THE CHEMICAL EVOLUTION
OF THE ATMOSPHERE
AND OCEANS

THE CHEMICAL EVOLUTION
OF THE ATMOSPHERE
AND OCEANS

HEINRICH D. HOLLAND

PRINCETON UNIVERSITY PRESS
PRINCETON, N.J.

THIS BOOK IS DEDICATED
WITH AFFECTION
TO MY MOTHER
Jeanette B. Holland

Contents

9 The Composition of the Atmosphere and Oceans during the Phanerozoic Eon 441

Preface

This volume is the sequel to *The Chemistry of the Atmosphere and Oceans*. The two books were originally planned as parts of a single volume, but separation into two volumes turned out to be the best way to complete the work. The first book dealt with the present state of the atmosphere-ocean system, and attempted to define quantitatively the processes that control the composition of these reservoirs today. The present volume seeks to define the evolution of the atmosphere-ocean system. It draws heavily on the contents of the first book, since the present-day operation of the system supplies important clues to its operation in the past.

The book progresses forward in time, and hence begins with the most speculative parts of the subject. The first four chapters deal with the atmosphere and oceans during the first half billion years of Earth history. While writing these chapters, I tried to keep in mind Mark Twain's comment in *Life on the Mississippi* that "There is something fascinating about science; one gets such wholesale returns of conjecture for such a trifling investment of fact." A rock record from the first half billion years of Earth history is surely needed to test the conjectures in these chapters.

The next four chapters are devoted largely to the atmosphere and oceans from 3.9 to 0.6 b.y.b.p. A major outcome of the writing of these chapters was a reaffirmation of belief in the rather conservative nature of Earth history: the Earth has surely changed with time, but the changes during the past 3.9 b.y. have not been dramatic.

The last chapter of the book deals with the Phanerozoic Eon. The availability of an extensive stratigraphic record from this era allows much more definite statements to be made about the composition of the atmosphere and oceans during the last 0.6 b.y. than during the preceding 3.9 b.y. The mineralogy of marine evaporites indicates that the composition of seawater during the Phanerozoic cannot have varied greatly. However, the changes in the isotopic composition of sulfur and strontium in seawater during this period demonstrate that there have been changes, and that these have not been trivial. A number of important questions remain to be answered. Among these, the effects of large extraterrestrial impacts, and the effects of changes in the Earth's orbital parameters on the ocean-atmosphere system are of particular interest.

The book was written as a monograph. I made no serious efforts to bridge the gap between the practitioners of the several disciplines whose approaches and data have been brought to bear on the evolution of the atmosphere and oceans. I can only hope that aeronomists, among others, will not find the nomenclature of mineralogy and petrology too obscure.

The range of topics considered in the book is uncomfortably large. Fortuantely, many friends have been willing to read and to criticize the manuscript. I would especially like to thank the following reviewers for taking the time to comment on the several chapters: Chapter 1: E. Anders, A. G. W. Cameron, A. E. Ringwood, G. W. Wetherill, and J. A. Wood. Chapter 2: A. Muan, M. Sato, and D. H. Speidel. Chapter 3: H. Craig, F. P. Fanale, and T. Kirsten. Chapter 4: D. M. Hunten, J. S. Levine, S. L. Miller, J. P. Pinto, and Y. L. Yung. Chapter 5: R. F. Dymek, R. B. Hargraves, and F. T. Mackenzie. Chapter 6: T. W. Donnelly, J. I. Drever, M. J. Mottl, and K. Muehlenbachs. Chapter 7: R. M. Garrels, D. E. Grandstaff, J. Kalliokoski, W. S. Meddaugh, G. J. Retallack, and M. Schidlowski. Chapter 8: E. S. Barghoorn, P. G. Brewer, E. Dimroth, C. Klein, A. H. Knoll, J. W. Schopf, and J. Veizer. Chapter 9: W. Alvarez, M. A. Arthur, H. P. Eugster, W. T. Holser, J. S. Leventhal, and E. S. Saltzman.

Many of my colleagues at several universities helped and sustained me during the preparation of the manuscript. Parts of the book were written while I was on sabbatical leave at the University of Hawaii in 1968–1969 and during the spring of 1981, and I gratefully acknowledge help from R. Moberly, K. E. Chave, S. O. Schlanger, and E. Zbinden. Parts of the book were written at Harvard University during a sabbatical leave in 1975–1976, during which I was supported by a Guggenheim Fellowship. Discussions with M. B. McElroy, A.G.W. Cameron, E. S. Barghoorn, J. A. Wood, J. J. McCarthy, and R. Siever were particularly helpful. Much of the remainder of the book was written at Heidelberg University in 1980 and during the summer of 1981, while I held a Senior Scientist Award from the Alexander von Humboldt-Stiftung. I am particularly grateful to Prof. Dr. K. O. Münnich for his unfailing hospitality at the Institut für Umweltphysik; to fellow visitors W. S. Broecker and H. E. Suess for stimulating discussions; and to Dr. H. Hanle of the Alexander von Humboldt-Stiftung for his warm and enthusiastic support. In addition, H. Hummel's suggestions during the spring of 1983 did much to improve the readability of the final manuscript. My debt to the staffs of all three academic institutions—Hawaii, Harvard, and Heidelberg—is also happily acknowledged. My greatest indebtedness, however, is surely to R. F. Quirk, who presided over the assembly of the manuscript, retyped large parts of it, prepared the indexes, and maintained order where chaos was a continual threat.

Cambridge, Massachusetts   **Heinrich D. Holland**
June 1983

THE CHEMICAL EVOLUTION
OF THE ATMOSPHERE
AND OCEANS

The Origin of the
Solar System

During the past three decades a great deal of progress has been made in our understanding of the origin of the solar system. Constraints on speculation have come from two lines of evidence. The first of these is physical: any theory of the origin of the solar system must be able to account for, or at least be consonant with, the present mass and motions of the bodies in the solar system. The second set of constraints is chemical and consists of a formidable array of isotopic and elemental analyses, determinations of the nature of solid phases, and analyses of organic compounds in meteorites, lunar samples, and the Earth. Both sets of constraints are described briefly in this chapter. The descriptions are somewhat superficial, because the events with which they deal predate those that are the main concern of this book.

1. Physical Constraints

Until quite recently, problems in mechanics dominated research into the origin of the solar system. Kant (1755) suggested that the solar system formed from a cloud of gas and dust, that the Sun condensed from the denser, central part of the cloud, and that the planets condensed from the less dense, outer portions of the gas cloud. Laplace (1796) proposed that the contracting mass of gas and dust which became the Sun threw off successive rings of matter from its equator, and that the condensation of matter in these rings formed the planets. This hypothesis and its variants failed to account for the distribution of angular momentum in the solar system. The Sun contains 98.85% of the mass of the solar system but only about 2% of its angular momentum (see, for instance, Urey 1952). The concentration of so much angular momentum in such a small fraction of the mass of the solar system was difficult to explain in terms of the ejection of mass from a single central body. This problem occupied a large number of astronomers prior to ca. 1950. It is no longer considered an obstacle in theories of the formation of the solar system that include a stage in which at least as much material as is required to form the planets was present in the form of a rotating disk of more or less the dimensions of the present solar system (Wetherill 1978). There are currently two competing explanations for the evolution of such a rotating disk into the present state of the solar system (see Kaula 1982). One mechanism involves the formation of giant protoplanets, i.e., gaseous blobs, by the gravitational collapse of portions of the disk. The other envisages a gravitationally stable disk in which larger bodies grew by the collision and coalescence of a swarm of smaller bodies (Wetherill 1980).

A great deal of computational work has been carried forward in the exploration of both mechanisms. Cameron and Pine's (1973) model, which neglected the effects of dissipation and angular momentum transfer during the accretion of the primitive solar nebula, proved to be close to the point of instability as judged by the ratio of the thermal to the rotational energy, and yielded characteristic cooling and angular momentum transfer times that have proved to be too short. Since then a series of papers have applied Lynden-Bell and Pringle's (1974) theory for the effects of dissipation in a viscous accretion disk to the problem of the formation of the Sun (see, for instance, Cameron 1978). Accretion is treated in terms of the inward flow of matter onto a central object in hydrostatic and rotational equilibrium. The process itself is apparently rather chaotic, but it is possible that the primitive solar accretion disk was repeatedly unstable against axisymmetric perturbation, and that within these perturbations, rings tended to form and to collapse on themselves, leading to the formation of giant gaseous protoplanets. The mass of these objects was apparently large enough to disturb the axisymmetry of the gravitational potential of the accretion disk. Their presence therefore complicates the computation of the evolutionary paths that can be taken by a collapsing solar nebula. They are, however, important as possible protoplanets. They would have consisted largely of hydrogen and helium, together with a small quantity of heavy elements that were present as convectively mixed grains. Complete melting and reduction of these grains in the presence of H_2 was probably followed by rapid coalescence and rain-out (DeCampli and Cameron 1979). The rained-out material could then have become either the core of a Jovian planet or one of the terrestrial planets after the loss of its gaseous envelope (Slattery, 1978; Slattery, DeCampli, and Cameron 1980). The effects of the rain-out process on the thermal evolution of gaseous protoplanets is complicated, because it involves the release of sizeable quantities of gravitational energy and a decrease in the opacity of the gas clouds. These effects have not been explored in detail.

The earlier papers by Cameron and his coworkers neglected the effect of the surrounding nebula on the evolution of gaseous protoplanets. Recently, Cameron, DeCampli, and Bodenheimer (1982) have shown that the evolution of the solar nebula can have a substantial influence on the evolution, and even on the existence, of gaseous protoplanets embedded in it. In particular, the time rate of increase of temperature in the nebula is of considerable importance for the evolution of protoplanets. These calculations again demonstrate that models of the type developed by

Cameron and his colleagues are very sensitive to the assumptions made and to the details of the computations. Gaseous protoplanets may well have existed during the early history of the solar system, but at present it seems at least equally likely that the nebular disk was gravitationally stable and that the planets formed by the accretion of particles at large in the nebula.

In the Cameron models, planetary formation tends to take place in approximately 10^5 to 10^6 years. On the other hand, the growth of the terrestrial planets from a flattened, rotating mixture of dust and gas by collision and coalescence of a swarm of smaller bodies may well have required on the order of 10^7 to 10^8 years. The initial growth of solid bodies could have occurred as a consequence of physical or chemical sticking of dust grains (Urey 1952), or due to the effects of small-scale gravitational instabilities in a thin central dust layer (see, for instance, Goldreich and Ward 1973). The accumulation of small planetesimals into planets could then have occurred as a consequence of collisions between individual bodies. It seems likely that this accumulation occurred, at least in part, in the presence of gas. However, it is quite possible that the gas was removed from the inner part of the solar system by solar UV or by corpuscular radiation on a time scale of 10^4 to 10^6 years, i.e., in periods of time shorter than those required for the terrestrial planets to grow by mutual collisions (Wetherill 1980). If so, the significant body of computational data that has been developed for the mechanics of gas-free accretion can be used to describe the growth of the terrestrial planets (see, for instance, Safronov 1972, and Wetherill 1978). Although this has been done in considerable detail, many uncertainties remain. Some of these can apparently be eliminated in a straightforward manner. Other unresolved questions may prove intractable.

It is clear from the above that the difference between the two most likely models for the condensation of the solar system is quite extreme. The time scale of accretion in the two models differs considerably, and the thermal history of the planets during their accretion is apt to be different. Until there is more certainty about the reality of giant gaseous protoplanets in the early solar system and about the role of collisional processes in the growth of the inner planets, the physical constraints currently imposed on our views of the early history of the Earth will be rather mild.

2. Chemical and Mineralogical Constraints

Measurements of the physical properties, including the motions of many of the larger bodies in the solar system, can be made precisely and repeatedly. However, the use of such measurements in the reconstruction of the early history of the solar system is difficult because so much has happened to these bodies in the meantime. The range of objects on which detailed chemical measurements can be made is much smaller, but some of these objects contain a largely undisturbed chemical record and have therefore turned out to be valuable sources of information for the recon-struction of the early history of the solar system. Among these objects, meteorites have been the most important. The Moon has also retained a reasonable record of the first half billion years of its history, a time interval for which only rather fragmentary and problematical evidence exists on Earth.

The Age and Origin of Meteorites

Meteorites, as Harold Urey was fond of saying, are the only pieces of extraterrestrial material that have come to us free of charge. His enthu-siasm for meteorites has proved to be more than justified, and the intensive study of these objects continues to yield new and surprising insights into the early history of the solar system.

THE AGE OF METEORITES

Meteorites have had a relatively uneventful history and still carry many of their birthmarks. An impressive number of age determinations have shown that most meteorites were formed some 4.55 ± 0.1 b.y. ago. The earliest measurements were reviewed by Anders (1962, 1964) and by Wood (1968). Kirsten (1978) and Podosek (1978) have provided recent reviews. The potassium-argon and $^{40}Ar-^{39}Ar$ ages of stony meteorites range from a high near 4.55 b.y. to values below 1 b.y. The relatively low K/Ar age of many stony meteorites is clearly the result of argon loss due to reheating prior to or during breakup of the parent meteorite bodies. The distribution of (U, Th)-He ages in chondrites is similar to that of their K/Ar ages, although the gas retention ages are not always the same (see, for instance, Zähringer 1968).

One of the most intriguing recent discoveries is that of K/Ar ages in excess of 4.6 b.y. for several of the white inclusions in the Allende meteorite (Kirsten 1980, p. 78). These include apparent ages of 5.27, 5.37, and 5.43 b.y., i.e., K/Ar ages that are almost certainly greater than the age of the solar system (see below). Several elements in white inclusions in Allende are known to have peculiar isotopic compositions, and it is not known whether the high K/Ar age of some of these inclusions is "real" or whether they are artifacts of potassium loss in the solar nebula (see, for instance, Podosek 1978).

Radiometric age determinations establish meteorites as the oldest objects, or at least as some of the oldest objects in the solar system. Their age does not define the age of the solar system. However, the distribution of the products of extinct radionuclides indicates that the solar system is almost certainly not much older than 4.6 b.y. The variable excess of ^{129}Xe relative to the other xenon isotopes in meteorites is largely due to differences in the ratio of ^{129}I to Xe in meteorites at the time of their formation and to the subsequent decay of the 15.7 m.y. ^{129}I to ^{129}Xe. The fact that ^{129}I was present at all during the formation of meteorites shows that the time interval between the end of nucleosynthesis in our part of the galaxy and the formation of meteorites cannot have been more than 10 to 15 half-lives of ^{129}I. Similarly, the presence of excess $^{131-136}$Xe in some meteorites due to the fission of ^{244}Pu indicates that ^{244}Pu was present during the accretion of meteoritic material. If the ratio ^{129}I/^{127}I was ca. 2×10^{-3} and the ^{244}Pu/^{238}U ratio ca. 3×10^{-2} at the end of nucleosynthesis, the time between the end of nucleosynthesis and the "freezing" of these radio-isotopes in meteorites was ca. 10^8 years.

Evidence for the former presence of extinct nuclides of much shorter half-life has been discovered recently. ^{26}Mg produced by the decay of ^{26}Al ($t_{1/2} = 0.76 \times 10^6$ yr) has been found in the carbonaceous chondrite Allende (Lee et al. 1976, 1977; Hutcheon et al. 1978) and in Leoville (Lorin and Christophe Michel-Levy 1978). ^{107}Ag produced by the decay of ^{107}Pd ($t_{1/2} = 6.5$ m.y.) has been found in the meteorite Santa Clara by Kelly and Wasserburg (1978) as well as in other ataxites (Kaiser and Wasserburg 1981). The short half-life of ^{26}Al implies that it was not produced during the same period of nucleosynthesis as ^{129}I and ^{244}Pu. The simplest explanation for the presence of extinct ^{26}Al is a second nucleosynthetic event a few million years before formation of the Allende inclusions (Cameron and Truran 1977). The presence of ^{26}Al at that time reinforces the notion that nucleosynthesis was involved in, and possibly acted as a trigger for, the collapse of the solar nebula. If sufficient ^{26}Al was produced, it may have been an important source of heat during the first few million years of solar system history, and may account—at least in part—for the high temperatures that were reached even in some of its smaller bodies.

THE COMPOSITION AND STRUCTURE OF METEORITES

The composition and the structure of meteorites have been subjects of active investigation since the beginning of the nineteenth century; since 1950 the volume of detailed petrographic, mineralogic, and chemical data has reached flood proportions. Most of the meteorites whose fall has been observed are stony meteorites (see table 1.1). Iron meteorites and stony-irons together account for only 7% of the observed falls. Among the stony meteorites the majority are chondrites, that is, they contain millimeter-sized globules of silicate minerals and/or silicate glass. Achondrites are stony meteorites without chondrules. Iron meteorites consist largely of

Table 1.1.

Relative proportions of chondrites, achondrites, stony-irons, and irons among meteorites actually seen to fall and among those collected on the Antarctic ice cap (in percentages).

	Meteorites seen to fall (Wood 1968)	*Meteorites collected on the Antarctic Ice Cap (Kusunoki 1975, and Yanai 1979)*
Stones		
Chondrites	85.7	94.9
Achondrites	7.1	1.5
Stony Irons	1.5	0
Irons	5.7	2.4
Dubious	—	1.2

metallic iron and nickel. These metals are present in large part as components of two alloys: kamacite, a low-Ni phase, and taenite, a high-Ni phase. Stony-irons contain both metallic and nonmetallic phases.

Each of the major meteorite groups has been subdivided variously (see, for instance, Van Schmus and Wood 1967; Anders 1968; Van Schmus 1969; and McSween 1979), but there is no unique way of assigning individual meteorites, each with its own name, to membership in families, clans, and tribes. The classification of Van Schmus and Wood (1967) (see figure 1.1), with its emphasis on chemical groups and petrological types and

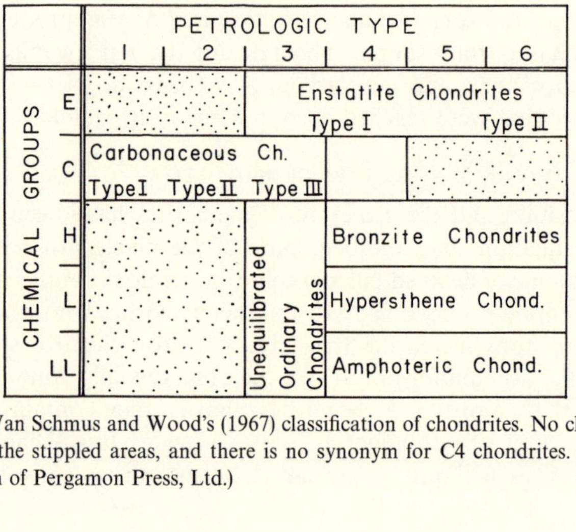

FIGURE 1.1. Van Schmus and Wood's (1967) classification of chondrites. No chondrites are known from the stippled areas, and there is no synonym for C4 chondrites. (Reproduced by permission of Pergamon Press, Ltd.)

the use of letters and numbers rather than proper names, is appealing and has been widely accepted. Assignment to chemical groups is based on the ratio by weight of total Fe to SiO_2 in the bulk analysis, on the SiO_2/MgO ratio in the bulk analysis, on the mole fraction $FeO/(FeO + MgO)$ in olivine and pyroxene, and on the ratio of metallic iron to total iron. Assignment to petrologic types is based on the pyroxene(s) and feldspar, on the presence or absence of silicate glass, on the nature of the metallic and sulfide minerals, on the basis of textures, and on the content of C and H_2O. Many of the observed chemical and petrological trends vary sympathetically, and hence a relatively simple classification can embrace most of the variability observed in stony meteorites.

Van Schmus and Wood's (1967) petrological types 3 to 6 are arranged in an order that could reflect changes imposed by the progressive thermal metamorphism of initially identical meteorites. The textures in type 2 chondrites are sharply defined; glass is present, as are sulfides and non-equilibrium mixtures of silicate minerals. In type 6 chondrites, glass is absent, chondrules are rare and almost invisible, feldspar is present, silicate minerals are in equilibrium with each other, and both metal and troilite are present (Wood 1968, pp. 43–44). The properties of intermediate meteorite types are easily interpolated.

Type 1 includes meteorites consisting wholly of very fine-grained material with a high volatile content. This corresponds to Wiik's (1956) type 1 carbonaceous chondrites. Meteorites of type 2 consist of two distinct fractions: chondrules and fine-grained matrix. The chondrules frequently contain silicate glass. The composition of glasses in chondrules is such that their presence can only be explained by a high-temperature history ($T > 1500°K$) followed by cooling, perhaps in a matter of minutes (see, for instance, Ringwood 1966a), and subsequent refrigeration below a temperature ceiling of a few hundred degrees Kelvin. The fine-grained matrix, on the other hand, frequently contains organic compounds (see Hayatsu and Anders 1981) and a variety of minerals, including phyllosilicates. These were probably formed hydrothermally; they could not have survived heating above a few hundred degrees Kelvin (see, for instance, Kvenvolden et al. 1970, 1971). Chemically, the chondrules and their matrix are quite distinct. The relative abundance of many elements in the matrix is surprisingly similar to their relative abundance in the Sun (see, for instance, Anders and Ebihara 1983). The chondrules, however, are strongly depleted in the more volatile elements. The bulk composition of any particular carbonaceous chondrite is largely dependent on the relative proportion of matrix and chondrules.

The similarity of the composition of the Sun and the matrix of carbonaceous chondrites suggests that the composition of the matrix was determined by condensation from a nebula of roughly solar composition at temperatures low enough to allow most elements to be captured in solar proportions. The lack of volatile elements in the high-temperature components (excluding chondrules) of carbonaceous chondrites can be explained in terms of the condensation of these "refractories" at temperatures that were too high for the incorporation of volatiles. On the other hand, the lack of volatile elements in chondrites of types 4, 5, and 6 and also in the refractory inclusions in carbonaceous chondrites can be ascribed to the loss of volatiles during post accretion metamorphism.

These notions have been integrated into the equilibrium condensation sequence of Larimer (1967), Grossman (1972), and Grossman and Larimer (1974). In this scheme a nebula of solar composition cools at a pressure of about 10^{-4} atm from $\geq 2000°K$ to temperatures below $400°K$. The condensation sequence and the reactions of condensates with the remaining nebula at lower temperatures are outlined in figures 1.2 and 1.3. The early condensates, which are formed between ca. $1700°K$ and $1440°K$, consist largely of a few rare metals, together with Al_2O_3, $CaTiO_3$, $Ca_2Al_2SiO_7$, and $MgAlO_4$. The next major component consists largely of metallic iron and nickel. The third major component consists of magnesium silicates. Together these three materials comprise about 70% of the potentially condensable "rocky" material of the nebula. On further cooling, these condensates collect the more volatile constituents, including water and organic compounds. During this stage the metals tend to react with H_2S to yield sulfides and with H_2O to yield oxides, which may in turn react further with silicate minerals.

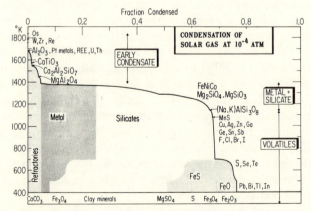

FIGURE 1.2a. The equilibrium condensation sequence of the solar nebula as proposed by Ganapathy and Anders (1974). (Reproduced by permission of the authors.)

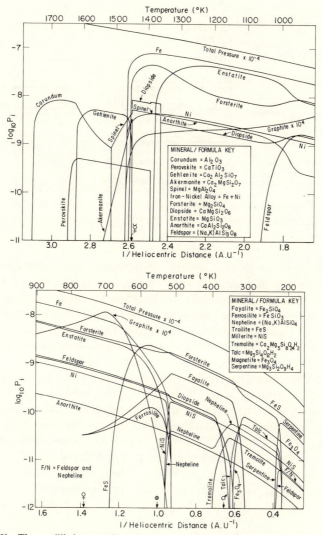

FIGURE 1.2b. The equilibrium condensation sequence of the solar nebula as proposed by Barshay (1981).

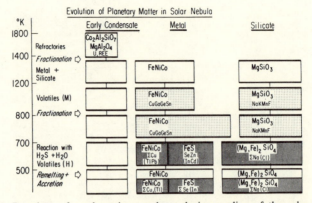

FIGURE 1.3. Reactions of condensation products during cooling of the solar nebula as proposed by Ganapathy and Anders (1974). (Reproduced by permission of the authors.)

Although this scheme is capable of explaining much of the chemistry and mineralogy of meteorites and can serve as an interesting framework for discussions of the chemistry of the larger bodies of the solar system, it has recently come under severe attack (see, for instance, Ringwood 1979, chap. 9; and rebuttals by Anders 1978, Wolf et al. 1979, and Anders and Wolf 1980). There are, at present, no compelling astrophysical reasons for the very high initial temperatures that are demanded by the proposed condensation sequence (but see Cameron and Fegley 1982). Furthermore, the observed mineralogy of the high-temperature phases has been reinterpreted in terms of the presence of localized high temperature events that partially evaporated previously accreted silicate-metal material (see, for instance, Wood 1981, and Cohen, Kornacki, and Wood 1983). However, it still remains to be seen whether this interpretation can really do justice to the chemical and isotopic properties of the high-temperature inclusions in meteorites. At present, the matter seems to be unsettled.

THE STRUCTURE AND CHRONOLOGY OF THE MOON

Before the first Apollo mission, lunar chronology was beset by uncertainties and engulfed in controversy. The outpouring of literature since then has been prodigious. More than 50,000 pages were published in the decade following 1969, and many of the important questions regarding the history and the present state of the Moon have been answered. A few major problems remain unsolved; unfortunately, the origin of the Moon is one of these.

THE EVOLUTION AND INTERNAL STRUCTURE OF THE MOON

Figure 1.4 shows what is generally accepted to be a reasonable represent-ation of the internal structure of the Moon. The lunar crust is less dense than the lunar interior and is chemically distinct from it; the thickness of the crust varies between 60 km and 100 km. Anorthositic and basaltic rocks dominate the geology of the lunar surface. The mare basalts, although visually very important, account for less than 1% of the lunar volume. The overall composition of the lunar highlands is given in table 1.2. They are rich in calcium and aluminum and were probably derived by crystallization from a magma ocean shortly after the formation of the Moon. Detailed models for the crystallization of such a magma ocean indicate that this process was very complex, and that it probably involved cycles of fractional crystallization, assimilation, and mixing (Herbert et al. 1978; Longhi and Boudreau 1979; Warren and Wasson 1979).

The Moon's upper mantle extends from the base of the crust to a depth of 400–500 km. There is some inconclusive evidence that this part of the

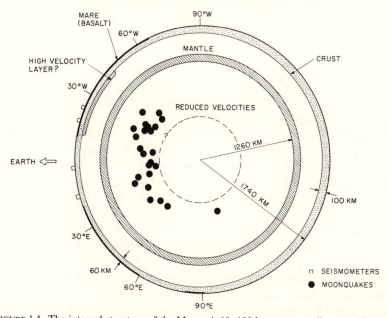

FIGURE 1.4. The internal structure of the Moon. A 60–100 km crust overlies a two-layered mantle. The transition zone between the two layers occurs between a depth of 400 and 480 km. The location of the epicenters of typically observed moonquakes is indicated by solid circles. P and S wave velocities increase sharply below the level indicated by the dashed cir-cle (Goins, Toksöz, and Dainty 1978 and 1979). (Reproduced by permission of the authors.)

TABLE 1.2.

Chemical composition of the lunar highlands, the bulk Moon and the Earth's mantle, after Taylor (1979). See also Morgan and Anders (1980). There is considerable controversy regarding the CaO and Al_2O_3 content of the Moon.

	1	2	3	4
$SiO_2\%$	45	42	46.1	45
$TiO_2\%$	0.56	0.4	0.2	0.16
$Al_2O_3\%$	24.6	8.0	4.3	3.3
$FeO\%$	6.6	12	8.2	8.0
$MgO\%$	8.6	31	37.6	40
$CaO\%$	14.2	6.0	3.1	2.6
$Na_2O\%$	0.45	0.1	0.4	0.2
K (ppm)	600	80–100	250	150
U (ppb)	240	30–40	—	15
Th (ppb)	900	120–150	—	60
K/U	2,500	2,500	—	10,000

[1] Lunar highlands surface; 2, Bulk Moon; 3, Earth's mantle; 4, Earth's mantle (NiO = 0.26%, Cr_2O_3 = 0.44%).

mantle is laterally heterogeneous. A minor discontinuity exists at a depth close to 480 km. The change to lower seismic velocities at that depth extends down to about 1000 km. Most moonquakes occur in the lower part of this zone. The nature of the lunar interior below 1100 km is rather uncertain, but its properties are consistent with the presence of ca. 1% of melt. The evidence for a lunar core remains inconclusive; seismic data restrict the size of the core—if one is present at all—to a radius of 170–360 km. At most, such a core could account for 2% of the volume of the Moon.

Numerous age determinations have suggested that the large-scale initial differentiation that produced most of the lunar crust took place between 4.4 and 4.6 b.y. ago (for a summary, see Kirsten 1978). Recent lead isotope investigations (Oberli, Huneke, and Wasserburg 1979) have refined this estimate, and have yielded a best estimate of 4.46 b.y. for the formation of the lunar highland crust. The following 0.5 b.y. of lunar history were dominated by bolide impacts. The rate at which the influx of extralunar material decreased between 4.5 and 3.9 b.y. ago is reasonably well known. The melting of the lunar interior that yielded the mare basalts produced flows between 3.9 b.y. and ca. 2.6 b.y. ago. However, mare volcanism may well have begun before the last major collisions. The impacts that formed the large basins produced some local melting but did not generate the floods of basalt that filled mare basins. These basalts were probably derived from the cumulate region produced during the early large-scale melting of the lunar mantle (Taylor 1979). Since the end of flooding by basalts, the face of the Moon has been altered mildly by the infall of material that produced craters such as Copernicus, Kepler, Aristarchus, and Tycho.

THE ORIGIN OF THE MOON

The three theories that dominated pre-Apollo discussions of the origin of the Moon still have their defenders. One of these, lunar capture, i.e., the capture of the Moon into an orbit around the Earth, is, however, most unlikely, because the mechanics of lunar capture into Earth orbit are extremely difficult. Kaula (1971) has assessed the likelihood of lunar capture as "horrendously improbable." The similarity of the isotopic composition of oxygen in lunar and terrestrial materials suggests a common origin for the Earth and the Moon; this does not, however, rule out the capture hypothesis, because similar $\delta^{18}O$ values have also been found in some meteorites that are probably related neither to the Earth nor to the Moon.

The other two theories both involve the formation of the Moon in orbit—one by the condensation of material ejected from the Earth, the other by condensation of material accumulated in Earth orbit during or after the accretion of the Earth. The hypothesis of a fission-origin of the Moon has not been popular for dynamic reasons, but there is no doubt that the composition of the Moon has much in common with that of the Earth's mantle (Brown 1978). Ringwood (1979, chap. 12) has made a reasonably strong case for a hypothesis that requires the material in the Moon to have been completely evaporated from the Earth's mantle and then to have been selectively recondensed. The spectacular depletion of volatiles, including Au, Ge, As, and Sb in the Moon compared to the Earth's mantle (see figure 1.5) can be explained partly by the absence of metallic iron in the system during the volatilization-recondensation process. The dynamics of the fission process are still poorly understood. Fragments that are knocked off the Earth by a glancing blow and that do not escape altogether from the Earth will describe elliptical orbits about the Earth's center of mass and will eventually return to the point on the ellipse where they started, i.e., at the surface of the Earth. Continued acceleration after fragments have been knocked out of the Earth is required to put them into Earth orbit. Proponents of the fission theory invoke the presence of a huge cloud of vaporized Earth material produced by the same planetesimal impact, and suggest that the expansion of this

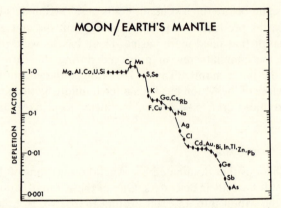

FIGURE 1.5. Comparison of the abundance of (mainly) volatile elements in the source regions of low-Ti mare basalts with corresponding abundances in the Earth's mantle (from Ringwood 1979). (Reproduced by permission of Springer-Verlag, New York, Inc.)

cloud tends to accelerate the debris fragments that are for a time present in the cloud. It remains to be seen whether this model is mechanically sound.

On dynamical grounds some variation of Ruskol's (1977) binary planet-coagulation hypothesis for the origin of the Moon seems most plausible; in this scheme the Moon forms from a ring of differentiated debris in terrestrial orbit. Ringwood (1979), however, believes that the geochemical problems created by this theory are insuperable. At present it seems certain that the origin of the Moon was nearly synchronous with that of meteorites and the Earth, and that the process was essentially complete ca. 4.50 b.y. ago. The relationship between the origin of the Moon and the Earth is still uncertain; it may turn out that all of the processes invoked in the three theories operated in concert, and that the origin of the Moon owes something to all of them.

The Age and Composition of the Earth

The age of the Earth is indistinguishable from that of the Moon and meteorites (see, for instance, Patterson 1956), although the available data still leave an uncertainty of 50–100 m.y. in the early chronology of the solar system. The accretion of the Earth may not have been essentially complete until ca. 4.45 b.y. ago, i.e., some 100 m.y. after the formation of many meteorites. The Moon and many meteorites reached temperatures close to or in excess of 1000°C within 100 m.y. of their formation. It is therefore likely, but not certain, that the Earth, or at least its outer portions, reached temperatures at least as high as 1000°C during or shortly after accretion.

The energy sources for the early heating of meteorites and the Moon have not been completely identified. In the Earth the release of gravitational energy and the decay of radioelements of geologically short and geologically long half-lives were probably of major importance, but heating during the T-Tauri phase of the Sun may not have been negligible. A "hot" origin for the Earth has become generally accepted, but the sequence of events during and shortly after accretion are still matters of debate. There is no unanimity regarding either the composition of the accreted materials or the order of their accretion. These uncertainties extend to the nature of the Earth's atmosphere as well, and only very broad limits can be set on the pressure and composition of the protoatmosphere.

Two types of models have been proposed recently for the accretion of the Earth: homogeneous accretion models and heterogeneous accretion models. Models of the first class explain the present state of the Earth in terms of the accretion of a single type of material; models of the second class invoke the sequential accretion of several types of material. Among the homogeneous accretion models the most widely supported variant (see, for instance, Urey 1962; Wood 1962) involves the accretion of an intimate mixture of silicate particles and metal particles that resembles ordinary chondritic meteorites; accretion is followed by the sinking of metal particles and by their coalescence into the Earth's core. This model has the distinct advantage of explaining the observation that the Earth's mantle is nearly homogeneous. The model has been criticized, particularly by Ringwood (1966a, 1966b, 1971, and 1975, pp. 550–552), for its apparent failure to account for the relatively high content of Ni, Co, Cu, and Au and for the rather high oxidation state of the upper mantle. Ringwood (1979, chap. 2) has argued that the abundance of the siderophile elements including Ni, Co, Cu, Ir, and Au in the upper mantle is one or two orders of magnitude greater than can be explained by equilibrium partition between a metallic phase and silicates.

This problem has been restudied by Jagoutz et al. (1979), who have determined the major, minor, and trace-element composition of spinel-lherzolite inclusions in alkali basalts (see also Basaltic Volcanism Study Project 1981, section 1.2.11). Such inclusions from the mantle are widespread. They have the great advantage over basalts of being much closer in composition to the mantle as a whole. Their composition is, however, variable, presumably as a consequence of magmatic events that have modified the mantle during the past 4.5 b.y.

Figure 1.6 shows the variation of the Mg/Si ratio in ultramafic nodules with their Al/Si ratio. The trend line has a slope opposite to that of the

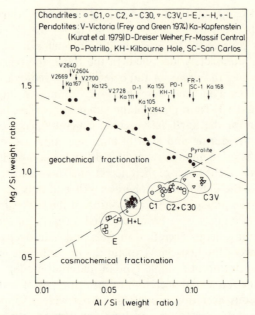

FIGURE 1.6. Plot of the Mg/Si vs. the Al/Si ratio of meteorites and terrestrial ultramafic nodules (Jagoutz et al. 1979). (Reproduced by permission of the authors.)

Mg/Si-Al/Si line for unfractionated meteorites. The ratios of the abundance of compatible refractory elements (Yb, Sc, Ca, Al) become increasingly more chondritic along the terrestrial trend line and are essentially chondritic at the intersection of the two trend lines. The Mg/Si ratio at this intersection is equal, within the stated uncertainties, to that of the solar photosphere. All these observations suggest that the composition of ultramafic nodules close to the intersection of the trend lines in figure 1.6 is quite close to that of the primitive mantle. It is reassuring that the composition of these nodules is close to the composition of Ringwood's pyrolite.

Major and trace element data for the most primitive nodules have been normalized to Si and C1 chondrite abundances, and have been plotted in figures 1.7 and 1.8. There is a large spread in the normalized abundance of the incompatible refractory lithophile elements compared to the uniform concentration of compatible refractory elements. The decrease in relative abundance from Mg to Zn is similar to that in C3 chondrites and may be related to the condensation temperatures of these elements. The relatively high concentration of Ni and Co in the Earth's mantle is emphasized by the small depletion of these elements relative to C1 chondrites and by the

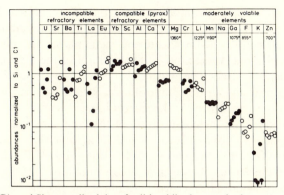

FIGURE 1.7. C1- and Si- normalized data for lithophile elements in the most primitive ultramafic nodules (Jagoutz et al. 1979). Open rings and solid circles have been alternated to allow better visual resolution. (Reproduced by permission of the authors.)

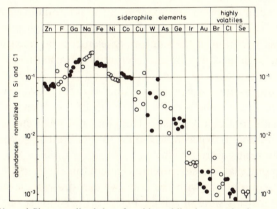

FIGURE 1.8. C1- and Si- normalized data for siderophile elements in the most primitive ultramafic nodules (Jagoutz et al. 1979). (Reproduced by permission of the authors.)

chondritic value of the Ni/Co ratio in primitive ultramafic inclusions. The Fe/Ni ratio in the inclusions is about 1.5 times the chondritic ratio. The normalized abundance of Ir and Au is more than a factor of 10 lower than that of Ni and Co; however, the Ir/Au ratios are nearly chondritic, as are the Os/Pd ratios (Morgan and Wandless 1979). Although the abundances in the paper by Jagoutz et al. (1979) differ somewhat from those in Ringwood's (1979) compilation, the differences do not alter the conclusion that the abundance of the siderophile elements in the mantle is not compatible with equilibrium between the mantle and the Earth's core. Provided the Earth as a whole is chondritic in composition, the core can be considered to have thermodynamic properties equal to that of the Fe-Ni phase in meteorites, and pressure effects on the distribution of the siderophile elements between silicate and metal phases are small (see also Kimura, Lewis, and Anders 1974). The validity of these conditions has been debated extensively. The evidence in favor of the proposition that the Earth has a near-chondritic composition is strong (see Ringwood 1979, chap. 2) but not conclusive for the siderophile elements.

Evidence regarding the composition of the core, especially the nature of its light element component(s) is still in a somewhat unsatisfactory state (see, for instance, Stevenson 1981). Ringwood (1979, chap. 3) has argued for the presence of oxygen as a constituent of "FeO" in the core, but the experimental data needed to confirm the reasonableness of this suggestion are not available, and the effects of dissolved O^{-2} on the distribution of the siderophile elements between a silicate phase and a metal phase are not known. It seems unlikely, however, that the effects of O^{-2} would be large enough to account for the observed differences between the predicted and observed abundances of the siderophile elements in the mantle. Brett (1971) has suggested that the effect of high pressure in the mantle on the partition of the siderophile elements between silicates and metal phases might account for the differences between their observed and predicted abundance in the mantle. Ringwood (1971, 1979) has criticized this suggestion, because he considers it unlikely that pressure would act in the same direction on the distribution of all of the siderophile elements between silicate and metal phases. This does not seem entirely unlikely; however, there are no experimental data to test Brett's proposal, and the problem is currently unresolved.

If it turns out that the siderophile elements in the Earth's mantle are really out of equilibrium with the core, we can draw on several possible explanations for this state of things. Perhaps the simplest of these is that complete equilibrium was not established between silicate and metal phases during the accretion of the Earth. The problem of homogenizing the trace element content of large parts of the mantle is quite severe, and it may have been difficult to equilibrate large volumes of silicate and metal phases during infall and core formation. Some slight evidence in favor of this view is offered by the heterogeneity of the concentration of siderophile elements in achondritic meteorites. Baedeker (1971) reported a range of 0.8 to 150 ppb for the Ir content of achondrites; Laul et al. (1972) reported a range of 0.06 to 31 ppb. The Au content of achondrites was reported to range from 2 to 60 ppb by Ehmann (1971) and from 0.08 to 78 ppb by Laul et al. (1972). The Ir content and the Au content suggested for the Earth's mantle by the data in table 1.3 fall within the wide concentration ranges for those elements in achondrites; their wide dispersion in these

TABLE 1.3.

Iridium and gold in ultramafic nodules and in carbonaceous chondrites (Jagoutz et al. 1979).

Data from ultramafic nodules:	Ir ppb	Au ppb	Ir/Au
Morgan and Wandless (1979) (average 18 nodules)	4.2	0.77	5.45
Jagoutz et al. (1979, Table 1) (average 6 nodules)	3.4	0.5	6.80
Carbonaceous chondrites: (data from Jagoutz laboratory)			
Orgueil C1	480	153	3.1
Murchison C2	660	150	4.4
Vigarano C3	763	151	5.1
Previous estimates for the upper mantle:			
Chou (1978)	2.4	3.0	0.80
Ringwood and Kesson (1977)	2.4	4.2	0.57

meteorites was probably produced by magmatic events very early in the history of the solar system.

A very different view of the present core-mantle relationship has been offered by the heterogeneous accretion model of Turekian and Clark (1969, 1975). In this model the inner portions of the planets are considered to consist of high-temperature condensates; lower temperature condensates are thought to become progressively more abundant outward. The high-temperature condensates would have been devoid of high vapor pressure materials. The lowest temperature condensates, which are thought to have been added to the planet in the manner of a veneer (see, for instance, Larimer and Anders 1967), would be material such as the volatiles in C1 carbonaceous chondrites. This model is attractive, because it offers a simple explanation for core-mantle disequilibrium within the framework of the accepted condensation sequence. However, the hypothesis has serious drawbacks, which have been summarized well by Ringwood (1979, chap. 7). It offers no ready explanation for the light element content of the core, for the apparently uniform FeO/(MgO + FeO) ratio of the mantle, for the presence of high-temperature condensates in the mantle, and for the nonprimordial abundance ratios of the siderophile and volatile elements. These objections could perhaps be overcome by calling on large-scale postaccretion homogenization of the mantle accompanied by interaction between the mantle and the core across the core-mantle boundary. It is not clear that such extensive homogenization is possible, and the need for such a process detracts a great deal from the simplicity of the original model.

Perhaps the most reasonable accretion models are those that involve neither completely homogeneous nor completely heterogeneous accretion. The two extremes are somewhat unlikely in any event; the accretion process must have been extremely complex. What seems to be ruled out most clearly by the composition of the Earth is complete condensation at very high temperatures in the center of a giant gaseous protoplanet; formation under such conditions offers no reasonable explanation for the presence of elements that condense at low temperatures. Accretion via planetesimals in the manner envisaged by Wetherill and by Safronov seems quite acceptable from a chemical point of view. The explanation for the relatively high abundance of the siderophile elements in the mantle may be found in one or more of the possibilities outlined above (see, for instance, Sun 1982).

During accretion, metallic Fe-Ni was almost certainly among the materials which fell on the Earth. The presence of such materials in the mantle during accretion is not ruled out by the oxidation state of present-day basalts or by the oxidation state of present-day volcanic gases as suggested by Ringwood (1979, chap. 2). This is an issue of considerable importance for the nature of the postaccretion atmosphere and is discussed at some length in the next chapter. The mass and chemical composition of the Earth's atmosphere during the accretion of the planet is still ill-defined. Its composition must have been such that reduction of iron and nickel in the accreting Earth was not complete, and its mass must have been sufficiently insubstantial so that removal was possible either during or after the accretion process under the influence of the Sun during its earliest evolutionary stages. These presumably included a T-Tauri phase and possibly other phases that were even more violent but probably short-lived.

References

Anders, E. 1962. Meteorite ages. *Rev. Mod. Phys.* 34:287–325.

———. 1964. Origin, age, and composition of meteorites. *Space Sci. Rev.* 3:583–714.

———. 1968. Chemical processes in the early solar system, as inferred from meteorites. *Accounts Chem. Res.* 1:289–298.

———. 1978. Procrustean science: Indigenous siderophiles in the lunar highlands, according to Delano and Ringwood. In *Proc. Lunar Planet. Sci. Conf. 9th*, 161–184. New York: Pergamon.

Anders, E., and Wolf, R. 1980. Moon and Earth: Compositional differences inferred from siderophiles, volatiles, and alkalis in basalts. *Geochim. Cosmochim. Acta* 44:2111–2124.

Anders, E., and Ebihara, M., 1983. Solar-system abundances of the elements. *Geochim. Cosmochim. Acta.* In press.

Baedeker, P. A. 1971. Iridium. In *Handbook of Elemental Abundances in Meteorites*, edited by B. Mason, 463–472. New York: Gordon and Breach Science Publishers.

Barshay, S. S. 1981. Combined condensation-accretion models of the terrestrial planets. Ph.D. diss., Massachusetts Institute of Technology.

Basaltic Volcanism Study Project. 1981. *Basaltic Volcanism on the Terrestrial Planets*. New York: Pergamon.

Brett, R. 1971. The Earth's core: Speculations on its chemical equilibrium with the mantle. *Geochim. Cosmochim. Acta* 35:203–221.

Brown, G. M. 1978. Chemical evidence for the origin, melting, and differentiation of the Moon. In *The Origin of the Solar System*, edited by S. F. Dermott, 597–609. New York: Wiley.

Cameron, A.G.W. 1978. Physics of the primitive solar accretion disc. *The Moon and Planets* 18:5–40.

Cameron, A.G.W., and Pine, M. R. 1973. Numerical models of the primitive solar nebula. *Icarus* 18:377–406.

Cameron, A.G.W., and Truran, J. W. 1977. The supernova trigger for formation of the solar system. *Icarus* 30:447–461.

Cameron, A.G.W., DeCampli, W. M., and Bodenheimer, P. 1982. Evolution of giant gaseous protoplanets embedded in the primitive solar nebula. Harvard-Smithsonian Center for Astrophysics, Cambridge, Mass.

Cameron, A.G.W., and Fegley, M. B. 1982. Nucleation and condensation in the primitive solar nebula. Cambridge, Mass.: Harvard-Smithsonian Center for Astrophysics.

Chou, C.-L. 1978. Fractionation of siderophile element ratios in the Earth's upper mantle and lunar samples (abstract). In *Lunar Planet. Sci.* 9:163–165. Houston, Tex.: Lunar and Planetary Institute.

Cohen, R. E., Kornacki, A. S., and Wood, J. A. 1983. Mineralogy and petrology of the Mokoia C3(V) chondrite. *Geochim. Cosmochim. Acta* (in press).

DeCampli, W. M., and Cameron, A.G.W. 1979. Structure and evolution of isolated giant gaseous protoplanets. *Icarus* 38:367–391.

Ehmann, W. D. 1971. Gold. In *Handbook of Elemental Abundances in Meteorites*, edited by B. Mason, 479–485. New York: Gordon and Breach Science Publishers.

Ganapathy, R., and Anders, E. 1974. Bulk compositions of the Moon and Earth, estimated from meteorites. In *Proc. Fifth Lunar Sci. Conf., Geochim. Cosmochim. Acta Suppl.* 5:1181–1206. New York: Pergamon.

Goins, N. R., Toksöz, M. N., and Dainty, A. M. 1978. Seismic structure of the lunar mantle: An overview. In *Proc. Lunar Planet. Sci. Conf. 9th*, 3575–3588. New York: Pergamon.

———. 1979. The lunar interior: A summary report. In *Lunar Planet. Sci.* 10:437–439. Houston, Tex.: Lunar and Planetary Institute.

Goldreich, P., and Ward, W. R. 1973. The formation of planetesimals. *Astrophys. J.* 183:1051–1061.

Grossman, L. 1972. Condensation in the primitive solar nebula. *Geochim. Cosmochim. Acta* 36:597–619.

Grossman, L., and Larimer, J. W. 1974. Early chemical history of the solar system. *Rev. Geophys. Space Phys.* 12:71–101.

Hayatsu, R., and Anders, E. 1981. Organic compounds in meteorites and their origins. In *Topics in Current Chem.*, no. 99, 1–37. New York: Springer-Verlag.

Herbert, F., Drake, M. J., and Sonett, C. P. 1978. Geophysical and geochemical evolution of the lunar magma ocean. In *Proc. Lunar Planet. Sci. Conf. 9th*, 249–262. New York: Pergamon.

Hutcheon, I. D., Steele, I. M., Smith, J. V., and Clayton, R. N. 1978. Ion microprobe, electron microprobe and cathodoluminescence data for Allende inclusions with emphasis on plagioclase chemistry. In *Proc. Lunar Planet. Sci. Conf. 9th*, 1345–1368. New York: Pergamon.

Jagoutz, E., Palme, H., Baddenhausen, H., Blum, K., Cendales, M., Dreibus, G., Spettel, B., Lorenz, V., and Wänke, H. 1979. The abundances of major, minor, and trace elements in the Earth's mantle as derived from primitive ultramafic nodules. In *Proc. Lunar Planet. Sci. Conf. 10th*, 2031–2050. New York: Pergamon.

Kaiser, T., and Wasserburg, G. J. 1981. What is the origin of [107]Ag and [109]Ag in nickel-rich ataxites. In *Proc. Lunar Planet. Sci. Conf. 12th*, 525–527. New York: Pergamon.

Kant, I. 1755. *Allgemeine Naturgeschichte and Theorie des Himmels.*

Kaula, W. M. 1971. Dynamical aspects of lunar origin. *Rev. Geophys. Space Phys.* 9:217–238.

———. 1982. The formation of the Sun and its planets. In *The Physics of the Sun*, edited by P. A. Sturrock, Chap. 13. Space Science Board Monograph.

Kelly, W. R., and Wasserburg, G. J. 1978. Evidence for the existence of [107]Pd in the early solar system. *Geophys. Res. Letters* 5:1079–1082.

Kimura, K., Lewis, R. S., and Anders, E. 1974. Distribution of gold and rhenium between nickel-iron and silicate melts: Implications for the abundance of siderophile elements on the Earth and Moon. *Geochim. Cosmochim. Acta* 38:683–701.

Kirsten, T. 1978. Time and the solar system. In *The Origin of the Solar System*, edited by S. F. Dermott, 267–346. New York: Wiley.

———. 1980. Geophysik in Heidelberg. *Sitzungsber. Heidelberger Akad. d. Wiss., Math.-naturw. Kl. Abh.* 4 (1979/80).

Kusunoki, K. 1975. A note on the Yamato meteorites collected in December 1969. In *Yamato Meteorites Collected in Antarctica in 1969*, edited by T. Nagata, 1–8. Mem. Nat. Inst. of Polar Res., Special Issue no. 5.

Kvenvolden, K., Lawless, J., Pering, K., Peterson, E., Flores, J., Ponnamperuma, C., Kaplan, I. R., and Moore, C. 1970. Evidence for extraterrestrial amino-acids and hydrocarbons in the Murchison meteorite. *Nature* 228:923–926.

Kvenvolden, K. A., Lawless, J. G., and Ponnamperuma, C. 1971. Nonprotein amino acids in the Murchison meteorite. *Proc. Nat. Acad. Sci. USA* 68:486–490.

Laplace, P. S. 1976. Exposition de Système du Monde. Paris.

Larimer, J. W. 1967. Chemical fractionations in meteorites: I. Condensation of the elements. *Geochim. Cosmochim. Acta* 31:1215–1238.

Larimer, J. W., and Anders, E., 1967. Chemical fractionations in meteorites: II. Abundance patterns and their interpretation. *Geochim. Cosmochim. Acta* 31:1239–1270.

Laul, J. C., Keays, R. R., Ganapathy, R., Anders, E., and Morgan, J. W. 1972. Chemical fractionations in meteorites: V. Volatile and siderophile elements in achondrites and ocean ridge basalts. *Geochim. Cosmochim. Acta* 36:329–345.

Lee, T., Papanastassiou, D. A., and Wasserburg, G. J. 1976. Demonstration of ^{26}Mg excess in Allende and evidence for ^{26}Al. *Geophys. Res. Letters* 3:109–112.

———. 1977. Aluminum-26 in the early solar system: Fossil or fuel? *Astrophys. J.* 211:L107–110.

Longhi, J., and Boudreau, A. E. 1979. Complex igneous process and the formation of the primitive lunar crustal rocks. In *Proc. Lunar Planet. Sci. Conf. 10th*, 2085–2105. New York: Pergamon.

Lorin, J. C., and Christophe Michel-Levy, M. 1978. Radiogenic ^{26}Mg fine-scale distribution in Ca-Al inclusions of the Allende and Leoville meteorites. In *Short Papers of the Fourth Int. Conf., Geochronology, Cosmochronology, Isotope Geology*, edited by R. E. Zartman, 257–259. U.S. Geol. Surv. Open-File Report no. 78-701.

Lynden-Bell, D., and Pringle, J. E. 1974. The evolution of viscous discs and the origin of the nebular variables. *Month. Not. Roy. Astron. Soc.* 168:603–637.

McSween, H. Y., Jr. 1979. Are carbonaceous chondrites primitive or processed? A review. *Rev. Geophys. Space Phys.* 17:1059–1078.

Morgan, J. W., and Wandless, G. A. 1979. Terrestrial upper mantle: Siderophile and volatile trace element abundances (abstract). In *Proc. Lunar Planet. Sci. Conf., 10th*, 855–857, Houston, Tex.: Lunar and Planetary Institute.

Morgan, J. W., and Anders, E. 1980. Chemical composition of Earth, Venus, and Mercury. *Proc. Nat. Acad. Sci. USA* 77:6973–6977.

Oberli, F., Huneke, J. C., and Wasserburg, G. J. 1979. U-Pb and K-Ar systematics of cataclysm and precataclysm lunar impactites (abstract). In *Proc. Lunar Planet. Sci. Conf., 10th*, 940–942. Houston, Tex.: Lunar and Planetary Institute.

Patterson, C. 1956. Age of meteorites and the Earth. *Geochim. Cosmochim. Acta* 10:230–237.

Podosek, F. A. 1978. Isotopic structures in solar system materials. *Ann. Rev. Astron. Astrophys.* 16:293–334.

Ringwood, A. E. 1966a. Genesis of chondritic meteorites. *Rev. Geophys.* 4:113–174.

————. 1966b. Chemical evolution of the terrestrial planets. *Geochim. Cosmochim. Acta* 30:41–104.

————. 1971. Core-mantle equilibrium: Comments on a paper by R. Brett. *Geochim. Cosmochim. Acta* 35:223–230.

————. 1975. *Composition and Petrology of the Earth's Mantle.* New York: McGraw-Hill.

————. 1979. *Origin of the Earth and Moon.* New York: Springer-Verlag.

Ringwood, A. E., and Kesson, S. E. 1977. Basaltic magmatism and the bulk composition of the Moon, II. Siderophile and volatile elements in Moon, Earth and chondrites: Implications for lunar origin. *The Moon* 16:425–464.

Ruskol, E. L. 1977. The origin of the Moon. In *Proceedings of the Soviet-American Conference on Cosmochemistry of the Moon and Planets,* edited by J. Pomeroy and N. Hubbard, 815–822. Washington, D.C.: NASA SP-370.

Safronov, V. S. 1972. *Evolution of the Protoplanetary Cloud and Formation of the Earth and the Planets.* Israel Program for Scientific Translations, Jerusalem (translated from Russian, 1969).

Slattery, W. L. 1978. Protoplanetary core formation by rain-out of iron drops. *The Moon and Planets* 19:443–457.

Slattery, W. L., DeCampli, W. M., and Cameron, A.G.W., 1980. Protoplanetary core formation by rain-out of minerals. *The Moon and Planets* 23:381–390.

Stevenson, D. J. 1981. Models of the Earth's core. *Science* 214:611–619.

Sun, S.-S. 1982. Chemical composition and origin of the Earth's primitive mantle. *Geochim. Cosmochim. Acta* 46:179–192.

Taylor, S. R. 1979. Structure and evolution of the Moon. *Nature* 281: 105–110.

Turekian, K. K., and Clark, S. P., Jr. 1969. Inhomogeneous accumulation of the Earth from the primitive solar nebula. *Earth Planet. Sci. Lett.* 6:346–348.

――――. 1975. The non-homogeneous accumulation model for terrestrial planet formation and the consequences for the atmosphere of Venus. *J. Atm. Sciences* 32:1257–1261.

Urey, H. C. 1952. *The Planets.* New Haven, Conn.: Yale University Press.

――――. 1962. Evidence regarding the origin of the Earth. *Geochim. Cosmochim. Acta* 26:1–13.

Van Schmus, W. R. 1969. The mineralogy and petrology of chrondritic meteorites. Earth-Science Reviews 5:145–184.

Van Schmus, W. R., and Wood, J. A. 1967. A chemical-petrologic classification for the chondritic meteorites. *Geochim. Cosmochim. Acta* 31: 747–765.

Warren, P. H., and Wasson, J. T. 1979. Effects of pressure on the crystallization of a moon-sized "chondritic" magma ocean (abstract). In *Proc. Lunar Planet. Sci. Conf., 10th,* 1304–1306. Houston, Tex.: Lunar and Planetary Institute.

Wetherill, G. W. 1978. Accumulation of the terrestrial planets. In *Protostars and Planets,* edited by T. Gehrels, 565–598. Tucson: University of Arizona Press.

――――. 1980. Formation of the terrestrial planets, *Ann. Rev. Astron. Astrophys.* 18:77–113.

Wiik, H. B. 1956. The chemical composition of some stony meteorites. *Geochim. Cosmochim. Acta* 9:279–289.

Wolf, R., Woodrow, A., and Anders, E. 1979. Lunar basalts and pristine highland rocks: Comparison of siderophile and volatile elements. In *Proc. Lunar Planet. Sci. Conf. 10th,* 2107–2130. New York: Pergamon.

Wood, J. A. 1962. Chondrules and the origin of the terrestrial planets. *Nature* 194:127–130.

――――. 1968. *Meteorites and the Origin of Planets.* New York: McGraw-Hill.

――――. 1981. The interstellar dust as a precursor of Ca,Al-rich inclusions in carbonaceous chondrites. *Earth Planet. Sci. Lett.* 56:32–44.

Yanai, K. 1979. Meteorite search in Victoria Land, Antarctica in 1977–1978 austral summer. In *Proceedings of the Third Symposium on Antarctic Meteorites*, edited by T. Nagata, 1–8. Tokyo: National Institute of Polar Research.

Zähringer, J. 1968. Rare gases in stony meteorites. *Geochim. Cosmochim. Acta* 32:209–237.

2

Inputs to the Earliest
Atmosphere

The history of the solar system that was pieced together in chapter 1 is still fragmentary and has a somewhat uncertain future. It seems likely, however, that the Earth did not condense directly from a solar nebula by a single-stage process. Direct condensation at high temperature alone seems to be ruled out by the relatively high oxidation state of the Earth's mantle and by the presence of low-temperature condensates in the Earth. Direct condensation during the entire temperature range of solar nebula condensation is unlikely, because it is difficult to visualize the required later homogenization of an initially heterogeneous, gravitationally stratified Earth.

The inference that the Earth accreted by a multistage process is supported by the abundance and isotopic composition of the rare gases in the Earth today. This chapter will deal first with the abundance data for these elements; it will then explore their implications for the nature of the earliest atmosphere. These can be extracted from the rare-gas data and from the compositional relationships between gases and the terrestrial silicate melts that probably existed during and shortly after the end of the Earth's accretion.

1. Abundance and Distribution of the Rare Gases

Table 2.1 lists the abundance of the rare gases in the Earth's atmosphere; table 2.2 details their isotopic composition. The abundance of He in the Earth's atmosphere is determined largely by its release from the Earth's interior and by its escape rate from the atmosphere into interplanetary space (see, for instance, Jenkins 1980). ^{3}He that is released from the Earth's mantle is nearly entirely primordial, whereas most of the released ^{4}He has been generated within the Earth by radioactive decay of nuclides within the ^{235}U, ^{238}U, and ^{232}Th series.

All of the rare gases other than He are too heavy to escape from the Earth's atmosphere in any but trace quantities. A very small amount of Ne is produced by nuclear reactions within the Earth; however, most of the Ne that is now in the atmosphere is primordial. Among the isotopes of Ar, ^{40}Ar is dominant by virtue of its generation within the crust and

TABLE 2.1.

Concentration of rare gases in the Earth's atmosphere at ground level, after Holland (1978).

Element	Concentration (ppm by volume)	Residence Time
Helium	5.24 ± 0.04	2×10^6 yr for escape
Neon	18.18 ± 0.04	largely accumulating
Argon	9340 ± 10	largely accumulating
Krypton	1.14 ± 0.01	largely accumulating
Xenon	0.087 ± 0.001	largely accumulating

TABLE 2.2.

Isotopic composition of the noble gases in the atmosphere (Holland) 1978).

Element	Abundance (%)	Element	Abundance (%)
Helium		Krypton	
^{3}He	1.4×10^{-4}	^{78}Kr	0.354 ± 0.002
^{4}He	~ 100	^{80}Kr	2.27 ± 0.01
		^{82}Kr	11.56 ± 0.02
Neon		^{83}Kr	11.55 ± 0.02
^{20}Ne	90.5 ± 0.07	^{84}Kr	56.90 ± 0.1
^{21}Ne	0.268 ± 0.002	^{86}Kr	17.37 ± 0.2
^{22}Ne	9.23 ± 0.07		
		Xenon	
Argon		^{124}Xe	0.096
^{36}Ar	0.337	^{126}Xe	0.090
^{38}Ar	0.063	^{128}Xe	1.919
^{40}Ar	99.600	^{129}Xe	26.44
		^{130}Xe	4.08
		^{131}Xe	21.18
		^{132}Xe	26.89
		^{134}Xe	10.44
		^{136}Xe	8.87

mantle by ^{40}K decay and by its subsequent release to the atmosphere. ^{36}Ar and ^{38}Ar are present in essentially primordial proportions. Kr and Xe are similar to Ne in that the production of these elements within the Earth by radioactive decay has been minor.

The pattern of rare gas abundances in the atmosphere is similar to that in meteorites. The terrestrial abundances of ^{20}Ne, ^{36}Ar, ^{84}Kr, and ^{132}Xe in figure 2.1 have been calculated by dividing the content of these gases in the atmosphere by the mass of the Earth and expressing the ratio in units of cm^3 STP of each isotope/gm of Earth. The rise from ^{20}Ne to ^{36}Ar and the decrease from ^{36}Ar to ^{84}Kr and ^{132}Xe is matched by the pattern of the abundance of these isotopes in carbonaceous chondrites (C1, C2, C30, and C3V) as well as in average ordinary high-iron chondrites (H) and in the atmosphere of Mars.

The patterns in figure 2.1 are, or course, not strictly comparable. Rare gases in meteorites are present largely within solid phases, whereas the terrestrial and Martian abundance patterns in figure 2.1 only account for that part of the planetary rare gas inventories that is currently residing in the atmosphere. A truly meaningful comparison between the rare gas patterns in meteorites and the Earth would have to include the rare gas content of the solid Earth as well. There has been and continues to be considerable controversy regarding the degree to which the Earth has lost its rare gases to the atmosphere. Recent estimates range from a few tens of percent to nearly complete degassing. The data in figures 2.2 and 2.3 for the rare gas content of a number of samples that have been derived from the mantle show a good deal of scatter in their absolute rare gas content and a fair excess of Xe relative to the abundance of the lighter rare gases compared to their proportion in the Earth's atmosphere. If these samples are representative of the mantle as a whole, the Earth has been degassed significantly but not completely. The best current estimate places the fraction degassed at approximately 1/2 (see chapter 3). The apparent deficit of Xe in the atmosphere may be due to the preferential incorporation of this gas in shales (Canalas, Alexander, and Manuel 1968; Podosek, Honda, and Ozima 1980), but there are not enough analyses of rare gases in shales to establish this explanation, and the validity of the hypothesis is widely doubted.

The addition of rare gases within the crust and mantle of the Earth to those in the atmosphere would probably roughly double the ^{20}Ne, ^{36}Ar and ^{84}Kr inventory, and might increase the ^{132}Xe inventory by as much as a factor of ten. If such an inventory had been used in figure 2.1, the similarity of the rare-gas pattern of the Earth and meteorites would have been enhanced rather than reduced.

The relative proportion of the rare gases in meteorites, the Earth, and Mars is quite different from their proportion in the Sun (see figure 2.4). The most pronounced difference is in the $^{20}Ne/^{36}Ar$ ratios. In the Sun ^{20}Ne is some forty times more abundant than ^{36}Ar, whereas in meteorites, the Earth, and Mars ^{20}Ne is considerably less abundant than ^{36}Ar. This difference virtually rules out the possibility that the major part of the rare

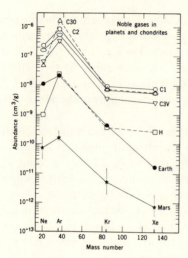

FIGURE 2.1 The abundance of the noble gases in chondritic meteorites, the Earth, and Mars. The terrestrial and Martian abundances were calculated by dividing the quantity of the rare gases in the two atmospheres by the mass of the respective planets (Anders and Owen 1977). (Copyright 1977 by the American Association for the Advancement of Science.)

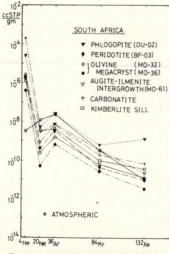

FIGURE 2.2a. Rare gas abundances in South African samples. "Atmospheric" is defined as the amount of a rare gas in the atmosphere divided by the mass of the Earth. Errors in the measurement of rare gas concentrations are estimated to be about 10% (Kaneoka, Takaoka, and Aoki 1978). (Reproduced by permission of Japan Scientific Societies Press.)

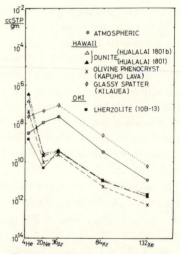

FIGURE 2.2b Rare gas abundances in Hawaii and Oki Island samples (Kaneoka, Takaoka, and Aoki 1978). (Reproduced by permission of Japan Scientific Societies Press.)

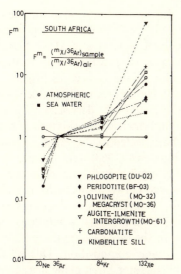

FIGURE 2.3a. Rare gas abundance patterns in South African samples relative to atmospheric abundances (Kaneoka, Takaoka, and Aoki 1978). Reproduced by permission of Japan Scientific Societies Press.)

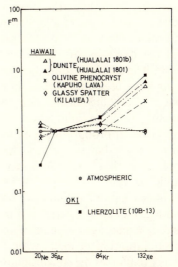

FIGURE 2.3b. Rare gas abundance patterns in Hawaii and Oki Island samples relative to atmospheric abundances (Kaneoka, Takaoka, and Aoki 1978). (Reproduced by permission of Japan Scientific Societies Press.)

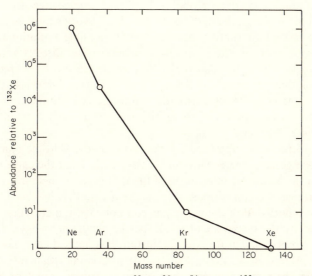

FIGURE 2.4. The relative abundance of ^{20}Ne, ^{36}Ar, ^{84}Kr, and ^{132}Xe in the Sun (Anders and Owen 1977).

gases was incorporated in these planets during equilibration at high temperatures with a gas of solar composition, although it may still be possible to explain the low $^{20}Ne/^{36}Ar$ ratios in the atmosphere of Mars, Venus, and the Earth by strong preferential loss of Ne (McElroy and Prather 1980). The available data for the solubility of the rare gases in silicate melts have been summarized in table 2.3. The figures in this table show that the solubility of the rare gases in silicate melts decreases with increasing atomic weight. Silicate melts equilibrated with a mixture of rare gases in solar proportions can therefore be expected to have a Ne/Ar ratio greater than that in the gas phase. If Kirsten's (1968) data are directly applicable, the $^{20}Ne/^{36}Ar$ ratio in melts equilibrated with solar gases would be approximately 120. The difference between this ratio and the ratios (see figure 2.1) that have been found in meteorites, in the terrestrial and Martian atmospheres, and in samples from the Earth's mantle is too large to encourage thoughts of an origin for the rare gases in these bodies by equilibration of melts with a gas of solar composition. This result simply reinforces the conclusion reached in chapter 1 that planetary accretion is unlikely to have taken place at high temperatures in the center of giant gaseous protoplanets of essentially solar composition.

At the same time, the similarity of the rare gas abundances in the Earth and in meteorites suggests that the Earth obtained its rare gases either in a manner similar to that by which meteorites obtained theirs, or that the Earth obtained its rare gases from an assemblage of meteorites. The first of these alternatives probably implies a single-stage accumulation history for the Earth; the likelihood of such a history was questioned earlier. The second alternative includes several possible evolutionary paths. There is, first of all, the assembling of meteorites directly into terrestrial planets. There are also paths that are much more complex: meteorites may have been assembled first into objects of lunar size that were then disrupted and finally reassembled into planets. On physical grounds alone, the more complex of these histories seem much more likely (see, for instance, Wetherill 1976).

This inference is supported by the data for the isotopic composition of the rare gases in meteorites and in the Earth. Even the early measurements of the isotopic composition of the rare gases in meteorites revealed a large range of compositions, and the great care that has been taken recently to identify all of the many gas components in meteorites has been rewarded by the discovery of a number of new components, each with its own isotopic signature (see, for instance, Alaerts, Lewis, and Anders 1979a,b; Alaerts et al. 1980). Much of the variability that has been discovered in the isotopic composition of the rare gases in meteorites is well understood, and is surely due to differences in the relative proportions of

TABLE 2.3.

Solubility* of rare gases in silicate melts.

	He	Ne	Ar	Kr	Xe
Enstatite melt at 1500°C	1.2×10^{-4}(a)	0.7×10^{-4} (a)	0.2×10^{-4} (a)	$[0.13 \times 10^{-4}]$ (b)	$[0.06 \times 10^{-4}]$ (b)
Gabbro-diabase melt at 1300°C	27×10^{-4} (c)	—	—	—	—
Na-Ca glass melt at 1200°C	9.2×10^{-4} (d)	—	—	—	—
K-Ca glass melt at 1480°C	25×10^{-4} (d)	—	—	—	—

* Solubility defined as the concentration of each gas in units of cc(STP)/gm in melts equilibrated with a gas phase containing the particular gas at a pressure of 1 atm.

(a) Kirsten (1968; measured).
(b) Kirsten (1968; estimated).
(c) Gerling (1940).
(d) Mulfinger and Scholze (1962).

several well-known gas constituents: trapped primordial gases, implanted solar gases, rare gases which are the decay products of relatively short-lived radioisotopes that were present in the solar nebula, and rare gases produced by cosmic ray bombardment of meteorites in space. However, the source of the isotopic signature of several of the rare gas fractions is still poorly understood.

The mechanisms of incorporation and even the phases in which the rare gases are located are still subjects of lively debate. Lewis, Srinivasan, and Anders (1975) showed that most of the trapped noble gases in the C3V carbonaceous chondrite Allende reside in a minor ($\leq 0.1\%$) phase that is resistant to attack by HCl and HF, soluble in HNO_3, and can be destroyed by other oxidizing agents, such as atomic oxygen and H_2O_2 (Frick 1977). Following the tradition of constructing slightly fey nomenclatures in this and related fields, the discoverers of the gas-rich phase have christened it "Q" for "quintessence." Anders et al. (1975) and Alaerts et al. (1980) suspect that it is a rare metallic oxide or sulfide, while Reynolds et al. (1978) favor the interpretation that "Q" is an ill-characterized, kerogenlike carbonaceous material. Whatever its true character, "Q" has now been demonstrated to be important as a carrier of rare gases in many classes of meteorites (Moniot 1980), and it seems likely that the phase was formed throughout the solar nebula during an early part of its condensation.

In spite of the extreme complexity of the present isotopic composition of the rare gases in meteorites, it does seem possible to extract some data regarding the isotopic composition of their trapped, primordial components and to compare these with the isotopic composition of the rare gases in the atmosphere. In Manuel's (1978) comparison of terrestrial and meteoritic noble gases, He and Ne were excluded, because interference from spallation products makes it impossible to identify unambiguously the isotopic composition of the trapped primordial component of these gases in iron meteorites. His results for Xe and Kr are shown in figures 2.5 and 2.6. In these, each of the spectra has been normalized to the spectrum observed in average carbonaceous chondrites (AVCC) so that

$$g_m{}^i = \frac{(^iX/^mX)_{\text{sample}}}{(^iX/^mX)_{\text{AVCC}}}. \qquad (2.1)$$

In figure 2.5, ^{132}Xe has been used as the reference isotope. The composition of AVCC is based on the data of Krummenacher et al. (1962). Abundances of ^{129}Xe have not been plotted, because of possible interference from the decay of extinct ^{129}I. The isotopic composition of Xe in the mantle samples is similar to that in air and in the Cape York iron meteorite. Atmospheric Xe is isotopically distinct from Xe in average carbo-

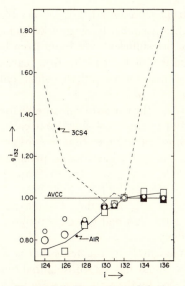

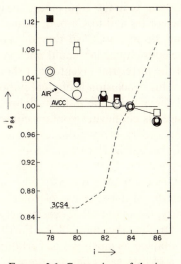

FIGURE 2.5. Comparison of the isotopic composition of Xe in samples from the Earth's mantle, in air, and in meteorites (Manuel 1978). The composition of Xe in average carbonaceous chondrites (AVCC) is used as a reference. (Reproduced by permission of Japan Scientific Societies Press.) 3CS4: carbon-rich residue of the Allende meteorite; large open square, $\square$; Kearsutitic amphibole inclusion; large open circle, $\bigcirc$: Josephinite, Oregon; large filled square, $\blacksquare$: Amphibole from a peridotite xenolith, Dish Hill, California; small open circle $\circ$: Cape York iron meteorite.

FIGURE 2.6. Comparison of the isotopic composition of Kr in samples from the Earth's mantle, air, and in meteorites (Manuel 1978). Symbols have the same significance as in figure 2.5. The small open squares refer to Kr in the Peña Blacea Springs achondrite. (Reproduced by permission of the Japan Scientific Societies Press.)

naceous chondrites and from the strange "Q" Xe found in carbon-rich residues in the Allende and in other carbonaceous meteorites. There is a suggestion that Xe in Ca-poor achondrites may be isotopically similar to that of air.

The data in figure 2.6 show a similar pattern. The isotopic composition of Kr in AVCC is taken from Eugster, Eberhardt, and Geiss (1967). Kr in the mantle is isotopically distinct from Kr in AVCC and shows an anomaly pattern opposite to that of the carbon-rich residue of Allende.

If the data in figures 2.5 and 2.6 are truly representative, it follows that Kr and Xe in the atmosphere are isotopically similar to Kr and Xe in iron meteorites and in Ca-poor achondrites, but sufficiently different from Kr and Xe in AVCC so that the contribution of Kr and Xe from carbonaceous chondrites could not have accounted for more than a small part of the terrestrial inventory of these gases.

This conclusion is of great importance for the earliest degassing history of the Earth. The data in figure 2.1 and in table 2.4 show that carbonaceous chondrites contain concentrations of volatiles that are roughly two orders of magnitude greater than the probable volatile inventory of the Earth. If carbonaceous chondrites were assembled to form the present-day Earth, a very large quantity of volatiles must have been lost either contemporaneously with or shortly after assembly. For physical reasons, a massive loss of volatiles after the Sun had passed through its early, disruptive stages seems very unlikely.

On the other hand, if the Earth is an assembly of fragments that never contained as much gas as carbonaceous chondrites, or whose initial burden of rare gases had been greatly reduced by the time they were incorporated into the Earth, there would be no need to invoke a massive loss of gas during the earliest part of Earth history to explain the small volatile inventory of our planet. The difference between the isotopic composition of Kr and Xe in the Earth and in trapped gases of meteorites was used above to argue for the presence of no more than a small component of gases from carbonaceous chondrites in the present-day Earth.

If this argument is valid, then it is likely that the Earth was initially poor in volatiles, and that there is no need to invoke an early, intense loss of volatiles from our planet. Ringwood (1979, chaps. 4 and 8), for instance, has pointed out that the overall composition of the Earth can be matched quite well by a mixture of about 10% of a low-temperature nebula condensate similar to C1 chondrites together with 90% of a high-temperature condensate that has been subjected to varying degrees of loss of volatile metals in amounts that are correlated with their volatilities. The rare-gas evidence seems to suggest that even the presence of 10% of an unmodified C1 component in the terrestrial mix may turn out to be inconsistent with the difference between the isotopic composition of the rare gases in C1 chondrites and those in the Earth's atmosphere.

Little can be said with assurance about the distribution of the rare gases between the atmosphere and the condensed parts of the growing planet. It is most likely, however, that primordial rare gases were somehow trapped during the formation of the Earth, and that they are still being released from the solid Earth. The discovery that the $^3He/^4He$ ratio in many volcanic gases, xenoliths diamonds, and volcanic glasses is higher

TABLE 2.4.

Comparison of some volatile elements in the Earth and meteorites, largely after Anders and Owen (1977).

	^{36}Ar ($10^{-8}cc(STP)/gm$)	C (ppm)	N (ppm)	H (ppm)	δD (‰)	$\delta^{13}C$ (‰)
Earth	4*	30†	2*	75*	−2	−6
Chondritic Meteorites						
Carbonaceous						
C1	87	36,000	2,800	7,900	+24	−7 to −11
C2	130	23,000	1,500	8,900	+5 to −17	−4 to −10
C3V	33	6,700	61	500	+18	−18
C3O	164	4,700	60	1,000	−12	−16
Ordinary						
H3	9.8	2,280	86	10‡		
H4	2.9	1,310	47	5‡		
H5	1.4	1,100	43	4‡		
H6	0.9	1,060	50	4‡		
H (weighted average)	2.0	1,190	47	5‡	?	−24

* Estimate for whole Earth (Holland 1978).
† Estimate for whole Earth except the mantle.
‡ These values are uncertain; they were estimated from the carbon content, assuming that 8% of the carbon is present as a polymer with C/H = 1.

than in the Earth's atmosphere (see, for instance, Craig, Clarke, and Beg 1975; Jenkins, Edmond, and Corliss 1978; Kurz et al. 1982; Ozima and Zashu 1983) is very hard to explain without calling on the trapping and subsequent release of primordial helium from the body of the Earth. It is likely that trapping occurred by the simple incorporation of solid rare-gas-containing objects. Some of the rare gases must have been released on impact, and were than reincorporated into the Earth by adsorption on solid particles and/or by dissolving in magmas at the surface of the growing planet. The partial pressure of the rare gases in the atmosphere that would be required to account for the presence of the estimated concentration of the rare gases in the solid Earth can be estimated from the data in figure 2.1 and table 2.3. The results are inconclusive for many reasons, including the large spread in the reported values for the solubility of He in silicate melts and the absence of comparable data for the solubility of the other rare gases in silicate melts. The most likely values of the required partial pressures of the rare gases as estimated with the presently available data are not impossibly large. However, if the growth period of the Earth was approximately 10^8 yrs, and if the half-life of ^{3}He escape from the atmosphere was less than 10^6 yrs, then unreasonably large quantities of ^{3}He would have had to have been available in the atmosphere to assure the presence of ^{3}He in its present concentration in mantle materials. On the whole, it therefore appears likely that the major portion of the rare gases, and of the other volatiles as well, was trapped directly together with the solids in infalling objects, and that reincorporation of volatiles in the body of the Earth after release on impact played a minor role in determining the volatile inventory of the Earth.

The view of the earliest history of the Earth that has been extracted from the accumulated data for the abundance and isotopic composition of the rare gases is therefore a rather classic one. It appears that the Earth was formed by the accumulation of preassembled materials that were rather poor in volatiles; a part of these volatiles was previously lost to an early atmosphere during accretion; since then, volatiles have been added gradually to the atmosphere and hydrosphere by degassing of the Earth's interior, and a number of these volatiles have been recycled repeatedly through the Earth's crust and upper mantle.

2. The Composition of the Earliest Volcanic Gases

We do not know what the temperature of the Earth was at the end of its main period of growth; but near-surface temperatures may well have been sufficiently high to allow wide-spread, if not pervasive, melting. Melting must have been pervasive if the Moon originated by fission from the

Earth; but high temperatures are also suggested by the results of calculations of the temperature history of the Earth during accretion. Even if accumulation of the Earth took as long as 10^8 yrs (Wetherill 1980), volcanism probably accompanied accretion or commenced during its waning stages. The intensity of volcanism and the composition of volcanic gases have always exerted a strong influence on the chemistry of the Earth's atmosphere and oceans. Thus a good deal could be learned about the near-surface chemistry of the earliest Earth if the quantity and the nature of the first volcanic gases were known.

Among the many compositional parameters of volcanic gases, their oxidation state is surely one of the most important. In the early Earth the oxidation of volcanic gases was probably determined in large part by the presence of metallic iron and FeO-containing silicates. The evidence for and against the presence of iron in the upper mantle during accretion was discussed in chapter 1. On balance, it seems likely that some iron was present in the upper mantle.

In heterogeneous accretion models metallic iron is allowed to condense as a core; the silicate mantle is considered to be a later addition. Such models run into potentially serious problems in accounting for the present degree of homogeneity of the mantle. They are also intuitively unlikely, since meteorites falling on the Earth today are a mixture of objects that includes irons and stony irons. The abundance of the siderophile elements in the upper mantle probably rules out equilibration of the mantle with large quantities of Fe^o (see chapter 1), unless Ringwood (1979, p. 128) is correct in his proposal that oxygen dissolved in metallic iron prevents the scavenging of siderophile elements from silicates at high pressures. However, the abundance of the siderophile elements is quite consonant with the presence of small amounts of Fe^o in the upper mantle early in Earth history and might simply be a consequence of nonequilibration between infalling Fe^o and surrounding silicates before the metallic iron was able to sink downward to the core.

We can now try to define the composition of volcanic gases released from the earliest mantle. A good deal of experimental work has been done to define the oxygen fugacity in gases equilibrated with silicates in the presence of metallic iron. Muan (1955), Muan and Osborn (1956), Speidel and Osborn (1967), and Speidel and Nafziger (1968) have defined the oxygen fugacity in a vapor phase in equilibrium with solids and melts in the system $Fe-O-MgO-SiO_2$ at temperatures near the liquidus. Figure 2.7 summarizes many of the data for this system at 1 atm total pressure between 1100°C and 1400°C. The phase assemblages at low oxygen fugacities in the lower stippled area include metallic iron (I). At intermediate oxygen fugacities, silicate phases either with or without wüstite (W)

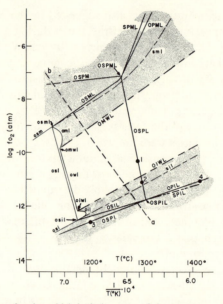

FIGURE 2.7. A 1/T-log f_{O_2} plot which includes most of the quaternary boundary curves of the system Fe-O-MgO-SiO$_2$ (Speidel and Nafziger 1968). Light lines are boundary curves in the ternary system Fe-O-SiO$_2$ (Muan 1955; Lindsley, Speidel, and Nafziger 1968); points 1 and 2 are from Speidel and Nafziger (1968); points 3 and 4 are from Nafziger (1966) and Nafziger and Muan (1967). Upper-case letters signify quaternary condensed phases; the lower case letters indicate condensed phases in the system Fe-O-SiO$_2$. Dash-dot lines indicate the additional ternary (light lines) and quaternary (heavy lines) boundary curves necessary to define T-f_{O_2} conditions of the quaternary assemblage silica-magnetite-liquid (SML), olivine-magnetite-liquid (OML), silica-iron-liquid (SIL), and olivine-iron-liquid (OIL). The position of the boundary curve silica-pyroxene-iron-liquid (SPIL) is not accurately known at present and is represented by a double dot-dot-dash line. All the possible univariant curves emanating from the invariant points omwl and oiwl are not shown in this diagram. The shaded areas indicate the T-f_{O_2} stability regions of magnetite-bearing and metallic iron-bearing assemblages. (Reproduced by permissionn of the *American Journal of Science*.)

are stable. At higher oxygen fugacities, in the upper stippled area, silicate phases and magnetite (M) are stable together. The stability region of hematite lies at oxygen fugacities beyond the limits of figure 2.7.

In the Earth's mantle the position of univariant curves involving silicate liquids almost certainly differs from that in the system Fe-O-MgO-SiO$_2$ The differences, however, are probably small. Walker et al. (1973) have determined the relationship between f_{O_2} in a gas phase and the composition of lunar basaltic melts at 1400°C equilibrated with an iron capsule.

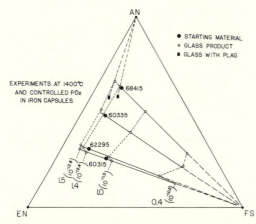

FIGURE 2.8. The values of f_{O_2} (in brackets) in gases equilibrated with iron capsules and melts of compositions similar to those of lunar basalts (Walker et al. 1973). AN = anorthite; EN = enstatite; FS = ferrosilite. (Reproduced by permission of the authors.)

With increasing f_{O_2}, iron in the container becomes oxidized and incorporated into the silicate melt; with decreasing f_{O_2}, ferrous iron in the silicate melt is reduced, metallic iron is produced, and the silicate melt becomes poorer in iron. All of the values of f_{O_2} are well below that of the iron-wüstite buffer at 1400°C. Comparison of figure 2.8 with figure 2.7 shows that melts in which the Fe/Mg + Fe ratio is similar to that of the Earth's mantle (ca. 0.12) fall within the f_{O_2} range indicated for melts within the simple system Fe-O-MgO-SiO$_2$.

The oxygen fugacity along the several univariant lines changes somewhat with increasing total pressure, and ferrosilite (FeSiO$_3$) becomes stable at pressures greater than 17.5 kb. This corresponds to a depth of ca. 50 km in the mantle. The various effects of pressure on the system Fe-O-SiO$_2$ have been described by Lindsley, Speidel, and Nafziger (1968), and the approximate phase relations in the system Fe-O-MgO-SiO$_2$ at a total pressure of ca. 15 kb are shown in figure 2.9.

Phase relationships at still higher pressures, and the oxygen fugacity of gases in equilibrium with possible mantle phases are poorly known. The possibility that ferrous iron disproportionates into metallic iron and ferric iron at very high pressures has been raised by Bell and Mao (1975). If such a disproportionation actually takes place in the mantle, the process could have had an important effect on the chemistry of the mantle and on the evolution of the oxidation state of volcanic gases.

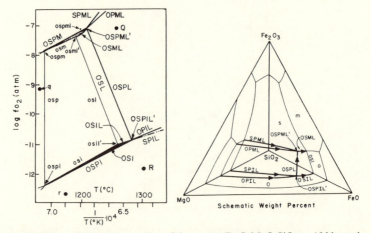

FIGURE 2.9. Phase relations in a portion of the system Fe-O-MgO-SiO$_2$ at 15 kb total pressure (Speidel and Nafziger 1968). Schematic liquidus relations are shown in the diagram on the right. The diagram on the left is a 1/T-log f_{O_2} plot of several of the boundary curves similar to those in figure 2.7. Abbreviations for the divariant fields and univariant curves are similar to those used in figure 2.7. The position of the invariant points at a total pressure of 15 kb is designated with a prime to distinguish these points from the corresponding points at a total pressure of 1 atm (figure 2.7). The latter points are plotted in this figure for comparison using the following symbols: Q = OSPML in figure 2.7; q = osml in figure 2.7; R = OSPIL in figure 2.7; and r = osil in figure 2.7. The shaded area is OSI. Ternary boundary curves are discussed by Lindsley, Speidel, and Nafziger (1968). (Reproduced by permission of the *American Journal of Science*.)

Volcanic gases in equilibrium with a basic magma saturated with respect to iron, olivine, pyroxene, and silica at 1300°C have an f_{O_2} value of about 10^{-12} atm. It is likely that the earliest volcanic gases were released at about this temperature and with approximately this value of f_{O_2} if iron was indeed present in the upper mantle during and shortly after the end of accretion. Even today the intrinsic oxygen fugacity of spinels from at least some upper mantle peridotites is close to that of the iron-wüstite boundary (Arculus and Delano 1981).

The value of f_{O_2} in a vapor phase in equilibrium with a magma at a given temperature defines a number of important ratios. Perhaps the most important of these is the ratio of the fugacity of hydrogen to that of water. The equilibrium constant for the reaction is

$$H_2O \rightleftharpoons H_2 + \tfrac{1}{2}O_2 \qquad (2.2)$$

$$K_{2.2} = \frac{f_{H_2} \cdot f_{O_2}^{1/2}}{f_{H_2O}}.$$

TABLE 2.5.

Equilibrium constants for some important gas reactions. Calculated on the basis of data in the JANAF tables (1965, 1966).

Temperature ($^\circ K$)	$\log K_{2.2}$	$\log K_{2.3}$	$\log K_{2.4}$	$\log K_{2.5}$	$\log K_{2.6}$
298	-40.06	-45.07	-140.3	-86.82	-57.22
1000	-10.07	-10.22	-41.83	-23.01	-18.33
1100	-8.89	-8.88	-38.03	-20.55	-16.84
1200	-7.90	-7.77	-34.85	-18.48	-15.58
1300	-7.07	-6.82	-32.17	-16.74	-14.50
1400	-6.35	-6.02	-28.86	-15.25	-13.59
1500	-5.73	-5.32	-27.86	-13.95	-12.80
1600	-5.18	-4.71	-26.12	-12.82	-12.10
1700	-4.70	-4.17	-24.57	-11.81	-11.48

$$K_{2.2} = \frac{f_{H_2} \cdot f_{O_2}^{1/2}}{f_{H_2O}};\ K_{2.3} = \frac{f_{CO} \cdot f_{O_2}^{1/2}}{f_{CO_2}};\ K_{2.4} = \frac{f_{CH_4} \cdot f_{O_2}^{2}}{f_{CO_2} \cdot f_{H_2O}^{2}};$$

$$K_{2.5.} = \frac{f_{H_2S} \cdot f_{O_2}^{3/2}}{f_{SO_2} \cdot f_{H_2O}};\ \text{and } K_{2.6} = \frac{f_{NH_3} \cdot f_{O_2}^{3/4}}{f_{N_2}^{1/2} \cdot f_{H_2O}^{3/2}}.$$

The value of $K_{2.2}$ is well known (see table 2.5) in the temperature range of particular interest for magma generation. Figure 2.10 shows the position of the univariant lines OSPI and OSPM at 1 atm total pressure and at a total pressure of 15 kb in relation to lines of constant f_{H_2}/f_{H_2O}. For a definition of the abbreviations see the legend in figure 2.7. At 1 atm the OSPI line lies between the lines $f_{H_2}/f_{H_2O} = 1$ and $f_{H_2}/f_{H_2O} = 10$, the OSPM line between the lines $f_{H_2}/f_{H_2O} = 10^{-2}$ and $f_{H_2}/f_{H_2O} = 10^{-3}$.

The oxygen fugacity along the univariant lines OSPI and OSPM also determines the ratio f_{CO}/f_{CO_2} in melts generated in a mantle containing metallic iron. Values for the equilibrium constant $K_{2.3}$ for the reaction,

$$CO_2 \rightleftharpoons CO + \tfrac{1}{2}O_2 \tag{2.3}$$

$$K_{2.3} = \frac{f_{CO} \cdot f_{O_2}^{1/2}}{f_{CO_2}}$$

have been calculated using the JANAF data, and are listed in table 2.5. Figure 2.11 shows that the ratio f_{CO}/f_{CO_2} in melts in equilibrium with the assemblage OSPI lies between 1 and 10; the f_{CO}/f_{CO_2} ratio in melts equilibrated with the assemblage OSPM is near 10^{-2} in the vicinity of 1200°C.

If graphite is present as a solid phase, the value of f_{CO} and of f_{CO_2} is fixed at any given temperature and f_{O_2} value. French (1966) has shown

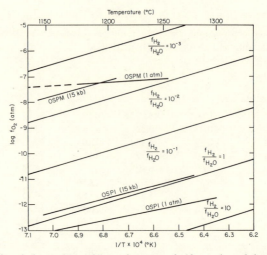

FIGURE 2.10. The relative position of lines of constant f_{H_2}/f_{H_2O} ratio and the univariant lines OSPI and OSPM in the system Fe-O-MgO-SiO$_2$. See text for sources of data.

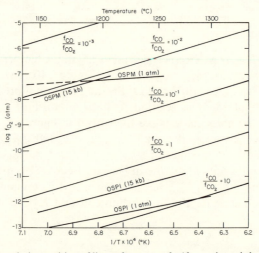

FIGURE 2.11. The relative position of lines of constant f_{CO}/f_{CO_2} ratio and the univariant lines OSLI and OSPM in the system Fe-O-MgO-SiO$_2$. See text for sources of data.

that the gas pressure in equilibrium with graphite and the f_{O_2}-buffer magnetite-wüstite (MW) exceeds 10 kb at 1000°C. Not enough carbon is present in the Earth to sustain an atmospheric $CO + CO_2$ pressure of this magnitude. The ($CO + CO_2$) pressure in equilibrium with graphite and the buffers wüstite-iron (WI) and silica-fayalite-iron (QFI) is very much lower than this at 1000°C, but even at the oxygen fugacity of these buffers the $CO + CO_2$ pressure at 1200°C in equilibrium with graphite exceeds the estimated maximum atmospheric pressure permitted by the probable carbon inventory of the Earth. It is unlikely, therefore, that graphite was a solid phase on the liquidus of early surface magmas, although graphite exists in the source regions of basalts today (Sato 1978).

The fugacity ratio f_{CH_4}/f_{CO_2} in a gas phase at high temperatures is determined by equilibrium in the reaction

$$CO_2 + 2H_2O \rightleftharpoons CH_4 + 2O_2 \qquad (2.4)$$

$$K_{2.4} = \frac{f_{CH_4} \cdot f_{O_2}^2}{f_{CO_2} \cdot f_{H_2O}^2}$$

and is fixed at any given temperature only if f_{O_2} and f_{H_2O} are known. In figure 2.12 the f_{CH_4}/f_{CO_2} ratio in a vapor phase for which $f_{H_2O} = 5$ atm has been plotted in the temperature and f_{O_2} range of interest. Under these conditions f_{CH_4} is trival compared to f_{CO_2} at magmatic temperatures, except along the boundary OSPI, where values of the ratio f_{CH_4}/f_{CO_2} are on the order of 0.1. At higher values of the water fugacity, f_{CH_4} can exceed f_{CO_2}. Such high water fugacities are unlikely in volcanic gases released reasonably unexplosively from volcanoes, but significant quantities of methane have been detected in fluid inclusions in diamonds (Melton and Giardini 1974, 1975, and 1981). The origin of these inclusions is not certain, but their methane content is consistent with a high-temperature,

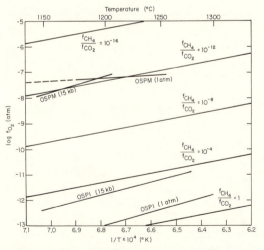

FIGURE 2.12. The relative position of the lines of constant f_{CH_4}/f_{CO_2} ratio and the invariant lines OSPI and OSPM in the system Fe-O-MgO-SiO$_2$ at $f_{H_2O} = 5$ atm. See text for sources of data.

high-pressure origin within the mantle. Volcanic gases released explosively from a carbon-rich part of the mantle can add and could have contributed significant quantities of CH_4 to the atmosphere.

The ratio of the fugacity of H_2S to that of SO_2 is determined by equilibrium in the reaction

$$SO_2 + H_2O \rightleftharpoons H_2S + \tfrac{3}{2}O_2 \qquad (2.5)$$

$$K_{2.5} = \frac{f_{H_2S} \cdot f_{O_2}^{3/2}}{f_{SO_2} \cdot f_{H_2O}}$$

Lines of constant f_{H_2S}/f_{SO_2} ratio in a vapor phase containing water at a fugacity of 5 atm are shown in figure 2.13 together with the log f_{O_2}-T lines for the OSPI and OSPM assemblages between ca. 1150°C and 1350°C. f_{H_2S} exceeds f_{SO_2} in a vapor phase in equilibrium with the assemblage OSPI, whereas f_{SO_2} is greater than f_{H_2S} in a vapor phase in equilibrium with the assemblage OSPM. In volcanic gases released at higher water pressures, the SO_2-H_2S equilibrium is shifted in the direction of H_2S. If pyrrhotite or a melt of pyrrhotite composition was present in the mantle in equilibrium with an assemblage whose f_{O_2} is buffered, then f_{S_2} and f_{SO_2} were, of course, both also buffered.

The oxidation state of nitrogen dissolved in basaltic melts depends on the fugacity of water and oxygen and on the total quantity of dissolved

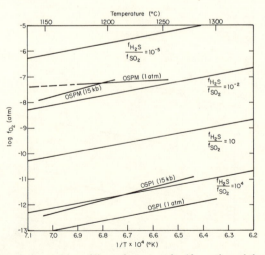

FIGURE 2.13. The relative position of lines of constant f_{H_2S}/f_{SO_2} ratio and the univariant lines OSPI and OSPM in the system Fe-O-MgO-SiO$_2$ at a water fugacity of 5 atm. See text for sources of data.

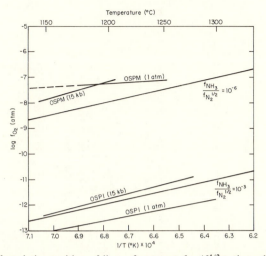

FIGURE 2.14. The relative position of lines of constant $f_{NH_3}/f_{N_2}^{1/2}$ ratio and the univariant lines OSPI and OSPM in the system Fe-O-MgO-SiO$_2$ at a water fugacity of 5 atm. See text for sources of data.

nitrogen and nitrogen compounds. In figure 2.14 lines of constant fugacity ratio $f_{NH_3}/f_{N_2}^{1/2}$ calculated from the equilibrium constant for the reaction,

$$\tfrac{1}{2}N_2 + \tfrac{3}{2}H_2O \rightleftharpoons NH_3 + \tfrac{3}{4}O_2 \qquad (2.6)$$

$$K_{2.6} = \frac{f_{NH_3} \cdot f_{O_2}^{3/4}}{f_{N_2}^{1/2} \cdot f_{H_2O}^{3/2}},$$

have been plotted for gases containing water vapor at a fugacity of 5 atm. The value of the ratio $f_{NH_3}/f_{N_2}^{1/2}$ is always much less than unity, even in equilibrium with metallic iron. NH$_3$ is therefore an important nitrogen species in volcanic gases only at values of $(P_{NH_3} + P_{N_2}) \leq 10^{-4}$ atm.

In table 2.6 some of the pertinent calculated fugacity ratios at $f_{O_2} = 10^{-12}$ and 10^{-8} atm are compared with those in present-day volcanic gases. Early volcanic gases released at low pressures in equilibrium with metallic iron would have consisted largely of H$_2$ and H$_2$O. CO would have been the dominant carbon species, H$_2$S the dominant sulfur species, and N$_2$ almost certainly the dominant nitrogen species.

The oxidation state of present-day volcanic gases is quite different. Their f_{O_2} value is roughly 10^{-8} atm, a little below that required for equilibrium with the assemblage OSPM between 1150° and 1200°C (Sato and Wright 1966; Sato and Moore 1973; Carmichael and Nicholls 1967; Gerlach and Nordlie 1975a,b,c), and in reasonable agreement with the

TABLE 2.6.

Comparison of calculated fugacity ratios in gases at $f_{O_2} = 10^{-12}$ and 10^{-8} with observed fugacity ratios in some present-day volcanic gases.

Fugacity Ratio	$f_{O_2} = 10^{-12.0}$ $f_{H_2O} = 5$ atm		$f_{O_2} = 10^{-8.0}$ $f_{H_2O} = 5$ atm		Volcanic Gases from Surtsey, Hawaii, and Erta'Ale (Holland 1978, p. 289)
	$T = 1400°K$	$T = 1500°K$	$T = 1400°K$	$T = 1500°K$	
f_{H_2}/f_{H_2O}	$10^{-0.3}$	$10^{+0.3}$	$10^{-2.3}$	$10^{-1.7}$	$10^{-2.1}$ to $10^{-1.3}$
f_{CO}/f_{CO_2}	1.0	$10^{+0.7}$	$10^{-2.0}$	$10^{-1.3}$	$10^{-1.5}$ to $10^{-1.3}$
f_{CH_4}/f_{CO_2}	$10^{-3.5}$	$10^{-2.5}$	$10^{-11.5}$	$10^{-10.5}$	no CH_4 observed
f_{H_2S}/f_{SO_2}	$10^{+3.4}$	$10^{+4.7}$	$10^{-2.6}$	$10^{-1.3}$	no H_2S observed*
$f_{NH_3}/f_{N_2}^{1/2}$	$10^{-3.6}$	$10^{-2.8}$	$10^{-6.6}$	$10^{-5.8}$	

* H_2S has been observed recently in Hawaiian volcanic gases.

mineralogy of basalts and diabases. The dominant gas species in present-day volcanic gases is H_2O; H_2 is quite subordinate. CO_2 is the dominant carbon species (see also Murck, Burruss, and Hollister 1978), SO_2 is the dominant sulfur species, and N_2 is the dominant nitrogen species.

It is very likely that the redox state of terrestrial volcanic gases has always fallen within the range represented by the values in table 2.6. No significant changes have been detected in the oxidation state of basaltic rocks during the past 2.5 b.y. (Steinthórssen 1974). However, some indirect evidence suggests that in Archean basaltic magmas f_{O_2} may have been lower than in more recent basaltic magmas. The Cr content of basaltic magmas is f_{O_2}-sensitive, since at low f_{O_2}, Cr^{+2} can contribute to the total Cr content of these magmas. The old layered intrusions—the Stillwater, Bushveld, and Great Dyke—are richer in Cr than their more recent counterparts, present MORBs and Phanerozoic ophiolites; D. Walker (personal communication, 1982) has pointed out that this difference might be related to a progressive increase of f_{O_2} in the mantle, and the matter surely deserves more careful study.

Contrary to some expressed opinions, the present oxidation state of basalts and of associated volcanic gases can tell us very little about the oxidation state of the mantle during and shortly after accretion; they do, however, impose the restriction that any explanation for the oxidation state of the early mantle and of early volcanic gases must be consistent with their later evolution to the present state of the Earth. An explanation that is reasonable but not compelling can be proposed for this evolution. Elemental iron must have been removed from the mantle soon after accretion if the early history of the Earth that was outlined above is roughly correct, because it is hard to imagine that a phase as dense as elemental iron would not have sunk rapidly. Thereafter, the oxidation state of magmas in the mantle must have been determined by the local bulk composition of the mantle and by the requirement of equilibrium or near-equilibrium between the several redox couples in the magma and in the residual solids.

Among the elements that are apt to exist in more than one oxidation state in the mantle iron, hydrogen, sulfur, and carbon are quantitatively the most important. After the removal of Fe°, iron was probably present dominantly in the divalent state, unless the disproportionation of "FeO" to Fe° and "Fe_2O_3" in the upper mantle turns out to be more important than seems likely at present.

Hydrogen was probably present largely as water of hydration in silicates (including OH^-), although some H_2 was presumably present in the sites that were also occupied by the rare gases. Sulfur presumably existed largely as a constituent of pyrrhotite. Carbon may have been present as graphite, as diamond, and as a constituent of carbonates (see, for instance, Eggler 1978). If very little H_2 was retained in the mantle, then H_2O released during melting acted as an oxidizing agent in magmas generated within the mantle. The reaction of H_2O with Fe^{+2} yielded some Fe^{+3}:

$$2Fe^{+2} + H_2O \rightleftharpoons 2Fe^{+3} + O^= + H_2, \tag{2.7}$$

reaction of H_2O with FeS yielded SO_2:

$$FeS + 3H_2O \rightleftharpoons \text{``FeO''} + SO_2 + 3H_2, \tag{2.8}$$

and reaction of H_2O with carbon yielded CO, CO_2, and CH_4:

$$C^\circ + H_2O \rightleftharpoons CO + H_2, \tag{2.9}$$

$$C^\circ + 2H_2O \rightleftharpoons CO_2 + 2H_2 \tag{2.10}$$

$$2C^\circ + 2H_2O \rightleftharpoons CH_4 + CO_2. \tag{2.11}$$

Decarbonation reactions that yield CO_2 and silicate minerals would not have involved a change in the oxidation state of carbon.

A detailed analysis of melting processes in the mantle involving all of the pertinent equilibria is not yet possible; however, we can show that the chemistry of some present-day volcanic gases and the redox state of some basaltic magmas are compatible with mantle melting under the proposed conditions.

Let us consider first reaction 2.7. Its equilibrium constant,

$$K_{2.7} = \left(\frac{a_{Fe^{+3}}}{a_{Fe^{+2}}}\right)^2 \cdot \frac{a_{O^=} f_{H_2}}{f_{H_2O}},$$

can be rewritten in the form

$$K'_{2.7} = \left(\frac{m_{Fe^{+3}}}{m_{Fe^{+2}}}\right)^2 \cdot \frac{f_{H_2}}{f_{H_2O}}$$

as long as $a_{O^=}$ and the overall activity coefficients $\gamma_{Fe^{+3}}$ and $\gamma_{Fe^{+2}}$ are roughly constant during the course of melting. The value of $K'_{2.7}$ for basalts can be extracted from Fudali's (1965) data. Figure 2.15 shows the value of $\log m_{Fe^{+3}}/m_{Fe^{+2}}$ in basaltic and andesitic magmas in equilibrium with a vapor phase of variable f_{O_2}. The slope of the lines of best fit should be 0.25. The fact that the slopes are closer to 0.20 suggests that the activity coefficient ratio $\gamma_{Fe^{+3}}/\gamma_{Fe^{+2}}$ is not quite constant, or that complexing of one or both species affects the slope somewhat. For our purposes it seems best to accept the measured value of the Fe^{+3}/Fe^{+2} ratio at 1200°C and $f_{O_2} = 10^{-8.0}$ atm; this value is close to the oxygen fugacity of present-day basaltic magmas. We can then calculate the value

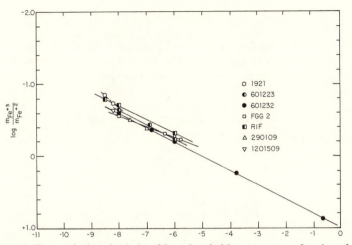

FIGURE 2.15. Ferrous-ferric ratios in basaltic and andesitic magmas as a function of f_{O_2} at 1200°C; after Fudali (1965).

of the equilibrium constant for the reaction,

$$4\text{"FeO"} + O_2 \rightleftharpoons 4\text{"FeO"}_{1.5} \tag{2.12}$$

$$K'_{2.12} \cong \left(\frac{m_{Fe^{+3}}}{m_{Fe^{+2}}}\right)^4 \cdot \frac{1}{f_{O_2}} = \frac{(10^{-0.65})^4}{10^{-8.0}} = 10^{5.4 \pm 0.2},$$

at this temperature and f_{O_2} value. The equilibrium constant $K'_{2.12}$ is related to $K'_{2.7}$ and $K_{2.1}$ by the expression,

$$K'_{2.7} = (K'_{2.12})^{1/2} \cdot (K_{2.1}).$$

$K'_{2.7}$ therefore has a value of $10^{-3.2 \pm 0.1}$ at 1200°C.

When melting takes place in the mantle, a vapor phase is probably absent; however, the presence of CO_2-rich fluid inclusions in mantle olivines indicates that basalt may be vesiculated even in its source regions. In the absence of a vapor phase, the degree to which Fe^{+2} and H_2O react in the manner suggested by equation 2.7 depends on the activity-concentration relationship of H_2 in basalts and andesites. The solubility of H_2O in basaltic and andesitic melts has been measured by Hamilton, Burnham, and Osborn (1964) (see table 2.7). These authors have shown that water dissolves in such melts largely via the reaction,

$$H_2O + O^= \rightleftharpoons 2OH^-. \tag{2.13}$$

However, molecular water is also present in the melts. The proportion of molecular water increases with increasing total water content and

TABLE 2.7.

The solubility of water in basaltic and andesitic melts at 1100°C (Hamilton, Burnham, and Osborn 1964).

		Solubility	
Run no.	Pressure (bars)	Wt. %	Mole %
Basalt			
204	1,034	3.09	10.15
196	2,000	4.59	14.37
203	3,000	5.93	17.82
200	4,000	7.30	21.07
209	5,343	8.51	23.73
205	6,067	9.37	25.52
Andesite			
A12	1,034	4.49	14.00
A 8	2,000	6.04	17.96
A 9	3,000	7.40	21.15
A11	4,000	8.70	23.98
A14	5,309	10.08	26.76

becomes equal to "OH^- water" when the total water content is equal to ca. 4 wt. percent (Stolper 1982). The equilibrium constant for reaction 2.13,

$$K_{2.13} = \frac{a_{OH^-}^2}{f_{H_2O} \cdot a_{O^=}},$$

can be approximated by the expression,

$$K'_{2.13} \simeq \frac{(2m_{H_2O})^2}{f_{H_2O}},$$

so long as $a_{O^=}$ is reasonably constant, and the total water content is small. From the experimental data it follows that for basalts at 1100°C,

$$m_{H_2O} \approx 5.3 \times 10^{-2} \, f_{H_2O}^{1/2} \text{ mol/kg melt},$$

and for andesites at the same temperature

$$m_{H_2O} \approx 8.0 \times 10^{-2} \, f_{H_2O}^{1/2} \text{ mol/kg melt}.$$

The solubility of water in these melts is not strongly dependent on temperature. Little error is therefore introduced by using these solubility data for calculations at 1200°C.

There are no comparable data for the solubility of H_2 in silicate melts. It seems reasonable to believe that H_2 behaves in a fashion similar to that of the rare gases, and that its concentration in silicate melts is proportional to its fugacity in a coexisting gas phase. The constant B_{H_2} in the equation.

$$m_{H_2} = B_{H_2} \cdot f_{H_2}, \tag{2.14}$$

is the Henry's Law constant of H_2. Blander et al. (1959) found that the Henry's Law constants for the solubility of the rare gases in salt melts follow the expression

$$B_i = a \exp(-br_i^2), \tag{2.15}$$

where a and b are constants and r_i is the gas kinetic radius of the rare gas atoms. Kirsten (1968) found this to be true as well for the relative solubility of the rare gases in enstatite melts. Since the molecular radius of H_2 is close to the atomic radius of He, the solubility of the two gases is probably similar.

Gerling's (1940) figure in table 2.3 for the solubility of He in a gabbro-diabase melt at $1300°C$ comes closest to the figure that is needed to solve the problem at hand. Roughly then,

$$B_{H_2} \approx B_{He} \approx \frac{27 \times 10^{-4} \text{ cc STP/gm atm}}{2.2 \times 10^4 \text{ cc STP/mol}} \times 10^3 \text{ gm/kg}$$

$$\approx 1.2 \times 10^{-4} \text{ mol/kg atm}$$

The expression for $K'_{2.7}$ for basalts at ca. $1200°C$ can now be recast in the form,

$$K'_{2.7} \simeq 10^{-3.2 \pm 0.1} \simeq \left(\frac{m_{Fe^{+3}}}{m_{Fe^{+2}}}\right)^2 \frac{m_{H_2}}{1.2 \times 10^{-4}} \cdot \frac{(5.3 \times 10^{-2})^2}{(m_{H_2O})^2},$$

and rearranged so that

$$\frac{m_{Fe^{+3}}}{m_{Fe^{+2}}} \cdot \frac{(m_{H_2})^{1/2}}{m_{H_2O}} \simeq 10^{-2.3}. \tag{2.16}$$

The uncertainty in the figure $10^{-2.3}$ is determined in large part by the very considerable uncertainty in the value of B_{H_2}.

The concentration of "FeO" in most basalts is $10 \pm 2\%$ by weight, i.e., approximately 1.5 mol/kg. Using this figure and equation 2.16, we can calculate the H_2/H_2O and the Fe^{+3}/Fe^{+2} ratio in basaltic melts as a function of their initial water content if no reaction takes place between the dissolved volatiles and the residual mantle minerals. As shown in figure 2.15, the oxidation of Fe^{+2} to Fe^{+3} by water dissolved in magmas does not go very far. If the initial water content is 1.8% by weight (i.e., $m_{H_2O} = 1.0$ mol/kg), the ratio m_{Fe+3}/m_{Fe+2} at equilibrium is 0.035. If the initial water content is 3.6% by weight, the m_{Fe+3}/m_{Fe+2} ratio is 0.055. The proportion of Fe^{+3} continues to rise with increasing water content of the magma, but concentrations of water as high as 4% in MOR basalts are unlikely or at least uncommon (see, for instance, Moore 1970; Killingsley and Muenow 1975; and Garcia, Liu, and Muenow 1979). The calculated m_{Fe+3}/m_{Fe+2} ratios would, of course, be too low if the estimated solubility of H_2 in magmas is too low; it would be too high if water in the melt reacts with residual mantle minerals.

The Fe^{+3}/Fe^{+2} ratio in almost unaltered, low alkali MOR basalts is less than 0.2 (see, for instance, Engel, Engel, and Havens 1965), but few basalts have Fe^{+3}/Fe^{+2} ratios as low as those suggested by the above calculations if the proposed mechanism for generating Fe^{+3} in mantle melts were entirely correct. Fortunately, the agreement between calculated and observed Fe^{+3}/Fe^{+2} ratios can be improved considerably by including the effects of reactions between magmas and their volatiles during their release close to the surface of the Earth. Subaerial volcanics have lost virtually all of their volatiles. Even submarine volcanics tend to be vesicular, although lavas extruded on the deeper parts of the ocean floor have frequently retained most, if not all, of their volatiles (see, for instance, Moore 1970).

If volatiles that have been released during vesiculation continue to react with their parent lava, additional Fe^{+3} will be produced. This can be shown in the following way. In the vapor phase within vesicles,

$$\frac{f_{H_2}}{f_{H_2O}} \approx \left(\frac{m_{H_2}}{m_{H_2O}}\right)_{vapor},$$

and therefore at equilibrium between a basalt and its vesicle-bound volatiles,

$$\left(\frac{m_{Fe+3}}{m_{Fe+2}}\right)^2 \left(\frac{f_{H_2}}{f_{H_2O}}\right) \approx \left(\frac{m_{Fe+3}}{m_{Fe+2}}\right)^2 \left(\frac{m_{H_2}}{m_{H_2O}}\right)_{vapor} \times 10^{-3.2}. \qquad (2.17)$$

If the initial Fe^{+2} content of the magma was 1.5 moles/kg, then the m_{H_2}/m_{H_2O} ratio in the final melt and the m_{H_2}/m_{H_2O} ratio in the associated volatiles can be calculated as a function of the initial water content of the magma. The results are shown in figure 2.15. m_{Fe+3}/m_{Fe+2} ratios of 0.10 can be obtained at geologically reasonable initial concentrations of water in basaltic magmas. However, Fe^{+3}/Fe^{+2} ratios of 0.2 are virtually impossible to explain by this mechanism. Such high ratios may well be due to preferential loss of H_2 during degassing followed by reaction of H_2O with Fe^{+3} beyond the limits set by the two closed system models. The

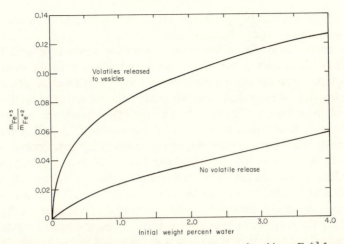

FIGURE 2.16. The calculated $m_{Fe^{+3}}/m_{Fe^{+2}}$ ratio in magmas produced in an Fe^{+3}-free mantle. The lower curve shows the relationship between $m_{Fe^{+3}}/m_{Fe^{+2}}$ and m_{H_2O} in a magma from which no volatiles have been released; the upper curve is drawn for a magma in which the volatiles have exsolved into vesicles.

even higher Fe^{+3}/Fe^{+2} ratios in many unaltered igneous rocks are presumably the result of the remelting of somewhat oxdized starting materials and the effects of fractional melting, fractional crystallization, and the oxidizing effects of meteoric and connate waters.

In the calculations that led up to figure 2.15, the effect of sulfur and of carbon on the redox state of magmas was neglected. This may not have been justified. The best estimate of the total inventory of sulfur in the oceans and crust is probably that of Holser and Kaplan (1966). Their figure of 5×10^{20} mol for the inventory of sulfur indicates that the abundance of this element is minor in the oceans and in sedimentary rocks compared to the abundance of water (1.1×10^{23} mol). If all of the near-surface sulfur was derived from the mantle as a constituent of SO_2, and if this SO_2 was generated by the reaction of water with pyrrhotite (see eqation 2.8), then approximately 1.5×10^{21} mol of H_2 were produced in this manner. This is approximately 1.3% of the quantity of near-surface water. Since water probably arrived at the surface together with such a percentage of H_2, the effect of SO_2 generation may not have been negligible compared to the effect of Fe^{+3} generation on the redox state of volcanic gases. In some volcanic gases the SO_2/H_2O ratio is much greater than might be expected on the basis of the relative abundance of S and H_2O in the oceans and in the Earth's crust. These high ratios are probably related to differences in the release rate of SO_2 and H_2O from volcanoes and/or to the recycling of crustal volatiles.

The effect of carbon on the redox state of magmas generated in the mantle depends on the oxidation state of this element in the mantle. If most of the carbon in the mantle is present as a constituent of carbonates, the redox state of magmas is only affected by the reduction of small quantities of the released CO_2 to CO. On the other hand, the presence of diamonds in kimberlites proves that at least some C^o is still present in the mantle. The best current estimate of carbon in the atmosphere, ocean, and crust is ca. 7.5×10^{21} mol (Holland 1978, p. 276). If all of this carbon was released from the mantle by reaction 2.10, then approximately 1.5×10^{22} mol of H_2 due to this reaction have been released; such a quantity of H_2 is considerably greater than the quantity that was presumably generated by the reaction of "FeO" with H_2O. Thus, if a sizable fraction of carbon in the mantle is or was present in elemental form, the calculations of the relationship between the Fe^{+3}/Fe^{+2} ratio in lavas and the H_2/H_2O ratio in associated volcanic gases are seriously in error. Since the oxidation of C^o in the mantle tends to reduce the extent to which H_2O reacts with Fe^{+2} to produce Fe^{+3} and H_2, C^o oxidation creates additional problems for the hypothesis that was advanced earlier to explain the observed range of Fe^{+3}/Fe^{+2} ratios in MOR basalts.

3. Summary

The abundance pattern of the rare gases in the Earth is similar to that in meteorites but differs from that in solar gases. This observation, together with the relatively homogeneous distribution of low-temperature condensates in the Earth, points to an origin of the Earth by the accumulation of previously assembled solids. The isotopic composition of Kr and Xe in the atmosphere differs significantly from the isotopic composition of these gases in the more volatile-rich, carbonaceous meteorites, but seems to be similar to those in the more volatile-poor meteorites. This suggests that the original inventory of volatiles in the materials that were assembled to form the Earth was not much larger than the current volatile inventory of the Earth, and that a major loss of volatiles from the Earth did not take place during or shortly after the end of accretion.

It is likely that the Earth was sufficiently hot toward the end of the period of accumulation so that melting of the mantle was widespread. Some metallic iron was probably present in the upper mantle during accretion, and must have influenced the oxidation state of the earliest volcanic gases. These probably consisted in large part of H_2, H_2O, CO, and H_2S. N_2 was almost certainly the dominant nitrogen species.

The oxidation state of volcanic gases after the removal of metallic iron from the upper mantle was probably determined in large part by the FeO and H_2O content of mantle silicates. If so, H_2O acted as an oxidizing agent for "FeO" and reacted with "FeO" to produce "Fe_2O_3" and H_2. The quantity of "Fe_2O_3" produced in this way depended on the evolution of volatiles from basaltic magmas. It is possible, though somewhat difficult, to account for the Fe_2O_3/FeO ratio in unaltered modern basalts by this mechanism alone. The reaction of FeS and elemental carbon with water in the regions of magma generation complicates matters. However, these reactions do not seem to rule out the proposed mechanism for the evolution of the gases associated with basaltic volcanism from a rather reduced state during and shortly after accretion to their present, considerably more oxidized composition.

References

Alaerts, L., Lewis, R. S., and Anders, E. 1979a. Isotopic anomalies of noble gases in meteorites and their origins: III. LL-chondrites. *Geochim. Cosmochim. Acta* 43:1399–1415.

————. 1979b. Isotopic anomalies of noble gases in meteorites and their origins: IV. C3 (Ornans) carbonaceous chondrites. *Geochim. Cosmochim. Acta* 43:1421–1432.

Alaerts, L., Lewis, R. S., Matsuda, J.-I., and Anders, E. 1980. Isotopic anomalies of noble gases in meteorites and their origins: VI. Presolar components in the Murchison C2 chondrite. *Geochim. Cosmochim. Acta* 44:189–209.

Anders, E., Higuchi, H., Gros, J., Takahashi, H., and Morgan. J. W. 1975. Extinct superheavy element in the Allende meteorite. *Science* 190:1262–1271.

Anders, E., and Owen, T. 1977. Mars and Earth: Origin and abundance of volatiles. *Science* 198:453–465.

Arculus, R. J., and Delano, J. W. 1981. Intrinsic oxygen fugacity measurements: Techniques and results for spinels from upper mantle peridotites and megacryst assemblages. *Geochim. Cosmochim. Acta* 45:899–913.

Bell, P. M., and Mao, H. K. 1975. Preliminary evidence of disproportionation of ferrous iron in silicates at high pressures and temperatures. *Carnegie Inst. Washington Yearbk.* 74:557–559.

Blander, M., Grimes, W. R., Smith, N. V., and Watson, G. M. 1959. Solubility of noble gases in molten fluorides: II. The LiF-NaF-KF eutectic mixture. *J. Phys. Chem.* 63:1164–1167.

Canalas, R. A., Alexander, E. C., Jr., and Manuel, O. K. 1968. Terrestrial abundance of noble gases. *J. Geophys. Res.* 73:3331–3334.

Carmichael, I.S.E., and Nicholls, J. 1967. Iron-titanium oxides and oxygen fugacities in volcanic rocks. *J. Geophys. Res.* 72:4665–4687.

Craig, H., Clarke, W. B., and Beg, M. A. 1975. Excess ^{3}He in deep water on the East Pacific Rise. *Earth Planet. Sci. Lett.* 26:125–132.

Eggler, D. H. 1978. Stability of dolomite in a hydrous mantle with implications for the mantle solidus. *Geology* 6:397–400.

Engel, A.E.J., Engel, C. G., and Havens, R. G. 1965. Chemical characteristics of oceanic basalts and the upper mantle. *Bull. Geol. Soc. Amer.* 76:719–734.

Eugster, O., Eberhardt, P., and Geiss, J. 1967. The isotopic composition of krypton in unequilibrated and gas-rich chondrites. *Earth Planet. Sci. Lett.* 2:385–393.

French, B. M. 1966. Some geological implications of equilibrium between graphite and a C-H-O gas phase at high temperatures and pressures. *Rev. Geophys.* 4:223–253.

Frick, U. 1977. Anomalous krypton in the Allende meteorite. *Proc. Lunar Sci. Conf. 8th*, 273–292. New York: Pergamon.

Fudali, R. F. 1965. Oxygen fugacities of basaltic and andesitic magmas. *Geochim. Cosmochim. Acta* 29:1063–1075.

Garcia, M. O., Liu, N.W.K., and Muenow, D. W. 1979. Volatiles in submarine volcanic rocks from the Mariana Island arc and trough. *Geochim. Cosmochim. Acta* 43:305–312.

Gerlach, T. M., and Nordlie, B. E. 1975a. The C-O-H-S gaseous system, Part I: Composition limits and trends in basaltic gases. *Amer. Jour. Sci.* 275:353–376.

———. 1975b. The C-O-H-S gaseous system, Part II: Temperature, atomic composition, and molecular equilibria in volcanic gases. *Amer. Jour Sci.* 275:377–394.

———. 1975c. The C-O-H-S gaseous system, Part III: Magmatic gases compatible with oxides and sulfides in basaltic magmas. *Amer. Jour Sci.* 275:395–410.

Gerling, E. K. 1940. The solubility of helium in melts. *Compt. Rend. Acad. Sci. USSR* 27:22–23.

Hamilton, D. L., Burnham, C. W., and Osborn, E. F. 1964. The solubility of water and effects of oxygen fugacity and water content on crystallization in mafic magmas. *Jour. Petrol.* 5:21–39.

Holland, H. D. 1978. *The Chemistry of the Atmosphere and Oceans.* New York: Wiley.

Holser, W. T., and Kaplan, I. R. 1966. Isotope geochemistry of sedimentary sulfates. *Chem. Geol.* 1:93–135.

JANAF. 1965. *Thermochemical Tables*. U.S. Dept. Commerce, National Bureau of Standards, Inst. for Applied Technology, PB 168 370D.

————. 1966. *Thermochemical Tables*. First Addendum, U.S. Dept. Commerce, National Bureau of Standards, Inst. for Applied Technology, PB 168 370D-1.

Jenkins, W. J. 1980. On the terrestrial budgets of helium and argon isotopes and the degassing of the Earth. Unpublished manuscript.

Jenkins, W. J., Edmond, J. M., and Corliss, J. B. 1978. Excess ^{3}He and ^{4}He in Galapagos submarine hydrothermal waters. *Nature* 272:156–158.

Kaneoka, I., Takaoka, N., and Aoki, K-I. 1978. Rare gases in mantle-derived rocks and minerals. In *Terrestrial Rare Gases*, edited by E. C. Alexander, Jr., and M. Ozima, 71–83. Japan Scientific Societies Press.

Killingsley, J. S., and Muenow, D. W. 1975. Volatiles from Hawaiian submarine basalts determined by dynamic high temperature mass spectrometry. *Geochim. Cosmochim. Acta* 39:1467–1473.

Kirsten, T. 1968. Incorporation of rare gases in solidifying enstatite melts. *J. Geophys. Res.* 73:2807–2810.

Krummenacher, D., Merrihue, C. M., Pepin, R. O., and Reynolds, J. H. 1962. Meteoritic krypton and barium versus the general isotopic anomalies in meteoritic xenon. *Geochim. Cosmochim. Acta* 26:231–249.

Kurz, M. D., Jenkins, W. J., Schilling, J. G., and Hart, S. R. 1982. Helium isotopic variations in the mantle beneath the central North Atlantic Ocean. *Earth Planet. Sci. Lett.* 58:1–14.

Lewis, R. S., Srinivasan, B., and Anders, E. 1975. Host phase of a strange xenon component in Allende. *Science* 190:1251–1262.

Lindsley, D. H., Speidel, D. H., and Nafziger, R. H. 1968. P-T-f_{O_2} relations for the system $Fe-O-SiO_2$. *Amer. Jour. Sci.* 266:342–360.

McElroy, M. B., and Prather, M. J. 1981. Noble gases in the terrestrial planets: Clues to evolution. *Nature* 293:535–539.

Manuel, O. K. 1978. A Comparison of terrestrial and meteoritic noble gases. In *Terrestrial Rare Gases*, edited by E. C. Alexander, Jr., and M. Ozima, 85–91. Japan Scientific Societies Press.

Melton, C. E., and Giardini, A. A. 1974. The composition and significance of gas released from natural diamonds from Africa and Brazil. *Amer. Mineral.* 59:775–782.

————. 1975. Experimental results and a theoretical interpretation of gaseous inclusions found in Arkansas natural diamonds. *Amer. Mineral.* 60:413–417.

————. 1981. The nature and significance of occluded fluids in three Indian diamonds. *Amer. Mineral.* 66:746–750.

Moore, J. G. 1970. Water content of basalt erupted on the ocean floor. *Contr. Mineral. Petrol.* 28:272–279.

Moniot, R. K. 1980. Noble-gas-rich separates from ordinary chondrites. *Geochim. Cosmochim. Acta* 44:253–271.

Muan, A. 1955. Phase equilibria in the system FeO-Fe_2O_3-SiO_2. *J. Metals* 7:965–976.

Muan, A., and Osborn, E. F. 1956. Phase equilibria at liquidus temperatures in the system MgO-FeO-Fe_2O_3-SiO_2. *Jour. Amer. Ceram. Soc.* 39:121–140.

Mulfinger, H. O., and Scholze, H. 1962. Löslichkeit und Diffusion von Helium in Glasschmelzen: I. Löslichkeit. *Glasstechn. Ber.* 35:466–478.

Murck, B. W., Burruss, R. C., and Hollister, L. S. 1978. Phase equilibria in fluid inclusions in ultramafic xenoliths. *Amer. Mineral.* 63:40–46.

Nafziger, R. H. 1966. Equilibrium phase compositions and thermodynamic properties of solid solutions in the system Mg-"FeO"-SiO_2. Ph.D. diss., Pennsylvania State University.

Nafziger, R. H., and Muan, A. 1967. Equilibrium phase compositions and thermodynamic properties of olivines and pyroxenes in the system MgO-"FeO"-SiO_2. *Amer. Mineral.* 52:1364–1385.

Ozima, M., and Zashu, S. 1983. Primitive helium in diamonds. *Science* 219:1067–1068.

Podosek, F. A., Honda, M., and Ozima, M. 1980. Sedimentary noble gases. *Geochim. Cosmochim. Acta* 44:1875–1884.

Reynolds, J. H., Frick, U., Neil, J. M., and Phinney, D. L. 1978. Rare-gas-rich separates from carbonaceous chondrites. *Geochim. Cosmochim. Acta* 42:1775–1798.

Ringwood, A. E. 1979. *Origin of the Earth and Moon.* New York: Springer-Verlag.

Sato, M. 1978. Oxygen fugacity of basaltic magmas and the role of gas-forming elements. *Geophys. Res. Letters* 5:447–449.

Sato, M., and Wright, T. L. 1966. Oxygen fugacities directly measured in magmatic gases. *Science* 153:1103–1105.

Sato, M., and Moore, J. G. 1973. Oxygen and sulfur fugacities of magmatic gases directly measured in active vents of Mount Etna. *Phil. Trans. R. Soc. Lond.* A274:137–146.

Speidel, D. H., and Osborn, E. F. 1967. Element distribution among coexisting phases in the system MgO-FeO-Fe_2O_3-SiO_2 as a function of temperature and oxygen fugacity. *Amer. Mineral.* 52:1139–1152.

Speidel, D. H., and Nafziger, R. H. 1968. P-T-f_{O_2} relations in the system Fe-O-MgO-SiO_2. *Amer. Jour. Sci.* 266:361–379.

Steinthórssen, S. 1974. The oxide mineralogy, initial oxidation state, and deuteric alteration in some Precambrian diabase dike swarms in Canada. Ph.D. diss., Princeton University.

Stolper, E. 1982. Water in silicate glasses: An infrared spectroscopic study. *Contrib. Mineral. Petrol.* 81:1–17.

Walker, D., Longhi, J., Grove, T. L., Stolper, E., and Hays, J. F. 1973. Experimental petrology and origin of rocks from the Descartes Highlands. *Proc. Lunar Science Conf. 4th* 1:1013–1032.

Wetherill, G. W. 1976. The role of large bodies in the formation of the Earth and Moon. *Proc. Lunar Science Conf. 7th* 3:3245–3257.

———. 1980. Formation of the terrestrial planets. *Ann. Rev. Astron. Astrophys.* 18:77–113.

3

Release and Recycling of Volatiles
during Earth History

1. Introduction

In the previous chapter a reasonably strong case was made for the proposition that the Earth never possessed a quantity of volatiles much larger than its present inventory. This chapter will seek to set some rather broad limits on the history of the release of these volatiles and on their recycling through the Earth's crust.

Numerous attempts have been made to reconstruct the degassing history of the Earth since Rubey's (1951) classic paper on the subject (see, for instance, Damon and Kulp 1958; Turekian 1959, 1964; Fanale 1971; Hamano and Ozima 1978; Fisher 1978; Hart and Hogan 1978; and Schwartzman 1978).

These reconstructions tend to rely heavily on the present-day distribution of the rare gases, particularly of ^{40}Ar, between the solid Earth and the atmosphere. Because ^{40}Ar has been generated in the Earth largely by the decay of ^{40}K since the Earth's formation, degassing could not have taken place entirely at or very close to 4.5 b.y. ago. It has, however, proved very difficult to extract detailed degassing schedules from the ^{40}Ar data; the construction of such schedules requires more precise data than are currently available for the total potassium content of the Earth, for the distribution of potassium between the Earth's crust and mantle, and for the retentivity of Ar in crustal rocks.

Really convincing evidence for the present-day addition of any primordial gases to the atmosphere was lacking until the discovery of excess ^{3}He in deep waters of the Pacific Ocean by Clarke, Beg, and Craig (1969) (see also Craig and Lupton 1981). The observed ^{3}He excess could only be explained by the injection of primordial helium at sea-floor spreading centers on mid-ocean ridges. Subsequent measurements have confirmed the presence of excess ^{3}He in deep ocean waters (Clarke, Beg, and Craig 1970; Jenkins et al. 1972; Craig, Clarke, and Beg 1975; Jenkins and Clarke 1976), and have demonstrated the presence of excess ^{3}He in rapidly quenched, glassy margins of oceanic basalts (Lupton and Craig 1975; Craig and Lupton 1976; Kurz et al. 1982) as well as in hydrothermal fluids associated with mid-ocean ridges (Lupton, Weiss and Craig 1977a,b; Jenkins, Edmond, and Corliss 1978). The discovery of large excesses of

[3]He in basaltic glasses has also shown that other rare gases in these basalts are largely primordial rather than due to atmospheric contamination, and this discovery has opened what promises to be a fruitful approach to the study of the degassing history of the Earth.

The rate of recycling of volatiles can probably be assessed best by a careful study of pertinent parts of the carbon cycle (see, for instance, Holland 1978, pp. 274–279). The last section of this chapter outlines the arguments on which the figure for the present-day rate of carbon cycling is based, and combines these data with the proposed primary degassing schedule to give at least a rough indication of the evolution of the volatile inventory in the atmosphere, oceans, and crust of the Earth.

2. Release of Primordial Gases

The helium content of the atmosphere is determined by the rate of release of He from the solid Earth and by the escape rate of He from the atmosphere into interplanetary space (for a recent discussion see, for instance, Holland 1978, pp. 253–259). The half-life of the atmospheric escape of [4]He is on the order of a few million years, and there seems to be little problem balancing the atmospheric budget of this isotope with the release of [4]He produced by the decay of U and Th in the solid Earth. Until recently, however, the [3]He budget of the atmosphere was not understood. Several processes were known to add [3]He: the decay of tritium produced by the interaction of cosmic rays with the atmosphere (Libby 1946; Craig and Lal 1961; Teegarden 1967), accretion from cosmic rays (Biswas, Ramadurai, and Sreenirasan 1967), and the decay of ^{3}H generated by the reaction of neutrons with lithium, $Li(n,\alpha)^3H$, in the solid Earth (Morrison and Pine 1955). However, the combined [3]He production rate due to all these processes is inadequate by about an order of magnitude to balance the atmospheric escape rate of [3]He (Kockarts and Nicolet 1962).

The problem of the apparent imbalance in the [3]He budget was finally resolved by the discovery that [3]He is released from mantle materials, particularly at mid-ocean spreading centers. [3]He released there is almost certainly primordial. Its flux can be estimated by multiplying the [3]He concentration in hydrothermal solutions in the Galapagos and in other parts of the East Pacific Rise (Lupton, Weiss, and Craig 1977b; Jenkins, Edmond, and Corliss 1978; Welhan 1981) by the total annual flow of such solutions which is required by the heat flow deficit in the vicinity of mid-ocean ridges (Wolery and Sleep 1976). If one adds to the marine flux the somewhat uncertain [3]He flux through the continents, one arrives at a total primordial [3]He flux of $(3.0 \pm 1.0) \times 10^{19}$ atoms/sec ($10^{3.2}$ mol/yr), a rate which is compatible with the estimated rate of escape of [3]He from the atmosphere (Jenkins 1980).

If we could obtain figures for the average ratio of the primordial non-radiogenic isotopes of the rare gases to ^{3}He in mantle-derived volatiles, we could arrive at a figure for the present-day release rate of these gases to the atmosphere. Unfortunately, such data are few. Kaneoka, Takaoka, and Aoki (1978) analyzed the rare gases in three samples from the island of Hawaii; Kaneoka and Takaoka (1980) have reported rare-gas analyses of samples from Maui and Oahu; and Kyser (1980) (see also Kyser and Rison 1982) has analyzed a collection of samples from Oahu, the island of Hawaii, South Africa, the Grand Canyon, and the Massif Central in France. Most of the samples described in these papers are listed in table 3.1.

There is a rather poor correlation between the ^{3}He and the ^{20}Ne content of the samples. However, figure 3.1 shows that their ^{3}He/^{36}Ar ratio

TABLE 3.1.

Samples for which ^{3}He and other rare gas analyses are available; the numbers are those used in figures 5.1 and 3.2.

No.	Sample	Source
1	Dunite 180lb. Hualalai	(K,T,A)
2	Dunite 1801, Hualalai	(K,T,A)
3	Olivine phenocrysts, Kapoho	(K,T,A)
4	Augite phenocryst HAa, Cpx ⎫ White Hill	
5	Olivine phenocryst HAa, Ol ⎬ Haleakala, Maui	(K,T)
6	Augite Crystal, W. of Puu Oili, Haleakala, Maui	(K,T)
7	Spinel lherzolite, SLC-45, Salt Lake Crater, Oahu	(K,T)
8	Spinel lherzolite, SLC-52, Salt Lake Crater, Oahu	(K,T)
9	Garnet pyroxenite, SLC-57, Salt Lake Crater, Oahu	(K,T)
10	Dunite, M-1, Hualalai	(K)
11	Dunite, KAP-10, Hualalai	(K)
12	Wehrlite, H-5, Hualalai	(K)
13	Dunite KAP-6, Hualalai	(K)
14	Spinel lherzolite, SLC-3, Salt Lake Crater	(K)
15	Garnet pyroxenite, SLC-26, Salt Lake Crater	(K)
16	Garnet pyroxenite, SAL-97, Salt Lake Crater	(K)
17	Cpx from spinel lherzolite, SLC-12, Salt Lake Crater	(K)
18	Garnet lherzolite, LBM-9, Matsoku, South Africa	(K)
19	Garnet lherzolite, KB 12-51, Jagersfontein, S. A.	(K)
20	Lherzolite, GC, Grand Canyon	(K)
21	Spinel lherzolite, MC-20, Massif Central, France	(K)

(K,T,A): Kaneoka, Takaoka and Aoki (1978).
(K,T): Kaneoka and Takaoka (1980).
(K): Kyser (1980).

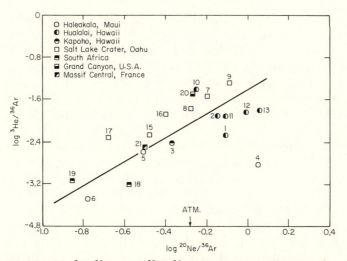

FIGURE 3.1. Plot of the ^{3}He/^{36}Ar vs. the ^{20}Ne/^{36}Ar ratio in the gas samples listed in table 3.1. The numbers next to data points in this figure identify the samples in table 3.1.

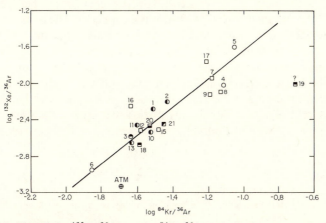

FIGURE 3.2. Plot of the ^{132}Xe/^{36}Ar vs. the ^{84}Kr/^{36}Ar ratio in the gas samples listed in table 3.1. The numbers next to the data points identify the samples in table 3.1; the symbols have the same meaning as in figure 3.1.

is reasonably well correlated with their ^{20}Ne/^{36}Ar ratio. The values of the ^{20}Ne/^{36}Ar ratio scatter approximately evenly around the ^{20}Ne/^{36}Ar ratio of the present-day atmosphere, $10^{-0.28}$. If the mean value of ^{20}Ne/^{36}Ar ratio in gases that are released from the mantle is equal to the atmospheric ratio, as would be expected if the atmosphere conserves these species, then the mean ^{3}He/^{36}Ar ratio of these gases is approximately $10^{-2.0}$. The ^{3}He flux is ca. $10^{3.2}$ moles/yr; the best estimate of the ^{36}Ar flux is therefore $10^{5.2}$ mol/yr and that of the ^{20}Ne flux is $10^{4.9}$ mol/yr. These values are uncertain, perhaps by as much as an order of magnitude. The ^{3}He flux is not known precisely, and the ^{20}Ne/^{36}Ar ratio of primordial gases released to the atmosphere today may not equal the current atmospheric ratio of these nuclides. If, for instance, the mean input ratio of ^{20}Ne/^{36}Ar were $10^{-0.7}$ and the mean ^{3}He/^{36}Ar ratio were $10^{-3.0}$, the ^{36}Ar degassing rate would be $10^{6.2}$ mol/yr and the ^{20}Ne degassing rate $10^{5.5}$ mol/yr.

The probable and reasonable minimum and maximum values of the ^{20}Ne and ^{36}Ar degassing rates based on the data in figure 3.1 are compared with the mean degassing rate of these nuclides in table 3.2. The mean degassing rate as used here is the total number of moles of each nuclide in the atmosphere divided by 4.5×10^9 yr. For both ^{20}Ne and ^{36}Ar, the most probable present-day degassing rate is nearly an order of magnitude less than the mean degassing rate; however, the maximum estimated values for the current degassing rates are indistinguishable from the mean degassing rate of these nuclides.

The ^{84}Kr content of the samples listed in table 3.1 is quite well correlated with their ^{132}Xe content, but the correlation between their ^{132}Xe/^{36}Ar and their ^{84}Kr/^{36}Ar ratios is even better. The data are plotted in figure 3.2 together with the value of these ratios in the atmosphere. The line of best fit through the analytical data does not pass through the point representing the composition of the atmosphere today. This is almost certainly due to a deficit of Xe in the atmosphere, which may be due to preferential trapping of the heavy rare gases in sediments and in sedimentary rocks. For recent discussions of this matter, see Podosek, Honda, and Ozima (1980), and Podosek, Bernatowicz, and Kramer (1981.)

TABLE 3.2.

Comparison of the estimated current degassing rate of ^{20}Ne, ^{36}Ar, ^{84}Kr, and ^{132}Xe with mean degassing rate of these nuclides.

| Nuclide | Degassing Rate (mol/yr) | | | |
	Minimum	Probable	Maximum	Mean
^{20}Ne	$10^{4.6}$	$10^{4.9}$	$10^{5.4}$	$10^{5.81}$
^{36}Ar	$10^{4.6}$	$10^{5.2}$	$10^{6.0}$	$10^{6.10}$
^{84}Kr	$10^{2.9}$	$10^{3.8}$	$10^{4.9}$	$10^{4.41}$
^{132}Xe	$10^{1.9}$	$10^{2.9}$	$10^{4.2}$	$10^{2.96}$

Note: See text for sources of data.

If Xe removal from the atmosphere were the only mechanism for shifting the composition of the atmosphere away from the correlation line in figure 3.2, the $^{84}Kr/^{36}Ar$ ratio of the atmosphere would be lower than the value of this ratio in all but one of the analyzed samples of mantle gases. This seems somewhat unlikely. It seems more likely that both Kr and Xe have been removed from the atmosphere in significant quantities, and that the mean $^{132}Xe/^{36}Ar$ and $^{84}Kr/^{36}Ar$ inputs to the atmosphere lie near the midpoint of the line segment in figure 3.2. The mean input rate of ^{84}Kr and ^{132}Xe in table 3.2 has been estimated on this basis; the probable minimum and maximum values correspond to $^{84}Kr/^{36}Ar$ values of $10^{-1.7}$ and $10^{-1.1}$, respectively.

The most probable input rates of ^{84}Kr and ^{132}Xe are, respectively, somewhat less and comparable to the mean input rate of these nuclides as defined in table 3.2. If significant quantities of Kr and Xe have been lost from the atmosphere, the calculated mean degassing rates from these gases are considerably less than the true degassing rates, and it is likely that the ratio of the present to the true mean degassing rate of Kr and Xe is similar to the ratio of the present and the mean degassing rate of Ne and Ar. It is obvious, however, that the available data are too incomplete to allow anything but tentative conclusions regarding the relative magnitude of their present-day flux and their mean flux to the atmosphere.

If the data in table 3.2 for the ^{20}Ne and ^{36}Ar flux from the Earth today turn out to be correct, the present flux of these gases is ca. 13% of their mean flux. If the degassing rate has remained constant during most of Earth history, ca. 87% of the two gases were released early in Earth history. Such an evolutionary model is rather unlikely. Heat generation in the Earth has been decreasing with time, and it is more than likely that the rate of degassing of the Earth has decreased concomitantly, both in response to the decreasing metabolic activity of the Earth and to the decreasing content of the nonradiogenic rare gases in the mantle. The functional relationships between the degassing rate, the rate of heat generation, and the rare gas content on the mantle are not known. However, at least some plausible models can be constructed to see whether and under what conditions a period of rapid degassing early in Earth history is required by the rare gas release data.

Heat loss from the Earth is related to processes of heat generation, largely by radioactive decay, and to the process of cooling itself. It seems likely that the rate of heat generation within the Earth has been more or less equal to the rate of heat loss. Turcotte (1980) has proposed that approximately 20% of the present heat flux from the Earth's interior is due to cooling, and that this percentage has probably remained roughly constant during the past 3 b.y. (see also McKenzie and Weiss 1980; Schubert,

TABLE 3.3.

The evolution of heat generation due to radioactive decay (q_{rad}), due to cooling (q_c), and the total heat flux (q_T); after Turcotte (1980).

Time (b.y. before present)	Heat Flow ($\mu cal/cm^2\ sec$)		
	q_{rad}	q_c	q_T
0	1.46	0.29	1.75[†]
1.0	1.78	0.40	2.18
2.0	2.29	0.52	2.81
3.0	3.19	0.71	3.90
3.6	4.16*		
4.6	6.67*		

* Based on Lambert (1976)
[†] Davies (1980) suggests that the present average heat flow is $80 \pm 8\ mW/m^2 \equiv 1.91 \pm 0.19$ $\mu cal/cm^2$ sec.

Stevenson, and Cassen 1980; Sleep and Windley 1982). Turcotte's values for q_{rad}, the rate of heat generation due to radioactive decay, and q_{total} are shown in table 3.3 and in figure 3.3. They tend to show that the heat flux 3 b.y. ago was approximately twice as great as today. The decrease in heat generation with time during the last 3 b.y. depends largely on the relative proportion of K, U, and Th, the three major long-lived radioisotopes in the Earth. Turcotte (1980) assumed that the Th/U ratio in the Earth is 4.0 and the K/U ratio is 1×10^4. If the radioelements are uniformly distributed throughout the crust and mantle, a uranium concentration of 31.6 ppb would be consistent with these ratios and with a present-day q_{rad} value of 1.46 $\mu cal/cm^2$ sec. This is in reasonable agreement with Jacobsen and Wasserburg's (1979) figure of 26 ppb for the mean uranium content of the crust and mantle. If the K/U ratio of the Earth turns out to differ from 1.0×10^4, the data in table 3.4 can be used to calculate more precisely the course of heat production by radioactive decay. Heat production decreases more rapidly if the K/U ratio of the Earth is greater than 1×10^4, but the effect of errors in estimates of the K/U ratio on the course of q_{rad} is large only during the early part of Earth history.

Estimation of the heat generated in the early Earth is complicated by the possible presence of short-lived radioisotopes and by the release of heat accompanying core formation. It seems fair to say that the proposed curve for q_{rad} in figure 3.3 is probably adequate for the past 3.5 b.y., but

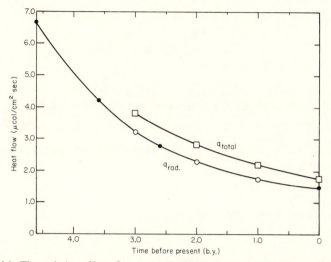

FIGURE 3.3. The variation of heat flow in the Earth through geologic time, q_{rad} = heat flow due to radioactive decay; open circles, Turcotte (1980); filled circles, calculated from Lambert (1976). q_{total} = total heat flow, Turcotte (1980).

TABLE 3.4.

Some data for heat generation by long-lived radioisotopes in the Earth.

Isotope	Half Life $t_{1/2}$ (yrs)	Heat Generation* (cal/gm yr)	Present Heat Generation in the Earth (10^{20} cal/yr)
^{235}U	0.87×10^9 ⎫	0.73	0.92
^{238}U	4.50×10^9 ⎭		
^{232}Th	13.20×10^9	0.20	1.00
^{40}K	1.29×10^9	2.7×10^{-5}†	0.34
^{244}Pu	8.20×10^7	14.07	0.00

* Voytkevitch and Yefanov (1979).
† Heat Generation in cal/gm K per year.

that it may be considerably in error for the first half billion years of Earth history.

The application of the heat production data to the problem of gas release is not straightforward. It seems likely that the gas release rate is proportional to the rate of heat dissipation and to the rare gas content of the portion of the mantle that is being degassed. If this is so, then

$$\frac{dM}{dt} \cong -\alpha m, \tag{3.1}$$

where

M = no. of moles of a nonradiogenic gas in the mantle

m = moles of the same gas per gm of mantle being degassed,

and where α is a proportionality constant that is linearly dependent on the rate of heat loss q_T:

$$\alpha \cong \beta q_T. \tag{3.2}$$

Since q_T is nearly proportional to q_{rad},

$$\alpha \cong \beta \eta q_{\text{rad}}, \tag{3.3}$$

where the value of η is approximately 1.2. It follows that

$$\frac{dM}{dt} \cong -1.2\beta q_{\text{rad}} m. \tag{3.4}$$

Two cases can now be distinguished. In the first, mantle degassing always takes place by the release of gases from the previously undegassed mantle. In this model the present mantle is heterogeneous and consists of a depleted portion from which gases have been released, and an undegassed portion that still contains much or all of its primordial complement of gases. This view of the mantle is supported by rare-gas data (Hart, Dymond, and Hogan 1979) and by the evolution of the Sm-Nd and Rb-Sr systems in the Earth (see, for instance, Jacobsen and Wasserburg 1979, 1981). In the second type of degassing, the mantle remains homogeneous during degassing with respect to the concentration of the gases in question. The real mantle may well behave in a manner intermediate between these two extreme cases.

If degassing has taken place only from undepleted portions of the mantle,

$$\frac{dM}{dt} \cong -1.2\beta q_{\text{rad}} m_0, \tag{3.5}$$

where m_0 is the concentration of a gas in undepleted mantle material. The present rate of change of gas in the mantle, $(dM/dt)_p$, is then given by the expression,

$$\left(\frac{dM}{dt}\right)_p \cong -1.2\beta (q_{\text{rad}})_p m_0, \tag{3.6}$$

where $(q_{rad})_p$ is the present value of q_{rad}; thus

$$\frac{dM}{dt} \cong \left(\frac{dM}{dt}\right)_p \cdot \frac{q_{rad}}{(q_{rad})_p}, \tag{3.7}$$

and the total quantity of gas transferred from the mantle to the atmosphere since a time t_1 is

$$M_p - M_1 = \int_{t_1}^{t_p} \left(\frac{dM}{dt}\right)_p \cdot \frac{q_{rad}}{(q_{rad})_p}\, dt. \tag{3.8}$$

The fraction of the present atmospheric complement of a gas that has accumulated in the time interval t_1 to t_p is

$$\frac{N_p - N_1}{N_p} = \left[\frac{1}{N_p}\left(\frac{dN}{dt}\right)_p\right]\int_{t_1}^{t_p} \frac{q_{rad}}{(q_{rad})_p}\, dt, \tag{3.9}$$

where N is the number of moles of the gas in the atmosphere. Given the heat generation data in figure 3.3, one can easily evaluate the integral in equation 3.9. The results have been plotted as curve 1 in figure 3.4. The course of this curve suggests that there was an early period of rapid degassing during which ca. 75% of the present complement of the non-radiogenic rare gases was released. The assumptions underlying the model calculation are that the mantle became homogeneous after the initial degassing event and that it has since become heterogeneous due to the release of gas from what is now the gas-depleted portion of the mantle.

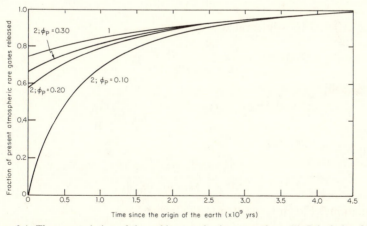

FIGURE 3.4. The accumulation of the noble gases in the atmosphere. (1) Calculation for a heterogeneous mantle; (2) calculation for a homogeneous mantle; ϕ_p = fraction of rare gases presently remaining in the mantle.

Early, large-scale melting of the mantle could have produced the required period of early gas loss and rehomogenization.

The inclusion in the above calculations of the probable effect of progressive changes in heat production on the gas release rate has not eliminated the need for a period of rapid early degassing. However, rapid early degassing is not required if the present-day rare gas release rate is much larger than the rate indicated by the available rare gas data, if the gas release rates turn out to be proportional to the heat generation rate raised to a power significantly greater than one, or if the concentration of the nonradiogenic gases in portions of the mantle undergoing degassing has varied considerably with time. The last of these possibilities is surely worth exploring further.

Consider a mantle within which a particular gas is homogeneously distributed. Equation 3.4 must then be written in the form,

$$\frac{dM}{dt} = -1.2\beta q_{rad}\frac{M}{W},\tag{3.10}$$

where

$$W = \text{the mass of the mantle in kg.}$$

If we again let

$$N = \text{no. of moles of the gas in the atmosphere,}$$

then

$$\frac{dN}{dt} = -\frac{dM}{dt}.\tag{3.11}$$

If equation 3.10 is integrated, we obtain

$$\ln\frac{M_p}{M_1} = -\frac{1.2\beta}{W}\int_{t_1}^{t_p} q_{rad}\,dt.\tag{3.12}$$

At present,

$$-\frac{dN_p}{dt} = \frac{dM_p}{dt} = -1.2\beta(q_{rad})_p\frac{M_p}{W}.\tag{3.13}$$

Thus,

$$\frac{1.2\beta}{W} = \frac{1}{M_p}\frac{dN_p}{dt}\frac{1}{(q_{rad})_p}.\tag{3.14}$$

Equation 3.12 can be written in the form

$$\ln\frac{M_p}{M_1} = -\left[\frac{1}{M_p}\left(\frac{dN_p}{dt}\right)\right]\int_{t_1}^{t_p}\frac{q_{rad}}{(q_{rad})_p}\,dt.\tag{3.15}$$

Now let the fraction of mantle degassing that has taken place during Earth history be defined by the parameter ϕ_p

$$\phi_p \equiv \frac{M_p}{M_0}. \tag{3.16}$$

If all of the atmosphere was generated by degassing of the mantle,

$$N_p = M_0 - M_p = M_p\left(\frac{1}{\phi_p} - 1\right), \tag{3.17}$$

so that equation 3.15 becomes

$$\ln\frac{M_p}{M_1} = -\left(\frac{1}{\phi_p} - 1\right)\left[\frac{1}{N_p}\left(\frac{dN_p}{dt}\right)\right]\int_{t_1}^{t_p}\frac{q_{rad}}{(q_{rad})_p}dt. \tag{3.18}$$

If we now define v_1 as

$$v_1 \equiv \frac{M_p}{M_1}, \tag{3.19}$$

then

$$N_1 = M_0 - M_1 = M_0 - \frac{M_p}{v_1}, \tag{3.20}$$

and

$$\frac{N_1}{N_p} = \frac{M_0 - \dfrac{M_p}{v_1}}{M_0 - M_p} = \frac{1 - \dfrac{\phi_p}{v_1}}{1 - \phi_p}. \tag{3.21}$$

Values of N_1/N_p computed using equations 3.18 and 3.21 are shown as the curves marked 2 in figure 3.4. If the fractional degassing of the mantle has been small, then the degassing of the mantle is essentially described by curve 1, and roughly 75% of the atmospheric burden of the nonradiogenic noble gases were released early in Earth history. If the fraction of gases presently retained in the mantle is small, then the quantity of gas released per year per unit of heat generated has become significantly less, the gas release rate earlier in Earth history was much higher, and the fraction of gas released very early in Earth history during a catastrophic gassing event was proportionally smaller. The effect of decreasing ϕ_p is minor until ϕ_p is smaller than 0.5 (see figure 3.4). Even if the mantle is presently 70% degassed, roughly two-thirds of the present atmospheric burden of nonradiogenic gases appears to have been released very early, perhaps during a catastrophic degassing event. Toward lower values of ϕ_p, the calculated value of N_1/N_p 4.5 b.y.b.p. decreases rapidly. If the mantle is now 90% degassed ($\phi_p = 0.10$), no early catastrophic degassing is required. For values of $\phi_p < 0.10$ there are no solutions to the physical problem, i.e., one cannot then account for the present degassing rate and for the present content of the nonradiogenic rare gases in the atmosphere within the framework of the proposed computational scheme.

These results show that a substantial early degassing event seems to be required, except for degrees of mantle degassing and rehomogenization that are in conflict with present evidence for the evolution of the mantle from the systematics of the Nd-Sm and Rb-Sr decay systems. Staudacher and Allègre (1982) have reached the same conclusion on the basis of differences between the isotopic composition of Xe in the atmosphere and in MORB gases. It must, however, be remembered that the data base for the proposed present-day release rate of the nonradiogenic rare gases is still very slim. If present-day release rates turn out to be more than a factor of four greater than the rates used in the above calculations, the case made above for a period of rapid early degassing becomes very weak.

3. Mechanisms of Rapid, Early Degassing

There are two obvious mechanisms for the rapid, early degassing of the Earth; both were mentioned in chapter 2. The first involves the loss of volatiles from accreting material during infall (see, for instance, Boslough, Weldon, and Ahrens 1980). The degree of volatile release by this mechanism alone may have been sufficient to account for the presence of one-quarter to one-half of the present complement of the nonradiogenic rare gases in the early atmosphere, although reincorporation of volatiles may have been quick enough to preclude the development of a thick atmosphere (Jakosky and Ahrens 1979). The release of the rare gases during infall was probably nonspecific. The relative proportion of the rare gases in the released volatiles was therefore probably similar to that in the rare gases that were incorporated into the growing Earth.

The second degassing mechanism involves large-scale melting of the Earth during and/or shortly after accretion. It is difficult to make a convincing analysis of the behavior of volatiles during such an episode. The available data are insufficient for describing the equilibrium distribution of volatiles between an atmosphere and a largely or completely molten Earth, and it seems likely that the actual distribution of volatiles would have been heavily influenced by the kinetics of accretion and of mantle mixing. Nevertheless, the extant experimental data do seem to permit some nontrivial observations regarding the behavior of a few volatile elements and compounds during large-scale early melting of the Earth.

Figure 3.5 shows the solubility of CO_2 and of water in diopside melts at pressures up to 30 kb. The concentration of CO_2 in these melts is

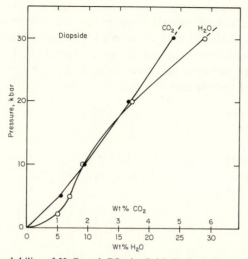

FIGURE 3.5. The solubility of H_2O and CO_2 in $CaMgSi_2O_6$ melt to 30 kbar. H_2O data along the diopside solidus from Rosenhauer and Eggler (1975). CO_2 data at 1625° from Mysen et al. (1976). (Reprinted by permission of the Oregon Department of Geology and Mineral Industries.)

roughly proportional to P_{CO_2}, so that

$$m_{CO_2} \approx 4.5 \times 10^{-5} P_{CO_2} \text{ mol/kg}, \tag{3.22}$$

where P_{CO_2} is in atmospheres. Consider now an Earth which grows by the accretion of material containing $m_{CO_2}^0$ moles of CO_2 per kg. Let the accreted material melt on impact, and let the atmosphere generated by the release of volatiles always equilibrate with the last batch of infalling material. A mass dW kg of infalling material brings with it $m_{CO_2}^0 \, dW$ mol of CO_2. After equilibration with the atmosphere, the concentration of CO_2, m_{CO_2}, will be determined by the expression

$$m_{CO_2} = B_{CO_2} \cdot P_{CO_2}, \tag{3.23}$$

where B_{CO_2} is the Henry's Law constant. The quantity of CO_2 added to the atmosphere will then be

$$dN_{CO_2} = (m_{CO_2}^0 - m_{CO_2}) \, dW \tag{3.24}$$

$$= (m_{CO_2}^0 - B_{CO_2} P_{CO_2}) \, dW. \tag{3.25}$$

The pressure P_{CO_2} is proportional to the total number of moles of CO_2 in the atmosphere:

$$P_{CO_2} = a_{CO_2} N_{CO_2}. \tag{3.26}$$

The proportionality constant a_{CO_2} varies during the process of accretion, because both the gravitational constant, g, and the surface area of the Earth increase with the size of the planet. If the accreting material is reasonably homogeneous, g will be proportional to the radius of the Earth; since the surface area of the Earth is proportional to the square of the radius, it follows that

$$a_{CO_2} = a^p_{CO_2} \left(\frac{W_p}{W}\right)^{1/3},\tag{3.27}$$

where W_p is the present mass of the Earth, W its mass during the accretion process, and $a^p_{CO_2}$ the present value of a_{CO_2}. Equation 3.25 can now be written in the form,

$$dN_{CO_2} = \left[m^0_{CO_2} - B_{CO_2} a^p_{CO_2} \left(\frac{W_p}{W}\right)^{1/3} N_{CO_2}\right] dW,\tag{3.28}$$

which should be valid except for values of W small enough so that CO_2 can escape into interplanetary space. If deviations of the term $(W_p/W)^{1/3}$ from unity are neglected, equation 3.28 can be integrated readily to yield

$$N_{CO_2} = \frac{m^0_{CO_2}}{B_{CO_2} a^p_{CO_2}} \left[1 - e^{-B_{CO_2} a^p_{CO_2} W}\right].\tag{3.29}$$

This equation satisfies the condition that N_{CO_2} is zero when W is zero; when W approaches infinity, N_{CO_2} approaches the value $m^0_{CO_2}/B_{CO_2} a_{CO_2}$. The physical significance of this is that as the Earth grows by accretion, the pressure of CO_2 in the atmosphere increases until it reaches a value at which atmospheric CO_2 is in equilibrium with the initial concentration, $m^0_{CO_2}$, in infalling material. After that point has been reached, no more CO_2 is lost from infalling material, and all of the CO_2 that is added to the Earth with newly accreted material becomes buried.

This is not strictly true, because a_{CO_2} continues to change as the mass of the Earth increases during accretion. However, the effect of neglecting the departure of W_p/W from unity is small, as can be shown by performing a numerical integration of equation 3.28. The value of $a^p_{CO_2}$ is the inverse of the number of moles of CO_2 needed to generate a pressure of one atmosphere; thus

$$a^p_{CO_2} = \frac{44 \text{ gm } CO_2/\text{mole}}{5.1 \times 10^{21} \text{ gm}} = 0.86 \times 10^{-20} \text{ mol}^{-1},$$

and the coefficient of W in equation (3.29) is

$$-B_{CO_2} a^p_{CO_2} = -4.5 \times 10^{-5} \times 0.86 \times 10^{-20} = -3.87 \times 10^{-25}.$$

The present mass of the Earth is 6.0×10^{24} kg; the calculated value of N_{CO_2} at the end of accretion is therefore

$$N_{CO_2} = \frac{m^0_{CO_2}}{3.87 \times 10^{-25}} (1 - e^{-2.23}) \, \text{mol}$$

$$= \frac{0.893 m^0_{CO_2}}{3.87 \times 10^{-25}} \, \text{mol}.$$

If this model is even roughly correct, the mass of CO_2 in the primordial atmosphere reached a value close to equilibrium with the CO_2 content of infalling material. This would also have been true for other gases whose solubility in a primordial molten Earth was equal to or greater than that of CO_2.

The fraction of CO_2 that accumulated in the atmosphere during the accretion process described above is approximately

$$\frac{N_{CO_2}}{m^0_{CO_2} W_p} \approx \frac{0.893}{3.87 \times 10^{-25} \times 6.0 \times 10^{24}} = 0.38.$$

This suggests that approximately one-third of the CO_2 in accreting material may have been released to the atmosphere.

Although the presence of one-third of the total terrestrial CO_2 in the atmosphere at the end of accretion is not unreasonable, the scheme outlined above may founder on the distribution of the rare gases. The data for their solubility in silicate melts is extremely limited (see table 2.3), and the data for the solubility of helium in silicate melts show an uncomfortably large scatter. If Kirsten's data in table 2.3 are the most appropriate for use in calculating the distribution of the rare gases between an atmosphere and a molten planet during accretion, then virtually all of the rare gases should have accumulated in the atmosphere. Since later reincorporation of ^{3}He is almost certainly ruled out by the rapid escape rate of this gas from the Earth's atmosphere, such an early, quasi- complete release is not compatible with the present distribution of the rare gases. On the other hand, if the solubility of the rare gases in the accreting Earth was as high as suggested by the data of Gerling and of Milfinger and Scholze in table 2.3, then these gases would have behaved in a manner similar to that of CO_2, and there may then be no serious conflict between the consequences of the proposed model and the present-day distribution of the rare gases.

The behavior of water differs from that of CO_2 and of the rare gases during the growth of a molten Earth, because it is considerably more soluble in silicate melts than these gases. Its concentration in silicate melts is nearly proportional to the square root of its partial pressure in the associated vapor phase if its concentration is less than ca. 0.5% (see chapter

2); thus,

$$m_{H_2O} \cong B_{H_2O} P_{H_2O}^{1/2}. \tag{3.30}$$

If we follow the procedure used above in calculating the distribution of CO_2 between the atmosphere and a molten Earth during accretion, then

$$dN_{H_2O} = [m_{H_2O}^0 - B_{H_2O} \cdot P_{H_2O}^{1/2}] \, dW, \tag{3.31}$$

and

$$dN_{H_2O} = \left[m_{H_2O}^0 - B_{H_2O}(a_{H_2O}^p)^{1/2} \left(\frac{W_p}{W} \right)^{1/6} (N_{H_2O})^{1/2} \right] dW, \tag{3.32}$$

where a^p, and W_p have the same significance. as before. The current best estimate for B_{H_2O} is probably 5.3×10^{-2}, the value used in chapter 2; the value of $a_{H_2O}^p$ is 3.5×10^{-21}; $m_{H_2O}^0$ can be approximated roughly by doubling the sum of the mass of water in the oceans (1.4×10^{24} kg) plus the mass of water in the crust (ca. 0.6×10^{24} kg), dividing by the mass of the Earth, and converting to molar concentration:

$$m_{H_2O}^0 \approx \frac{4 \times 10^{24} \text{ gm}}{18 \text{ gm/mol} \times 6 \times 10^{24} \text{ kg}} \approx 3.7 \times 10^{-2} \text{ mol/kg}.$$

If these values of B_{H_2O}, $a_{H_2O}^p$, and $m_{H_2O}^0$ are substituted into equation 3.32, and if a numerical integration is carried out, it turns out that the term in $(N_{H_2O})^{1/2}$ approaches $m_{H_2O}^0$ at values of W much smaller than W_p. This means that the water content of the atmosphere of an Earth accreting in the molten state rapidly reaches equilibrium with the water content of molten, infalling material. Thereafter the course of N_{H_2O} is defined very nearly by the expression,

$$N_{H_2O} \approx \left(\frac{m_{H_2O}^0}{B_{H_2O}} \right)^2 \frac{1}{a_{H_2O}^p} \left(\frac{W}{W_p} \right)^{1/3} \tag{3.33}$$

$$\approx 1.39 \times 10^{20} \left(\frac{W}{6 \times 10^{24}} \right)^{1/3}. \tag{3.34}$$

When the Earth had reached its present mass, the 1.39×10^{20} moles of water in the atmosphere would have exerted a pressure of about 0.5 atm and would have represented about 0.06% of the total water inventory of the planet. The water pressure would therefore have been minimal, and nearly all of the water would have been dissolved in the molten planet. These conclusions were already foreseen by Rubey (1951). They are rather insensitive to the uncertainties in the assumptions that were made above regarding the solubility of water in the primordial molten Earth and the water content of the accreting material. The quantitative difference between the distribution of CO_2 and that of water between the atmosphere and a molten, accreting Earth is therefore real and quite large; CH_4, SO_2, H_2S, N_2, and H_2 would have behaved more like CO_2 than like water.

The relative proportions of the volatiles in the Earth's atmosphere at the end of the period of accretion clearly depended on the quantity of melt generated during accretion, on the release of volatiles during infall without melting, and on the kinetics of the reaction of released volatiles with unmolten, infallen material. If large-scale melting did not occur during much of the accretion process, a considerable quantity of volatiles, including water, may have been released to the atmosphere from infalling material. If melting occurred during the last stages of accretion, a large proportion of the slightly soluble volatiles such as CO_2 and the rare gases would have remained in the primordial atmosphere. Water would, how-ever, have tended to be dissolved in the late silicate melt. The fraction of water that might have been lost to such a magma layer depends somewhat on the quantity of water present in the atmosphere before melting. Figure 3.6 shows that roughly 50% of a quantity of water equal to the present mass of the oceans $(1.4 \times 10^{24}$ gm) could have dissolved in a magma layer

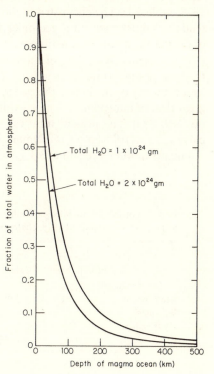

FIGURE 3.6. The fraction of water in an at-mosphere in equilibrium with a magma ocean of depth between 0 and 500 km.

only 50 km thick if the solubility of water in the melt was equal to that used above ($B_{H_2O} = 5.3 \times 10^{-2}$).

4. Recycling of Volatiles

Most of the Ne and Ar degassed from the Earth's interior during geologic time has accumulated in the atmosphere. The present quantity of these gases in the atmosphere is therefore nearly equal to their degassing rate integrated over the last 4.5 b.y. This is not true for most of the other volatiles. Table 3.5 shows that a sizable fraction of the degassed nitrogen and water, most of the sulfur, and nearly all of the carbon have been transferred from the atmosphere to the Earth's crust. Nitrogen is present there in large part as a constituent of organic compounds, water largely as a constituent of hydrous silicates, sulfur largely as a constituent of pyrite and anhydrite, and carbon as a constituent of organic compounds, graphite, and carbonates of calcium and magnesium. The residence times of carbon and sulfur for burial or reburial, with sedimentary rocks are geologically short. The residence time of carbon in the oceans is approximately 8×10^4 yrs, that of sulfur approximately 8×10^6 yrs. (Holland 1978, table 5.1). The chemistry of the ocean-atmosphere system therefore tends to adjust itself on a geologically short time scale, so that the rate of output of dissolved constituents is equal to their rate of input.

The present-day cycling of these volatiles tends to be rapid compared to their primary rate of degassing from the mantle. This is best illustrated by the operation of the geochemical cycle of carbon, for which a great deal of information is now available (see for instance Holland 1978, pp. 262–283). The present total rate of erosion and of sedimentation is close to 2×10^{16} gm/yr (Holeman 1968). The carbon buried with new sediments amounts to some $(3.4 \pm 0.6) \times 10^{14}$ gm/yr (Holland 1978, table

TABLE 3.5.

Distribution of carbon, nitrogen, sulfur, and water in the near-surface geochemical reservoirs (Holland 1978, chaps. 5 and 6).

	Carbon ($\times 10^{18}$ gm)	Nitrogen ($\times 10^{18}$ gm)	Sulfur ($\times 10^{18}$ gm)	Water ($\times 10^{21}$ gm)
Atmosphere	0.69	3,950	<0.001	≤0.2
Biosphere	1.1	0.02	<0.1	1.6
Hydrosphere	40	22	1,300	1,400
Crust	90,000	2,000	15,000	600

6.6). Most of this carbon is derived from the erosion of older sedimentary rocks. Elemental and organic carbon in old sedimentary rocks are oxidized to CO_2 during weathering; carbonate carbon is released by reaction with atmospheric and soil CO_2, and is transported to the oceans, largely as $HCO_3{}^-$ in river water. There seems to be very little change in the mean content of elemental carbon of sedimentary rocks with time (see chapter 5). Thus the fraction of carbon supplied to new sediments by the weathering of old sedimentary rocks is essentially equal to the fraction of old sedimentary rocks in the new sediments. Approximately 75% of recent sediments have been derived from the weathering of old sedimentary rocks, the remainder from the weathering of igneous and high-grade metamorphic rocks. Approximately 75% of the carbon in new sediments thus has been derived from old sedimentary rocks. Since igneous and high-grade metamorphic rocks contain very little carbon compared to modern sediments, approximately 25% of the carbon in modern sediments remains to be accounted for. The only reasonable sources for this carbon are the carbon released from sedimentary rocks during metamorphism and juvenile carbon. The missing carbon amounts to $(8 \pm 3) \times 10^{13}$ gm/yr.

This quantity is large compared to the quantity of juvenile carbon that is probably added annually to the atmosphere. The total inventory of carbon in the crust today is approximately 9×10^{22} gm (table 3.5). If this quantity of carbon has been released from the mantle at a constant rate, the rate of carbon addition from this source has been 2×10^{13} gm/yr, a figure that is almost certainly a strong upper limit for the actual present-day flux of juvenile carbon. The analysis of the present rate of degassing of ^{20}Ne and ^{36}Ar earlier in this chapter suggested that these gases are being released presently at about 13% of the mean rate calculated over geological time. There is no reason to believe that this rate is equal to the mean rate at which juvenile CO_2 is being released from the mantle; but if CO_2 and the rare gases are behaving roughly similarly, then juvenile carbon is being added at a rate of approximately 0.3×10^{13} gm/yr. This accounts for only about 4% of the carbon that must be degassed annually to account for the carbon content of modern sediments. Even if the proposed figure for the degassing rate of juvenile carbon is too low by a factor of 4, it still follows that carbon released during the metamorphism of sedimentary rocks accounts for most of the nonanthropogenic CO_2 that is currently being added to the atmosphere. This is probably true also for sulfur and nitrogen. Degassing during the operation of the geological cycle is therefore currently much more important for the geochemistry of these elements than the contribution of juvenile material from the mantle.

5. Summary

The magnitude of the release rate of primordial gases from the Earth has been debated for many years. The discovery that primordial ^{3}He is being released in easily measurable quantities has finally led to a technique for arriving at a potentially reliable estimate for the release rate of the primordial rare gases. The few data that are currently available indicate that the present release rate of ^{20}Ne and ^{36}Ar is approximately 13% of the mean release rate for these gases averaged over all of geologic time. If this figure is approximately correct, then it can serve as an anchor point for degassing models of the Earth. Three of these have been explored. In the first, the degassing rate has been set equal to the present rate during all of geologic time. In the second and third models the degassing rate in the past is assumed to have been proportional to the rate of heat loss from the Earth. The second and third models differ in the assumptions that were made regarding the present state of the mantle and the mechanism of mantle degassing. Model 2 assumes that gas release has taken place from packets of mantle, all of which contain the same amount of primordial gases. This leads to a heterogeneous mantle consisting of depleted and undepleted portions. Model 3 assumes that the mantle has remained homogeneous during degassing, and that the concentration of the primordial gases released per unit mass of mantle has decreased with time. All three models indicate that there was a period of rapid degassing early in Earth history, unless the mantle is now more depleted in volatiles than seems reasonable.

Early rapid degassing of the rare gases is easily explained both in terms of gas release accompanying the infall of material during the accretion of the Earth and in terms of gas release during melting on or shortly after impact. The two mechanisms probably lead to different ratios of the rare gases, CO_2, and water in the atmosphere toward the end of the process of accretion, but there are not enough experimental data for the solubility of the rare gases in silicate melts to choose between the two mechanisms or to propose intermediate schemes of volatile release. It is certain, however, that nearly all of the water in infalling material was trapped in the Earth if large-scale melting accompanied accretion. Even if a magma ocean only 50 km deep formed toward the end of the accretion process, much of the water that had accumulated in the atmosphere would have been dissolved in the magma.

Most of the volatiles that have been released from the mantle during Earth history have been removed into the crust and have been recycled by weathering and by degassing accompanying metamorphism. An analysis of the carbon cycle shows that the rate of degassing of juvenile carbon today is probably no greater than a few percent of the rate at which carbon is currently being recycled by the metamorphism of sedimentary rocks.

References

Biswas, S., Ramadurai, S., and Sreenirasan, N. 1967. $He^3/(He^3 + He^4)$ ratios in cosmic rays, path lengths in space, and energy spectrum of helium nuclei in local interstellar space. *Phys. Rev.* 159:1063–1069.

Boslough, M. B., Weldon, R. J., and Ahrens, T. J. 1980. Impact-induced water loss from serpentine, nontronite and kernite. *Proc. Lunar Planet. Sci. Conf. 11th*, 2145–2158. New York: Pergamon.

Clarke, W. B., Beg, M. A., and Craig, H. 1969. Excess ^{3}He in the sea: Evidence for terrestrial primordial helium. *Earth Planet. Sci. Lett.* 6:213–220.

———. 1970. Excess helium 3 at the North Pacific Geosecs Station. *J. Geophys. Res.* 75:7676–7678.

Craig, H., and Lal, D. 1961. The production rate of natural tritium. *Tellus* 13:85–105.

Craig, H., Clarke, W. B., and Beg, M. A. 1975. Excess ^{3}He in deep water on the East Pacific Rise. *Earth Planet. Sci. Lett.* 26:125–132.

Craig, H., and Lupton, J. E. 1976. Primordial neon, helium, and hydrogen in oceanic basalts. *Earth Planet. Sci. Lett.* 31:369–385.

———. 1981. Helium-3 and mantle volatiles in the oceans and the oceanic crust. In *The Oceanic Lithosphere*, vol. 7 of *The Sea*, chap. 11. New York: Wiley.

Damon, P. E., and Kulp, J. L. 1958. Inert gases and the evolution of the atmosphere. *Geochim. Cosmochim. Acta* 13:280–292.

Davies, G. F. 1980. Review of oceanic and global heat flow estimates. *Rev. Geophys. and Space Phys.* 18:718–722.

Fanale, F. P. 1971. A case for catastrophic early degassing of the Earth. *Chem. Geol.* 8:79–105.

Fisher, D. E. 1978. Terrestrial potassium and argon abundances as limits to models of atmospheric evolution. In *Terrestrial Rare Gases*, edited by E. C. Alexander, Jr., and M. Ozima, 173–183. Japan Scientific Societies Press.

Hamano, Y., and Ozima, M. 1978. Earth-atmosphere evolution model based on Ar isotopic data. In *Terrestrial Rare Gases*, edited by E. C. Alexander, Jr., and M. Ozima, 155–171. Japan Scientific Societies Press.

Hart, R., and Hogan, L. 1978. Earth degassing models and the heterogeneous vs. homogeneous mantle. In *Terrestrial Rare Gases*, edited by E. C. Alexander, Jr., and M. Ozima, 193–206. Japan Scientific Societies Press.

Hart, R., Dymond, J., and Hogan, L. 1979. Preferential formation of the atmospheric-sialic crust system from the upper mantle. *Nature* 278: 156–159.

Holeman, J. N. 1968. The sediment yield of major rivers of the world. *Water Resour. Res.* 4:737–747.

Holland, H. D. 1978. *The Chemistry of the Atmosphere and Oceans.* New York, Wiley.

Jacobsen, S. B., and Wasserburg, G. J. 1979. The mean age of mantle and crustal reservoirs. *J. Geophys. Res.* 84:7411–7427.

———. 1981. Transport models for crust and mantle evolution. *Tectonophys.* 75:163–179.

Jakosky, B. M., and Ahrens, T. J. 1979. The history of an atmosphere of impact origin. *Proc. Lunar Planet. Sci. Conf. 10th*, 2727–2739. New York: Pergamon.

Jenkins, W. J. 1980. On the terrestrial budgets of helium and argon isotopes and the degassing of the Earth. Unpublished manuscript.

Jenkins, W. J., Beg, M. A., Clarke, W. B., Wangersky, P. J., and Craig, H. 1972. Excess ^{3}He in the Atlantic Ocean. *Earth Planet. Sci. Lett.* 16:122–126.

Jenkins, W. J., and Clarke, W. B. 1976. The distribution of ^{3}He in the western Atlantic Ocean. *Deep Sea Res.* 23:481–494.

Jenkins, W. J., Edmond, J. M., and Corliss, J. B. 1978. Excess ^{3}He and ^{4}He in Galapagos submarine hydrothermal waters. *Nature* 272: 156–158.

Kaneoka, I., Takaoka, N., and Aoki, K-I. 1978. Rare gases in mantle-derived rocks and minerals. In *Terrestrial Rare Gases*, edited by E. C. Alexander, Jr., and M. Ozima, 71–83. Japan Scientific Societies Press.

Kaneoka, I., and Takaoka, N. 1980. Rare gas isotopes in Hawaiian ultra-mafic nodules and volcanic rocks: Constraint on genetic relationships. *Science* 208:1366–1368.

Kockarts, G., and Nicolet, M. 1962. Le problème aéronomique de l'hélium et de l'hydrogène neutres. *Ann. Geophys.* 18:269–290.

Kurz, M. D., Jenkins, W. J., Schilling, J. G., and Hart, S. R. 1982, Helium isotopic variations in the mantle beneath the central North Atlantic Ocean. *Earth Planet. Sci. Lett.* 58:1–14.

Kyser, T. K. 1980. Stable and rare gas isotopes and the genesis of basic lavas and mantle xenoliths. Ph.D. diss., University of California, Berkeley.

Kyser, T. K., and Rison, W. 1982. Systematics of rare gas isotopes in basic lavas and ultramafic xenoliths. *Jour. Geophys. Res.* 87:5611–5630.

Lambert, R. St. J. 1976. Archean thermal regimes, crustal and upper mantle temperatures and a progressive evolutionary model for the

Earth. In *The Early History of the Earth*, edited by B. F. Windley, 363–387. New York: Wiley.

Libby, W. F. 1946. Atmospheric helium-3 and radiocarbon from cosmic radiation. *Phys. Rev.* 69:671–672.

Lupton, J. E., and Craig, H. 1975. Excess ³He in oceanic basalts: Evidence for terrestrial primordial helium. *Earth Planet. Sci. Lett.* 26:133–139.

Lupton, J. E., Weiss, R. F., and Craig, H. 1977a. Mantle helium in the Red Sea brines. *Nature* 266:244–246.

———. 1977b. Mantle helium in hydrothermal plumes in the Galapagos Rift. *Nature* 267:603–604.

McKenzie, D., and Weiss, N. 1980. The thermal history of the Earth. In *The Continental Crust and Its Mineral Deposits*, edited by D. W. Strangway, 575–590. Geol. Assoc. Canada, Spec. Paper 20.

Morrison, P., and Pine, J. 1955. Radiogenic origin of helium isotopes in rock. *Ann. N.Y. Acad. Sci.* 62:71–92.

Mysen, B. O., Eggler, D. H., Seitz, M. G., and Holloway, J. R. 1976. Carbon dioxide in silicate melts and crystals: I. Solubility measurements. *Amer. Jour. Sci.* 276:455–479.

Podosek, F. A., Honda, M., and Ozima, M. 1980. Sedimentary noble gases. *Geochim. Cosmochim. Acta* 44:1875–1884.

Podosek, F. A., Bernatowicz, T. J., and Kramer, F. E. 1981. Adsorption of xenon and krypton on shales. *Geochim. Cosmochim. Acta* 45:2401–2415.

Rosenhauer, M., and Eggler, D. H. 1975. Solution of H_2O and CO_2 in diopside melt. *Carnegie Inst. Washington Yearbk.* 74:474–479.

Rubey, W. W. 1951. Geologic history of seawater: An attempt to state the problem. *Bull. Geol. Soc. Amer.* 62:1111–1147.

Schubert, G., Stevenson, D., and Cassen, P. 1980. Whole planet cooling and the radiogenic heat source contents of the Earth and Moon. *J. Geophys. Res.* 85:2531–2538.

Schwartzman, D. W. 1978. On the ambient mantle $^4He/^{40}Ar$ ratio and the coherent model of degassing of the Earth. In *Terrestrial Rare Gases*, edited by E. C. Alexander, Jr., and M. Ozima, 185–191. Japan Scientific Societies Press.

Sleep, N. H., and Windley, B. F. 1982. Archean plate tectonics: Constraints and inferences. *Jour. Geol.* 90:363–379.

Staudacher, T., and Allègre, C. J. 1982. Terrestrial xenology. *Earth Planet. Sci. Lett.* 60:389–406.

Teegarden, B. J. 1967. Cosmic-ray production of deuterium and tritium in the Earth's atmosphere. *J. Geophys. Res.* 72:4863–4868.

Turcotte, D. L. 1980. On the thermal evolution of the Earth. *Earth Planet. Sci. Lett.* 48:53–58.

Turekian, K. K. 1959. The terrestrial economy of helium and argon. *Geochim. Cosmochim. Acta* 17:37–43.

———. 1964. Degassing of argon and helium from the Earth. In *The Origin and Evolution of Atmospheres and Oceans*, edited by P. J. Brancazio and A.G.W. Cameron, 74–82. New York: Wiley.

Voytkevich, G. V., and Yefanov, K. N. 1979. Geothermal significance of plutonium-244 in the early life of the Earth. *Geochemistry International* 16:1–5.

Welhan, J. A. 1981. Carbon and hydrogen gases in hydrothermal systems: The search for a mantle source. Ph.D. diss, University of California, San Diego.

Wolery, T. J., and Sleep, N. H. 1976. Hydrothermal circulation and geochemical flux at mid-ocean ridges. *Jour. Geol.* 84:249–275.

4

The Chemistry of the Earliest Atmosphere and Oceans

1. Introduction

The previous three chapters have set some rather broad limits on the accretion history of the Earth, on the composition of the volatiles that were introduced into the atmosphere during the first few hundred million years of Earth history, and on the rate at which these volatiles were added. Armed with these limits, we can set about defining the probable range of the composition of the atmosphere and oceans during the earliest part of Earth history. This range turns out to be large, in part due to uncertainties in the rate and the composition of the inputs to the atmosphere during this period, and in part because we are still rather ill-informed about the processes by which these inputs were removed from the ocean-atmosphere system. All that can be claimed for the proposed history of the early atmosphere-ocean system is that it is consistent with the apparent requirements for the origin of life on Earth, and that it falls within the rather wide limits that are imposed by the later geologic record and by the present state of the Earth.

During the earliest part of Earth history the input of volatiles to the atmosphere was dominated by the release of primordial gases. It seems likely—but is not certain (see below)—that these gases were released from volcanoes at magmatic temperatures. Their oxidation state is not known, but it must have been equal to or intermediate between the oxidation state of volcanic gases today and the oxidation state of volcanic gas in equilibrium with metallic iron. The composition of the end members is shown in table 2.6. Volcanic gases today are dominated by H_2O. CO_2 is the major compound of carbon, SO_2 the major compound of sulfur, and N_2 the only nitrogen-containing molecule of importance. In volcanic gases equilibrated with metallic iron, H_2 is approximately as abundant as H_2O, CO is the dominant carbon compound, H_2S the dominant sulfur compound, and N_2 the dominant nitrogen species. The major reducing gas in such a mixture is H_2, because the ratio of the abundance of carbon and sulfur compounds to water is less than 0.1. The proportion of H_2 in volcanic gases was therefore almost certainly of major importance for the oxidation state of the early ocean-atmosphere system.

The total input rate of H_2 was the product of its proportion in volcanic gases and the flux of these gases to the atmosphere. We do not know this flux, but we can set fairly severe lower and upper limits on its value. The lower limit is imposed by the requirement that the early release rate of primordial gases was no less than the mean release rate during geologic time. The amount of water in the oceans together with the quantity in the crust is 2.0×10^{24} gm. The minimum H_2O injection rate early in Earth history was therefore approximately

$$\frac{2.0 \times 10^{24} \text{ gm}}{18 \text{ gm/mol} \times 4.5 \times 10^9 \text{ yrs}} \cong 2.5 \times 10^{13} \text{ mol/yr.}$$

The maximum injection rate was probably about 100 times this value, because at such a rate all of the water now present in the oceans and in the crust would have been released in 45 m.y., i.e. in a very small fraction of Earth history. If the existence of the early, rapid period of degassing proposed in chapter 3 is supported by additional rare gas data, then at the proposed maximum release rate only about 20 m.y. would have been required for the accumulation of half the present inventory of ^{20}Ne, ^{36}Ar, and ^{84}Kr in the atmosphere.

Table 4.1 summarizes the rather wide limits on the H_2 injection rates that are set by these limits on the rate of release of water to the atmosphere and by the constraints on the composition of volcanic gases. The permitted range of the injection rates is approximately four orders of magnitude. The actual injection rates were probably close to the upper limit during the earliest part of Earth history and had approached values not far above the lower limit well before the deposition of the earliest known sedimentary rocks approximately 3.9 b.y. ago.

TABLE 4.1.

Limits on the rate of injection of hydrogen into the early atmosphere.

	Limits Imposed by the Injection Rate of Volcanic Gases			
Limits Imposed by the Oxidation State of Volcanic Gases	Minimum Rate		Maximum Rate	
	(mol H_2/yr)	(atoms H/cm^2 sec)	(mol H_2/yr)	(atoms H/cm^2 sec)
Most Oxidized ($H_2/H_2O = 10^{-2}$)	2.5×10^{11}	1.9×10^9	2.5×10^{13}	1.9×10^{11}
Most Reduced ($H_2/H_2O = 1$)	2.5×10^{13}	1.9×10^{11}	2.5×10^{15}	1.9×10^{13}

2. The Escape of Hydrogen from the Atmosphere

Hydrogen is known to be escaping from the atmosphere today at a rate close to 3×10^8 atoms/cm^2 sec (Donahue 1966; Joseph 1967; Brinton and Mayr 1971; Vidal-Madjar, Blamont, and Phissamay 1973; Meier and Mange 1973; Hunten and Strobel 1974; Hunten and Donahue 1976). H_2 escape is preceded by the photodissociation of H_2O in the stratosphere and by the upward transport of hydrogen to the exosphere; the upward flux of biologically produced methane may also be adding significant quantities of hydrogen to the upper atmosphere (Wofsy, McConnell, and McElroy 1972) and is probably enhancing the hydrogen escape rate considerably.

In an oxygen-free atmosphere the escape rate of hydrogen is proportional to the mixing ratio of hydrogen in the lower parts of the atmosphere. The processes that determine the proportionality constant are complicated (Hunten 1973; Hunten and Strobel 1974; Walker 1977; Kasting, Liu, and Donahue 1979), but it seems likely that the escape flux, F_c, of hydrogen atoms is approximately related to the total mixing ratio, f_t, of hydrogen* by the expression (Walker 1977, p. 164):

$$F_c \approx 2.5 \times 10^{13} f_t \, \text{cm}^{-2} \, \text{sec}^{-1}, \tag{4.1}$$

where the mixing ratio, f_t, is related to the mixing ratio $f(H)$ of H and the mixing ratio $f(H_2)$ of H_2 by the expression

$$f_t = f(H) + 2f(H_2). \tag{4.2}$$

The value of the coefficient of f_t in equation 4.1 is close to that derived more recently by Kasting, Liu, and Donahue (1979). Its value is not particularly sensitive to the composition of the atmosphere. It is, however, sensitive to the temperature of the exosphere if this temperature drops below ca. 800°K. Such low temperatures might have prevailed in the early exosphere if the partial pressure of an atmospheric component active in the infrared was sufficiently high; if this was the case, then the warming effect due to the absorption of solar ultraviolet radiation was reduced considerably (Warneck 1980). If the exosphere temperature was less than 800°K, the H_2 content of the atmosphere would have been higher for a given flux rate than that predicted by equation 4.1. Such low temperatures are, however, unlikely in a hydrogenous atmosphere (Hunten, personal communication, 1982).

The residence time of hydrogen in the atmosphere is geologically very short. The hydrogen content of the early atmosphere must therefore have

---

* The mixing ratio of a component i is defined as the fraction of the number of atoms or molecules of i in a gas.

adjusted itself rapidly until the rate of hydrogen escape was equal to the net flux of hydrogen into the upper atmosphere. This was probably less than the injection rate of hydrogen from volcanoes, because a variety of reactions within the atmosphere almost certainly removed H_2 from the atmosphere (see below). Equation 4.1 can therefore only be used to suggest upper limits for the value of the mixing ratio of hydrogen in the early atmosphere.

The lowest value for the hydrogen injection rate listed in table 4.1 is 1.9×10^9 atoms H/cm^2 sec. This is about an order of magnitude greater than the escape rate of hydrogen at present. The hydrogen mixing ratio, f_t, would have been $\leq 0.8 \times 10^{-4}$ in dynamic equilibrium with this injection rate. The calculated equilibrium mixing ratio of hydrogen for an injection rate of 1.9×10^{13} atoms H/cm^2 sec is ≤ 0.8. It seems likely, however, that equation 4.2 breaks down at such high injection and escape rates. If all of the heat flux provided to the thermosphere today by solar ultraviolet radiation were used to carry atmospheric gases out of the Earth's gravitational potential well, the outflow of hydrogen would be ca. 2×10^{11} atoms/cm^2 sec (Walker 1982). A much larger thermospheric heat source would almost certainly have been required to sustain a hydrogen escape flux of 2×10^{13} atoms/cm^2 sec. The required quantity of thermospheric heat could have come either from a greatly enhanced solar luminosity in the extreme ultraviolet (see below), from a greatly enhanced solar wind, or from a combination of the two. The mixing ratio of hydrogen in an atmosphere fed by an Earth from which hydrogen was degassed at rates in excess of 10^{11} atoms/cm^2 sec is therefore uncertain, but could have been substantially greater than 0.8.

Free oxygen would have been virtually absent from such an atmosphere (Kasting, Liu, and Donahue 1979; Kasting and Walker 1981; for a dissenting opinion, see Towe 1981). As shown in figure 4.1a, the O_2 concen-

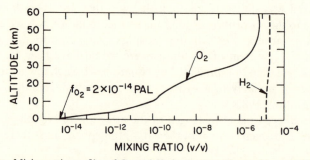

FIGURE 4.1a. Mixing ratio profiles of O_2 and H_2 in an atmosphere with a total H_2 mixing ratio of 17×10^{-6} (Kasting and Walker 1981). (Reproduced by permission of the authors.)

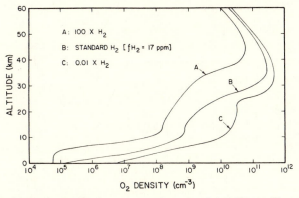

FIGURE 4.1b. O_2 density profiles for an atmosphere containing 17 ppm H_2 and for atmospheres containing 100 times more and 100 times less H_2 (Kasting and Walker 1981). (Reproduced by permission of the authors.)

trated at ground level would have been negligible even if the total H_2 mixing ratio was as low as 17×10^{-6} (17 ppm). The same conclusion applies even for an atmosphere with an H_2 mixing ratio of 17×10^{-8} (see Figure 4.1b).

3. Chemical Reactions in the Atmosphere

Gas mixtures containing H_2, H_2O, and CO in the proportions discussed above are thermodynamically stable at magmatic temperatures but not at temperatures between 0° and 100°C. It therefore seems worthwhile to explore the thermodynamics of such mixtures on cooling, although it will be shown later that kinetics, rather than thermodynamics, almost certainly controlled the chemistry of the early atmosphere. At equilibrium and in the presence of liquid water, CO should react nearly completely with H_2 to yield CH_4 and water, or graphite and water (see, for instance, chapter 4 in Miller and Orgel 1974, and Holland 1962). This point is illustrated by figure 4.2, which shows the equilibrium pressure of CH_4 and CO_2 as a function of P_{H_2} in a gas phase at 25°C in equilibrium with liquid water. At H_2 pressures in excess of ca. 1×10^{-4} atm, CH_4 is the dominant carbon compound, and its pressure depends solely on the total quantity of carbon in the atmosphere. The upper horizontal line in figure 4.2a gives the equilibrium CH_4 pressure in the atmosphere when ca. 1% of the total quantity of present-day crustal carbon is present in the atmosphere. For a variety of reasons this is probably a strong upper limit for P_{CH_4} in the early atmosphere.

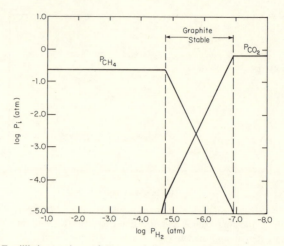

FIGURE 4.2a. Equilibrium values of P_{CH_4} and P_{CO_2} as a function of P_{H_2} when $\Sigma M_C = 15$ mol/cm².

The equivalent horizontal line in figure 4.2b gives the equilibrium CH_4 pressure in an atmosphere where ca. 0.1% of the total quantity of present-day crustal carbon is present in the atmosphere. Toward lower values of the hydrogen pressure, these lines end at their intersection with the line along which P_{CH_4} is defined by P_{H_2} and the presence of graphite (see also

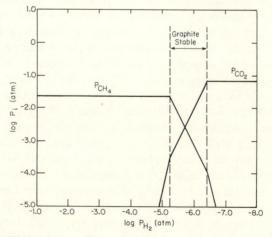

FIGURE 4.2b. Equilibrium values of P_{CH_4} and P_{CO_2} as a function of P_{H_2} when $\Sigma M_C = 1.5$ mol/cm².

Suess 1962). At 25°C the equilibrium constant for the reaction,

$$C_{graph.} + 2H_2 \rightleftharpoons CH_4, \tag{4.3}$$

is

$$K_{4.3} = \frac{P_{CH_4}}{P_{H_2}^2} = 7.9 \times 10^8.$$

The slope of the corresponding log P_{CH_4} line in figures 4.2a and 4.2b is -2, and negligibly small values of P_{CH_4} are reached when $P_{H_2} < 10^{-7}$ atm.

Before H_2 drops to such low values, the equilibrium pressure of CO_2 becomes substantial. Its value in the absence of graphite is determined by equilibrium in the reaction,

$$CO_2 + 4H_2 \rightleftharpoons CH_4 + 2(H_2O)_{Liq}, \tag{4.4}$$

for which

$$K_{4.4} = \frac{P_{CH_4}}{P_{CO_2} \cdot P_{H_2}^4} = 8.1 \times 10^{22}.$$

When graphite is present, the controlling equation becomes

$$CO_2 + 2H_2 \rightleftharpoons C_{graph.} + 2(H_2O)_{Liq}, \tag{4.5}$$

for which

$$K_{4.5} = \frac{1}{P_{CO_2} \cdot P_{H_2}^2} = 1.0 \times 10^{14}.$$

The slope of the line defining log P_{CO_2} in the presence of graphite is $+2$. The intersection of this line with the horizontal line which defines the atmospheric pressure that is exerted when the total initial carbon complement is present as CO_2 marks the maximum equilibrium CO_2 pressure in the atmosphere.

If 15 mol CH_4/cm^2 are initially present in an atmosphere equilibrated with liquid water at 25°C, and if the H_2 pressure is gradually lowered, graphite should appear when $P_{H_2} = 10^{-4.75}$ atm. As P_{H_2} drops below this value, P_{CH_4} and the total pressure of carbon compounds decreases. When $P_{H_2} = 10^{-5.7}$ atm, $P_{CH_4} = P_{CO_2} = 2.8 \times 10^{-3}$. Toward lower values of P_{H_2} the total pressure of carbon compounds increases. If the total number of moles of carbon in the solid and gaseous state remains constant, graphite is converted to CO_2 as P_{H_2} drops, until all of the graphite deposited during the destruction of CH_4 has been oxidized. A further decrease in P_{H_2} or the addition of O_2 does not change P_{CO_2} after the oxidation of graphite is complete.

If the CH_4 pressure in a highly reducing atmosphere is less than the value of P_{CH_4} at the crossover point in figure 4.2, graphite will not be formed at equilibrium as P_{H_2} is lowered. CH_4 will be converted directly to CO_2, and the pressure of the two gases will be equal when $P_{H_2} = 1.9 \times 10^{-6}$ atm.

The equilibrium pressure of CO will always be very small in atmospheres equilibrated with liquid water at 25°C. P_{CO} is related to P_{H_2} and P_{CH_4} by the expression,

$$CO + 3H_2 \rightleftharpoons CH_4 + (H_2O)_{Liq}, \tag{4.6}$$

for which

$$K_{4.6} = \frac{P_{CH_4}}{P_{CO} \cdot P_{H_2}^3} = 2.5 \times 10^{26}.$$

At equilibrium P_{CO} will always be less than 10^{-8} atm for the range of P_{H_2} and ΣM_C in figures 4.2a and 4.2b. CO in highly reduced volcanic gases should therefore be converted quantitatively to CH_4 in the model atmosphere that has been described above. Such a conversion would not violate any mass balance requirements. If the ratio of carbon compounds to water in early volcanic gases was equal to the ratio of C to H_2O in the oceans plus the crust today, then the ratio P_{CO}/P_{H_2O} was approximately

$$\frac{P_{CO}}{P_{H_2O}} \approx \frac{9 \times 10^{22} \text{ gm crustal C/12 gm per mol C}}{2 \times 10^{24} \text{ gm } H_2O \text{ in the oceans + crust/18 gm per mol } H_2O}$$

$$\approx 7 \times 10^{-2}.$$

If P_{H_2}/P_{H_2O} in these gases was approximately 1, then

$$P_{H_2} \approx 15 P_{CO}.$$

Since only 3 moles of H_2 are required to convert one mole of CO to CH_4 and H_2O, there would probably have been enough H_2 present for the complete conversion of CO to CH_4 and to sustain the required flux of H_2 from the atmosphere into interplanetary space if thermodynamic equilibrium had prevailed.

If, however, P_{H_2}/P_{H_2O} in volcanic gases was much less than 1, then this would no longer have been possible. Today P_{H_2}/P_{H_2O} in volcanic gases is on the order of 1×10^{-2}, and most of the carbon in volcanic gases is present as a constituent of CO_2. The maximum production of CH_4 by CO_2 reduction would therefore have been very minor. In an abiotic equilibrium atmosphere fed by such gases, most of the volcanic H_2 would have been removed by reaction with CO_2 and CO; the atmospheric H_2

pressure would therefore almost certainly have been considerably less than the pressure calculated in the previous section, where P_{H_2} was considered to have been controlled by hydrogen escape alone. It will be shown later that inorganic and biological reactions of H_2 with CO_2 to yield organic compounds have been of major importance during much of geologic time in maintaining the redox state of the crust and the distribution of crustal carbon between the reservoir of elemental carbon and that of carbonate carbon.

The ratio of N_2 to NH_3 in an equilibrated atmosphere is determined by the equation

$$N_2 + 3H_2 \rightleftharpoons 2NH_3. \tag{4.7}$$

The equilibrium constant for this reaction is

$$K_{4.7} = \frac{P_{NH_3}^2}{P_{N_2} \cdot P_{H_2}^3} = 6.7 \times 10^5.$$

The equilibrium N_2/NH_3 ratio at any given temperature therefore depends not only on the hydrogen pressure but on the total number of moles of nitrogen, ΣM_N, in the atmosphere. This is shown in figure 4.3, where the equilibrium value of P_{NH_3} is plotted as a function of P_{H_2} for two values of ΣM_N: half the total number of moles of nitrogen now in the crust plus atmosphere, and 1% of this value (see Holland 1978, p. 306).

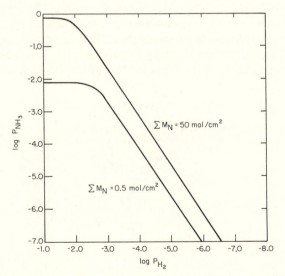

FIGURE 4.3. The equilibrium NH_3 pressure as a function of P_{H_2} for two values of the total number of moles of nitrogen in the atmosphere.

For both values of ΣM_N the equilibrium pressure of NH_3 varies considerably with P_{H_2} in the range of reasonable P_{H_2} values.

H_2S is the only sulfur compound of potential importance for the early atmosphere. It is, however, destroyed rapidly by solar ultraviolet radiation and was probably removed from the atmosphere by redox reactions and as a participant in the synthesis of organic compounds by electrical discharges in the atmosphere (see, for instance, Miller and Orgel 1974, p. 91).

Although the above description of possible equilibrium atmospheres is an interesting exercise in physical chemistry, it is probably not particularly useful in defining the state of the early atmosphere and oceans. The importance of reaction kinetics casts considerable doubt on the conclusions advanced earlier for the presence of a primordial methane-ammonia atmosphere (Holland 1962) and for the composition of the "primordial soup" (Sillén 1965a). The concentration of many of the minor components in the present atmosphere is controlled by photochemical reactions, by the effects of electrical discharges, and by biological reactions. The last factor was presumably missing very early in Earth history, but the first two were almost certainly of major importance.

Table 4.2 summarizes the intensity of the present-day and past sources of terrestrial energy. Until recently it was assumed that the intensity of the solar UV flux during the early stages of Earth history was roughly comparable to the intensity of the present-day flux. This view has now been challenged. Imhoff and Giampapa (1980) and Canuto et al. (1982, 1983) have reported that several T-Tauri stars emit strongly in the UV. The flux emitted per unit surface area of these stars is 100–5000 times greater than the present flux from the Sun. The Sun probably resembled these stars during the early stages of its evolution; it is quite possible therefore (see table 4.2) that the solar UV flux was much more intense

TABLE 4.2.

Present sources of external and internal energy (Gaustad and Vogel 1982).

Type of energy	Contemporary Earth Age 4.6×10^9 yr	Earth at age 5×10^8 yr	Earth at age 10^7 yr
Total solar radiation	265 000	170 000	132 000
Far UV (20–150 nm)	1.4	4–30	100–10 000
X-Ray (0.3–6 nm)	0.2	7	70–700
Radioactivity from crust (35 km)	15.5	47	—
Heat from volcanic emission	0.15	>0.15	—
Electric discharges	4	4	—

Note: Energy sources averaged over entire Earth (cal cm^{-2} y^{-1}).

during the first 10 m.y. and significantly more intense during the next 500 m.y. of Earth history than it is today (Gaustad and Vogel 1982).

The intensity of the other sources of energy that affected the chemistry of the early atmosphere is not known. The disparity between their present-day intensity and that of solar UV light is very large, and it seems likely that solar UV and lightning discharges were the only energy sources whose effects were important for the chemistry of the early atmosphere.

It is virtually certain that the combined effect of these energy sources on the composition of the atmosphere was dramatic. In Miller's (1953) classic experiments, mixtures of CH_4, NH_3, H_2, and H_2O were subjected to electric sparks generated across a spark gap by a Tesla coil in the apparatus shown in figure 4.4. The water in the 500 ml flask was boiled to bring the vapor into the region of the spark and to circulate the gases in the 5-liter flask. The products of the discharge were condensed and washed through the U-tube into the small flask. The nonvolatile products remained there, but the volatile products were recirculated past the spark, which was operated continuously for a week. At the end of the experiment the gases were pumped out, and the accumulated products in the small flask were analyzed by chromatography.

The results of the analysis are given in table 4.3. The identified compounds accounted for 15% of the total carbon. A substantial quantity of polymer or tar was formed; this was perhaps a cyanide, an aldehyde, or a mixed cyanide-aldehyde polymer. Some of the methane remained undecomposed and was found in the gas phase together with CO and N_2 formed in the spark. The results of the experiment were unexpected: a small number of relatively simple compounds accounted for a large proportion of the products, and these included a surprising number of substances that occur in living organisms.

The variation in the concentration of NH_3, HCN, and the aldehydes in the U-tube and of the concentration of amino acids in the 500 ml flask during the course of the experiment are shown in figure 4.5. Large amounts of cyanide and aldehydes were formed during the first 125 hours of sparking; the rate of synthesis of these compounds then fell off. The amino acids were formed at a more or less constant rate throughout the run.

Since 1953 a great many similar experiments have been performed with a large variety of starting mixtures. The results have been summarized by Miller and Orgel (1974), Miller, Urey, and Oró (1976), Dose and Rauchfuss (1975), and Fox and Dose (1972). The yields and the variety of organic compounds produced in these experiments decrease considerably as CH_4 is replaced by CO and as NH_3 is replaced by N_2 in the starting mixtures. No amino acids were found in the product of sparked

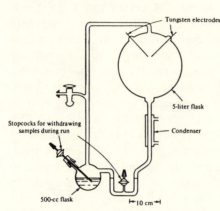

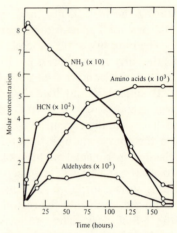

FIGURE 4.4. Apparatus used by Miller (1953) for sparking mixtures of CH_4, NH_3, H_2, and H_2O. (Copyright 1953 by the American Association for the Advancement of Science.)

FIGURE 4.5. The concentration of ammonia, hydrogen cyanide, and aldehyde in the U-tube, and the concentration of amino acids in the 500 ml flask while sparking a mixture of methane, ammonia, water, and hydrogen in the apparatus shown in figure 4.4 (Miller and Orgel 1974). (Reproduced by permission of the authors.)

mixtures of CO_2 + N_2 + H_2O (see Kawamoto and Akaboshi 1982, and Khare et al. 1981), but Abelson (1966) found HCN among the products of discharge experiments with mixtures of N_2, CO, and H_2 in a vessel with a cold trap, which maintained a low partial pressure of water vapor and removed complex molecules produced during the experiments.

These and similar experiments have been of great importance in attempts to construct a reasonable theory for the origin of life on Earth. They are of equal importance for understanding the evolution of the ocean-atmosphere system, because they suggest very strongly that electrical discharges and ultraviolet radiation tend to convert mixtures of reduced gases into mixtures of somewhat larger molecules and high polymers, and that these products tend to condense and to be washed out of the atmosphere onto land surfaces and into the oceans.

Few of the experimenters in this field have attempted to translate their results into rate equations that can be used to estimate the fate of the gases that were injected into the early atmosphere. Lasaga, Holland, and Dwyer (1971) studied the effect of ultraviolet radiation on methane; they developed a reasonable theoretical model to explain the observed polymerization in their experiments, and predicted that a methane atmosphere

TABLE 4.3.

Yields from sparking a mixture of CH_4, NH_3, H_2O, and H_2 (Miller and Orgel 1974).

Compound	Yield (μM)	Yield (%)
Glycine	630	2.1
Glycolic acid	560	1.9
Sarcosine	50	0.25
Alanine	340	1.7
Lactic acid	310	1.6
N-Methylalanine	10	0.07
α-Amino-n-butyric acid	50	0.34
α-Aminoisobutyric acid	1	0.007
α-Hydroxybutyric acid	50	0.34
β-Alanine	150	0.76
Succinic acid	40	0.27
Aspartic acid	4	0.024
Glutamic acid	6	0.051
Iminodiacetic acid	55	0.37
Iminoaceticpropionic acid	15	0.13
Formic acid	2330	4.0
Acetic acid	150	0.51
Propionic acid	130	0.66
Urea	20	0.034
N-Methyl urea	15	0.051

Note: 59 mmoles (710 mg) of carbon were added as CH_4. The percentage yields are based on the carbon.

would be polymerized in $< 10^8$ yrs to form a potentially thick primordial "oil slick" on the land and sea surface. This conclusion is in keeping with the nature of hydrocarbons in the atmosphere of Titan (Strobel 1981) and with the rapid formation of tarry deposits in spark and UV irradiation experiments; however, Lasaga, Holland, and Dwyer's (1971) model for the early atmosphere is obviously oversimplified (see, for instance, Allen, Pinto, and Yung 1980), and has been superseded by the calculations of Kasting, Zahnle, and Walker (1983).

The most complete treatment to date of the production rate of a "prebiologic" molecule in the early atmosphere is that given for formaldehyde by Pinto, Gladstone, and Yung (1980). In the absence of shielding by molecular oxygen, CO_2 and H_2O are photolyzed in the troposphere by the reactions,

$$CO_2 + hv \longrightarrow CO + O \tag{4.8}$$

and

$$H_2O + h\nu \longrightarrow H + OH. \tag{4.9}$$

These radicals participate in a complex series of reactions, the net effect of which may be written as

$$CO_2 + 2H_2 \xrightarrow{h\nu} H_2CO + H_2O. \tag{4.10}$$

Carbon monoxide participates in a similar reaction scheme, whose net effect can be expressed by the equation

$$2CO + H_2O \xrightarrow{h\nu} H_2CO + CO_2. \tag{4.11}$$

Pinto, Gladstone, and Yung (1980) (see also Bar-Nun and Chang 1983) computed the production rate of H_2CO via these reactions in an atmosphere with an N_2 pressure of 0.8 bar, an H_2O pressure of 0.013 bar, a CO pressure of 3.0×10^{-7} bar, and a variety of CO_2 and H_2 pressures.

Vertical profiles of the abundance of important minor constituents found in their standard photochemical model ($P_{CO_2} = 3 \times 10^{-4}$ bar, $P_{H_2} = 1 \times 10^{-3}$ bar) are shown in figure 4.6; the production and destruction rates of formyl radicals and formaldehyde are shown in figure 4.7.

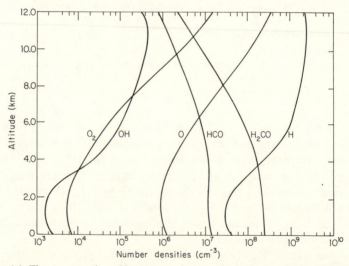

FIGURE 4.6. The concentration of important minor species in the standard model of Pinto, Gladstone, and Yung (1980). The surface mixing ratios of N_2, H_2O, CO_2, CO, and H_2 are 1, 1.5×10^{-2}, 3×10^{-4}, 3×10^{-7}, and 1×10^{-3}, respectively. These gases are well mixed throughout the troposphere, except for water vapor. The height independent eddy diffusion coefficient is 10^6 cm^{-1} sec^{-1}. (Copyright 1980 by the American Association for the Advancement of Science.)

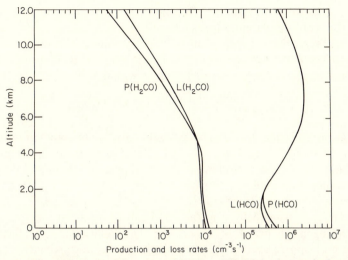

FIGURE 4.7. The production and loss rates of HCO and H_2CO in the standard model of Pinto, Gladstone, and Yung (1980). (Copyright 1980 by the American Association for the Advancement of Science.)

Approximately 99% of the H_2CO produced photochemically is destroyed by photolysis. However, a small fraction is incorporated into rain droplets and is delivered to the oceans. The removal rate of formaldehyde by this mechanism was calculated using the scavenging coefficients of Wofsy (1976) for H_2CO in the present atmosphere. The results are shown in table 4.4. Since the rain-out rate is sensitive to the thermal structure and humidity of the atmosphere, the figures in this table are only approximate.

TABLE 4.4.

Rainout rate of formaldehyde computed by Pinto, Gladstone, and Yung (1980) as a function of the CO_2 and H_2 pressure in their model atmosphere.

	Rainout Rate (molecules/cm² sec) P_{H_2} (bar)			
P_{CO_2} (bar)	1×10^{-4}	3×10^{-4}	1×10^{-3}	3×10^{-3}
6×10^{-4}			4.8×10^8	
3×10^{-4}	1.7×10^7	4.4×10^7	2.8×10^8	1.0×10^9
1.2×10^{-3}			1.2×10^9	
4×10^{-3}			3.1×10^9	

They are, however, fascinating. Despite its large destruction rate, formaldehyde would apparently be rained out at a rate of ca. $10^{8.3 \pm 1}$ molecules/cm^2 sec, which is equivalent to $10^{10.7 \pm 1}$ moles/yr. If formaldehyde were added to the present oceans at this rate and if it were allowed to accumulate, its concentration after 10 m.y. would be

$$\frac{10^{17.7 \pm 1} \text{ moles H}_2\text{CO}}{10^{21.1} \text{ kg seawater}} = 10^{-3.4 \pm 1} \text{ mol/kg.}$$

It seems unlikely that formaldehyde would have accumulated in the early oceans (Abelson 1966). The calculated rate of input is not spectacularly large, but the process could have played a significant role in the synthesis of prebiologic molecules. The effect of an early, more intense UV flux on the production rate of formaldehyde remains to be assessed.

The role of formaldehyde in the geochemistry of atmospheric carbon and hydrogen was probably fairly minor. The mean input rate of carbon to the atmosphere averaged over geologic time is approximately 1.7×10^{12} mol/yr. The rain-out rate of formaldehyde from Pinto, Gladstone, and Yung's (1980) standard atmosphere is 5% of this rate. Since the input rate of carbon compounds to the early atmosphere was probably considerably greater than the mean input rate, the percentage of carbon sequestered as formaldehyde from the standard atmosphere was probably even smaller. Only if the upper limit of H_2CO rain-out rates in table 4.4 is closer to the true value, and if the carbon input rate to the atmosphere was close to the geologic average, would the removal of carbon as a constituent of formaldehyde have represented a considerable fraction of the total removal rate of carbon from the early atmosphere.

The loss rate of H_2 from the atmosphere as a constituent of H_2CO is equal to the removal rate of H_2CO itself. In Pinto, Gladstone, and Yung's (1980) standard model the hydrogen pressure is 1×10^{-3} atm. Their calculated escape rate of hydrogen is 1.0×10^{10} atoms/cm^2 sec, which is ca. 25 times the computed rain-out of H_2CO. Since the rain-out rate of H_2CO and the H_2 escape rate are both roughly proportional to the hydrogen pressure in the atmosphere, it is likely that the rain-out of H_2CO played only a minor part in the H_2 budget of the early atmosphere, unless the CO_2 pressure in the early atmosphere was considerably in excess of 3×10^{-4} atm, or if the effect of a higher UV flux turns out to have been dramatic.

4. The Interaction of the Atmosphere with Continental Surfaces

Little is known about the chemical composition of the early crust. Many of the high-grade metamorphic rocks formed between 3.5 and 3.8 b.y. ago have a composition close to that of average crustal rocks (see chapter 5), and it seems likely that similar rocks were present more than 3.8 b.y. ago. Basaltic and ultramafic rocks must also have been present, but nothing is known about the relative proportions of these rock types. Fortunately, our present ignorance is not a serious deterrent to a discussion of the interaction of the early atmosphere with continental surfaces, because this interaction was not influenced strongly by the differences in the composition of the igneous rocks that are likely to have been common on the early Earth.

Chemical weathering today is dominated by carbonation, hydration, and oxidation reactions (see, for instance, Holland 1978, chap. 2). CO_2 and H_2O were almost certainly important constituents of the earliest volcanic gases; carbonation and hydration reactions were therefore probably important during weathering on the early Earth. The earliest known metasediments contain strong evidence for the operation of carbonation and hydration reactions about 3.8 b.y. ago (see chapter 5). During carbonation reactions, H_2CO_3 that is dissolved in water reacts with minerals to yield dilute bicarbonate solutions and solid residues that are frequently rich in clays. The carbonation of anorthite is a suitable example of such reactions:

$$CaAl_2Si_2O_8 + 2H_2CO_3 + H_2O \longrightarrow Ca^{+2} + 2HCO_3^{-} + Al_2Si_2O_5(OH)_4.$$
anorthite   kaolinite

Ca^{+2} and HCO_3^{-} tend to travel to the oceans in solution, and are removed from seawater by the deposition of $CaCO_3$. Kaolinite tends to be transported mechanically to the oceans and to accumulate there as a constituent of marine sediments.

The evolution of present-day river waters is now reasonably well understood (see, for instance, Holland 1978, chap. 4), and we can apply this understanding to reconstruct the evolution of river waters on the early Earth. The reaction of CO_2 with silicate rocks on early continental surfaces must have released many cations, among which Ca^{+2}, Mg^{+2}, Na^{+}, K^{+}, and Fe^{+2} were almost certainly the most abundant. Today the first four of these are the major cations in nearly all river waters. Fe^{+2} is oxidized rapidly, and is precipitated, largely within soils, as an oxide or hydroxide of iron. The concentration of dissolved iron in most river waters is well below 1 ppm (Durum and Haffty 1963); its concentration in seawater ranges from ca. 0.1 to 60 ppb (see, for instance, Brewer 1975). In an anoxic atmosphere with a hydrogen mixing ratio of 1×10^{-3}, Fe^{+2} is overwhelmingly stable with respect to Fe^{+3} (see, however, Baur 1978, for a dissenting view). In the presence of atmospheric and dissolved H_2S, a variety of iron sulfides are apt to precipitate. The behavior of iron in modern anoxic sediments is probably a useful guide to the behavior of iron on the early Earth (see chapter 9). FeS, Fe_3S_4, and FeS_2 probably precipitated in the earliest weathering horizons and river courses (see, for instance, Berner 1980), and the concentration of dissolved iron in rivers was probably no greater than it is today.

Iron sulfides in such settings could have been quantitatively important sinks for volcanic H_2S. If the ratio of sulfur to carbon in the earliest volcanic gases was close to its average during Earth history, then the ratio of the input of H_2S to CO_2 was approximately equal to the ratio of the crustal inventory of sulfur to that of carbon. If we take 16×10^{21} gm as the crustal inventory of sulfur (Holser and Kaplan 1966) and 90×10^{21} gm as the crustal inventory of carbon (Holland 1978, p. 276) then

$$\left[\frac{d_v M_{H_2S}/dt}{d_v M_{CO_2}/dt} \right] \approx \frac{16 \times 10^{21} \text{ gm S}/32 \text{ gm S/mol}}{90 \times 10^{21} \text{ gm C}/12 \text{ gm C/mol}} = 0.067 \frac{\text{mole S}}{\text{mole C}}. \quad (4.13)$$

If most of the volcanic CO_2 was removed from the atmosphere by carbonation reactions involving basaltic rocks, then the ratio of Fe^{+2} to HCO_3^- was determined by the cationic composition of basalts and by the completeness of chemical weathering. From table 2.10 in Holland (1978), it follows that in average basalts the ratio,

$$\frac{2m_{Fe^{+2}}}{(2m_{Fe^{+2}} + 2m_{Mg^{+2}} + 2m_{Ca^{+2}} + m_{Na^+} + m_{K^+})},$$

is approximately 0.25. During weathering involving the complete removal of these cations, the ratio $m_{Fe^{+2}}/m_{HCO_3^-}$ is therefore approximately 0.12 After comparable weathering of average granite, this ratio has a value close to 0.07 (see table 2.9 in Holland 1978). The minimum value of the S/Fe ratio in iron sulfides is 1.0. Thus, if all of the iron released during weathering were precipitated as a constituent of one or more iron sulfides, then each mole of volcanic CO_2 could lead to the precipitation of ca. 0.1 mol of sulfur. Since the removal ratio of sulfur to carbon is potentially greater than the crustal ratio of these elements, the precipitation of iron sulfides could have acted as a major H_2S sink on the early Earth. This removal mechanism must have been in competition with the photochemical destruction of H_2S and the loss of H_2S by reactions in lightning discharges. The relative importance of these several removal mechanisms is not known, but it seems likely that their combined effect kept the partial pressure of H_2S in the atmosphere at a very modest level as soon as a steady state between the input of gases to the atmosphere and their removal as a component of sediments had been established. The H_2S pressure in the early atmosphere is hard to gauge. The equilibrium between pyrite and siderite at 25°C in the presence of liquid water can be written in the form,

$$FeCO_3 + 2H_2S \rightleftharpoons FeS_2 + CO_2 + H_2 + (H_2O)_{Liq}. \quad (4.14)$$

The equilibrium partial pressure of CO_2, H_2, and H_2S is therefore related by the expression,

$$K_{4.14} = \frac{P_{CO_2} \cdot P_{H_2}}{P_{H_2S}^2} \approx 10^{+10}$$

(data from Robie, Hemingway, and Fisher 1978). For all likely values of P_{CO_2} and P_{H_2} in the early atmosphere, the equilibrium value of P_{H_2S} is therefore $\leq 10^{-5}$ atm. Removal of H_2S by iron sulfide precipitation alone might therefore have maintained the H_2S content of the atmosphere at the ppm level.

It could, however, be argued that the rate of release of volatiles early in Earth history was so rapid that neutralization by weathering did not keep up with the input of acid volatiles. Lafon and Mackenzie (1974) have calculated the evolution of the composition of ocean water on an Earth on which all of Rubey's (1951) excess volatiles were degassed instantaneously and were then permitted to react with slightly simplified average igneous rock. Their results are shown in figures 4.8 and 4.9. The calculated initial pH of the oceans is approximately 0.1 and approaches present values only when more than 0.5 kg of average igneous rock have been

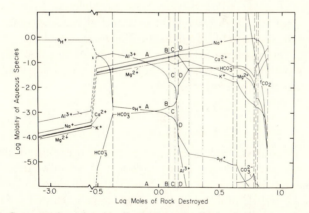

FIGURE 4.8. Concentration of major dissolved species in the simulated early ocean as a function of the progress of the reaction (measured by the amount of igneous rock destroyed). The vertical dashed lines separate distinct phase assemblages of product minerals (see figure 4.9). The values of f_{CO_2} are calculated by assuming equilibrium between the gas phase and the aqueous phase. All concentrations are relative to 1 kg of water (Lafon and Mackenzie 1974). (Reproduced by permission of the Society of Economic Paleontologists and Mineralogists.)

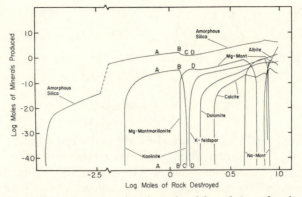

FIGURE 4.9. Amounts of sedimentary minerals produced (in moles) as a function of the progress of the reaction in Figure 4.8 (Lafon and Mackenzie 1974). (Reproduced by permission of the Society of Economic Paleontologists and Mineralogists.)

destroyed by reaction with each kilogram of water released. The low initial pH is largely due to the presence of the 0.946 moles of HCl per kg H_2O in the unreacted volatiles. The presence of such an acid ocean in the earliest Earth seems unlikely but cannot be ruled out. It was pointed out earlier that the maximum injection rate of water into the early atmosphere was probably $\leq 2.5 \times 10^{15}$ mol/yr. There is no reason to believe that the ratio of the other volatiles to water was very different from that in the outer layers of the Earth today. The quantity of NaCl currently present in evaporites is probably less than the NaCl content of the oceans (see chapter 9). The number of moles of Cl^- in the oceans and in the crust together is therefore $\leq 1.4 \times 10^{21}$ mol. The HCl injection rate early in Earth history was therefore probably

$$\leq \frac{1.4 \times 10^{21} \text{ mol} \times 100}{4.5 \times 10^9 \text{ yrs}} = 3 \times 10^{13} \text{ mol/yr.}$$

The total inventory of sulfur and carbon in the ocean-atmosphere-crust system was discussed above. From these figures it follows that the probable maximum injection rate of H_2S was 1×10^{13} mol/yr and that of CO_2 was 17×10^{13} mol/yr. Today rivers annually carry ca. 4.6×10^{13} mol of HCO_3^- to the sea. The annual removal of 3×10^{13} mol of highly reactive HCl, together with the removal of a smaller amount of H_2S and approximately six times as much CO_2, does not, therefore, seem to demand extraordinarily extreme atmospheric conditions. The fraction of the Earth's surface covered by water was probably smaller on the early Earth than it is today. This in turn probably implies a less rapid hydrologic cycle but a somewhat larger land area. The "continents" would have consisted almost entirely of igneous rocks; these tend to weather rather rapidly even in the absence of a plant cover (Holland 1978, pp. 103–106). If volcanic features dominated the earliest landscapes, slopes were probably steep enough to insure the mechanical removal of soil zones and the continuous exposure of fresh rock. Thus, unless the rates of volatile input used above are seriously in error, there seems to be no reason to believe that the primordial ocean was ever very acid.

Siever (1977) has suggested that life might have originated in icy oceans surrounded by smoking volcanoes. This is an intriguing thought. A complete freeze-over, leading to what might be called the "golf-ball Earth," can probably be ruled out on the grounds that CO_2 and some of the other gases that were injected into the atmosphere would have accumulated there in the absence of weathering reactions mediated by the hydrologic cycle; ultimately their presence would probably have produced a greenhouse effect sufficiently strong to raise the surface temperature of the Earth above 0°C. Melting would then have taken place, followed by the establishment of a hydrologic cycle and the removal of much of the accumulated atmospheric inventory of gases.

5. The Primordial Oceans

The oceans must have grown apace during the earliest part of Earth history. Liquid water could have existed at temperatures well above 100°C. Whether it actually did so depended, among other factors, on the rate of evolution of water from the Earth's interior, on the rate at which this water reacted with surface rocks to produce hydrated minerals, and on the rate at which water percolated underground. If we take 2.5×10^{13} mol/yr as the minimum water release rate from the Earth's mantle during this period of Earth history, and if 5% by weight water of hydration was added to surface rocks during weathering, then the complete removal of the injected water would have required a weathering rate of roughly

$$2.5 \times 10^{13} \frac{\text{mol/H}_2\text{O}}{\text{yr}} \times 18 \frac{\text{gm H}_2\text{O}}{\text{mol H}_2\text{O}} \times 20 \frac{\text{gm rock}}{\text{gm H}_2\text{O}} = 0.9 \times 10^{16} \text{ gm/yr.}$$

This is about half the current figure of 2×10^{16} gm/yr for sediment transport to the oceans (Holeman 1968). In the absence of an ocean, the hydrologic cycle would have been very modest; thus weathering reactions probably did not keep up with the injection of water even at its minimum injection rate. Removal of water from the surface as interstitial water in mechanically produced sediments probably delayed the appearance of standing bodies of water, but if the inferences drawn in chapter 3 in favor of a rapid early degassing of the Earth are correct, sizable bodies of water were almost certainly present on the Earth more than 4.0 b.y.b.p.

The salinity of the early oceans was determined largely by the HCl/H_2O ratio in the released volatiles. We do not know the value of this ratio, but it seems likely that it was not far from the present crustal ratio of Cl^- to H_2O. If we accept the figures from the previous section of this chapter for the Cl^- and H_2O content of the crust, then the chloride concentration, $_0m_{Cl^-}$, in the earliest oceans was on the order of

$$_0m_{Cl^-} \approx \frac{1.4 \times 10^{21} \text{ mol } Cl^-}{2 \times 10^{21} \text{ kg } H_2O} = 0.7 \text{ mol/kg}.$$

The precision of this figure should not be overrated (see chapter 9) and its applicability is clouded by our ignorance of the first few hundred million years of Earth history, but the calculation does indicate that the earliest oceans almost certainly contained a significant concentration of Cl^-. Na^+, K^+, Ca^{+2}, and Mg^{+2} must have been the dominant cations. Their relative proportions are difficult to gauge. We are now coming close to a quantitative understanding of the factors that control the ratio of the major cations in modern seawater (Holland 1978, chap. 5); a fairly detailed proposal for the chemical evolution of seawater during the last 1 b.y. of Earth history is given in chapters 8 and 9. It is shown there that the ratio of the major cations in seawater has probably been somewhat variable, and that their concentration has been determined by the cation inputs to the oceans and by the mix of mechanisms by which they were removed.

Na^+ has almost certainly been the dominant cation in seawater throughout Earth history, the ratio $_0m_{K^+}/_0m_{Na^+}$ has been less than unity, and Ca^{+2} and Mg^{+2} have been present in small but not negligible concentrations. The HCO_3^- concentration was probably low in the early oceans, perhaps smaller than it is today. The concentration of SO_4^{-2} depended largely on the oxidation state of the atmosphere and on the manner in which H_2S was removed from the atmosphere. At equilibrium, the SO_4^{-2} concentration is determined by the reaction,

$$H_2S + 4(H_2O)_{Liq} \rightleftharpoons SO_4^{-2} + 2H^+ + 4H_2, \tag{4.15}$$

for which

$$K_{4.15} = \frac{a_{SO_4^{-2}} \cdot a_{H^+}^2 \cdot P_{H_2}^4}{P_{H_2S}} \cong 10^{-41.7}.$$

When $P_{H_2} = 10^{-3}$ atm and $P_{H_2S} = 10^{-4}$ atm, the equilibrium value of $a_{SO_4^{-2}}$ is negligibly small, even compared to a_{HS^-}. This is illustrated by figure 4.10, which shows the equilibrium concentration or partial pressure

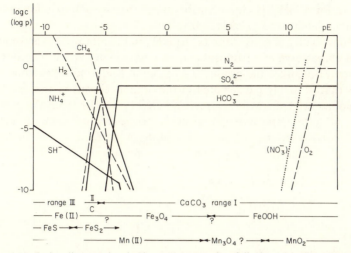

FIGURE 4.10. Redox diagram showing log c (concentration; fully drawn) and log p (pressure; broken lines) for various species as a function of pE in a model atmosphere-ocean system. At the bottom of the diagram the stability range of various solid phases is indicated (Sillén 1965b). (Reproduced by permission of the Royal Swedish Academy of Sciences.)

of several constituents of the atmosphere-ocean system as a function of its oxidation state (Sillén 1965b). The diagram confirms that at H_2 pressures in excess of about 10^{-4} atm, sulfur, carbon, and nitrogen should all have existed predominantly in their most reduced states. It is doubtful, however, that they obeyed the dictates of thermodynamics.

The same doubts extend to the oxidation state of nitrogen. It was shown above that N_2 was almost certainly the dominant nitrogen species in volcanic gases, and that N_2 should have been converted to NH_3 in the atmosphere if P_{H_2} was $\geq 10^{-3}$ atm. It is extremely unlikely, however, that significant quantities of NH_3 were ever produced in the early atmosphere. If NH_3 had been produced, it would have been decomposed by the flux of solar ultraviolet radiation (see, for instance, Levine 1982, and Levine, Augustsson, and Natarajan 1982). In an atmosphere containing only N gas species, a quantity of N_2 equal to the current amount of atmospheric N_2 can be produced by the photodissociation of NH_3 in 10^7 years even if the NH_3 pressure is maintained at values as low as 10^{-5} atm during the course of the conversion (Kuhn and Atreya 1979). The photochemical destruction of NH_3 may have been slowed by the presence of H_2 in the atmosphere and of H_2O in the upper atmosphere (Pollack and Yung 1980). On the other hand, the NH_3 abundance in the upper troposphere and in the higher atmosphere may have been reduced by its reaction with

H_2S to form NH_4SH clouds in the upper troposphere (see, for instance, Weidenschilling and Lewis 1973).

The difficulty of converting N_2 to NH_3 in the atmosphere is somewhat of an embarrassment. Atmospheric ammonia is not an absolutely necessary constituent for the production of amino acids and other small molecules that probably participated in the organic chemistry that led to the development of life (Miller, Urey, and Oró 1976). However, the preservations of some important amino acids in the oceans seems to demand NH_4^+ concentrations $\geq 1 \times 10^{-3}$ mol/kg. Aspartic acid deaminates reversibly to fumaric acid and NH_4^+ by the reaction,

$$^-OOC - CH_2 - \underset{\underset{NH_3^+}{|}}{CH} - COO^- \rightleftharpoons$$

$$^-OOC - CH = CH - COO^- + NH_4^+, \quad (4.16)$$

for which the equilibrium constant is

$$K_{4.16} = \frac{a_{NH_4^+} \cdot a_{Fumarate}}{a_{DL\text{-aspartate}}}.$$

The value of $K_{4.16}$ is 1.0×10^{-3} at $0°C$ and 2.7×10^{-3} at $24°C$ (Bada and Miller 1968). The half-life for attainment of equilibrium is ca. 29 m.y. at $0°C$ and ca. 96,000 years at $25°C$. If aspartic acid was needed for the evolution of life, and if the aspartate/fumarate ratio was not allowed to fall substantially below 1.0, then $_0m_{NH_4^+}$, the concentration of NH_4^+ in the oceans, must have been $\geq 1 \times 10^{-3}$ m. Bada and Miller (1968) and Miller and Orgel (1974, p. 48) suggest that NH_4^+ concentration between 10^{-3} and 10^{-2} m seem reasonable for the prebiotic synthesis of most organic compounds of biological interest. Miller and van Trump (1981) have shown that such an NH_4^+ concentration is consistent with the synthesis of amino acids through the amino nitriles (Strecker synthesis) in the primitive ocean.

An NH_4^+ concentration of 1×10^{-2} m in the primordial oceans probably implies that several percent of the total quantity of nitrogen in the atmosphere and oceans at that time were present in the oceans as a constituent of NH_4^+. If the ratio of N/H_2O in early volatiles was equal to their present ratio in the atmosphere-ocean-crust reservoir (see Holland 1978, p. 306), then

$$\left(\frac{dM_N/dt}{dM_{H_2O}/dt} \right)_{volc.} \approx \frac{7 \times 10^{21} \text{ gm N}/14 \text{ gm N/mol}}{2 \times 10^{24} \text{ gm } H_2O/18 \text{ gm } H_2O/\text{mol}} = 4.5 \times 10^{-3}.$$

Thus, if all of the early nitrogen was present in the primordial ocean,

$$_0m_{NH_4^+} \approx 4.5 \times 10^{-3} \frac{mol\ N}{mol\ H_2O} \times 55 \frac{mol\ H_2O}{kg\ H_2O} = 0.25\ mol/kg.$$

An actual concentration of 0.01 mol/kg in an early ocean would have implied that approximately 4% of the available nitrogen was present as marine NH_4^+.

The source of this ammonia is not immediately obvious. It has been suggested by S. L. Miller that the release of volatiles at lower than magmatic temperatures could supply the necessary amount of NH_3, because with falling temperature the NH_3-N_2 ratio of important oxide buffers is shifted in the direction of NH_3. The release of volatiles in submarine settings is an intriguing possibility, but it is difficult to imagine that this process was important on an early Earth on which the quantity of liquid water was a small fraction of that in the present oceans.

The cycling of seawater through volcanics is a more promising method for generating ammonia. If the N_2 pressure was 0.1 atm, the concentration of dissolved N_2 in seawater at 25°C was approximately 6×10^{-5} mol N_2/ kg H_2O. Reaction of this seawater with volcanics equilibrated with metallic iron and magnetite at temperatures of 100°–200°C should have converted a sizable fraction of the dissolved N_2 to NH_3 via the reaction,

$$3Fe^0 + 4H_2O + N_2 \longrightarrow Fe_3O_4 + 2NH_3 + H_2, \qquad (4.17)$$

If the conversion was 50% complete, approximately 6×10^{-5} mol NH_3 were generated per kg of seawater that reacted with volcanics in this manner. The volume of seawater that reacted with hot volcanics is extremely difficult to estimate. The total quantity of seawater was almost certainly less than today, and the quantity of hot volcanics almost certainly much greater. If we use a rate ten times as large as the best estimate for the current rate of seawater cycling through midocean ridges (see chapter 6), the potential rate of generation of ammonia by the reaction of seawater with hot volcanics was approximately

$$\approx 6 \times 10^{-5} \frac{mol\ NH_4^+}{kg\ H_2O} \times 0.2 \times 10^{15} \frac{kg\ H_2O}{yr}$$

$$\approx 1.2 \times 10^{10} \frac{mol\ NH_4^+}{yr}.$$

If the volume of seawater was 10% of the present volume, the time interval, Δt, required to raise the NH_4^+ concentration of seawater to 0.01 mol/kg was defined by the expression,

$$\frac{1.2 \times 10^{10} \dfrac{\text{mol NH}_4^+}{\text{yr}} \times \Delta t}{1.4 \times 10^{20} \text{ kg H}_2\text{O}} \cong 0.01 \frac{\text{mol NH}_4^+}{\text{kg}};$$

hence

$$\Delta t \cong 1.1 \times 10^8 \text{ yrs.}$$

NH_4^+ generation by this mechanism could therefore have been of considerable importance for the early ocean-atmosphere system. If the mechanism was indeed important, the production of CH_4 by the reduction of CO_2 and HCO_3^- by iron and the generation of H_2 by the reaction of water with iron were probably also quantitatively significant.

The reaction of a "primordial oil slick" or of a layer of organic polymer containing nitrogen with hot volcanics should probably be counted among possible additional mechanisms for generating ammonia. However, we know so little about the nature and the quantity of such organic polymers that a feasibility calculation is not warranted. Finally, the hydrolysis of HCN to formate and ammonia in the oceans may have been a significant source of NH_4^+. Irradiation of a variety of mixtures of CO, N_2, and H_2 produces HCN as the major reaction product (Abelson 1966), and it is likely that HCN was an important starting material for the synthesis of biomolecules on the primitive Earth (see, for instance, Sanchez, Ferris, and Orgel 1967; Matthews and Moser 1967; Ferris and Edelson 1978; Ferris et al. 1978). Oligomerization of HCN in aqueous solutions containing more than 0.1 mol/kg HCN can lead to the synthesis of amino acids, purines, and pyrimidines (see, for instance, Schwartz 1979). In solutions where the concentration of HCN is ≤ 0.01 mol/kg, the principal reaction is the hydrolysis of HCN to formate and ammonia. NH_3 is produced at high HCN concentrations. The concentration of HCN in the primordial oceans is not known, but a concentration low enough to favor the hydrolysis of HCN to formate and ammonia is probably not unreasonable.

Localized environments with the desired chemical and physical properties are an attractive alternative to a uniformly ammoniacal ocean as a source area of prebiotic organic compounds. Charles Darwin's "warm little pond with all sorts of ammonia and phosphoric salts, light, heat, electricity, etc." in which "a protein compound was chemically formed, ready to undergo still more complex changes" (see Ponnamperuma and Caren 1968) is just such an environment. The high concentration of ammonia and phosphoric salts could be due either to the particularly extensive operation of reactions that produce these constituents, or to the local addition of these desiderata by, say, the infall of carbonaceous meteorites (see, for instance, Kvenvolden et al. 1970; Kvenvolden, Lawless, and Ponnamperuma 1971), or of cometary material (Delsemme 1979), or by the evaporative concentration of seawater in areas of low rainfall and high evaporation rate. A high evaporation rate could have been due to the local climate, as in the Persian Gulf today; it could also have been due to the heating effect of lava flows in shallow, restricted parts of the primordial ocean. Both of the latter possibilities are interesting, because they leave open the possibility that compounds formed in such special environments were mixed back into larger, more permanent bodies of water, where they were available for further syntheses. Polypeptides may well have formed in such environments (see, for instance, Fox 1964; Saunders and Rohlfing 1972; Rohlfing 1976).

Special environments may also be required for the phosphorylation reactions that were probably important in prebiologic chemistry. The concentration of dissolved phosphorus in rivers and in the oceans today is uniformly low (see, for instance, Turekian 1969; Holland 1978, pp. 212–219; Meybeck 1982; Froelich et al. 1982). This is due in part to the small solubility of the mineral apatite, $Ca_5(PO_4)_3(OH,F)$, and in part to the uptake of phosphorus by plants during photosynthesis. If the model of the earliest oceans that was proposed earlier in this chapter is roughly correct, the pH and the concentration of calcium were probably sufficiently high, so that the concentration of dissolved phosphate was not much greater than its concentration in the present oceans (3 to 4 × 10^{-6} mol/kg); figure 4.11 shows that in the absence of strong complexing agents the pH of seawater has to be dropped by approximately 3 pH units before the concentration of phosphate reaches 1×10^{-4} mol/l (Schwartz 1971). Oxalate is one of the more interesting complexing agents

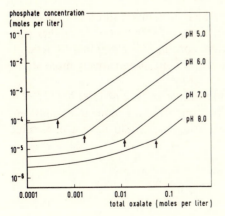

FIGURE 4.11. Solubility of fluorapatite in the presence of various quantities of total oxalate. The arrows indicate the point at which saturation with calcium oxalate is reached for each pH (Schwartz 1971).

for Ca^{+2}, but figure 4.11 shows that rather large concentrations of oxalate are required to depress the Ca^{+2} activity significantly at pH values close to neutrality. It also seems likely that in an ocean saturated with respect to $CaCO_3$, the activity of free Ca^{+2} was independent of the oxalate concentration.

Other means of increasing the concentration of phosphate and/or polyphosphate have therefore been sought. Ponnamperuma and Gabel (1968) reported that the mono-, di-, tri-, and tetraphosphates of adenosine were formed when a dilute solution of adenine and ribose was irradiated with ultraviolet light in the presence of ethylmetaphosphate. Rabinowitz et al. (1969) found that the addition of phosphine, PH_3, to methane, ammonia, and water in electric discharge experiments produced organic phosphorus derivatives such as aminoalkyl phosphates and aminoalkane phosphonates in addition to the normal array of organic compounds.

Although the pathways of a prebiologic synthesis of biochemically important phosphorus molecules are still very imperfectly understood, reduced phosphorus compounds such as phosphine and phosphide minerals might have served as interesting starting materials. Volcanic gases and iron meteorites are the most likely sources of such compounds in the early Earth. Phosphorus in igneous rocks is present largely as a constituent of apatite. The fugacity of phosphine, PH_3, in volcanic gases associated with basaltic flows is therefore controlled by equilibria such as

$$5CaMg(SiO_3)_2 + 3PH_3 + 8H_2O \rightleftharpoons$$
$$Ca_5(PO_4)_3(OH) + 5MgSiO_3 + 5SiO_2 + 12H_2. \quad (4.18)$$

The value of the equilibrium constant,

$$K_{4.18} = \frac{f_{H_2}^{12}}{f_{PH_3}^3 \cdot f_{H_2O}^8},$$

can be calculated approximately at magmatic temperatures. The necessary thermochemical data and their sources are listed in table 4.5 together with the calculated value of the equilibrium constant for reaction 4.18. At constant values of f_{H_2} and f_{H_2O}, phosphine becomes progressively more stable toward higher temperatures.

Basaltic magmas have temperatures in the vicinity of $1500°K$ ($1227°C$). Volcanic gases emitted at that temperature from a magma in equilibrium with diopside, enstatite, apatite, and tridymite would have a phosphine fugacity of

$$f_{PH_3} \cong 10^{-8.9} \times \frac{f_{H_2}^4}{f_{H_2O}^{8/3}} \text{ atm.} \quad (4.19)$$

TABLE 4.5.

Thermochemical data for the calculation of $K_{4.18}$ at magmatic temperatures.

| Compound | ΔG_f^0 (kcal/mol) | | | | | | Source of Data |
	1000°K	1100°K	1200°K	1300°K	1400°K	1500°K	
PH_3	+8.641	+11.255	+13.864	+16.465	+19.059	+21.645	(1)
$CaMgSi_2O_6$ (diopside)	−628.652	−614.831	−600.886	−586.997	−572.315	−556.327	(2)
H_2O	−46.023	−44.690	−43.348	−42.000	−40.641	−39.274	(2)
$Ca_5(PO_4)_3(OH)$ (apatite)	−1307.758	−1272.497	−1236.572	−1200.790	−1165.256	−1129.948	(2)
$MgSiO_3$ (clinoenstatite)	−300.293	−293.220	−286.167	−279.141	−271.277	−262.067	(2)
SiO_2 (tridymite)	−174.380	−170.248	−166.127	−162.029	−157.947	−153.874	(2)
log $K_{4.18}$	42.7	38.1	34.2	31.1	28.4	26.7	

(1) JANAF tables (1965).
(2) Robie, Hemingway, and Fisher (1978).

In present-day volcanic gases, f_{H_2} is much smaller than f_{H_2O}. It is therefore understandable that f_{PH_3} is negligibly small under such conditions. In early volcanic gases f_{H_2} was probably on the same order as f_{H_2O}. Even if both were equal to 1 atm, f_{PH_3} was only on the order of 10^{-9} atm, and only a negligible fraction of the phosphorus typically present in basaltic melts would have been carried off as phosphine.

Even the maximum likely input of PH_3 to the early atmosphere would have been small compared to the probable flux of PO_4^{-3} sustained by the weathering of apatite. From equation 4.19 it follows that

$$\frac{f_{PH_3}}{f_{H_2O}} \approx 10^{-8.9} \frac{f_{H_2}^4}{f_{H_2O}^{11/3}}. \tag{4.20}$$

The exponents of f_{H_2} and f_{H_2O} on the right side of equation 4.20 are so nearly the same, that the ratio $f_{H_2}^4 / f_{H_2O}^{11/3}$ can be set equal to unity when $f_{H_2} \cong f_{H_2O}$. If we again use 2.5×10^{15} moles/yr as the maximum input of water to the early atmosphere, the maximum input of volcanic PH_3 was approximately 3×10^6 moles/yr. This is trivial compared to the river input of PO_4^{-3} today (ca. 7×10^{10} moles/yr; Holland 1978, p. 217) and was probably of no significance for the course of prebiologic chemistry unless volcanic phosphine was deposited and used very locally.

Criticisms can of course be leveled at the above calculation; the pyroxenes in basalts are hardly pure enstatite and diopside; F^- frequently substitutes extensively for OH^- in apatite; basalts are not always silica-saturated; the thermochemical data, especially those for phosphine and apatite, are not very precise. The errors that are introduced into the calculation by these uncertainties are difficult to evaluate; nevertheless it is unlikely that phosphine was ever more than a trace constituent of volcanic gases.

Other gaseous phosphorus compounds appear to have been present in even smaller quantities than phosphine. The equilibrium constant $K_{4.21}$ for the reaction,

$$PH_3 + 3HCl \rightleftharpoons PCl_3 + 3H_2, \tag{4.21}$$

is

$$K_{4.21} = \frac{f_{PCl_3} \cdot f_{H_2}^3}{f_{PH_3} \cdot f_{HCl}^3},$$

and has a value of $10^{-2.3}$ at $1500°K$ (data from JANAF tables (1965) and Robie and Waldbaum (1968). The fugacity of HCl in the earliest volcanic gases is not known, but it is likely that the ratio f_{HCl}/f_{H_2O} was considerably less than unity. If f_{H_2} was roughly equal to f_{H_2O}, then f_{HCl}/f_{H_2} was also considerably smaller than one, and

$$\frac{f_{PCl_3}}{f_{PH_3}} = 10^{-2.3} \left[\frac{f_{HCl}}{f_{H_2}}\right]^3 \ll 1. \tag{4.22}$$

Similar arguments indicate that at 1500°K,

$$\frac{f_{POCl_3}}{f_{PH_3}} < 10^{-2.5},$$

$$\frac{f_{PCl_5}}{f_{PH_3}} < 10^{-15},$$

and

$$\frac{f_{PF_3}}{f_{PH_3}} \leq 10^{-2}.$$

Unless these calculations are more seriously in error than the stated uncertainties in the thermochemical data tend to indicate, hypotheses for the abiologic synthesis of organic compounds should not count on the availability of more than trace amounts of volcanic phosphine, or of phosphorus halogen compounds.

Iron meteorites hold out a somewhat brighter hope than volcanic gases for the production of phosphine. They contain ca. 0.13% phosphorus (Mason 1962); most of this phosphorus is present as a constituent of the mineral schreibersite $(Fe,Ni)_3P$. Ablation of large iron meteorites during passage through the Earth's atmosphere can be quite severe. Volatilized schreibersite would probably react with a mildly reducing atmosphere to form a variety of compounds, including phosphine and organic phosphorus compounds. The total quantity of phosphorus compounds formed in this manner could have been significant. The volatilization of 1 kg of iron meteorite per cm^2 of Earth surface would have liberated ca. 1.3 gm/cm^2 of phosphorus. If all of this phosphorus had been converted to phosphine, the partial pressure of PH_3 would have been 1.5×10^{-3} atm. Reaction with water and water vapor would gradually have converted phosphine to phosphoric acid via intermediates that might have been of prebiologic importance. The phosphoric acid would have been neutralized rapidly by a variety of reactions.

Phosphide minerals that survive passage through the atmosphere today are oxidized at the Earth's surface. Fe_3P reacts slowly with hot water, but is reported to give no reaction with cold water. Ni_3P does not react with cold or hot water on the time scale of a laboratory chemist. Phosphine is probably an intermediate product in its oxidation, but the quantity of phosphine generated in this manner surely depends on the chemistry of the environment in which the oxidation takes place.

The rate of phosphine generation by all processes related to meteorite infall is difficult to estimate. If one accepts a meteorite infall rate of 3×10^{12} gm/yr (Keil 1969) on the early Earth, if 7% of this mass consisted of iron meteorites; and if these iron meteorites contained 1% P, then the infall of P from this source was ca. 2×10^9 gm/yr. This figure is much greater than the estimated volcanic input of PH_3, but it is three orders of magnitude less than the current river flux of dissolved phosphorus to the oceans. Schwartz (1972b) therefore suggests that meteoritic phosphides could not have played a significant role in the phosphorus chemistry of the primitive Earth. This may not be true if abnormal quantities of schreibersite managed to find their way into restricted environments where the chemistry, both organic and inorganic, was quite different from that of the contemporary "world ocean."

The repeated return to special environments in the above discussion may, of course, be considered special pleading; however, strict adherents to the notion that only the major environments were important locales for prebiologic chemistry could be accused of simplemindedness. As Siever (1977) has pointed out, there were probably many intermontane basins on the early Earth, many of them filled with playa lakes, whose chemical composition depended on a variety of local factors. Some of these lakes could well have been so long-lived, that significant prebiologic evolution took place within them. Schwartz (1972a), for instance, has shown that an "evaporating-pond" model in which dilute solutions of uridine and ammonium oxalate in contact with synthetic hydroxyapatite were evaporated under relatively mild conditions led to the synthesis of uridine phosphates. In later experiments Schwartz et al. (1975) showed that even larger yields and a higher proportion of multiply phosphorylated products were obtained with a fluorapatite of igneous origin as the source of phosphate. Such syntheses are among the first that are geologically plausible. Handschuh and Orgel (1973) have pointed out that the mineral struvite, $MgNH_4PO_4 \cdot 6H_2O$, may also have precipitated in such environments, and that phosphorylation would certainly have occurred at the temperatures reached by dark-colored surfaces in strong sunlight; this would probably have led to the synthesis of nucleotides such as uridine 5'-phosphate and of nucleoside diphosphates such as uridine 5'-diphosphate. Corliss, Baross, and Hoffman (1981) have pointed out that submarine hot springs could have provided interesting thermal environments for the synthesis of prebiologic organic compounds. It seems likely that laboratory experiments designed to illuminate the origin of life will gradually demonstrate which, if any, of the special environments suggested to date are likely to have played a significant role in the prebiologic organic chemistry of the early Earth (Lazcano, Oró, and Miller 1983).

6. Summary

The chemistry of the earliest atmosphere and oceans was determined by high-temperature processes for which a thermodynamic treatment is appropriate, by low-temperature processes that were in part kinetically determined, and by atmospheric processes for which kinetics were clearly dominant. The high-temperature processes included the injection of volcanic gases at magmatic temperatures and the reaction of these gases, as well as aqueous solutions, with volcanic rocks at temperatures up to several hundred degrees centigrade. The low temperature processes included weathering reactions of atmospheric constituents with surface rocks and a multitude of reactions within and at the boundaries of the earliest oceans. Atmospheric processes included photochemical reactions driven largely by solar UV radiation and probably reactions driven by electrical discharges in the atmosphere. None of these processes are well defined. All that can be done at present is to reconstruct the state of the early Earth as imaginatively as possible, to set reasonable limits on the range of the parameters that seem to be important, to treat this current vision of the early Earth with as much rigor as our present understanding permits, and to remain skeptical of the results.

The oxidation state of the earliest volcanic gases must have been within the limits set by present-day volcanic gases and by volcanic gases in equilibrium with metallic iron. Their rate of input to the early atmosphere was probably more rapid than the mean degassing rate of the Earth (see chapter 3). If so, the injection rate of H_2 was sufficiently rapid so that the hydrogen pressure was significant, possibly on the order of 10^{-3} atm; this conclusion can only be upset if the removal of H_2 by photochemical reactions in the atmosphere was much more rapid than seems likely. O_2 is virtually absent from such an atmosphere. Methane should have been produced if thermodynamic equilibrium had prevailed, but the presence of this gas and of NH_3 in more than trace quantities is hard to reconcile with their almost certain destruction by photochemical reactions in the early atmosphere. Both gases were probably generated during the reaction of seawater containing dissolved carbon compounds, nitrogen, and nitrogen compounds with volcanic rocks at temperatures of a few hundred degrees centigrade; but it seems unlikely that these reactions generated enough CH_4 and NH_3 to produce highly reducing atmospheres such as those that had been considered the most likely candidates for the Earth's earliest atmosphere.

Weathering almost certainly involved carbonation and hydration reactions but not oxidation reactions. It is likely that weathering reactions were sufficiently rapid to keep up with the injection of acid volcanic gases, and that the pH of the early oceans was not very much lower than that of the present oceans. The major cations, then as now, were probably Na^+, Mg^{+2}, K^+, and Ca^{+2}; the major anion was surely Cl^-. The cation ratios are difficult to estimate, but it is hard to see how Na^+ could fail to have been the dominant cation.

The organic chemistry of the early oceans was probably extraordinarily complex. Organic compounds were almost certainly generated in the atmosphere and rained out into the oceans. It is possible that the oceans were covered by a layer of highly polymerized carbon compounds. A host of organic reactions must have proceeded within the primordial ocean, at its edges, in contact with hot igneous rocks in and near volcanic centers, and with infalling meteoritic and cometary material. It seems likely that life originated during the first 500 m.y. of Earth history. We believe that we understand the first steps in the synthesis of the first organisms. The nature of the later steps is still quite obscure, but it is probable that some of them were taken in microenvironments.

References

Abelson, P. H. 1966. Chemical events on the primitive Earth. *Proc. Nat. Acad. Sci. USA* 55:1365–1372.

Allen, M., Pinto J. P., and Yung, Y. L. 1980. Titan: Aerosol photochemistry and variations related to the sunspot cycle. *Astrophys. J.* 242: L125–L128.

Bada, J. L., and Miller, S. L. 1968. Ammonium ion concentration in the primitive ocean. *Science* 159:423–425.

Bar-Nun, A., and Chang, S. 1983. Photochemical reactions of water and carbon monoxide in Earth's primitive atmosphere. *Jour. Geophys. Res.* 88:6662–6672.

Baur, M. E. 1978. Thermodynamics of heterogeneous iron-carbon systems: Implications for the terrestrial primitive reducing atmosphere. *Chem. Geol.* 22:189–206.

Berner, R. A. 1980. *Early Diagenesis: A Theoretical Approach.* Princeton N.J.: Princeton University Press.

Brewer, P. G. 1975. Minor elements in seawater. In *Chemical Oceanography.* vol. 1, edited by J. P. Riley and G. Skirrow, chap. 7. New York: Academic Press.

Brinton, H. C., and Mayr, H. G. 1971. Temporal variations of thermospheric hydrogen derived from in situ measurements. *J. Geophys. Res.* 76:6198–6201.

Canuto, V. M., Levine, J. S., Augustsson, T. R., and Imhoff, C. L. 1982. UV radiation from the young Sun and oxygen and ozone levels in the prebiological paleoatmosphere. *Nature* 296:816–820.

————. 1983. Oxygen and ozone in the early Earth's atmosphere. *Precamb. Res.* 20:109–120.

Corliss, J. B., Baross, J. A., and Hoffman S. E. 1981. Submarine hydrothermal systems: A probable site for the origin of life. In *Geology of Oceans*, vol. C4 of *Proc. XXVI Int'l Congress of Geology*, Paris.

Delsemme, A. H. 1979. Scientific returns from a program of space missions to comets. In *Space Missions to Comets*, edited by M. Neugebauer et al., 139–178. NASA Conf. Publ. 2089.

Donahue, T. M. 1966. The problem of atomic hydrogen. *Ann. Geophys.* 22:175–188.

Dose, K., and Rauchfuss, H. 1975. *Chemische Evolution und der Ursprung lebender Systeme.* Stuttgart: Wissenschaftliche Verlagsgesellschaft.

Durum, W. H., and Haffty, J. 1963. Implications of the minor element content of some major streams of the world. *Geochim. Cosmochim. Acta* 27:1–11.

Ferris, J. P., and Edelson, E. H. 1978. Chemical evolution: 31. Mechanism of the condensation of cyanide to HCN oligomers. *J. Organic Chem.* 43:3989–3991.

Ferris, J. P., Joshi, P. C., Edelson, E. H., and Lawless, J. G. 1978. HCN: A plausible source of purines, pyrimidines and amino acids on the primitive Earth. *Jour. Mol. Evol.* 11:293–311.

Fox, S. W. 1964. Experiments in molecular evolution and criteria of extraterrestrial life. *Bio. Science* 14(12):13–21.

Fox, S. W., and Dose, K. 1972. *Molecular Evolution and the Origin of Life.* San Francisco: Freeman.

Froelich, P. N., Bender, M. L., Luedtka, N. A., Heath, G. R., and DeVries, T. 1982. The marine phosphorus cycle. *Amer. Jour. Sci.* 282:474–511.

Gaustad, J. E., and Vogel, S. N. 1982. High energy solar radiation and the origin of life. *Origins of Life* 12:3–8.

Handschuh, G. J., and Orgel, L. E. 1973. Struvite and prebiotic phosphorylation. *Science* 179:483–484.

Holeman, J. N. 1968. The sediment yield of major rivers of the world. *Water Resour. Res.* 4:737–747.

Holland, H. D. 1962. Model for the evolution of the Earth's atmosphere. In *Peptrologic Studies: A Volume to Honor A. F. Buddington*, edited by A.E.J. Engel et al., 447–477. Boulder: The Geological Society of America.

————. 1978. *The Chemistry of the Atmosphere and Oceans.* New York: Wiley.

Holser, W. T., and Kaplan, I. R. 1966. Isotope geochemistry of sedimentary sulfates. *Chem. Geol.* 1:93–135.

Hunten, D. M. 1973. The escape of light gases from planetary atmospheres. *J. Atmos. Sciences* 30:1481–1494.

Hunten, D. M., and Strobel, D. F. 1974. Production and escape of terrestrial hydrogen. *J. Atmos. Sciences* 31:305–317.

Hunten, D. M., and Donahue, T. M. 1976. Hydrogen loss from the terrestrial planets. *Ann. Rev. Earth Planet. Sci.* 4:265–292.

Imhoff, C. L., and Giampapa, M. S. 1980. The ultraviolet spectrum of the T-Tauri star RW Aurigar. *Astrophys. J.* 239:L115–L119.

JANAF. 1965. *Thermochemical Tables.* U.S. Dept. of Commerce, National Bureau of Standards, Inst. for Applied Technology, PBS 168 370D.

Joseph, J. H. 1967. Diurnal and solar variations of neutral hydrogen in the thermosphere. *Ann. Geophys.* 23:365–373.

Kasting, J. F., Liu, S. C., and Donahue, T. M. 1979. Oxygen levels in the prebiological atmosphere. *J. Geophys. Res.* 84:3097–3107.

Kasting, J. F., and Walker, J.C.G. 1981. Limits on oxygen concentration in the prebiological atmosphere and the rate of abiotic fixation of nitrogen. *J. Geophys. Res.* 86:1147–1158.

Kasting, J. F., Zahnle, K. J., and Walker, J. C. G. 1983. Photochemistry of methane in the early Earth's atmosphere. *Precamb. Res.* 20: 121–148.

Kawamoto, K., and Akaboshi, M. 1982. Study on the chemical evolution of low molecular weight compounds in a highly oxidized atmosphere using electric discharges. *Origins of Life* 12:133–141.

Keil, K. 1969. Meteorite composition. In *Handbook of Geochemistry,* vol. 1, edited by K. H. Wedepohl, chap. 4. New York: Springer-Verlag.

Khare, B. N., Sagan, C., Zumberge, J. E.. Sklarew, D. S., and Nagy, B. 1981. Organic solids produced by electrical discharge in reducing atmospheres: Tholin molecular analysis. *Icarus* 48:290–297.

Kuhn, W. R., and Atreya, S. K. 1979. Ammonia photolysis and the greenhouse effect in the primordial atmosphere of the Earth. *Icarus* 37: 207–213.

Kvenvolden, K., Lawless, J., Pering, K., Peterson, E., Flores, J., Ponnamperuma, C., Kaplan, I. R., and Moore, C. 1970. Evidence for extraterrestrial amino acids and hydrocarbons in the Murchison meteorite. *Nature* 228:923–926.

Kvenvolden, K. A., Lawless, J. G., and Ponnamperuma, C. 1971. Nonprotein amino acids in the Murchison meteorite. *Proc. Nat. Acad. Sci. USA* 68:486–490.

Lafon, G. M., and Mackenzie, F. T. 1974. Early evolution of the oceans: A weathering model. In *Studies in Paleo-Oceanography,* edited by W. W. Hay, 205–218. Soc. of Econ. Paleont. and Min., Spec. Publ. no. 20, Tulsa, Okla.

Lasaga, A. C., Holland, H. D., and Dwyer, M. J. 1971. Primordial oil slick. *Science* 174:53–55.

Lazcano, A., Oró, J., and Miller, S. L. 1983. Primitive Earth environments: Organic syntheses and the origin and early evolution of life. *Precamb. Res.* 20:259–282.

Levine, J. S. 1982. The photochemistry of the paleoatmosphere. *Jour. Molec. Evol.* 18:161–172.

Levine, J. S., Augustsson, T. R., and Natarajan, M. 1982. The prebiological paleoatmosphere: Stability and composition. *Origins of Life* 12:245–259.

Mason, B. 1962. *Meteorites.* New York: Wiley.

Matthews, C. N., and Moser, R. E. 1966. Prebiological protein synthesis. *Proc. Nat. Acad. Sci. USA* 56:1087–1094.

Meier, R. R., and Mange, P. 1973. Spatial and temporal variations of the Lyman-alpha airglow and related atomic hydrogen distributions. *Planet. Space Sci.* 21:309–327.

Meybeck, M. 1982. Carbon, nitrogen, and phosphorus transport by world rivers. *Amer. Jour. Sci.* 282:401–450.

Miller, S. L. 1953. A production of amino acids under possible primitive Earth conditions. *Science* 117:528–529.

Miller, S. L., and Orgel, L. E. 1974. *The Origins of Life on the Earth.* Englewood Cliffs, N.J.: Prentice-Hall.

Miller, S. L., Urey, H. C., and Oró, J. 1976. Origin of organic compounds on the primitive Earth and in meteorites. *Jour. Mol. Evol.* 9:59–72.

Miller, S. L., and van Trump, J. E. 1981. The Strecker synthesis in the primitive ocean. In *Origin of Life*, edited by Y. Wolman, 135–141. Dortrecht, Holland: Reidel.

Pinto, J. P., Gladstone, G. R., and Yung, Y. L. 1980. Photochemical production of formaldehyde in the Earth's primitive atmosphere. *Science* 210:183–185.

Pollack, J. B., and Yung, Y. L. 1980. Origin and evolution of planetary atmospheres. *Ann. Rev. Earth and Planet. Sci.* 8:425–487.

Ponnamperuma, C., and Caren, L. 1968. Chemical studies on the origin of life. In *Encyclopedia of Polymer Science and Technology*, vol. 9, edited by H. F. Mark, N. G. Gaylord, and N. M. Bikayes, 649–659. New York: Wiley-Interscience.

Ponnamperuma, C., and Gabel, N. W. 1968. Current status of chemical studies on the origin of life. *Space Life Sci.* 1:64–96.

Rabinowitz, J., Woeller, F., Flores, J., and Krebsbach, R. 1969. Electric discharge reactions in mixtures of phosphine, methane, ammonia and water. *Nature* 224:796–798.

Robie, R. A., and Waldbaum, D. R. 1968. Thermodynamic properties of minerals and related substances at 298.15°K (25.0°C) and one atmosphere (1.013 bars) pressure and at higher temperatures. *U.S. Geol. Survey Bull.* 1259.

Robie, R. A., Hemingway, B. S., and Fisher, J. R. 1978. Thermodynamic properties of minerals and related substances at 298.15°K and 1 bar (10^5 pascals) pressure and at higher temperatures. *U.S. Geol. Survey Bull.* 1452.

Rohlfing, D. L. 1976. Thermal polyamino acids: Synthesis at less than 100°C. *Science* 193:68–70.

Rubey, W. W. 1951. Geologic history of seawater, an attempt to state the problem. *Bull. Geol. Soc. Amer.* 62:1111–1147.

Sanchez, R. A., Ferris, J. P., and Orgel, L. E. 1967. Studies in prebiotic synthesis: 2. Synthesis of purine precursors and amino acids from aqueous hydrogen cyanide. *J. Molec. Biol.* 30:223–253.

Saunders, M. A., and Rohlfing, D. L. 1972. Polyamino acids: Preparation from reported proportions of "prebiotic" and extraterrestrial amino acids. *Science* 176:172–173.

Schwartz, A. W. 1971. Phosphate: Solubilization and activation on the primitive Earth. In *Chemical Evolution and the Origin of Life*, edited by R. Buvet and C. Ponnamperuma, 207–215. Amsterdam: North-Holland.

Schwartz, A. W. 1972a. Prebiotic phosphorylation-nucleotide synthesis with apatite. *Biochim. Biophys. Acta* 281:477–480.

———. 1972b. The sources of phosphorus on the primitive Earth: An inquiry. In *Molecular Evolution: Prebiological and Biological*, edited by D. L. Rohlfing and A. I. Oparin, 129–140. New York: Plenum.

———. 1979. Chemical evolution: The genesis of the first organic compounds. In *Organic Chemistry of Sea Water*, edited by E. K. Duursma and R. Dawson. New York: Elsevier.

Schwartz, A. W., van der Veen, M., Bisseling, T., and Chittenden, G.J.F. 1975. Prebiotic nucleotide synthesis: Demonstration of a geologically plausible pathway. *Origins of Life* 6:163–168.

Siever, R. 1977. Early Precambrian weathering and sedimentation: An impressionistic view. In *Chemical Evolution of the Early Precambrian*, edited by C. Ponnamperuma, 13–23. New York: Academic Press.

Sillén, L. G. 1965a. Oxidation state of Earth's ocean and atmosphere: 1. A model calculation on earlier states. The myth of the "probiotic soup." *Arkiv Kemi* 24:431–456.

———. 1965b. Oxidation state of the Earth's ocean and atmosphere: 2. The behavior of Fe, S, and Mn in earlier states. Regulating mechanisms for O_2 and N_2. *Arkiv Kemi* 25:159–176.

Strobel, D. F. 1981. Chemistry and evolution of Titan's atmosphere (abstract). In *Program and Abstracts IAMAP*, Third Scientific Assembly, 20. Hamburg, Federal Republic of Germany.

Suess, H. E. 1962. Thermodynamic data on the formation of solid carbon and organic compounds in primitive planetary atmospheres. *J. Geophys. Res.* 67:2029–2034.

Towe, K. M. 1981. Environmental conditions surrounding the origin and early Archean evolution of life: A hypothesis. *Precamb. Geol.* 16:1–10.

Turekian, K. K. 1969. The oceans, streams, and atmosphere. In *Handbook of Geochemistry*, vol. 1, edited by K. H. Wedepohl, chap. 10. New York: Springer-Verlag.

Vidal-Madjar, A., Blamont, J. E., and Phissamay, B. 1973. Solar Lyman alpha changes and related hydrogen density distribution at the Earth's exobase (1969–1970). *J. Geophys. Res.* 78:1115–1144.

Walker, J.C.G. 1977. *Evolution of the Atmosphere.* New York: Macmillan.

———. 1982. The earliest atmosphere of the Earth. *Precamb. Res.* 17: 147–171.

Warneck, P. 1980. Was wissen wir über die primitive Erd-Atmosphäre? Unpublished manuscript.

Weidenschilling, S. J., and Lewis, J. S. 1973. Atmospheric and cloud structures of the Jovian planets. *Icarus* 20:465–476.

Wofsy, S. C. 1976. Interactions of CH_4 and CO in the Earth's atmosphere. *Ann. Rev. Earth and Planet. Sci.* 4:441–469.

Wofsy, S. C., McConnell, J. C., and McElroy, M. B. 1972. Atmospheric CH_4, CO, and CO_2. *J. Geophys. Res.* 77:4477–4493.

5

The Acid-Base Balance of the Atmosphere-Ocean-Crust System

1. Introduction

The conversion of igneous rocks into sedimentary rocks involves several types of chemical reactions. Among these the titration reactions of bases with acids are probably the most important and their effects the most dramatic. Nearly all silicate minerals are attacked by weathering acids. With the exception of Fe^{+2}, which is usually oxidized and reprecipitated in weathering horizons today, all of the major mono- and divalent cations are released from their parent minerals during intense weathering and pass out of soil zones in solution. The charge of these cations is balanced by that of the anions of the weathering acids. There are many of these acids, but only three are quantitatively important today: H_2CO_3, H_2SO_4, and HCl. H_2S may have been important very early in Earth history (see chapter 4). Table 5.1 is an inventory of the mass and the H^+-equivalents

TABLE 5.1.

Approximate mass and H^+-equivalents of carbon, sulfur and chlorine compounds in the atmosphere, oceans, and crust.

	Mass (gm)	H^+-equivalents
Carbon (Hunt 1972)		
Elemental carbon	2×10^{22}	
Carbonate carbon	7×10^{22}	
	9×10^{22}	12×10^{21}
Sulfur (Holser and Kaplan 1966)		
Oceans	1.3×10^{21}	
Evaporites	5×10^{21}	
Other sedimentary rocks	2.7×10^{21}	
Metamorphic and igneous rocks	7×10^{21}	
	16×10^{21}	1×10^{21}
Chlorine (Holland 1978, p. 211)		
Oceans	3×10^{22}	
Evaporites	$(3 \pm 2) \times 10^{22}$	
	$(6 \pm 2) \times 10^{22}$	$(1.7 \pm 0.6) \times 10^{21}$

of carbonate carbon, sulfur, and chlorine in the present-day atmosphere, oceans, and crust. The figures in this compilation are rather rough, but they are probably correct within a factor or two. Carbonic acid has clearly been the most abundant of the weathering acids. The H^+-equivalents of HCl seem to be comparable to those of the sulfur-containing acids.

I argued in chapter 3 that the release of new volatiles from the interior of the Earth is slow today compared to the mean degassing rate of the Earth during the past 4.5 b.y., and that most of the present-day weathering acids are recycled volatiles derived from the metamorphism and melting of crustal rocks. The degassing of the Earth during its first half billion years is difficult to reconstruct in the absence of a rock record. However, the mineralogy and the composition of even the earliest metasediments allow some reasonably firm inferences to be drawn regarding weathering at the time of their deposition. The more complete rock record of more recent times can be used to reconstruct fairly quantitatively the relationship between the supply of acids and the supply of bases by the exposure of rocks to weathering. This chapter attempts to reconstruct the evolution of the acid-base balance of the crust and to explore the implications of this evolution for the CO_2 content of the atmosphere during the past 3.8 b.y.

2. The Metasediments at Isua, Greenland

The oldest known sequence of metasedimentary rocks is that at Isua, Greenland. The area lies to the northeast of Godthaab, within the Archean craton of Greenland (figure 5.1). The Isua supracrustals form a belt of medium-grade metavolcanics and metasediments that is enclosed in regional gneisses (figure 5.2). The belt is a crescent-shaped raft approximately 35 km long and 2.5 km wide, which has been folded about a central mass of Amitsoq tonalite gneiss (Bridgwater and McGregor 1974; Bridgwater, Watson, and Windley 1973, 1978; Allaart 1976; Bridgwater, Collerson, and Myers 1978). Pb/Pb, Rb/Sr, U/Pb, and Sm/Nd dating has yielded consistent dates of 3780 m.y. for the supracrustal rocks (Moorbath, O'Nions, and Pankhurst 1975; Baadsgaard 1976; Vitrac et al. 1977; Hamilton et al. 1978). The sequence was already highly deformed before the injection of the Amitsoq gneisses circa 3700 m.y. ago. Since then, both the gneisses and the supracrustal rocks have been strongly deformed, and contacts that were originally discordant are preserved only locally. The sequence was intruded by the Ameralik dikes between 2800 and 3700 m.y.b.p. and by at least two sets of Proterozoic dikes between 1800 and 2500 m.y.b.p. Pre-Ameralik metamorphism reached the amphibolite facies. Semipelitic to pelitic garnet-biotite schists in the sequence are characterized by the mineral assemblage quartz + plagioclase + biotite + garnet + ilmenite ± muscovite ± graphite ± epidote ± carbonate (see, for

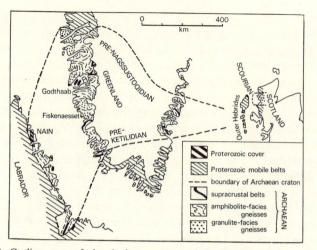

FIGURE 5.1. Outline map of the Archean craton of the North Atlantic region (after Bridgwater, Watson, and Windley 1973). (Reproduced by permission of the Royal Society, London.)

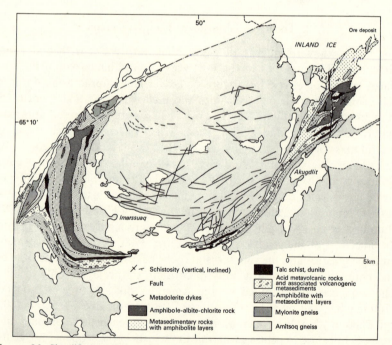

FIGURE 5.2. Simplified geologic map of the geology of Isua, Greenland (Allaart 1976). (Reproduced by permission of John Wiley & Sons, Ltd.)

instance, Boak and Dymek 1982); secondary chlorite and white mica are locally abundant. Anthophyllite is present in metamorphosed basic rocks within the series. The metamorphism and the extreme deformation have destroyed virtually all sedimentary and igneous structures with the exception of gross lithologic layering (Gill, Bridgwater, and Allaart 1983).

Biotite schist derived in part from tuffaceous (?) felsic volcanic rocks contains the assemblage quartz $\pm$ plagioclase $\pm$ muscovite $\pm$ biotite $\pm$ ilmenite $\pm$ epidote $\pm$ K-feldspar $\pm$ garnet. Kyanite has been observed in one sample. Fe-Mg partitioning between coexisting biotites and the core of garnets indicates metamorphic equilibration at temperatures of $500°–550°C$. The development of kyanite indicates a minimum pressure of 3.5 kb during metamorphism. These figures are consistent with temperatures and pressures during normal, moderate-pressure regional metamorphism (Boak and Dymek 1980).

The metasediments contain both chemical and clastic components. The chemical sediments consist of cherts, ironstones, marbles, calc-silicates, and biotite-bearing amphibolites that are thought to have been derived from impure calcareous sediments. The major clastic sediments are pelites and a suite of rather variable rocks; these range from impure calcareous units to quartz-feldspar biotite rocks that contain clasts of K-rich, felsic, acid volcanic rocks. Major layers and pods of dunite, clinopyroxene-peridotite, and their talc-schist amphibolitic derivatives appear to have been emplaced as tectonic pods, possibly at the time of thrust slicing.

Analyses of a few rather pure and several impure carbonates from the Isua supracrustals are compiled in table 5.2. Measurements of the CO_2 content are available for only two of the samples, and one of these is highly siliceous. Nevertheless, a good indication of the bulk composition of the carbonates in these rocks can be obtained, because most of the weight loss on ignition during analysis must have been due to loss of CO_2. This is illustrated by the figures in table 5.3. The sum of the number of moles of $(FeO + MgO + CaO)$ is essentially equal to that of the calculated number of moles of CO_2 in samples such as 175568J, which contain very little SiO_2. In siliceous samples the difference between the sum $(FeO + MgO + CaO)$ and their CO_2 content is generally well correlated with the number of moles of SiO_2. Sample 173070 is an exception. Decarbonation has apparently been less intense in this than in the other samples of siliceous carbonates. The data indicate that all, or nearly all, of the FeO, MgO, and CaO in these rocks was present at one time as a constituent of carbonate phases, and that decarbonation reactions are responsible for their present CO_2 deficit. All of the carbonates seem to have been highly magnesian. In most of the samples the mole ratio $(FeO + MgO)/CaO$ is equal to or greater than 1.0. Thus the initial carbonates

TABLE 5.2.

Composition in weight percentage of calcareous metasediments from Isua, Greenland (Gill, Bridgwater, and Allaart 1983).

Sample No.	175568B	175568E	175568F	175568I	175568J	170728	173070	158496	158497
SiO_2	17.99	33.49	1.96	4.73	1.78	3.19	27.29	47.85	13.74
TiO_2	0.31	0.19	0.06	0.03	0.03	0.02	0.11	0.00	0.00
Al_2O_3	6.69	4.29	1.15	2.59	1.04	0.04	6.38	0.49	0.05
Fe_2O_3	0.43	1.24	2.13	0.00	0.00	0.06	0.81	1.10	1.18
FeO	8.84	7.73	4.69	6.55	7.96	3.47	11.30	8.53	4.07
MnO	0.33	0.18	0.21	0.22	0.28	0.24	0.61	0.57	0.59
MgO	13.60	27.00	25.13	19.00	16.18	22.84	8.34	14.72	16.80
CaO	20.60	6.00	18.14	23.94	27.30	23.51	16.41	18.68	28.22
Na_2O	0.43						0.06	0.00	0.00
K_2O	1.49	0.01			0.08		2.28	0.00	0.01
H_2O^+ volat.	27.18	19.06	44.64	40.30	43.61	44.95	25.48	1.61	0.22
H_2O^- LOI								0.00	0.00
P_2O_5	0.03	0.01	0.01	0.01	0.01		0.02	0.09	0.17
CO_2								6.33	35.06
Total	97.92	99.19	98.12	97.38	98.27	98.32	99.09	99.97	100.11

TABLE 5.3.

Moles of FeO, MgO, CaO, CO_2, and SiO_2 per Kg of the calcareous metasediments of table 5.2.

Sample No.	175568B	175568E	175568F	175568I	175568J	170728	173070	158496	158497
FeO	1.23	1.07	0.65	0.91	1.11	0.48	1.57	1.18	0.56
MgO	3.37	6.69	6.23	4.71	4.01	5.66	2.07	3.65	4.17
CaO	3.68	1.07	3.24	4.27	4.87	4.20	2.93	3.34	5.04
[FeO + MgO + CaO]	8.28	8.83	10.12	9.89	9.99	10.34	6.57	8.17	9.77
CO_2*	6.18	4.33	10.14	9.17	9.91	10.22	5.79	1.44	7.97
[FeO + MgO + CaO] − CO_2*	2.10	4.50	−0.02	0.72	0.08	0.12	0.78	6.73	1.80
SiO_2	3.00	5.58	0.33	0.79	0.30	0.53	4.55	7.97	2.29

* Calculated either from direct CO_2 measurements or from the figures for LOI.

Table 5.4.

Composition in weight percentage of metapelites from Isua, Greenland (Gill, Bridgwater, and Allaart 1983).

Sample No.	Schists							Amphibolites	
	173164	173153	168255	170531	168294	173140	177705	172976	173030
SiO_2	52.68	47.40	44.42	41.09	40.56	44.00	47.91	48.72	45.34
TiO_2	1.22	1.31	0.82	0.16	0.91	0.89	0.20	1.24	0.90
Al_2O_3	13.02	14.41	13.02	8.04	13.29	10.51	7.14	15.23	11.19
Fe_2O_3	2.53	2.09	1.48	1.16	2.77	2.18	0.65	2.21	3.12
FeO	11.42	11.76	9.88	8.65	9.45	9.42	5.07	11.20	11.32
MnO	0.23	0.27	0.23	0.30	0.34	0.37	0.73	0.50	0.30
MgO	5.35	3.52	6.65	13.42	6.69	5.87	8.29	3.70	3.43
CaO	4.94	7.50	9.54	15.90	13.05	12.03	16.37	11.00	13.35
Na_2O	1.06	0.48	0.05	0.34	0.24	0.27	0.18	0.68	1.77
K_2O	3.89	4.60	5.67	1.52	3.61	3.62	2.32	0.83	0.87
H_2O^+		1.18	1.01	0.80	0.98	1.06		0.94	0.90
H_2O^-	1.08								
P_2O_5	0.18	0.22	0.05	0.04	0.02	0.19	0.09	0.18	0.21
CO_2	2.05	4.90	6.58	8.13	8.45	9.45	11.28	3.25	7.10
Total	99.65	99.64	99.40	99.51	100.22	99.86	100.74	99.68	99.80

probably consisted of a member of the dolomite-ankerite series and in some samples included a member of the magnesite-siderite series as well. This conclusion is borne out by mineralogical work (Allaart 1976).

A number of the schists and amphibolites in the Isua supracrustals also contain significant quantities of carbonate. Analyses of such calcareous rock samples are collected in table 5.4. Figure 5.3 shows that their CaO content is reasonably well correlated with their CO_2 content. The correlation between MgO and CaO is rather poor. Most of the CaO in these rocks is present as a constituent of $CaCO_3$. It is not clear how much of the MgO and FeO in these rocks was once present as a constituent of carbonates that were destroyed during metamorphism.

The presence of nearly pure carbonates and of calcareous metapelites in which virtually all of the calcium is present in a carbonate phase indicates that intense carbonation was taking place during weathering more than 3780 m.y. ago. The low Na_2O content of the metapelites in table 5.4 supports this conclusion. The Na_2O content of these rocks is comparable to that of more recent, unmetamorphosed pelites; their Na_2O content is much lower than that of all reasonable parent rocks. Na-plagioclase is easily destroyed by carbonation reactions during terrestrial weathering (see, for instance, Holland 1978, chap. 2), and this was almost certainly the mechanism of sodium loss from the parent rocks of the Isua metasediments. Intense carbonation during weathering implies the presence of an atmosphere containing a significant quantity of CO_2; this is consistent with the model for the early atmosphere that was developed in chapter 4.

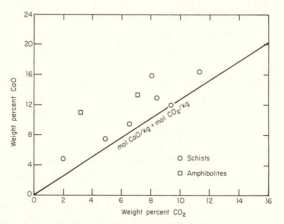

FIGURE 5.3. The relationship between the CaO and the CO_2 content of metapelites from Isua, Greenland; data from table 5.4.

The presence of nearly pure carbonates and of calcareous metapelites in the Isua supracrustals shows that the body of water in which these sediments were deposited was saturated with respect to one or more of the alkaline earth carbonates. If conditions at Isua were representative of the state of the world some 3800 m.y. ago, then the nature of weathering, solution transport, and sedimentation was already similar to that at the present time. The data are consistent with an Earth in which weathering was largely subaerial. Hargraves's (1976) suggestion that a shallow, globe-encircling sea covered the Earth during Archean time is somewhat unlikely. The notion is based on the presence of high geothermal gradients during this era and the presence of a thin continental crust. The pressure-temperature conditions during the metamorphism of the Isua supracrustals indicate that neither of these conditions prevailed there, and studies of other Archean terrains (see, for instance, Windley 1977, pp. 330–333) suggest that the thickness of Archean continents was comparable to those of the present day.

The chemical and petrologic data for the Isua metasediments cannot be used for a quantitative analysis of the acid-base balance of the Earth 3800 m.y. ago, because decarbonation reactions have destroyed much of the necessary evidence. For rock units that have retained many of their sedimentary features we must therefore go to somewhat younger sedimentary series.

3. The Barberton Mountain Land, South Africa

Archean sedimentary rocks of surprisingly low metamorphic grade are preserved in the greenstone belts of several Archean cratons. A great deal of attention has been paid to these belts during the past twenty-five years, and the results of many relatively detailed studies have been summarized in various publications (see, for instance, Lowe 1980). Greenstone belts occur as highly deformed volcanic and sedimentary keels that are steeply infolded into surrounding, coeval to younger, metamorphic and granitoid plutonic rocks. Greenstone sequences commonly have narrow metamorphic aureoles, but most of the rock units within such belts are altered only to the lowest greenschist facies.

Anhaeusser et al. (1969) have emphasized the remarkable similarity of the rock association, stratigraphy, metamorphism, and tectonic evolution of the known Archean greenstone belts. Their stratigraphic sections are usually between 10 and 25 km in thickness and include distinct volcanic and sedimentary sequences. In belts formed prior to 3000 m.y.b.p., such as the Barberton Mountain Land and the Pilbara Block of Australia (see, for instance, Cooper, James, and Rutland 1982), a thick volcanic sequence is overlain by a sedimentary succession. Belts formed less than 3000 m.y. ago commonly consist of several volcanic cycles, each capped by a thick sedimentary unit (Lowe 1980).

The most thoroughly studied and chemically analyzed of the greenstone belts is probably that of the Barberton Mountain Land in the eastern Transvaal, some 300 km east of Pretoria. The geology of the area was described by Visser (1956), and has since been the subject of numerous petrologic, mineralogic, sedimentological and geochemical studies. Although the analytical data are still somewhat inadequate, they do give a fair degree of insight into the nature of near-surface processes between 3000 and 3500 m.y.b.p. A geological map of the area is shown in figure 5.4 and a stratigraphic section in figure 5.5. The basal Onverwacht Group is approximately 14,300 m thick. It includes a lower ultramafic unit, 7500 m in thickness, consisting largely of ultramafic and mafic lavas, and an overlying unit, some 6800 m thick, that is composed of mafic lavas, but which also includes significant thicknesses of felsic tuff and lavas, chert, and carbonate rock (Viljoen and Viljoen 1969a and 1969b). Most of the sequence has suffered low-grade regional thermal metamorphism. The sedimentary rocks of the Onverwacht Group contain a diverse assemblage of sediments, including volcaniclastic sand and silt, coarse-grained detrital carbonate, BIF, and thin chert units that were originally deposited as fine-grained carbonate or other orthochemical sediments.

The base of the upper division of the Onverwacht Group is defined by a persistent sedimentary unit, the Middle Marker. This unit averages 10 m in thickness and is composed largely of fine-grained chert that contains significant amounts of dolomite, hematite, pyrite, and carbonaceous matter. Similar units are not uncommon at major breaks within volcanic sequences in other greenstone belts. Calcite and dolomite are the most common carbonates in such units (Lowe 1980). There is much evidence to suggest that the upper part of the Onverwacht Group was formed in a variety of shallow water, possibly evaporitic environments. Toward the close of Onverwacht deposition, a complex association of shallow water to subaerial environments probably existed in the southern part of the Mountain Land, while deep, quiet-water conditions developed to the north (Lowe and Knauth 1977).

Above the Onverwacht Group, volcanic materials are rather insignificant components of the sedimentary sequence in the Barberton Mountain area. The overlying Fig Tree Series is some 2000 m thick. Greywackes and banded cherts dominate its lower division; reworked, uncomformable grit horizons, greywackes, and a volcanic horizon dominate its upper division (Anhaeusser et al. 1968).

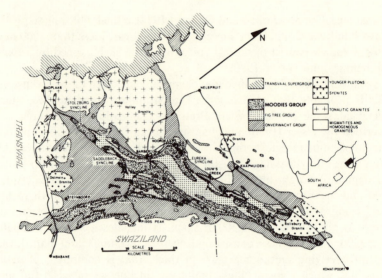

FIGURE 5.4. Geological map of the Barberton Mountain Land (Eriksson 1977). (Reproduced by permission of Elsevier Scientific Publishing Company.)

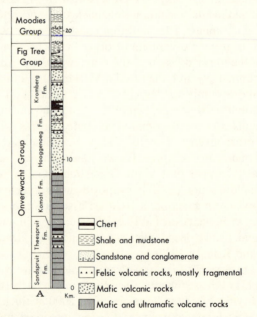

FIGURE 5.5. Stratigraphic section of the Barberton Mountain Land (Lowe 1980; after Viljoen and Viljoen 1969a,b). (Reproduced by permission from the *Annual Review of Earth and Planetary Sciences*, vol. 8, © 1980 by Annual Reviews, Inc.)

The Fig Tree Group is mainly a deep-water turbiditic assemblage in northern outcrops; southward it includes oxide-facies banded iron formations, volcaniclastic units, bedded barite (see, for instance, Reimer 1982), and shale-greywacke units. This association reflects deposition in an area of volcanic and tectonic instability, mainly under quiet subaqueous conditions. The lowest unit of the Fig Tree Series consists of shale, banded iron formation, and banded chert. In the Sheba and Hlambanyati hills the shales of this zone are generally carbonaceous. Ferruginous shales are also common and are frequently associated with banded iron formation. Banded cherts a few inches to several feet in thickness consist of alternating layers of white chert, iron formation, and iron-rich shale. The passage from shale to iron formation is frequently gradational. Oölitic chert that is apparently crossbedded has been found in one area. Greywackes and grits overlie the basal units of the Fig Tree Series. The greywackes are dark-colored to black sandstones derived from an area of acid, basic, and ultramafic volcanic rocks, granites, metamorphic rocks, and sediments (Reimer 1975). Fragments of various feldspars, black shale, black chert, carbonates, muscovite, and biotite are contained in a matrix of chlorite clay minerals, and small grains of chert and quartz. This unit is followed by a shale with numerous narrow bands of chert. Both are similar to their counterparts in the basal unit of the series.

The next unit consists of grits and conglomerate grits, in which rounded grains of clear quartz and fragments of chert, microcline, and micropegmatite are set in a matrix of quartz, feldspar, sericite, carbonate, and some secondary silica. Above these sedimentary horizons the Fig Tree Series contains a volcanic sequence. Trachytic lavas are the dominant rock types, but pyroclastics have been found in several areas. The youngest sequence in the Fig Tree Series consists of shales and greywackes, which are very similar to the shales and greywackes in the lower parts of the series.

The uppermost group of the Swaziland System is the Moodies Group, a widely distributed unit that is some 3500 m thick where the succession is most complete. A detailed account of the stratigraphy of the Moodies Group has been given by Visser (1956). A basal conglomerate is almost universally present. In some areas it could be called a boulder bed; in others it passes laterally into grit or even into sandstone. The boulders and pebbles in the conglomerate are well rounded and consist of black, chert banded iron formation, red, banded jasper, grit, quartzite, quartz porphyry, and granite. Some carbonate has been observed as a constituent of the Basal Conglomerate, which is overlain by predominantly arenaceous rocks. These are composed mainly of quartzite that is rich in calcium carbonate; locally there are intercalated shale lenses and bands of

TABLE 5.5.

Chemical analyses of greywackes from the Sheba Formation, Fig Tree Series, Barberton Mountain Land (Reimer 1975).

	1 K1	*2* V1	*3* C4	*4* C6	*5* C7	*6* C8	*7* C9	*8* C10	*9* C11	*10* C12	*11* C13	*12* C14	*13* C15	*14* C16	*15* C17	*16* C18	*17* C25	*18* SF4B	*19* SF3	*20* SF2	*21* Sh7	*22* Fg15
SiO_2	63.88	63.32	65.1	66.0	63.8	72.0	70.6	66.0	66.2	70.6	68.0	67.0	66.8	65.2	67.0	66.0	57.2					
TiO_2	0.59	0.59	0.50	0.45	0.56	0.60	0.54	0.69	0.51	0.47	0.46	0.49	0.49	0.51	0.43	0.53	0.42					
Al_2O_3	10.97	7.9	9.48	11.5	10.5	9.34	10.3	12.5	10.3	9.97	10.1	10.0	9.61	10.7	10.3	10.3	9.48					
Fe_2O_3	6.67	0.95	7.88	6.26	7.74	6.28	8.55	8.23	7.02	6.40	7.05	9.14	6.07	7.21	5.42	6.87	4.95					
FeO		6.01																				
MgO	5.50	5.67	3.35							3.40												
CaO	1.70	1.95	3.19	2.05	1.51	0.80	1.67	1.34	2.29	1.51	2.47	1.47	3.00	2.21	2.43	2.39	1.60					
Na_2O	1.27	2.09	1.77	2.80	1.30	1.22	1.71	2.13	2.06	1.84	1.50	1.34	1.76	1.55	1.64	1.70	2.90					
K_2O	1.63	1.44	1.35	0.90	1.61	1.54	0.97	1.13	1.26	1.26	1.55	1.32	1.35	1.98	2.14	1.45	1.18					
H_2O^-	LOI	0.06																				
H_2O^+	6.92	3.51																				
CO_2	L.O.I.	3.19																4.6	6.3	4.3 L.O.I.	6.2	2.2
P_2O_5		0.11																				
C			1.010	0.728	1.342	0.267	0.897	0.371	0.869	0.610	0.884	0.510	1.046	0.697	0.826	0.833	0.516					
S			0.007	0.013	0.089	0.082	0.087	0.011	0.035	0.023	0.229	0.063	0.015	0.020	0.017	0.040	0.008					

Trace elements (ppm)

	1 K1	*2* V1	*3* C4	*4* C6	*5* C7	*6* C8	*7* C9	*8* C10	*9* C11	*10* C12	*11* C13	*12* C14	*13* C15	*14* C16	*15* C17	*16* C18	*17* C25	*18* SF4B	*19* SF3	*20* SF2	*21* Sh7	*22* Fg15
Mn	1770		1090	800	600	610	610	490	1010	670	910	750	1040	710	860	810	600					
Rb			53	46	70	73	36	49	43	52	49	45	47	67	83	43	61	88	53	31	28	89
Sr			142	129	81	41	80	59	126	78	109	50	101	97	115	104	164	23	99	27	35	
Ba			408	156	286	331	243	274	285	207	315	224	231	526	502	284	519					
Zr			113	140	150	166	123	206	120	124	127	103	123	109	115	131	157	91	112	105	111	
Cr																		768	820	824	694	706
Ni			264	232	362	280	291	342	283	287	266	306	280	317	257	318	266	570	444	443	666	277
Sc			13	10	14	13	13	15	13	12	12	13	12	13	11	14	11	15	12	15	21	15

Average values   Be = 1.2   Cu = 43   Co = 32   V = 77   Nb = 7   Pb = 6
Ga = 11   Zn = 82   Y = 22   Yb = 1.2   Th = 6

TABLE 5.5. (continued)

	1 STIV4	2 STIV6a	3 STIV7	4 STIV9	5 STIV8a	2 B4	3 17c
SiO_2	69.91	66.80	67.33	68.89	69.14	68.84	64.50
TiO_2	0.49	0.47	0.60	0.47	0.58	0.41	0.48
Al_2O_3	10.05	10.11	11.15	9.48	9.96	9.20	10.46
Fe_2O_3	0.05	0.52	2.84	2.65	0.73	1.45	2.37
FeO	4.53	5.67	4.17	3.42	4.86	4.54	3.99
MgO	2.73	4.27	4.29	3.89	4.21	3.67	5.00
CaO	2.78	1.94	0.99	1.99	1.82	2.17	2.36
Na_2O	1.17	1.76	1.88	1.77	1.13	1.78	2.00
K_2O	3.18	1.73	1.50	1.38	2.02	1.80	2.30
H_2O	0.09	0.17	0.16	0.18	0.12	0.07	0.09
H_2O	2.65	3.08	3.45	2.89	2.75	2.73	3.17
CO_2	2.97	3.02	1.64	3.10	2.68	3.69	3.57
P_2O_5	0.07	0.17	0.09	0.07	0.07	0.06	0.08
MnO	0.12	0.08	0.07	0.14	0.12	0.13	0.20
Total	100.79	99.79	100.16	100.32	100.19	100.54	100.57

Trace elements (ppm)

	1 STIV4	2 STIV6a	3 STIV7	4 STIV9	5 STIV8a	2 B4	3 17c
Rb					56		
Sr					53		
Cr	562	467	569	501	337	393	473
Ni	276	359	318	254	50	211	292
Cu	57	45	61	50		40	37
As	3	2	4	5		4	3
Co	55	45	55	55	65	65	45
Zn	55	160	115	100	145	90	85
Pb	5	10	15	10	5	5	5
Cd	<5	<5	<5	<5	<5	<5	<5
Ag	<1	<1	<1	<1	<1	<1	<1

pure, white marble. The carbonate in this horizon is apprently a primary constituent of the sediments. The overlying lower shale consists of reddish, thinly bedded shales and of sandy shales. Lenses of coarse quartzite are found near the top of the unit. The Middle Quartzite is a cross-bedded, feldspathic unit, whose matrix consists of very small grains of quartz, feldspar, sericite, carbonate, biotite, and some secondary quartz. A thin band of amygdaloidal lava follows the Middle Quartzite, and this in turn is followed by the Middle Shale Zone (a small thickness of shale with red jasper, and of sandy shale), the Upper Quartzite, and the Upper Shale.

The facies analysis of a portion of the Moodies Group has revealed the existence of both barred and nonbarred tidal environments (Eriksson 1977). Back-barrier muds and sands are considered to have formed on mid- to lower-tidal flats and flood-tidal deltas. A tidally influenced estuary is characterized by tidal channel and estuarine mud-flat sediments and subtidal through intertidal sand shoals oriented at right angles to the paleoshoreline. Implicit in this paleoenvironmental interpretation is the existence of laterally contiguous differences in tidal range. Along present-day coasts, increases in tidal range are often associated with embayments or bights where tidal currents become funneled. A similar situation may have existed during deposition of the Moodies Group. A back-barrier depositional setting is favored for these sands; the tidal range there was probably less than 5 m, and the water depth between 0 and 15 m. The reasonably common occurrence of shallow water sediments in the Swaziland System is impossible to reconcile with the notion of a world-encircling Archean ocean (Hargraves 1976), but is consistent with an Earth in which subaerial weathering and erosion were active, and where sedimentation took place both in deep and in shallow water environments.

Reimer (1975) has analyzed a relatively large number of samples of greywackes and some shales from the Sheba Formation, which is the lowest stratigraphic unit in the Fig Tree Series. His data (see table 5.5) give us some insight into the acid-base balance in this area approximately 3400 m.y. ago. The Sheba Formation consists of greywacke (84%), pelite (14%), and chert (2%). These sediments were derived in large part from igneous rocks, and it is possible to estimate, at least roughly, the quantity of acid that was required to effect the transformation (see, for instance, Garrels and Mackenzie 1971, chap. 9). The number of moles of acid required to convert a given quantity of igneous rock into a particular suite of sedimentary rocks depends both on the nature of the original igneous rock and on the nature of the sedimentary rocks that are produced. The importance of the nature of the parent igneous rock is illustrated by the data in figures 5.6, 5.7, and 5.8. In these, the MgO, CaO, K_2O, and Na_2O content of the igneous rocks in table 5.6 has been plotted as a function of

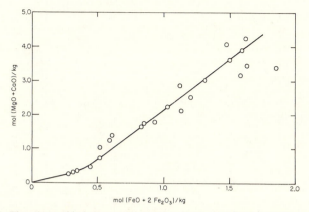

FIGURE 5.6. The relationship between the number of moles of (MgO + CaO) and the number of moles of $(FeO + 2Fe_2O_3)$ in igneous rocks; data from table 5.5.

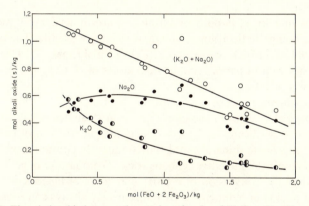

FIGURE 5.7. The relationship between the number of moles of $(K_2O + Na_2O)$ and the number of moles of $(FeO + 2Fe_2O_3)$ in igneous rocks; data from table 5.5.

their iron content. The compositions plotted are those of granitic to basaltic igneous rocks.

Basaltic rocks are considerably richer in total iron, MgO, and CaO than granitic rocks. Their K_2O content is much lower, and their Na_2O content is somewhat lower than that of granitic rocks. The sum of the number of moles of (MgO + CaO)/kg of igneous rocks increases rapidly with increasing iron content (see figure 5.6); the sum of the number of moles of $(MgO + CaO + K_2O + Na_2O)$ increases nearly linearly with increasing iron content (see figure 5.8). Thus the complete titration of these

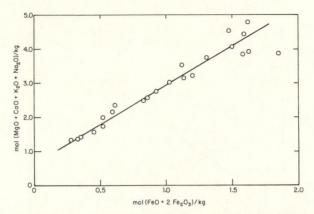

FIGURE 5.8. The relationship between the sum of the number of moles of (MgO + CaO + K_2O + Na_2O) and the number of moles of (FeO + $2Fe_2O_3$) in igneous rocks; data from table 5.5.

four bases in 1 kg of gabbro or basalt requires considerably more acid than their complete titration in an equal weight of granite. The complete titration of CaO, MgO, Na_2O, and K_2O in 1 kg of Clarke's (1924) average igneous rock requires approximately 2.8 moles of H_2CO_3 and/or H_2SO_4, or 5.6 moles of HCl.

Most sedimentary rocks are not products of the complete acid titration of igneous rocks. Most of the contained iron in shales is still present as a constituent of silicates and oxides; a significant fraction of their MgO is present in silicate minerals; and they frequently contain more K_2O in silicate minerals than average igneous rock. The quantity of acid that has been used in the conversion of a particular mass of igneous rock into a particular packet of sediments is therefore nearly always considerably less than the sum of the number of moles of titratable cations in the igneous rock. The quantity of acid that has been used to produce a particular group of sediments is difficult to determine accurately, because sedimentary rocks are almost always products of complex processes involving mixing and sorting. Nevertheless, meaningful computations of acid use can be made.

Consider an average igneous rock that undergoes subaerial weathering today. During this process Mg^{+2}, Ca^{+2}, K^+, and Na^+ are removed; the charge of these cations in soil water and in rivers is balanced largely by HCO_3^- (see, for instance, Holland 1978, chap. 2). Significant quantities of SiO_2 are also removed, and some water and oxygen are added. Al_2O_3 and total iron tend to be roughly conserved in the residuum. The changes in the composition of average igneous rock during such weathering are

TABLE 5.6.

Composition of granitic to basaltic igneous rocks (Poldervaart 1955).

Index	SiO_2	TiO_2	Al_2O_3	Fe_2O_3	FeO	MnO	MgO	CaO	Na_2O	K_2O	P_2O_5	CO_2	Reference
1.	74.2	0.2	13.6	1.3	0.8	—	0.3	1.1	3.0	5.4	0.1	—	22 calc-alkali rhyolites, Nockolds 1954, p. 1012.
2.	73.9	0.3	13.5	1.4	1.0	0.1	0.4	1.2	3.4	4.7	0.1	—	102 rhyolites, Daly 1933, p. 9.
3.	72.5	0.4	13.9	0.9	1.7	0.1	0.5	1.3	3.1	5.4	0.2	—	72 calc-alkali granites, Nockolds 1954, p. 1012.
4.	70.8	0.4	14.6	1.6	1.8	0.1	0.9	2.0	3.5	4.1	0.2	—	546 granites, Daly 1933, p. 9.
5.	69.2	0.5	14.7	1.7	2.2	0.1	1.1	2.6	3.9	3.8	0.2	—	794 silicic igneous rocks, Nockolds 1954, p. 1032.
6.	67.2	0.6	15.8	1.3	2.6	0.1	1.6	3.6	3.9	3.1	0.2	—	137 granodiorites, Nockolds 1954, p. 1014.
7.	65.7	0.6	16.1	1.7	2.7	0.1	1.9	4.5	3.7	2.8	0.2	—	40 granodiorites, Daly 1933, p. 15.
8.	65.1	0.5	15.6	2.1	2.5	—	2.5	4.3	3.5	3.7	0.2	—	average plutonic igneous rock, Vogt 1931, p. 47.
9.	62.4	0.7	15.9	3.0	3.3	0.2	3.0	5.1	3.4	2.7	0.3	—	average Cordilleran-Appalachian igneous rock, Knopf 1916, p. 622
10.	60.3	0.8	17.5	3.4	3.1	0.2	2.8	5.9	3.6	2.1	0.3	—	87 andesites, Daly 1933, p. 16.
11.	60.1	1.1	15.6	3.1	3.9	0.1	3.5	5.2	3.9	3.2	0.3	—	average igneous rock, Clarke 1924, p. 29.
12.	57.6	0.8	16.9	3.2	4.5	0.1	4.2	6.8	3.4	2.2	0.3	—	70 diorites, Daly 1933, p. 16.
13.	54.9	1.5	16.7	3.3	5.2	0.2	3.8	6.6	4.2	3.2	0.4	—	635 intermediate igneous rocks. Nockolds 1954, p. 1032.
14.	54.6	1.3	17.4	3.5	5.5	0.2	4.4	8.0	3.7	1.1	0.3	—	49 andesites, Nockolds 1954, p. 1019.
15.	54.1	0.9	18.1	2.5	5.8	0.1	5.5	8.4	3.4	1.0	0.2	—	10 parental calcalkaline magmas, Nockolds and Allen 1953, p. 139.
16.	52.4	1.5	16.5	2.7	5.8	0.1	5.5	8.5	3.4	1.3	0.2	—	50 diorites, Nockolds 1954, p. 1019.
17.	51.3	2.0	14.2	2.9	9.1	0.2	6.1	10.5	2.3	0.9	0.4	—	137 "normal" tholeiites, Nockolds 1954, p. 1021.
18.	51.0	1.4	15.6	1.1	9.8	0.2	6.4	10.5	2.2	1.0	0.2	—	average tholeiite, Green and Poldervaart 1955.
19.	49.9	1.4	16.0	5.4	6.5	0.3	7.0	9.1	3.2	1.5	0.4	—	198 basalts, Daly 1933, p. 17.
20.	49.7	2.2	14.2	3.7	10.0	0.2	6.3	9.6	2.6	0.7	0.3	—	43 plateau basalts, Daly 1933, p. 17.
21.	48.6	1.8	15.7	2.8	8.1	0.2	6.8	10.8	2.3	0.7	0.3	—	637 mafic igneous rocks, Nockolds 1954, p. 1032.
22.	47.0	3.0	15.1	3.7	8.1	0.2	8.7	10.9	2.7	1.0	0.3	—	aver. olivine basalt Pacific, Green and Poldervaart 1955.
23.	46.1	2.6	14.8	3.2	8.8	0.2	7.9	10.8	2.7	1.0	0.4	—	96 "normal" alkali basalts, Nockolds 1954, p. 1021.

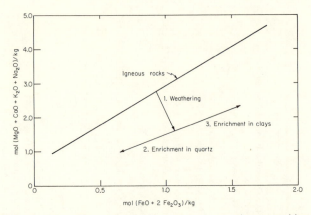

FIGURE 5.9. Some of the effects of weathering and sorting on the composition of igneous rocks during their conversion into sediments.

indicated by arrow 1 in figure 5.9. During transport and deposition, soils tend to be separated by grain size. Residual sand-sized particles are apt to be separated from silt and from clay minerals; the several fractions are normally deposited as sands, silts, and muds. Sands tend to be enriched in quartz; fine-grained sediments tend to be enriched in minerals that contain untitrated cations. The chemical changes associated with the addition of quartz or of other minerals that contain neither iron nor acid-titratable cations are indicated by arrow 2; changes associated with the removal of such minerals and enrichment in clays are indicated by arrow 3.

Cation exchange and reverse weathering reactions during interaction with seawater tend to modify the cation content of sediments (see, for instance, Holland 1978, chap. 5), and the addition of carbonate and sulfate minerals can increase the quantity of MgO and CaO dramatically. The effects of these minerals can be removed by subtracting the number of moles of combined CO_2 and SO_3 from the total number of moles of MgO and CaO:

$$mol(MgO + CaO)_{sil.}/kg \approx mol(MgO + CaO - CO_2 - SO_3)/kg, \quad (5.1)$$

where the subscript sil. indicates silicate.

This relationship is somewhat imprecise, because the presence of iron in carbonate minerals has been neglected; however, the quantity of siderite and of the ankeritic component in dolomite is usually small, and the error is normally not serious. Expression 5.1 does not distinguish between the distribution of MgO and CaO individually between carbonates and silicates. This distinction is not important when we seek to estimate the quantity of acid that has been used in the conversion of igneous to sedimentary rocks.

The effects of the separation of sediments into fractions that are rich and fractions that are poor in acid titratable cations can be compensated approximately by normalizing the Al_2O_3 content of sediments to that of igneous rocks. Most common igneous rocks contain ca. 15.6% Al_2O_3 by weight (see table 5.6). In the calculations that are described below, the quantity of acid-titratable cations in sedimentary rocks has been corrected by multiplying the measured concentration of these cations by the ratio, $15.6/wt\% Al_2O_3$:

$$mol(MgO + CaO)_{sil.corr.} = mol(MgO + CaO)_{sil.} \frac{15.6}{wt\% Al_2O_3}, \quad (5.2)$$

$$mol(K_2O + Na_2O)_{corr.} = mol(K_2O + Na_2O) \frac{15.6}{wt\% Al_2O_3}. \quad (5.3)$$

The total iron content of sedimentary rocks has been corrected in the same manner, because aluminum and iron cohere sufficiently well in most sediments to make this a reasonable procedure.

These operations have been performed on analyses of grcywackes and shales of the Sheba Formation (see tables 5.5, 5.7, and 5.8), and the results are shown in figures 5.10 and 5.11. The number of data points in figure 5.10 is smaller than in figure 5.11, because not all of the available analyses included CO_2 determinations. Without such measurements the value of $(MgO + CaO)_{sil.}$ cannot be calculated. The results indicate that the rocks of the Sheba Formation are derived from igneous rocks that are on average more iron-rich than Clarke's (1924) average igneous rock $(FeO + 2Fe_2O_3 = 0.93$ mol/kg). The values of $(K_2O + Na_2O)_{corr.}$ of the greywackes in figure 5.11 scatter around the igneous rock line of figure 5.7, indicating that weathering of the feldspars in these sedimentary rocks has been minor. Plots of the analyses of representative greywackes of all ages (Pettijohn 1975, p. 228) also scatter about the igneous rock line in figure 5.11, but at lower total iron values. With one exception the shales plot significantly below the igneous rock line.

The sum $(MgO + CaO)_{sil. corr.}$ for the single Sheba shale and for the few greywackes for which CO_2 analyses are available plots well below the igneous rock line in figure 5.10. This shows that significant quantities of $(MgO + CaO)$ were removed during the formation of these sedimentary rocks from their parent igneous rocks. Presumably, Mg^{+2} and Ca^{+2} released during weathering were precipitated as constituents of carbonate minerals within or beyond the confines of the Barberton area (Reimer 1975). The average loss of $(MgO + CaO)$ during the production of the

TABLE 5.7.

Analyses of pelites from the Sheba Formation, Fig Tree Series, Barberton Mountain Land (Reimer 1975).

	1 C3	2 C5	3 V1
SiO_2	56.0	63.7	54.11
TiO_2	0.46	0.72	1.00
Al_2O_3	10.3	12.3	17.64
Fe_2O_3	14.4	11.0	1.43
FeO			7.45
MgO	3.35	3.35	6.97
CaO	2.17	0.13	0.22
Na_2O	0.28	0.81	2.14
K_2O	0.72	2.70	2.76
H_2O^-			0.24
H_2O^+			5.14
CO_2			0.09
P_2O_5			0.14
C	1.19	0.16	0.33
S	0.009	0.009	

Trace elements (ppm)			
Mn	1260	290	510
Rb	28	83	
Sr	85	16	
Ba	516	316	
Zr	34	112	
Cr			
Ni	958	400	
Sc	28	21	

Sheba greywackes from parent igneous rocks was approximately 1.7 mol/kg igneous rock. There are too few data for the Sheba shales, but the figure for cation loss during the formation of these shales is probably comparable.

The scheme developed above for estimating the degree to which igneous rocks were titrated by acids during their conversion to sediments can be faulted on at least two grounds. It could be argued that in the presence of insufficient atmospheric oxygen iron is not conserved in soils but removed in solution as Fe^{+2} (see chapter 7), and should therefore be included in the roster of acid titratable cations. This argument is quite reasonable, but it can be shown that the inclusion of iron among the ti-

TABLE 5.8.

Analyses of shales from the Fig Tree Series.

SiO_2	62.11	57.35	57.25	52.73
Al_2O_3	16.26	15.73	12.00	16.60
Fe_2O_3 \ FeO	9.10*	9.70*	10.20*	11.00*
MgO	5.17	4.70	5.66	5.67
CaO	0.33	0.95	3.81	2.54
Na_2O	1.38	1.08	1.84	2.50
K_2O	2.34	2.02	1.94	1.40
H_2O^+	n.d.	n.d.	n.d.	n.d.
H_2O^-	n.d.	n.d.	n.d.	n.d.
C	2.21	3.14	1.47	1.19
CO_2	n.d.	n.d.	n.d.	n.d.
TiO_2	n.d.	n.d.	n.d.	n.d.
P_2O_5	n.d.	n.d.	n.d.	n.d.
MnO_2	n.d.	n.d.	n.d.	n.d.

n.d. = not determined.
* Total iron expressed as FeO.
Analyst: M. J. Borcsik.

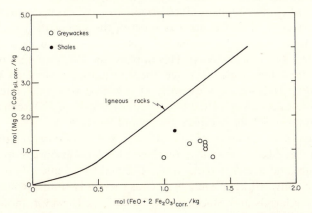

FIGURE 5.10. The $(MgO + CaO)_{sil.\ corr.}$ of sedimentary rocks of the Sheba Formation, Fig Tree Series, Barberton Mountain Land, plotted as a function of their $(FeO + 2Fe_2O_3)_{corr.}$ content.

tratable ions is unnecessary so long as this element is reprecipitated in sediments as a constituent of oxides and silicates, whose formation does not involve the consumption of acid, and as long as these minerals occur together with the clay fraction of sedimentary rocks. This seems to be sufficiently true of most sedimentary rocks.

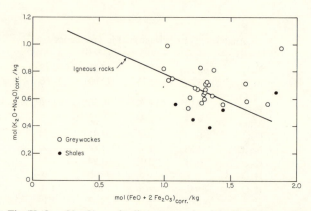

FIGURE 5.11. The $(K_2O + Na_2O)_{corr.}$ of sedimentary rocks of the Sheba Formation, Fig Tree Series, Barberton Mountain Land, plotted as a function of their $(FeO + Fe_2O_3)_{corr.}$ content.

A second objection to the proposed computational scheme concerns the presence of sulfides in sedimentary rocks. If these have been formed at the expense of original iron oxides and silicates, the number of moles of sulfur per kg of rock should be added to the number of moles of consumed acid computed as outlined above. In most cases this represents a small correction. The S-content of most of the sedimentary rocks of the Swaziland Series is much less than 0.5% (see tables 5.5 and 5.7, and Moore, Lewis, and Kvenvolden 1974). This implies that the formation of 1 kg of these sedimentary rocks involved the consumption of much less than 0.15 mol/kg H_2S; this is a quantity of acid that is a small fraction and within the uncertainty of the estimated 1.7 moles of CO_2 consumed in the formation of 1 kg of typical Swaziland Series rocks.

The use of ca. 1.7 moles of CO_2 to convert 1 kg of parent igneous rock into average Sheba greywacke or shale plus an appropriate quantity of carbonates and ancillary rocks is nearly equal to the quantity of CO_2 required to convert 1 kg of average igneous rock today into average modern sediment (Garrels and Mackenzie 1971, p. 248). However, modern sediments consist largely of older sediments that have been eroded and redeposited; only about 25% of modern sediments are derived directly from the weathering of igneous and/or high grade metamorphic rocks. The total rate of acid use in weathering today is therefore only about one quarter of the rate that would be required if all modern sediments were first-cycle sediments. We do not know what fraction of the sediments deposited between 3000 and 3500 m.y. ago were first cycle sediments. It is certain that on average they had gone through fewer cycles than modern sediments, and it is likely that many of them were first cycle, since the available isotopic evidence suggests that continental materials produced earlier in Earth history were recycled rapidly through the mantle (see, for instance, O'Nions et al. 1979; De Paolo 1980; Allègre and Othman 1980; Veizer 1983). All of this suggests that the rate of acid usage per kg of sediment formed per year during mid-Archean time was considerably greater than the rate at which acid is used in the production of 1 kg of sediment today. If the total rate of sediment production was higher during the mid-Archean than it is today, the evolution of acid and its use in the titration of igneous rocks was much more rapid then than today. The release of juvenile volatiles was probably more rapid during the Archean Era (see chapter 3), and enough volatiles had already been released, so that the degassing of earlier sedimentary and altered volcanic rocks probably contributed significantly to the flux of CO_2 from the Earth's interior. A return to such an early Earth could be arranged by increasing the rate of tectonism and plutonism in such a fashion that nearly all sedimentary rocks become highly metamorphosed and largely remelted on a time scale of a few hundred million years. Most future sediments would then become first-cycle sediments, and the CO_2 released during prior metamorphism and melting would be available to produce new generations of sediments that are acid-titrated approximately as completely as their predecessors.

Quantitative modeling of this process is impossible at present, because we know too little about the sedimentary rocks that were formed during mid-Archean time. Our knowledge of the geologic record of the mid-Archean is confined almost entirely to greenstone belts; although these do give a reasonably consistent picture of events in such settings, they are probably not representative of mid-Archean sedimentary environments as a whole (see, for instance, von Brunn and Hobday 1976; Pretorius 1975; and Vos 1975).

4. Archean and Lower Proterozoic Shales of the Canadian Shield

The analytical data for the Barberton Mountain Land are extremely useful; they are, however, somewhat lacking, because they are rather few in number and because they were collected in a relatively small area. A more extensive coverage of a rather larger area is to be found in Cameron and Garrels' (1980) data for Archean and Aphebian (early Proterozoic) shales from the Canadian Shield. The Archean samples in their sets of analyses are more than 2500 m.y. old, but they are probably younger than the Fig Tree Series. The Aphebian shales span the age range from ca. 1600 to 2500 m.y.b.p. The 406 samples of Archean shales collected at 153 localities in the Superior Province of the Canadian Shield were drawn into eight composites representing the districts within which the samples were

Helikian sediments, volcanics

Archean gneisses, granites

Aphebian sediments, volcanics

Archean sediments, volcanics

Boundary, Superior Province........
Boundary, Canadian Shield..........
Sampled areas, Proterozoic.................■
Sampled areas, Archean.......................▲

FIGURE 5.12. Location map for the Archean and Aphebian shale composites of Cameron and Garrels (1980). (Reproduced by permission of Elsevier Scientific Publishing Company.)

collected. The 396 samples of Aphebian shales from 54 localities within or flanking the Superior Province were collected into nine composites, each representing a particular stratigraphic unit. The location of the composites is shown in figure 5.12, their chemical composition in tables 5.9 and 5.10. The geology of the units sampled has been described briefly by Cameron and Baumann (1972) and by Cameron and Jonasson (1972). The samples are variably but weakly metamorphosed.

The parameters that were developed for determining acid consumption in the formation of sediments from the Fig Tree Series were calculated for the Archean and Aphebian composites of table 5.9, with the exception of the composites of Aphebian iron formations. The results are shown in figures 5.13 and 5.14. On average, the corrected iron content of the shales indicates a derivation from igneous rocks that were somewhat less basaltic than the precursors of the Fig Tree shales. Composite 4 is an exception. Cameron and Garrels (1980) have pointed out the rather anomalous composition of this sample.

The value of $(MgO + CaO)_{sil.\,corr.}$ of the composites falls well below that of igneous rocks; values of $(K_2O + Na_2O)_{corr.}$ fall below the igneous-rock line with the exception of the value for composite 11; the samples

Major element composition of Archean and Aphebian shale composites from the Canadian Shield (Cameron and Garrels 1980).

Sample description		SiO_2	TiO_2	Al_2O_3	Fe_2O_3[e]	MnO	MgO	CaO	Na_2O	K_2O	CO_2	C	S	P_2O_5	H_2O	LOI[f]	$\delta^{34}S$ (‰)
Archean:																	
(1)[a] Red Lake[b]	12c[c]	56.3	0.65	18.1	10.4	0.15	2.60	1.08	1.0	4.05	0.0	2.1	2.24	0.07	3.0	5.8	+1.22
(2) Atikokan	27o[d]	59.0	0.70	17.3	7.2	0.09	3.71	1.54	2.9	3.13	0.6	0.1	0.16	0.20	3.2	3.7	+0.6
(3) Lac des Iles and Beardmore	103o	55.9	0.77	17.8	8.5	0.10	4.43	2.17	2.5	3.18	1.1	0.1	0.14	0.17	4.1	4.4	—
(4) Michipicoten and Oba	6c/6o	57.2	0.54	12.7	11.7	0.03	1.46	0.64	2.6	1.40	0.0	5.0	4.70	0.10	3.6	12.0	-0.46
(5) Timmins	110c/31o	59.4	0.69	16.4	6.3	0.08	3.20	2.61	3.2	2.38	1.5	1.4	1.06	0.14	2.9	5.3	+1.44
(6) Matachewan and Larder Lake and Kenojevis River	67o	57.5	0.83	18.4	8.1	0.09	3.83	1.39	2.5	3.20	0.9	0.2	0.14	0.19	3.5	4.4	—
(8) Desmeloizes and Amos-Barraute	15o	59.4	0.86	18.3	7.2	0.09	3.44	1.42	2.3	2.50	0.6	0.2	0.11	0.16	3.6	4.1	—
(9) Chapais	29o	62.6	0.61	16.6	6.2	0.06	2.58	1.44	2.6	3.48	0.8	0.0	0.12	0.19	2.8	3.6	+2.1
Total	406																
Weighted average		58.0	0.73	16.9	7.6	0.09	3.33	2.07	2.5	2.73	1.1	0.7	0.66	0.16	3.4	4.5	+1.31 ±0.32
Aphebian:																	
(10) Whitewater	16o	59.4	0.90	18.7	6.0	0.04	2.66	0.35	2.3	4.24	0.0	1.8	0.05	0.16	3.8	5.7	—
(11) Gowganda	24o	58.8	0.81	16.5	7.6	0.13	4.23	1.63	4.5	2.18	0.0	0.0	0.10	0.19	3.6	2.9	+2.4
(13) Rove	137o	60.7	0.87	17.6	6.7	0.04	2.72	0.42	1.7	4.05	0.0	0.6	0.21	0.15	4.2	5.1	+16.5
(14) Albanel	2c/36o	67.7	0.54	11.1	3.3	0.01	1.23	0.89	0.5	5.38	0.6	5.1	1.84	0.09	1.3	9.0	-0.34
(16) Attikamagen	50o	67.0	0.59	13.5	4.4	0.05	1.56	0.07	0.3	6.95	0.0	2.0	0.94	0.07	3.0	6.2	+13.61
(18) Menihek	61o	62.2	0.55	14.5	7.5	0.13	2.41	0.40	0.7	4.76	0.3	2.0	0.94	0.12	3.9	6.5	+10.99
Total	326																
Weighted average		62.6	0.73	15.6	6.1	0.06	2.42	0.50	1.4	4.65	0.1	1.6	0.63	0.13	3.6	5.9	+11.79 ±4.37
Aphebian iron formation:																	
(12) Gunflint	46o	54.9	0.43	7.7	21.3	0.15	2.10	1.18	0.1	1.86	1.7	0.3	0.98	0.32	5.4	9.9	+6.18
(13) Temiscamie	14o	42.2	0.20	5.9	28.2	0.04	5.01	1.83	0.2	2.80	7.0	2.0	0.62	0.19	3.6	12.9	+15.8
(17) Ruth	10o	51.3	0.91	9.3	20.0	3.92	1.09	0.13	0.0	4.44	0.2	1.1	0.09	0.15	6.0	8.6	—
Total	70																

[a] Composite code; [b] sample localities; [c] number of samples taken from cores; [d] number of samples from outcrops; [e] total Fe as Fe_2O_3; [f] loss on ignition.

TABLE 5.10.

Minor element composition of Archean and Aphebian shale composites from the Canadian Shield (Cameron and Garrels 1980).

Composite code[a]	Li	Be	B	F	Cl	Sc	V	Cr	Co	Ni	Cu	Zn	Ga	Ge	As	Rb	Sr	Y	Mo	Ag	Sn	Sb	Ba	La	Hg[b]	Pb	U
Archean:																											
1	43	3.6	70	400	400	21	150	141	45	147	87	195	21	0.3	32	112	50	16	2.0	0.17	3.8	1.1	470	21	81	13	1.7
2	57	3.5	7	500	100	15	130	119	30	91	47	96	16	0.3	23	117	280	18	5.0	0.08	1.5	0.9	730	26	69	13	2.3
3	51	3.3	30	400	100	19	150	147	35	125	59	100	14	0.3	17	113	190	18	3.5	0.06	1.4	1.2	600	25	151	12	2.1
4	18	2.9	10	300	100	9	75	55	35	82	99	435	17	1.5	75	43	530	14	4.0	0.19	5.9	5.9	320	25	241	12	1.3
5	27	2.5	30	400	100	15	110	130	38	147	82	702	16	0.3	73	68	180	18	2.0	0.95	6.2	1.4	460	21	206	51	1.4
6	61	3.2	50	400	100	18	150	168	35	140	52	114	17	0.3	61	99	160	21	1.5	0.14	1.4	0.7	680	28	94	16	1.7
8	49	3.4	30	300	200	18	150	140	30	120	48	102	22	0.7	13	73	230	22	1.0	0.10	1.9	0.3	510	21	67	12	1.3
9	28	3.1	20	400	100	12	110	62	20	52	56	93	18	0.3	18	99	240	11	0.5	0.10	1.2	0.1	560	22	73	14	1.8
Weighted average	42	3.0	33	400	100	16	130	133	35	127	66	323	16	0.3	46	91	199	18	2.2	0.40	3.28	1.2	550	24	147	26	1.7
Aphebian:																											
10	69	3.4	20	700	100	17	130	147	17	69	67	78	25	0.3	17	190	90	23	10.0	0.26	2.3	3.4	850	23	618	12	8.6
11	44	5.4	5	300	100	19	210	104	30	82	40	108	17	0.3	5	94	250	22	1.0	0.10	1.7	0.1	430	29	126	48	3.3
13	85	4.0	30	700	200	20	170	145	19	69	70	173	23	2.7	14	178	70	26	6.0	0.46	2.8	0.9	610	37	299	22	6.5
14	17	3.8	20	300	200	9	89	60	69	61	222	17	19	0.3	81	134	30	19	2.0	0.35	2.0	0.4	2100	73	360	12	4.2
16	41	4.5	30	1500	100	10	240	68	8	25	30	47	14	1.7	21	211	40	23	4.5	0.31	2.1	1.4	450	48	585	23	5.1
18	50	5.4	20	1200	100	13	260	63	11	41	50	111	20	1.5	19	188	40	34	6.0	0.56	3.5	2.8	500	45	719	20	7.0
Weighted average	60	4.4	25	840	150	16	188	105	22	57	75	114	20	1.7	23	174	69	25	4.5	0.41	2.6	1.3	492	43	431	22	5.9
Aphebian iron formation:																											
12	67	4.6	20	300	200	9	170	48	14	22	34	36	13	2.8	57	56	10	28	12.5	0.08	1.1	0.6	220	33	611	12	5.9
15	25	3.9	30	700	200	7	130	26	6	13	10	42	7	9.4	13	129	20	23	1.0	0.05	8.5	0.3	180	36	475	11	1.1
17	79	8.4	10	1400	100	11	330	43	20	22	26	47	15	9.7	49	188	20	88	12.5	0.16	5.4	0.1	400	140	815	12	10.8

[a] Composite code from table 5.9.
[b] Hg in ppb.

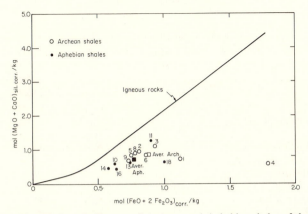

FIGURE 5.13. The $(MgO + CaO)_{sil.\ corr.}$ of Archean and Aphebian shales of the Canadian Shield; data from Cameron and Garrels (1980).

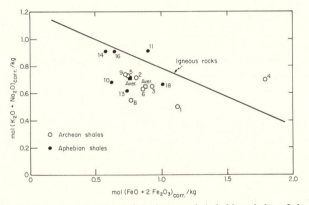

FIGURE 5.14. The $(K_2O + Na_2O)_{corr.}$ of Archean and Aphebian shales of the Canadian Shield; data from Cameron and Garrels (1980).

in this composite are thought to have had a glacial origin and to be less weathered than the components of the other composites. The loss of $(MgO + CaO)$ and of $(K_2O + Na_2O)$ during the formation of the average Archean and average Aphebian shales from the Canadian Shield are shown in table 5.11. Since the average Archean shale in these composites was apparently derived from igneous rocks that were slightly richer in iron than the average Aphebian shale, the $(MgO + CaO)$ content of the parent rocks of the Archean shales was slightly higher, and their $(K_2O + Na_2O)$ content slightly lower than those of the parent rocks of the Aphebian shales. The calculated total acid use in moles of CO_2/kg of igneous

TABLE 5.11.

Acid consumption in the formation of Archean and Aphebian shales from the Canadian Shield.

	Archean Shales	Aphebian Shales
$(FeO + 2Fe_2O_3)_{corr.}$	0.88 mol/kg	0.76 mol/kg
$(MgO + CaO)_{ign.}$	1.82	1.46
$(MgO + CaO)_{sil.corr.}$	0.88	0.67
$\Delta(MgO + CaO)_{sil.corr.}$	0.94	0.79
$(K_2O + Na_2O)_{ign.}$	0.84	0.89
$(K_2O + Na_2O)_{corr.}$	0.64	0.72
$\Delta(K_2O + Na_2O)_{corr.}$	0.20	0.17
$\Delta(MgO + CaO + K_2O + Na_2O)_{corr.}$	1.14	0.96

rock seems to have been somewhat higher for the Archean than for the Aphebian shales, but the difference is probably not significant.

The difference between both figures and the value of 1.7 moles CO_2/kg, which was estimated for acid use in the production of the Fig Tree grey-wackes, is probably real. Whether this difference is simply due to the difference between ancient greywackes and shales is not clear. There are too few analyses of Fig Tree shales on which to base a meaningful comparison between the composition of shales from the two areas.

The Canadian shales are clearly not the product of intense acid titration. This point is well illustrated by figure 5.15, in which the mole ratio, $Na_2O/[Al_2O_3 + \frac{1}{3}(MgO + CaO)_{sil.}]$, in the Archean and Aphebian shales has been plotted against the value of their mole ratio, $K_2O/[Al_2O_3 + \frac{1}{3}(MgO + CaO)_{sil.}]$. The mole ratios of nearly all of the composites falls within a narrow band, for which $0.30 \leq (K_2O + Na_2O)/[Al_2O_3 + \frac{1}{3}(MgO + CaO)_{sil.}] \leq 0.45$.

The $CaO_{sil.}$ content of these shales is small. Hence the values of $(MgO + CaO)_{sil.}$ are close to those of $(MgO)_{sil.}$ alone. In micas where the octahedral sites are occupied by Al^{+3} and/or Mg^{+2}, and where the interlayer cation sites are occupied by K^+ and/or Na^+, the ratio $(K_2O + Na_2O)/(Al_2O_3 + \frac{1}{3}MgO)$ is 1/3. In the Archean and Aphebian shales, K^+ and Na^+ are the only quantitatively important interlayer cations. The shales do, however, contain sizable quantities of iron, and it is likely that Fe^{+2} and Fe^{+3} occupy a significant proportion of the octahedral sites of phyllosilicate minerals in these shales. There are no mineralogical data to

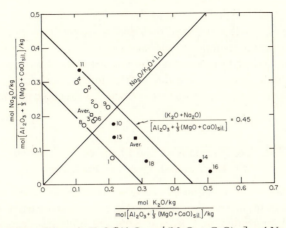

FIGURE 5.15. Values of the ratio $K_2O/[Al_2O_3 + \frac{1}{3}(MgO + CaO)_{sil.}]$ and $Na_2O/[Al_2O_3 + \frac{1}{3}(MgO + CaO)_{sil.}]$ in Archean and Aphebian shales of the Canadian Shield; data from Cameron and Garrels (1980).

indicate how much of the iron in these shales is present in the phyllosilicates. Iron would exert its maximum effect on the ratio of interlayer cations to octahedral cations if all of the iron were trivalent and if all iron in the shales substituted for Al^{+3} in octahedral sites of the phyllosilicates. The value for the ratio, $(K_2O + Na_2O)/[Al_2O_3 + Fe_2O_3 + \frac{1}{3}(MgO + CaO)_{sil.}]$, in average Archean shale would then be 0.28; its value in average Aphebian shale of the Canadian Shield would be 0.34. These ratios correspond to those in rocks in which the content of K_2O and Na_2O is enough, so that the phyllosilicates are dominantly micas rather than illites and chlorites.

In all but one of the Archean composites the mole ratio Na_2O/K_2O is greater than 1.0, whereas in all but one of the Aphebian composites the Na_2O/K_2O ratio is less than 1.0. It was pointed out above that the odd Aphebian composite (11) is probably of glacial origin, and is presumably the product of less intense chemical weathering than the other Aphebian shales. The marked difference between the Na_2O/K_2O ratio of the Archean and Aphebian shales may be due to differences in the Na_2O/K_2O ratio of their source rocks (Cameron and Garrels 1980; Eade and Fahrig 1971). This is not certain, however, because the chemical composition of shales reflects not only the composition of their source rocks but also the local intensity of weathering, the effects of reactions between clays and seawater, and those of later diagenetic reactions. Until sets of analyses from other parts of the Archean and Aphebian world become available that are comparable to Cameron and Garrels's (1980) analyses of shales

from the Canadian Shield, it is difficult to be certain whether the differences in the Na_2O/K_2O ratio of their shales are due to local changes, or
whether they are a response to global changes in the composition of the
source areas and/or in the composition of seawater. As the next section
will show, these questions are difficult to answer even for sedimentary
rocks from periods of Earth history for which a much larger data base is
available.

5. The Late Precambrian and Phanerozoic Sediments of the Russian and American Platforms

By far the most thorough and extensive study of the composition of sedi-
mentary rocks in any portion of the Earth's crust was carried out in the
USSR by Vinogradov and Ronov (1956a,b) and has been continued
since then by Ronov and several of his collaborators (Ronov 1956; Ronov,
Migdisov, and Barskaya 1969; Ronov and Migdisov 1971; Ronov and
Yaroshevsky 1967 and 1976).

The papers published by Vinogradov and Ronov in 1956 are based on
198 documented average samples of carbonate rocks composed of 8847
core samples, on 252 average samples of shale composed of 6804 core
samples, and on 158 average samples of sandstones and siltstones composed of 3709 core samples. The location of each of the average samples
is shown in figure 5.16; their composition by stages, series and systems is
shown in tables 5.12, 5.13, and 5.14. Most of the data which have been
accumulated since then were summarized by Ronov and Migdisov (1971).
Their compilation also includes data for the North American Platform
and is reproduced in table 5.15.

The data in these tables have been subjected to the analysis described
in the previous two sections of this chapter. Figure 5.17 is a plot of the
$mol(MgO + CaO)_{sil.\ corr.}/kg$ of shales of the Russian Platform plotted as
a function of their iron content. The data points show that all of the shales
are depleted in $(MgO + CaO)_{sil.}$ compared to their parent igneous rocks.
The depletion is rather variable. Triassic shales are least depleted, Sinian
(late Proterozoic) shales are most depleted, and shales from other periods
are depleted to an intermediate degree. The variable degree of depletion
in $(MgO + CaO)_{sil.\ corr.}$ is related to the carbonate content of the shales.
Figure 5.18 shows that $(MgO + CaO)_{sil.\ corr.}$ increases significantly with
increasing carbonate content of the shales, and that this increase is es-
sentially independent of the age of the shales. The low value of $(MgO + CaO)_{sil.\ corr.}$ in the Sinian shales of the Russian Platform is apparently
related to their low carbonate content, the high value of $(MgO + CaO)_{sil.\ corr.}$ of the Permian shales to their high carbonate content. The

FIGURE 5.16. Location of the documented average samples of carbonates, shales, and sand-stones of the Russian Platform (Vinogradov and Ronov 1956b). (Reproduced by permission from Pergamon Press, Ltd.)

reason or reasons for this trend are not clear. The Al_2O_3 content of the shales decreases with increasing carbonate content more or less in proportion to the dilution of the clay by the carbonate component. Curiously, the value of $(MgO + CaO)_{sil. corr.}$ of sandstones of the Russian Platform does not increase with increasing carbonate content, and the effect is apparently not universal in shales (see, for instance, Perry and Hower 1970, and Hower et al. 1976).

The relationship in figure 5.18 creates problems in assessing the quantity of acid used in the formation of shales of a given period in Earth

TABLE 5.12.

Chemical composition (in weight percentage) of composite samples of shales of the Russian Platform (Vinogradov and Ronov 1956b).

System, series, stage		Number of samples and analyses	Number of specimens in the samples	SiO_2	Al_2O_3	Fe_2O_3	TiO_2	CaO	MgO	K_2O	Na_2O	SO_3	CO_2	Loss on ignition*	Total	10^{-4} %		10^{-3} %		
																Th	U	Co	Ni	Cu
Proterozoic (Sinian)	(Sn₂)	17	765	58.21	18.34	8.37	0.88	0.55	2.26	3.96	0.76	0.11	1.17	5.65	100.26	13.7	4.4	2.0	1.5	4.8
Cambrian	(Cm)	14	436	60.36	17.83	6.71	0.84	0.83	2.17	4.82	0.49	0.15	0.70	5.03	99.93	13.2	3.7	1.6	2.5	5.9
Ordovician	(S₁)	8	198	43.77	11.82	5.25	0.64	12.82	3.40	3.95	0.79	0.16	11.13	6.21	99.94	10.1	3.2	1.2	1.3	2.9
Gothlandian	(S₂)	6	531	49.58	12.62	5.46	0.60	8.99	4.99	4.42	0.69	0.15	7.37	5.38	100.25	9.5	3.5	2.4	1.0	8.5
Silurian	(S)	14	729	46.24	12.16	5.34	0.62	11.18	4.08	4.15	0.75	0.15	9.52	5.85	100.04	9.9	3.3	1.4	1.2	4.0
Middle Devonian	(D₂)	41	822	49.93	16.26	6.69	0.80	5.79	3.89	3.55	0.60	0.50	5.61	6.29	99.91	10.5	3.9	1.0	1.5	3.3
Frasnian stage	(D₃)	41	1433	46.48	14.52	6.99	1.02	10.92	2.55	3.54	0.58	0.16	8.77	4.80	100.33	10.1	4.4	1.1	1.6	3.9
Famennian stage	(D₃)	9	253	40.01	11.18	5.15	0.57	11.87	6.42	4.15	0.64	6.62	7.99	5.13	99.75	9.9	3.2	1.3	3.0	4.9
Upper Devonian	(D₃)	53†	1721†	45.34	14.02	6.59	0.93	10.94	3.33	3.65	0.59	1.47	8.48	4.89	100.23	10.2	4.1	1.1	1.9	4.1
Devonian	(D)	94	2543	47.34	15.00	6.64	0.87	8.69	3.58	3.61	0.59	1.05	7.23	5.50	100.10	10.4	4.0	1.1	1.7	3.7
Lower Carboniferous	(C₁)	17	267	54.09	18.49	6.39	0.99	2.99	1.53	2.61	0.53	0.53	2.47	9.73	100.35	14.2	5.9	0.8	1.8	2.0
Middle Carboniferous	(C₂)	19	241	50.99	16.11	6.62	0.70	5.64	4.16	5.00	0.73	0.11	5.59	4.61	100.26	12.7	5.0	1.0	3.8	2.9
Carboniferous	(C)	36	508	52.45	17.24	6.51	0.84	4.32	2.92	3.87	0.63	0.31	4.12	7.04	100.25	13.8	5.5	0.9	2.9	2.5
Upper Permian	(P₂)	27	708	42.48	11.22	5.65	0.55	13.24	5.81	2.71	1.31	1.60	11.52	3.99	100.08	7.7	3.5	0.9	2.8	3.0
Permian	(P)	28‡	730‡	42.97	11.45	5.78	0.57	12.94	5.67	2.71	1.30	1.55	11.34	3.90	100.18	7.7	3.5	0.9	2.8	3.0
Paleozoic	(Pz)	203	5720	49.38	15.21	6.56	0.80	7.41	3.57	3.68	0.72	0.78	6.44	5.56	100.11	11.0	4.1	1.0	2.3	3.4
Triassic	(T)	4	121	57.40	14.72	6.12	0.71	10.10	3.68	2.29	1.03	0.09	7.42	3.73	100.63	10.0	4.1	1.4	3.6	4.0
Jurassic	(J)	16	328	55.12	16.55	6.24	0.77	4.78	1.98	3.12	1.40	0.15	3.69	6.03	99.83	14.0	3.4	1.1	2.3	3.4
Lower Cretaceous	(Cr₁)	8	183	58.24	18.76	6.54	0.86	1.65	1.92	2.64	1.34	0.24	0.94	6.85	99.98	10.1	4.1	1.3	2.2	3.3
Upper Cretaceous	(Cr₂)	4	72	51.52	10.49	4.61	0.50	13.46	1.55	1.98	1.15	0.01	11.33	3.08	99.68	10.1	4.0	0.7	1.4	2.8
Cretaceous	(Cr)	12	255	56.09	16.00	5.90	0.74	5.59	1.80	2.42	1.28	0.17	4.40	5.60	99.99	10.0	4.1	1.1	1.9	3.1
Mesozoic	(Mz)	32	704	54.94	16.11	6.09	0.75	5.81	2.12	2.76	1.31	1.15	4.42	5.58	100.04	12.1	3.7	1.1	2.3	3.4
Tertiary	(Tr)	12	352	57.10	11.59	6.14	0.59	7.25	2.16	2.32	1.18	0.32	5.69	5.67	100.01	8.6	4.3	0.8	1.5	2.8
Quaternary	(Q)	5	28	59.65	13.54	5.97	0.67	4.93	3.07	3.15	0.74	0.00	4.06	4.31	100.09	11.2	4.9	0.6	1.3	1.7
Cenozoic	(Kz)	17	380	57.85	12.16	6.09	0.62	6.57	2.42	2.56	1.05	2.22	5.21	5.27	100.02	9.4	4.5	0.7	1.4	2.5
Total and average		252	6804	50.65	15.10	6.47	0.78	7.19	3.31	3.49	0.81	0.63	6.10	5.58	100.11	11.0	4.1	1.1	2.2	3.5

* Includes organic matter, water of crystallization and constitutional water, halides and sulfur of the sulfides.
† These figures include three samples composed of 35 specimens dated no more accurately than D_3.
‡ These figures include one sample composed of 22 specimens of the Lower Permian clays from the Keltminskii borehole.

Chemical composition (in weight percentage) of composite samples of sandstones of the Russian Platform (Vinogradov and Ronov 1956b).*

System, series, stage		Number of samples and analyses	Number of specimens in the samples	SiO_2	Al_2O_3	Fe_2O_3	TiO_2	CaO	MgO	K_2O	Na_2O	SO_3	CO_2	Loss on ignition**	Total
Upper Proterozoic (Sinian)	(Sn_2)	14	551	73.90	10.28	4.83	0.61	0.93	1.56	2.88	0.75	0.12	0.86	2.95	99.67
Cambrian	(Cm)	13	251	78.02	8.82	3.62	0.54	1.02	1.04	2.21	0.57	0.09	1.32	2.42	99.67
Ordovician	(S_1)	2	133	52.79	9.73	6.34	0.46	10.04	3.86	2.16	1.21	0.29	9.55	3.28	99.71
Gothlandian	(S_2)	1	5	66.58	1.96	3.52	0.15	12.51	1.19	1.30	0.62	0.04	9.53	1.18	98.58
Silurian	(S)	3	138	57.39	7.14	5.40	0.36	10.86	2.97	1.88	1.01	0.21	9.54	2.58	99.34
Middle Devonian	(D_2)	31	829	73.50	7.21	4.31	0.70	2.86	2.05	1.76	0.28	0.43	3.75	2.97	99.82
Frasnian stage	(D_3^1)	26	725	77.74	7.69	5.16	0.70	1.06	0.91	1.87	0.37	0.04	1.53	2.72	99.79
Upper Devonian	(D_3)	6	110	71.21	7.40	3.75	0.51	3.92	2.22	2.57	0.33	1.50	4.06	2.43	99.90
Devonian	(D)	32	835	76.51	7.64	4.89	0.66	1.60	1.15	2.00	0.37	0.31	2.01	2.67	99.81
Lower Carboniferous	(C_1)	63	1664	75.03	7.42	4.60	0.68	2.22	1.60	1.88	0.32	0.37	2.86	2.82	99.80
Middle Carboniferous	(C_2)	13	111	76.41	6.46	3.09	0.50	3.71	0.85	0.89	0.28	0.77	2.89	3.74	99.59
Carboniferous	(C)	8	73	55.73	12.72	4.66	0.61	7.83	2.50	4.55	0.75	0.09	7.51	2.59	99.54
Upper Permian	(P_2)	21	184	68.53	8.84	3.69	0.57	5.28	1.48	2.29	0.46	0.51	4.65	3.30	99.60
Permian	(P)	16	564	45.71	8.96	4.12	0.39	13.97	5.29	1.89	1.17	3.98	11.04	3.18	99.70
Paleozoic	(Pz)	130	3352	70.14	8.28	4.34	0.60	4.10	2.03	2.09	0.54	0.78	3.94	2.91	99.75
Triassic	(T)	1	9	59.63	7.78	4.54	0.44	11.35	1.45	4.26	0.00	0.00	6.46	4.14	100.05
Jurassic	(J)	8	67	63.61	8.75	5.88	0.46	6.99	1.54	1.84	1.09	0.15	5.83	3.54	99.68
Lower Cretaceous	(Cr_1)	9	114	73.03	8.72	6.72	0.56	1.49	1.13	1.76	0.67	0.52	1.72	3.93	100.25
Upper Cretaceous	(Cr_2)	5	46	70.16	5.54	4.14	0.36	7.10	1.12	1.92	0.76	0.26	4.80	3.38	99.54
Cretaceous	(Cr)	14	160	72.00	7.58	5.80	0.49	3.49	1.13	1.82	0.70	0.43	2.82	3.73	99.99
Mesozoic	(Mz)	23	236	68.54	7.80	5.77	0.48	5.05	1.29	1.93	0.81	0.31	4.02	3.70	99.70
Tertiary	(Tr)	3	93	74.64	8.03	4.26	0.52	2.72	1.29	1.74	0.68	0.38	1.64	3.60	99.50
Quaternary	(Q)	2	28	70.46	6.87	2.88	0.38	7.00	2.07	2.50	0.53	0.00	5.36	1.95	100.00
Cenozoic	(Kz)	5	121	72.97	7.56	3.71	0.46	4.43	1.60	2.05	0.62	0.23	3.13	2.98	99.74
Total or average		158	3709	70.00	8.22	4.53	0.58	4.25	1.89	2.06	0.58	0.69	3.87	3.02	99.69

* Preliminary data. The study of sands continues.
** See note to table 2, Vinogradov and Ronov 1956b.

TABLE 5.14.

Chemical composition (in weight percentage) of composite samples of carbonate rocks of the Russian Platform (Vinogradov and Ronov 1956b).

Column groups: *Analyses of documented samples* (No. of samples … U); *Previously published analyses* (CaO … SO_3, each with Total No. of analyses and Average); *Arithmetic mean of published analyses and analyses of documented samples* (CaO, MgO, SO_3).

System, series, stage		No. of samples	No. of specimens in a sample	Insoluble residue	R_2O_3	CaO	MgO	SO_3	Sr	Ba	Th (in 10^{-4}%)	U*** (in 10^{-4}%)	CaO No.	CaO Avg	MgO No.	MgO Avg	Fe_2O_3 No.	Fe_2O_3 Avg	Al_2O_3 No.	Al_2O_3 Avg	R_2O_3 No.	R_2O_3 Avg	SiO_2 No.	SiO_2 Avg	SO_3 No.	SO_3 Avg	CaO mean	MgO mean	SO_3 mean
Upper Proterozoic (Sinian)	(Sn_2)	8*	20	11.60	0.37	27.10	19.60	0.04	0.0033	0.0012	n.d.	n.d.	10	29.87	10	21.24	10	0.76	12	1.41	15	4.58	10	1.92	—	—	28.48	20.42	0.04
Cambrian	(Cm_2)								n.d.	n.d.	n.d.	n.d.	16	34.00	16	11.53	12	3.48	—	—	—	—	13	7.54	1	0.32	34.00	11.53	0.32
Ordovician	(S_1)	6	293	21.33	1.53	33.61	7.88	0.08	0.0067	0.0093	4.9	2.0	398	34.72	398	9.16	160	5.11	104	3.09	211	3.41	189	9.34	24	0.30	34.16	8.52	0.25
Upper Silurian	(S_2)	4	177	19.94	2.07	35.61	7.19	1.79	0.0490	0.0240	4.2	2.7	14	43.40	14	7.55	3	0.11	—	—	5	1.21	3	1.93	2	0.22	39.50	7.37	1.00
Silurian	(S)	10	470	20.80	1.75	34.41	7.60	0.76	0.0173	0.0130	4.6	2.3	412	35.01	412	9.10	163	5.02	104	3.09	216	3.36	192	9.22	26	0.29	34.71	8.35	0.52
Middle Devonian	(D_2)	15	297	18.46	1.34	33.74	8.05	5.45	0.0400	0.0153	2.3	1.2	97	30.78	97	8.40	18	3.16	18	4.48	70	2.86	18	19.98	46	4.25	32.26	8.22	4.85
Frasnian stage	(D_3)	37	2245	13.61	1.00	41.61	4.00	0.70	0.0280	0.0051	3.7	1.3	497	42.42	497	4.26	69	2.48	68	3.65	112	3.24	93	8.35	47	0.35	42.01	4.13	0.52
Famennian stage	(D_3^2)	31	1963	8.00	0.37	36.68	11.54	8.69	0.0730	0.0036	2.2	1.3	151	37.37	151	8.77	30	1.22	28	2.09	31	3.67	31	10.70	34	1.38	37.02	10.15	5.03
Upper Devonian	(D_3)	68	4208	11.05	0.71	39.36	7.44	4.34	0.0454	0.0044	3.0	1.3	871	39.39	870	6.19	133	2.10	132	3.54	189	3.96	159	11.70	—	—	39.37	6.81	2.93
Devonian**	(D)	83	4505	12.39	0.83	38.34	7.55	4.55	0.0447	0.0056	2.8	1.6	1040	38.80	1038	6.52	175	2.11	170	3.54	291	3.66	204	10.23	196	1.53	38.61	7.03	3.41
Lower Carboniferous	(C_1)	26	814	4.83	0.16	40.68	10.52	2.26	0.0403	0.0005	1.3	2.3	1157	47.18	1103	4.83	408	1.53	318	0.66	666	1.13	529	2.70	530	2.28	43.93	7.67	1.99
Middle Carboniferous	(C_2)	24	1229	7.30	0.28	40.25	9.27	1.56	0.0226	0.0016	1.2	2.7	1411	43.18	1359	7.55	337	1.02	282	1.98	684	1.65	598	5.69	483	0.72	41.71	8.41	1.43
Upper Carboniferous	(C_3)	14	520	2.12	0.12	37.37	13.21	6.06	0.0129	0.0019	3.4	3.4	946	39.04	945	12.42	65	1.06	66	3.25	626	2.28	97	6.76	755	1.30	38.20	12.81	4.91
Carboniferous**	(C)	64	2563	5.17	0.19	39.79	10.64	2.83	0.0252	0.0014	1.2	2.7	3815	44.18	3661	7.75	878	1.28	720	1.47	2211	1.63	1404	4.26	—	3.76	41.98	9.19	2.46
Lower Permian	(P_1)	17	922	4.86	0.20	34.82	12.49	14.17	0.0863	0.0019	1.0	2.4	988	35.79	988	9.50	3	3.66	2	0.63	655	1.12	48	6.38	917	6.86	35.30	11.00	10.51

Stratigraphic unit																													
Upper Permian	(P$_2$)	4	66	13.04	0.83	39.28	7.33	1.26	0.0150	0.0067	1.8	5.4	1488	37.86	1484	8.55	56	1.54	53	2.99	1050	4.84	285	13.64	1043	2.33	38.57	7.94	1.80
Permian**	(P)	21	988	6.42	0.32	35.67	11.50	11.67	0.0790	0.0029	1.3	3.3	2593	35.50	2584	8.78	104	1.52	101	3.02	1798	3.38	379	10.60	1985	4.40	35.58	10.14	8.03
Paleozoic	(Pz)	178	8526	9.56	0.59	38.33	9.13	4.56	0.0400	0.0042	2.2	2.1	7876	40.12	7711	8.01	1332	13.04	1107	2.08	4531	2.55	2192	6.36	4134	3.19	39.22	8.57	3.87
Jurassic	(J)	2	25	35.20	9.65	27.10	4.60	0.21	0.0192	0.0086	4.7	2.3	4	24.57	4	1.12	3	12.66	2	8.80	4	11.79	2	46.66	2	0.07	25.83	2.86	0.14
Lower Cretaceous	(Cr$_1$)	1	14	12.60	0.78	45.80	1.80	0.28	0.0280	0.0050	2.0	2.8	—	—	—	—	—	—	—	—	—	—	—	—	—	—	45.80	1.80	0.28
Upper Cretaceous	(Cr$_2$)	6	161	14.20	0.70	46.40	0.40	0.17	0.0826	0.069	2.7	1.6	266	45.00	166	0.60	66	1.29	83	2.51	215	3.64	199	17.51	48	0.35	45.70	0.50	0.26
Cretaceous	(Cr)	7	175	14.00	0.71	46.30	0.60	0.21	0.0735	0.0087	2.5	1.9	266	45.00	166	0.60	66	1.29	83	2.51	215	3.64	199	17.51	48	0.35	45.65	0.60	0.28
Mesozoic	(Mz)	9	200	18.71	2.70	42.00	1.49	0.21	0.0614	0.0087	3.3	2.0	270	44.70	170	0.61	69	1.78	85	2.66	219	3.79	201	17.80	50	0.34	43.35	1.05	0.27
Tertiary	(Tr)	3	101	49.00	2.55	24.00	0.70	0.22	0.1500	0.0220	7.1	3.6	176	45.60	145	1.74	65	1.13	46	2.16	128	3.39	114	13.58	68	0.21	34.84	1.22	0.21
Quaternary	(Q)	—	—	—	—	—	—	—	—	—	n.d.	n.d.	19	50.32	19	0.82	7	1.00	4	0.58	13	1.53	15	3.08	7	0.24	50.32	0.82	2.24
Cenozoic	(Kz)	3	101	49.00	2.55	24.00	0.70	0.22	0.1500	0.0220	7.1	3.6	195	46.14	164	1.63	72	1.11	50	2.04	141	3.22	129	12.36	75	0.22	35.07	1.16	0.22
Totals of analyses and average content by series and stages**		198	8847	10.66	0.71	37.83	9.08	4.11	0.0447	0.0047	2.4	2.1	7636	40.78	7396	7.90	1312	2.17	1086	2.16	4485	2.78	2214	8.53	4016	3.21	39.30	8.80	3.66
Totals of analyses and average content by systems**		198	8847	10.66	0.71	37.83	9.08	4.11	0.0477	0.0047	2.4	2.1	8351	40.50	8050	7.74	1483	2.09	1242	2.24	4894	2.74	2448	8.17	4268	3.10	39.16	8.62	3.60

* Analyses of common, not documented average samples.

** The total number of published data for stages and series is less than for the corresponding systems (D,C,P), for additional samples dated only as Devonian, Carboniferous or Permian were used for the latter. For this reason different averages are given at the bottom of the table: 1—for series and stages, and 2—for systems.

TABLE 5.15.

The abundance and average composition of sedimentary rocks of the Russian and North American platforms (Ronov and Migdisov 1971).

Stratigraphic Interval	Platforms	Rocks and complexes	Abundance in %, of the total volume of deposits	Number of analyses	Refs.	SiO_2	TiO_2	Al_2O_3	e_2O_3	FeO	MnO	MgO	CaO	K_2O	Na_2O	P_2O_5	C_{org}	CO_2	SO_3 total	Cl	H_2O	Sum	$S_{sulf.}$	$S_{pyr.}$	Th. $n \cdot 10^{-4}$%	U. $n \cdot 10^{-4}$%
LATE PROTEROZOIC (Pt_3)	RUSSIAN PLATFORM	Sands and aleurites	58.0	34 samples from 1350 specimens	1	72.78	0.56	9.79	2.63	2.04	0.102	1.67	1.37	2.94	0.69	0.073	0.08	1.39	0.53	0.105	3.33	100.07	0.056	0.156	7.9	1.5
		Clays and shales	37.6	34 samples from 1226 specimens	1	57.65	0.86	17.04	4.26	3.17	0.113	2.38	1.20	4.18	0.93	0.090	0.35	1.10	1.05	0.118	6.19	100.68	0.100	0.300	11.6	2.9
		Carbonates	2.4	3 samples from 106 specimens	1	28.73	0.36	6.52	1.41	1.63	0.112	6.52	25.96	2.23	0.56	0.064	0.06	23.21	0.52	0.081	1.97	99.33	0.048	0.160	3.3	1.3
		Effusives and tuffs	2.0	12 samples from 12 specimens	2	49.23	1.66	5.55	5.57	5.60	0.192	5.16	7.83	0.83	2.74	0.331	—	—	0	—	5.48	100.17	—	—	2.2	0.6
		Average composition of rocks	100.0	83 samples from 2694 specimens	1	65.56	0.68	12.56	3.26	2.52	0.108	1.75	2.02	3.34	0.81	0.084	0.18	1.78	0.71	0.107	4.56	100.03	0.071	0.207	9.1	2.0
		Average composition of rocks excluding carbonate matter		"	1	68.61	0.71	13.14	3.41	2.64	0.113	1.80		3.49	0.85	0.088	0.19	—	0.22‡	—	4.66	99.92	—	—	9.5	2.1

PALEOZOIC (Pz)	RUSSIAN PLATFORM	Sands and aleurites	17.5	319 samples from 4187 specimens	1	68.28	0.47	7.44	2.49	1.29	0.065	2.38	5.58	1.94	0.57	0.123	0.23	4.79	1.60	0.086	3.38	100.71	0.402	0.240	5.4	3.0
		Clays and shales	27.6	401 samples from 6734 specimens	1	47.94	0.78	14.26	4.10	1.85	0.064	3.85	7.53	3.70	0.65	0.112	0.70	6.80	2.18	0.148	6.07	100.73	0.351	0.473	10.0	4.0
		Carbonates	44.1	257 samples from 9951 specimens	1	6.12	0.12	1.23	0.75	0.35	0.037	8.95	39.53	0.49	0.21	0.039	0.26	36.75	3.82	0.074	2.22	100.95	1.320	0.209	1.7	2.1
		Gypsums and anhydrites	4.5	14 samples from 303 specimens	1	0.53	0	0.07	0.08	0	0.006	1.56	38.96	0.04	0.08	0.004	—	3.07	52.64	—	3.51	100.55	21.08	—	—	1.4
		Salts	5.6	9 analyses	2	0.71	0	0.17	0.10	0	0	0.19	0.60	0.94	44.22	—	—	—	0.86	51.650	0.57	100.01	0.344	—	—	—
		Effusives and tuffs	0.7	584 analyses		44.96	4.39	13.80	7.22	5.80	0.100	6.36	7.13	1.68	3.00	0.690	—	0.73	0.23	—	3.61	99.70	—	—	2.2	0.6
		Average composition of rocks	100.0	1584 samples from 21768 specimens	1	28.25	0.37	5.89	1.96	0.93	0.046	5.55	22.32	1.64	2.87	0.074	0.34	19.07	4.98	2.987	3.48	100.76	1.717	0.265	4.5	2.6
		Average composition of rocks excluding carbonate matter, etc*			1	60.72	0.84	13.15	4.37	2.09	0.118	5.30		3.53	0.88	0.165	0.78	—	0.57‡	—	7.31	99.82	—	—	9.0	5.2
	NORTH AMERICAN PLATFORM	Sands**	~17	30 analyses	3	85.83	0.80	3.72	1.26	0.13	—	1.14	1.43	2.14	1.15	0.150	—	—	—	—	2.26	100.01	—	—	—	—
		Clays and shales	~28	473 analyses	3,4,5	53.68	0.88	15.98	4.02	1.40	—	3.61	5.48	4.33	0.51	0.101	—	5.27	0.57	—	4.61	100.44	—	—	14.9	9.9
		Carbonates	~45	792 analyses	3,5	4.39	0.10	0.95	0.72	0.42	—	8.56	42.38	0.17	0.15	0.208	—	42.15	0.26	—	—	100.46	—	—	1.0	1.7
		Gypsums and anhydrites	~4.5	28 analyses	3	1.65	—		0.39	—	—	0.71	33.35	—	—	—	—	2.52	44.44	—	17.13	100.19	—	—	0.7	1.3
		Salts	~5.5	48 analyses	3		0.39		0.05	—	—	0.18	1.14	0.47	43.48	—	—	—	1.47	51.450	0.46	99.09	—	—	—	—
		Average composition of rocks	100.0	1371 analyses	3,4, 5,6	31.69	0.43	5.55	1.68	0.60	—	5.10	22.41	1.68	2.80	0.147	0.31	20.56	2.36	2.830	2.47	100.62	—	—	5.1	3.6
		Average composition of rocks excluding carbonate matter, etc.*			3,4, 5.6	69.99	0.94	12.27	3.69	1.34		2.97		3.67	0.90	0.327	0.69	—	0.18‡	—	3.71	100.68	—	—	11.2	7.8

(continued)

TABLE 5.15 (continued)

Stratigraphic Interval	Platforms	Rocks and complexes	Abundance in % of the total volume of deposits	Number of analyses	Refs.	SiO_2	TiO_2	Al_2O_3	Fe_2O_3	FeO	MnO	MgO	CaO	K_2O	Na_2O	P_2O_5	C_{org}	CO_2	SO_3 total	Cl	H_2O	Sum	$S_{sulf.}$	$S_{pyr.}$	$Th·n·10^{-4}\%$	$U·n·10^{-4}\%$
Mesozoic (Mz) and Cenozoic (Cz)	Russian Platform	Sands and aleurites	35.0	305 samples from 2250 specimens	1	70.01	0.46	7.45	3.23	1.33	0.059	1.34	4.90	1.87	0.87	0.147	0.36	3.34	0.90	0.069	4.46	100.80	0.177	0.185	5.7	2.3
		Clays and shales	43.3	259 samples from 2764 specimens	1	55.61	0.72	14.45	3.65	1.95	0.057	2.20	4.90	2.39	1.10	0.107	0.94	3.62	1.40	0.101	7.67	100.87	0.137	0.377	10.9	2.7
		Carbonates	20.5	109 samples from 880 specimens	1	16.61	0.13	3.22	1.06	0.67	0.048	1.02	41.20	0.80	0.62	0.124	0.47	31.56	0.60	0.078	2.56	100.77	0.049	0.190	3.4	2.1
		Siliceous rocks	1.3	128 analyses	2	81.77	0.35	7.19	2.00	0.73	0.016	0.93	1.15	1.13	0.36	0.035	—	—	0.50	—	4.39	100.55	—	—	—	—
		Average composition of rocks	100.0	801 samples from 6022 specimens	1	52.98	0.50	9.60	2.96	1.45	0.056	1.64	12.29	1.85	0.90	0.122	0.64	9.20	1.05	0.084	5.46	100.78	0.131	0.267	7.4	2.4
		Average composition of rocks excluding carbonate matter			1	66.67	0.63	12.08	3.72	1.82	0.070	2.59 (MgO+CaO)		2.33	1.13	0.153	0.81	—	0.34‡	0.106	6.87	99.32	—	—	9.3	3.0
	N. Am. Platform	Sands	≀ 30	122 analyses	3	78.79	0.68	6.86	1.52	—	—	1.08	4.02	1.73	0.88	0.043	—	2.35	0.13	—	1.98	100.06	—	—	—	—
		Clays and shales†	≀ 50	959 analyses	3,4,5	60.10	0.62	14.71	3.55	0.45	—	3.00	3.74	2.16	0.87	0.093	—	4.35	0.55	—	6.52	100.71	—	—	13.6	3.7
		Carbonates†	≀ 20	279 analyses	3,5	10.12	—	1.89	0.88	—	—	2.86	44.42	0.43	0.19	0.036	—	36.50	1.19	—	2.01	100.53	—	—	1.5	3.5
		Average composition of rocks	100	1360 analyses	3,4,5,6	55.71	0.51	9.79	2.41	—	—	2.40	11.96	1.69	0.74	0.067	0.76	10.18	0.55	—	4.20	100.97	—	—	7.1	2.6
		Average composition of rocks excluding carbonate matter			3,4,5,6	72.60	0.66	12.76	3.14	—	—	2.03 (MgO+CaO)		2.20	0.96	0.087	0.99	—		—	5.55	100.98	—	—	9.3	3.4

		5199 analyses (33215 specimens)																					
$\bar{O}$	Average composition of sedimentary cover of Russian and North American platforms	—	41.16	0.45	7.57	2.18	0.86	0.057	3.82	17.63	1.83	1.94	0.089	0.42	15.15	2.18	1.670	3.56	100.57	—	—	9.2	4.8
	Average composition of sedimentary cover of Russian and North American platforms excluding carbonate matter, etc. *	—	68.21	0.75	12.55	3.62	1.34	0.092	\{ 3.25		3.00	0.93	0.163	0.70	—	0.43‡	—	5.53	100.56	—	—	—	—

* When evaluating the average composition of sedimentary rocks, the carbonate and sulfate content and the soluble salts were excluded.

** We had at our disposal only data on the composition of sands of the Middle and Upper Paleozoic (C_{2-3}, P_1, and P_2).

† When the average composition of Mesozoic carbonate rocks was deduced, Jurassic carbonates were not taken into account because there were no data.

‡ Data are given only for pyritic sulfur.

References:
1. Our data.
2. Evaluations from literature.
3. Evaluations from Hill et al. 1967.
4. Evaluations from Adams and Weaver 1958.
5. Evaluations from White 1959.
6. Evaluations from Trask and Patnode 1942.

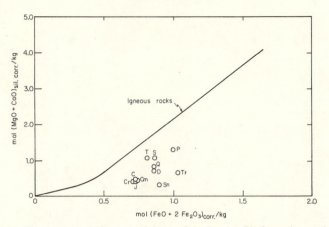

FIGURE 5.17. The (MgO + CaO)$_{sil.\,corr.}$ of shales of the Russian Platform; data and symbols from Vinogradov and Ronov (1956b).

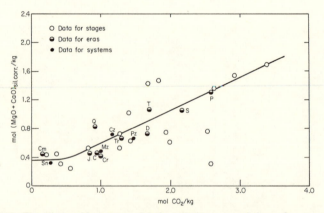

FIGURE 5.18. The relationship between (MgO + CaO)$_{sil.\,corr.}$ and the carbonate content of shales of the Russian Platform; data from Vinogradov and Ronov (1956b).

history. Their carbonate content is quite variable and must reflect paleogeographic conditions during sedimentation. Since the carbonate budget of the oceans is a global matter, the carbonates deposited with particular shales of the Russian Platform may have been derived from quite distant parts of the world, and their quantity does not necessarily reflect the local CO_2 budget.

Fortunately, the validity of the values of (MgO + CaO)$_{sil.\,corr.}$ of the Russian shales can be checked, because values of (MgO + CaO)$_{sil.\,corr.}$ can also be computed for the associated sandstones of the Russian Platform. These sandstones are rather dirty. Their Al_2O_3 and their total iron content are quite high, and this circumstance permits the reconstruction of their parent igneous rocks without a very long extrapolation. Table 5.16 shows a comparison of the values of $(FeO + 2Fe_2O_3)_{corr.}$, $(MgO + CaO)_{sil.\ corr.}$, $(K_2O + Na_2O)_{corr.}$, and acid use (ΔCO_2) of sandstones and shales of the Russian Platform for the Late Proterozoic, Paleozoic, and Mesozoic + Cenozoic.

The agreement between the sets of data for the two rock types is quite reasonable for the late Proterozoic and for the Paleozoic, but rather poor for the Mesozoic and Cenozoic Eras. The calculations based on the composition of shales always seem to indicate derivation from somewhat more granitic rocks and the use of less acid for the conversion of igneous rocks to sedimentary rocks than the calculations based on the composition of sandstones.

These differences are probably due in part to the simplicity of the model of sediment formation which underlies the calculations. However, the reasonable consistency of the two sets of data suggests that the model is not grossly wrong. The two sets of data together indicate that the iron content of the source rocks of the Russian Platform has been nearly constant during the last billion years, and that acid use per kg of igneous rock converted to sedimentary rock may have passed through a shallow minimum during the Paleozoic Era.

The data for the North American Platform are rather scant compared to those for the Russian Platform. A good many of them were compiled

TABLE 5.16.

Comparison of calculated acid use in the production of sandstones and shales of the Russian Platform.

	$(FeO + 2Fe_2O_3)_{corr.}$	$(MgO + CaO)_{sil.corr.}$	$(K_2O + Na_2O)_{corr.}$	ΔCO_2
Late Proterozoic				
Sands and aleurites	0.98 mol/kg	0.55 mol/kg	0.42 mol/kg	1.80 mol/kg
Clays and shales	0.89	0.51	0.59	1.56
Paleozoic				
Sands and aleurites	1.03	1.04	0.62	1.34
Clays and shales	0.84	0.81	0.55	1.20
Mesozoic and Cenozoic				
Sands and aleurites	1.23	0.94	0.71	1.83
Clays and shales	0.79	0.64	0.46	1.33

Note: Data from table 5.15.

in tables 4.6, 4.7 and 4.8 of Holland (1978). Calculated values of $(MgO + CaO)_{sil.\,corr.}$ for these analyses are shown in figure 5.19, together with the curve drawn through the corresponding Russian data in figure 5.18. There are too few points to test goodness of fit; the scatter of the few points in figure 5.20 around the curve is comparable to that for the Russian data.

Two striking changes in the nature of sedimentary rocks occurred during the Phanerozoic. One of these is the reduction in the proportion of dolomite in carbonate rocks. This decrease is well illustrated by the

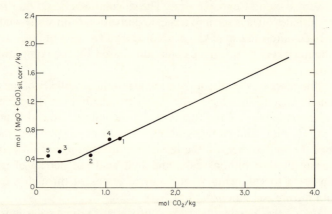

FIGURE 5.19. Relationship between $(MgO + CaO)_{sil.\,corr.}$ and the carbonate content of shales of the North American Platform; data points from the following sources: (1) Paleozoic shales of the North American Platform (Ronov and Migdisov 1971); (2) Mesozoic and Cenozoic shales of the North American Platform (Ronov and Migdisov 1971); (3) Composite analysis of 51 Paleozoic shales (Clarke 1924); (4) Composite analysis of 27 Mesozoic and Cenozoic shales (Clarke 1924); (5) Precambrian lutites (Nanz 1953).

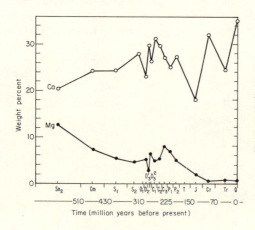

FIGURE 5.20. Variation in the average percentage of calcium and magnesium in the carbonate rocks of the Russian Platform (Vinogradov and Ronov 1956b). (Reproduced by permission of Pergamon Press, Ltd.)

data of figure 5.20. The reasons for the change have been warmly debated during most of this century. It is likely that most sedimentary dolomites were formed during or shortly after sedimentation (see chapter 6). Their relative rarity in Mesozoic and Cenozoic sediments therefore raises questions concerning progressive changes in the composition of seawater, the nature of carbonate environments, and the supply of CO_2 for the conversion of MgO in magnesium silicates to the $MgCO_3$ component in dolomites. Garrels and Mackenzie (1971, p. 238) have suggested that there is a reciprocal relationship between the MgO/Al_2O_3 ratio in shale rocks of the Russian Platform and the Mg/Ca ratio in carbonates during the last 2000 m.y., and that this is due to the redistribution of Mg between shales and carbonate rocks if there has been a fairly constant total amount of Mg deposited. Unfortunately, an arithmetic error seems to have crept into their computations. The data in table 5.17 show that the MgO/Al_2O_3 ratio in Mesozoic and Cenozoic shales of the Russian Platform is not significantly higher than the value of the MgO/Al_2O_3 ratio during the Proterozoic, and that it is distinctly lower than the MgO/Al_2O_3 ratio in the Paleozoic shales of the Russian Platform.

There does not, therefore, seem to be a correlation between the demise of dolomite and the MgO/Al_2O_3 ratio of shales of the Russian Platform.

TABLE 5.17.

The MgO/Al_2O_3 ratio in Proterozoic and Phanerozoic shales.

	Russian Platform		
	Vinogradov and Ronov (1956b)	*Ronov and Migdisov (1971)*	*Garrels and Mackenzie (1971, p. 238), based on Vinogradov and Ronov (1956b)*
Proterozoic	0.12	0.14	0.12
Paleozoic	0.23	0.27	0.14
Mesozoic	0.13	} 0.15	0.19
Cenozoic	0.20		0.20

	North American Platform			
	Ronov and Migdisov (1971)	*Clarke (1924)*	*Nanz (1953)*	*Perry and Hower (1970) and Hower et al. (1976)*
Proterozoic	—	—	0.12	—
Paleozoic	0.23	0.14	—	—
Mesozoic	} 0.20	} 0.19	—	—
Cenozoic			—	0.12

The situation on the North American Platform appears to have been rather similar. The only serious discrepancy between the several data sets for the North American Platform is in the figures for the MgO/Al_2O_3 ratio in Paleozoic shales. This difference is probably related to differences between the carbonate content of the shale composite of Clarke (1924) and that of Ronov and Migdisov (1971) (see figure 5.19). The MgO/Al_2O_3 ratio of shales therefore gives no indication that the decrease in the abundance of dolomite during the past 200 m.y. is related to a transfer of MgO from carbonate to silicate minerals in sedimentary rocks. The data in table 5.16 may mean that the availability of acid to titrate igneous rocks was actually greater during the past 200 m.y. than during the preceding 400 m.y. A lack of CO_2 to convert magnesium silicates to carbonates does not, therefore, seem to have been responsible for the increasing scarcity of dolomite during Mesozoic and Cenozoic time. An alternative hypothesis is suggested in chapter 6.

The second major change in the nature of sedimentary rocks during the Phanerozoic Era is not as dramatic as the decrease in the dolomite/limestone ratio. Vinogradov and Ronov (1956a) discovered that the potassium content of shales of the Russian Platform decreased during the Phanerozoic Era while their sodium content increased (see figure 5.21). The increase in the Na/K ratio of the shales is particularly pronounced at the end of the Carboniferous period. This is shown quite clearly by the

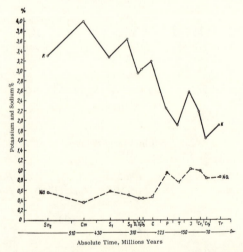

FIGURE 5.21. Changes in the potassium and sodium content of shales of the Russian Platform (Vinogradov and Ronov 1956a). (Reproduced by permission of Pergamon Press, Ltd.)

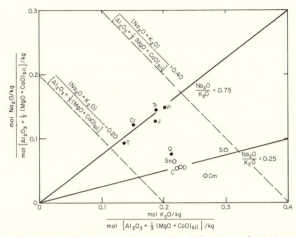

FIGURE 5.22. The relationship between the ratio $Na_2O/[Al_2O_3 + \frac{1}{3}(MgO + CaO)_{sil.}]$ and the ratio $K_2O/[Al_2O_3 + \frac{1}{3}(MgO + CaO)_{sil.}]$ in shales of the Russian Platform; data from Vinogradov and Ronov (1956a,b).

plot of figure 5.22. The mole ratio of Na_2O/K_2O in late Precambrian through Carboniferous shales of the Russian Platform is close to 0.25. The mole ratio of Na_2O/K_2O of younger shales is approximately 0.75. The Quaternary shales of the Russian Platform seem to be a throwback to those of the early Paleozoic.

Weaver (1967) has shown that the shales of the North American Platform behave in much the same way. His analysis of some 50,000 X-ray diffraction patterns of North American shales demonstrated that the proportion of illite decreased dramatically during the Carboniferous at the expense of expandable clays and kaolinite. His results, as replotted by Garrels and Mackenzie (1971, 235), are shown in figure 5.23. The chemical signature of this striking change in mineralogy can be calculated, and has been confirmed by the direct chemical analyses that are plotted in figure 5.24. The change in the Na/K ratio of shales on both platforms is not accompanied by a discernible trend in the degree of acid titration of the parent igneous rocks. This is demonstrated most convincingly by figure 5.25 and 5.26. The most undepleted shales are those of the Silurian, Permian, and Tertiary of the Russian Platform. On the average, both sets of shales are considerably more depleted in alkalis than the Archean and Aphebian shales of the Canadian Shield that were discussed earlier in this chapter. This trend parallels the gradual disappearance of greywackes in sedimentary suites at the expense of quartz arenites (Pettijohn 1975, p. 594).

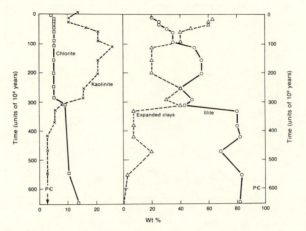

FIGURE 5.23. Variations in the relative percentage of the clay minerals in shales of the North American Platform during the Phanerozoic Era (Garrels and Mackenzie 1971, p. 235; data from Weaver 1967). (Reproduced by permission of W. W. Norton & Co., Inc., copyright 1971.)

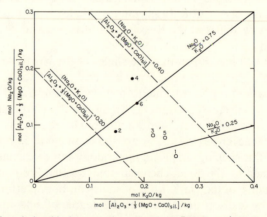

FIGURE 5.24. The relationship between the ratio $Na_2O/[Al_2O_3 + \frac{1}{3}(MgO + CaO)_{sil.}]$ and the ratio $K_2O/[Al_2O_3 + \frac{1}{3}(MgO + CaO)_{sil.}]$ in shales of the North American Platform; sources of data same as for figure 5.19.

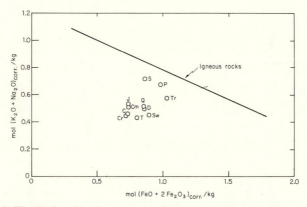

FIGURE 5.25. The $(K_2O + Na_2O)_{corr.}$ content of late Proterozoic and Phanerozoic shales of the Russian Platform; data from Vinogradov and Ronov (1956a,b).

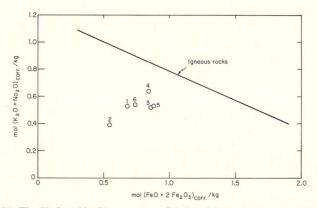

FIGURE 5.26. The $(K_2O + Na_2O)_{corr.}$ content of shales of the North American Platform; the sources of data are the same as those for figure 5.19.

6. Recent Sediments

Recent sediments are much easier to sample than ancient sediments; nevertheless, the number of sufficiently complete chemical analyses of Recent sediments is still far from sufficient to derive a precise figure for the present-day acid-base balance. Many of the available data sets are listed in table 5.18; the functions $(MgO + CaO)_{sil.\ corr.}$ and $(K_2O + Na_2O)_{corr.}$

TABLE 5.18.

Analyses of modern sediments.

	1	2	3	4	5	6	7	8	9	10	11	12
SiO_2	69.96	55.34	63.91	26.96	48.73	55.02	32.50	27.40	30.81	63.8	63.3	61.5
TiO_2	0.59	0.84	0.65	0.38	0.98	0.72	0.29	0.64	0.70	0.6	0.8	1.4
Al_2O_3	10.52	17.84	13.30	7.97	19.12	16.61	8.97	9.31	11.00	13.7	16.2	18.5
Fe_2O_3	3.47	7.04	5.66	3.00	9.48	7.26	4.82	4.89	5.50	9.5	8.3	5.5
FeO		1.13	0.67	0.87								2.5
MnO	0.06	0.48	0.50	0.33	0.094	0.26	0.066	0.07	0.10	1.2	0.3	0.1
MgO	1.41	3.42	1.95	1.29	3.17	2.19	2.21	3.17	2.87	3.3	2.2	2.3
CaO	2.17	0.93	0.75	0.30	0.89	1.57	20.92	24.67	19.10	3.1	2.6	2.2
Na_2O	1.51	1.53	0.94	0.80	—	2.76	1.61	1.73	1.97	1.0	2.4	2.4
K_2O	2.30	3.26	1.90	1.48	3.49	2.32	1.72	1.14	1.63	1.7	1.7	1.1
H_2O^+	1.96	6.54	7.13	3.91								
H_2O^-	3.78											
P_2O_5	0.18	0.14	0.27	0.15	0.37	0.18	0.27	0.11	0.11	0.3	0.2	0.2
CO_2	1.40				0.63	2.25	16.23	18.12	15.22	1.8	2.0	2.2
SO_3	0.03											
Cl	0.30											
C_{org}	0.66	0.24	0.22	0.31			2.25	0.8	0.65			
$CaCO_3$		0.79	1.09	50.09								
$MgCO_3$		0.83	1.04	2.16								
S							2.90					
Misc.	0.32											

1. Composite analysis of 235 samples of Mississippi Delta sediments; Clarke 1924, p. 509.
2. Lithogenous pelagic clays (Chester and Aston 1976).
3. Siliceous pelagic sediments (Chester and Aston 1976).
4. Calcareous pelagic sediments (Chester and Aston 1976).
5. Average of 60 clays from the Barents Sea (Calvert 1976. p. 192).
6. Average of 6 basin clays from the Gulf of Paria (Calvert 1976, p. 193).
7. Sediments from the Black Sea (Calvert 1976, p. 254).
8. Five aleuritic-pelitic muds from the Mediterranean Sea (Stanley 1972, p. 362).
9. Eight pelitic muds from the Mediterranean Sea (Stanley 1972, p. 362).
10. Average radiolarian ooze (Poldervaart 1955).
11. Average blue mud (Poldervaart 1955).
12. Average terrigenous mud (Poldervaart 1955).

for these sets are shown in figures 5.27 and 5.28, and the ratio $Na_2O/[Al_2O_3 + \frac{1}{3}(MgO+CaO)_{sil.}]$ is plotted against the ratio $K_2O/[Al_2O_3 + \frac{1}{3}(MgO + CaO)_{sil.}]$ in figure 5.29. Most of the data in these figures are for clay-rich, calcareous, and siliceous sediments. Data for silts and sands were excluded, as well as data for unusual sediments, such as those containing abnormally high concentrations of manganese.

The scatter of data in figures 5.27, 5.28, and 5.29 is roughly comparable to that in the figures for Phanerozoic shales of the Russian and North American Platforms, and there is nothing startlingly different about the acid use indicated by the composition of Recent sediments. The average

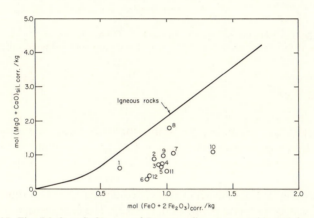

FIGURE 5.27. The $(MgO + CaO)_{sil.\,corr.}$ of Recent sediments; see table 5.18 for sources of data.

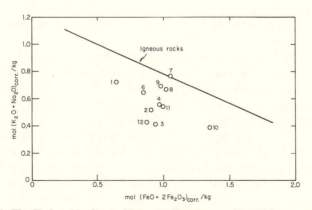

FIGURE 5.28. The $(K_2O + Na_2O)_{corr.}$ of Recent sediments; see table 5.18 for sources of data.

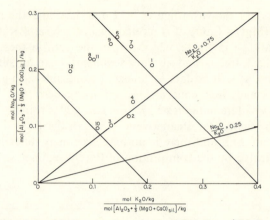

FIGURE 5.29. The relationship between the ratio $Na_2O/[Al_2O_3 + \frac{1}{3}(MgO + CaO)_{sil.}]$ and the ratio $K_2O/[Al_2O_3 + \frac{1}{3}(MgO + CaO)_{sil.}]$ of Recent sediments; see table 5.18 for sources of data.

Mississippi Delta sediment (1) is slightly unusual because its iron content is quite low, not because its $(MgO + CaO)_{sil. corr.}$ content is unusually low. The small amount of $(MgO + CaO)_{sil.}$ loss of aleuritic-pelitic sediments from the Mediterranean Sea (8) is probably due to a large volcanic component in these sediments; Mediterranean pelitic muds (9) fall within the cluster of data points in figure 5.27. The composition of average radiolarian ooze (10) is probably set apart more by an abnormally high iron content than by an abnormally low content of $(MgO + CaO)_{sil. corr.}$. The $(K_2O + Na_2O)_{corr.}$ values of the sets of Recent sediments correlate reasonably well with their $(MgO + CaO)_{sil. corr.}$ values. Only the abnormally high value of the $(K_2O + Na_2O)_{corr.}$ content of Black Sea sediments (7) looks somewhat unusual.

Some of the values of the ratios $Na_2O/[Al_2O_3 + \frac{1}{3}(MgO + CaO)_{sil.}]$ and $K_2O/[Al_2O_3 + \frac{1}{3}(MgO + CaO)_{sil.}]$ in figure 5.29 are slightly unusual for Phanerozoic sediments. Their sum falls within the normal range, but their ratio in several of the sets of data (6, 7, 8, 9, 10, 11, and 12) is distinctly higher than that of the Phanerozoic sediments discussed in the previous section. The high-Na_2O sediments include those that are rather immature; the Mediterranean sediments 8 and 9 are cases in point. None of the sets have Na_2O/K_2O ratios as low as those that are characteristic of the Paleozoic sediments of the Russian and North American platforms. The data for Recent sediments place them firmly within the group of Mesozoic and Cenozoic sediments described in the previous section.

7. The Implications of the Acid-Base Balance for the History of Atmospheric Carbon Dioxide

The data for the acid-base balance of sediments during Earth history are still fragmentary. They show that carbonation reactions were taking place at least as early as 3800 m.y.b.p., and they indicate that the intensity of these reactions 3400 m.y. ago was comparable to their intensity today. They are consistent with the observation that Phanerozoic sedimentary rocks tend to be more mature than Precambrian sedimentary rocks, but they are not sufficiently detailed to document the fluctuations in the acid-base balance that almost certainly occurred during the past 3800 m.y. They set some limits on P_{CO_2} in the past, but they do not permit a quantitative reconstruction of the variations in the CO_2 content of the atmosphere.

The CO_2 content of the atmosphere today (2.5×10^{18} gm) is trivial compared to the quantity of combined CO_2 in the carbonates of sedimentary rocks (ca. 2.6×10^{23} gm). Even if the carbon content of the biosphere and the HCO_3^- content of the oceans are included, the quantity of carbon in the "exchangeable reservoirs" is less than 0.1% of the carbon in the Earth's crust (see, for instance, Holland 1978, p. 276). The residence time of carbon in the exchangeable reservoirs, i.e. in the atmosphere, biosphere, and hydrosphere is approximately 10^5 years, and is determined largely by the residence time of HCO_3^- in the oceans (see, for instance, Holland 1978, chap. 5). The fundamental theorem for a system such as the atmosphere-biosphere-ocean-crust system is that the state of the atmosphere, the biosphere, and the oceans tends to adjust itself until the rate of CO_2 output into sediments is equal to its rate of input when averaged over a few residence times of CO_2 in the exchangeable reservoirs. From a geologic point of view this is an extremely short period of time; a knowledge of time lags between changes of the rate of CO_2 input and output are probably of importance only for understanding processes that take place on the time scale of glacial periods and for predicting the effects of human intervention in the chemistry of the atmosphere, the biosphere, and the oceans.

The major inputs and outputs of CO_2 are shown schematically in the rather simple representation of the geochemical cycle in figure 5.30. Solid lines in this diagram indicate the flow of solid constituents, dashed lines the flow of volatiles. The volatile balance of the system is controlled by three cycles and by a more or less unidirectional process. Cycle A involves the weathering of sedimentary rocks, their redeposition as sediments, and the conversion of these new sediments into sedimentary rocks. Cycle B involves the metamorphism and melting of sedimentary rocks, the weathering of the metamorphic and igneous products and their deposition as

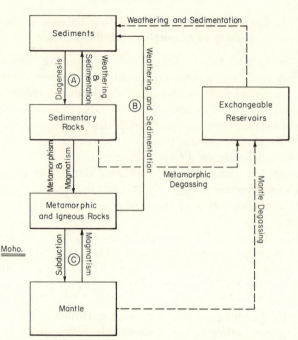

FIGURE 5.30. A simplified view of the geochemical cycle.

sediments, and the conversion of these sediments into new sedimentary rocks.

Cycle C, which can be regarded as part of cycle B, involves the subduction of crustal material into the mantle and the reappearance of this material as new igneous rocks. The more or less unidirectional process is the degassing of the mantle. This process is not completely unidirectional, because some degassed volatiles can be returned to the mantle during subduction.

All three cycles are in operation today. Old sedimentary rocks are the major source of new sediments; the weathering of metamorphic and igneous rocks probably contributes approximately 25% to the mass of new sediments (Holland 1978, chap. 4). The conversion of these volatile-poor rocks into average sediments requires the addition of volatiles; these are supplied by the degassing of sedimentary rocks that are undergoing metamorphism and melting and by the addition of juvenile volatiles from the mantle. The relative proportion of sedimentary, metamorphic, and igneous rocks undergoing weathering has probably changed with time (see section 3 of this chapter). It is hard to avoid the conclusion that among the rocks exposed to weathering, the proportion of igneous rocks was greater early in Earth history than it is today. It also seems likely that the fraction of sedimentary rocks destroyed by metamorphism has decreased since the Archean in favor of destruction by weathering. If this is the case, then the importance of cycles B and C has gradually decreased in relation to the importance of cycle A.

During most, if not all, of Earth history, CO_2 seems to have been the major weathering acid. Its neutralization has required the conversion of nearly all of the CaO and a part of the MgO in silicate rocks exposed to weathering into limestones and dolomites. Rocks containing sizable quantities of Mg- and Ca-silicates occur both on land and in the oceans. However, the rate of submarine weathering seems to be so much slower than the rate of subaerial weathering (see chapter 6) that most of the silicate Ca^{+2} and Mg^{+2} has probably always been released by subaerial weathering.

The rate of release of cations from silicate minerals is a complicated function of climatic, biologic, and topographic factors as well as of the CO_2 content of the atmosphere. High temperatures, heavy rainfall, and dense vegetation speed the rate of chemical weathering (see, for instance, Holland 1978, chap. 2). Topographic relief is of little importance except when relief is very low. Times of rapid sea-floor spreading are probably accompanied by a rise in sea level and flooding of the continents (Harrison 1980). Periods of maximum mantle degassing seem to coincide with periods during which the area of continental rock exposed to weathering is at a minimum. Since high topographic relief in continental areas does not, per se, lead to high rates of chemical weathering, the atmospheric CO_2 pressure during periods of rapid degassing is apt to rise until the rate of chemical weathering is sufficiently fast to balance the rate of CO_2 input into the atmosphere. Such increases in P_{CO_2} are probably accompanied by an increase in the Earth's mean annual temperature due to the operation of the greenhouse effect of CO_2. Our knowledge of the value of the coefficient dT/dP_{CO_2} is still uncertain. If the coefficient is large, one might expect that climates during periods of rapid degassing, and therefore of marine transgression, are warm, and that periods of marine regression tend to coincide with relatively cold climates. There is some support for the existence of such a relationship. Figure 5.31 shows a recent compilation by Ronov et al. (1980) of the percentage of the continents covered by seawater during the Phanerozoic Era. The late Cenozoic and the Permo-Triassic are the two periods of maximum emergence (see also figure 6.36). Figure 5.32 shows a generalized temperature and precipitation history of the Earth (Frakes 1979). The proposed decrease in mean global temperature since the end of the Cretaceous Era

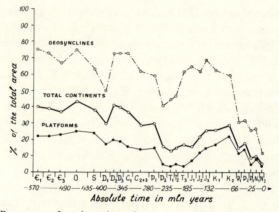

FIGURE 5.31. Percentage of total continental area covered by seas during the Phanerozoic Era (Ronov et al. 1980). (Reproduced by permission of Elsevier Scientific Publishing Company.)

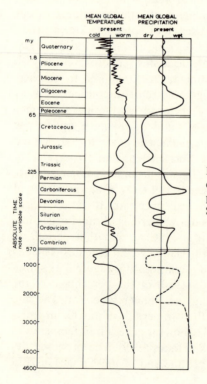

FIGURE 5.32. Generalized temperature and precipitation history of the Earth (Frakes 1979, p. 261). (Reproduced by permission of Elsevier Scientific Publishing Company.)

matches the emergence of the continents quite well. This match is confirmed by the $\delta^{18}O$ data in figure 9.49. However, the major previous glaciation took place during Permo-Carboniferous time, and seems to have occurred well before the period of continental emergence in the Triassic Era. The correlation between climate and degree of continental emergence is therefore hardly perfect; but there still appears to be some uncertainty concerning the history of sea level (see chapter 6) as well as of world climate during the Triassic, and the correlation may improve with time. Climate is known to be influenced by a great many factors other than the CO_2 content of the atmosphere, and it will be interesting to see just how much of a role the changes in atmospheric CO_2 have played in shaping the Earth's major, long-term climatic trends.

Land plants increase the rate of chemical weathering (see, for instance, Holland 1978, pp. 21–25), because they increase the CO_2 content of soil air, largely as a result of respiration and the microbial decay of recent plant matter in soils. This enhancement was minor or absent during most of geologic time, since vascular land plants did not develop until Silurian time and did not cover the majority of the exposed continents until the Carboniferous Era. The magnitude of the effect of the land plant cover is difficult to gauge. Studies of the rate of chemical weathering in northern Iceland, where the Skálfandafljót flows across a basaltic lava desert devoid of vegetation, have shown that the HCO_3^- concentration of river water reaches values as high as 25 ppm (0.4 meq/kg) (Cawley, Burruss, and Holland 1969; Holland 1978, pp. 103–106). Unless CO_2 is supplied by an unsuspected buried peat layer, the HCO_3^- concentration in this river reaches values about one-third as high as those in the more vegetated lower reaches of the river without benefit of plant decay in a soil horizon. The effect of plants on the rate of chemical weathering may therefore not be as great as suggested by the biological amplification of P_{CO_2} in soil air.

The release of Ca^{+2} and Mg^{+2} ions from silicates during chemical weathering must be followed by the precipitation of all or part of these cations as a constituent of carbonate minerals. Nearly all of the calcite and aragonite are precipitated in the oceans today as parts of marine organisms. The annual rate of precipitation of $CaCO_3$ is much more rapid today than the flux of Ca^{+2} into the oceans (see, for instance, Holland 1978, chap. 5); the Ca^{+2} budget of seawater is therefore balanced by the resolution of much of the precipitated $CaCO_3$ in the deeper parts of the oceans. Before the development of organisms with $CaCO_3$ shells, calcite and aragonite must have been precipitated inorganically and/or as a constituent of stromatolites. It is virtually certain that the surface layers of the oceans have been saturated or supersaturated with respect to calcite and aragonite since early Precambrian time. Unfortunately this, by itself, does not determine the partial pressure of CO_2 in the atmosphere. However, the requirement that $CaCO_3$ is stable with respect to calcium-silicates does place a lower limit on P_{CO_2}. The mineralogical aspects of this problem have been discussed elsewhere (Holland 1978, pp. 279–283). The thermodynamic data required to predict the CO_2 pressure at which $CaCO_3$ is no longer a stable phase in the oceans are not available, and the problem is complicated by the sluggishness of most reactions between seawater and silicate minerals at temperatures at and below 25°C; it is most likely, however, that the proportion of calcium removed from the oceans as a constituent of carbonates would decrease with decreasing CO_2 pressure. An increase in P_{CO_2} would have little effect on the removal of Ca^{+2} from the oceans, because nearly all of the river input of Ca^{+2} is already removed as a constituent of marine carbonates under present-day conditions.

Little of the river input of Mg^{+2} is removed as a constituent of carbonates today. It was shown above that this is apparently not due to a decrease in the rate of supply of CO_2 during the past 200 m.y., and it is suggested in the next chapter that ecological changes coupled with the reaction of seawater with MOR basalts are largely responsible for the decline in dolomite precipitation.

Current models for the evolution of the Sun require an increase in solar luminosity by 25% since the formation of the solar system (Newman and Rood 1977). Such an increase in the solar constant could have had profound effects on terrestrial climates, but there is no evidence from the geologic record for large changes in the Earth's mean global temperature. This conflict cannot be explained by the apparent inability of solar models to account for the low observed neutrino flux (Davis and Evans 1976). Even models that are forced to fit the present neutrino flux data require a similar increase in solar luminosity. Sagan and Mullen (1972) suggested that the solution must lie in a more efficient atmospheric greenhouse effect on Earth during the period when the solar luminosity was low. Their preferred greenhouse gas for the early earth was ammonia in a mixing ratio of 10^{-5} to 10^{-6}; this was thought to have been replaced by CO_2 at a later time (see also Hart 1978).

Owen, Cess, and Ramanathan (1979) have coupled the evolutionary model of Hart (1978) to a modification of the radiative-convective CO_2-climate model of Augustsson and Ramanathan (1977), and have calculated the CO_2 pressure required to maintain a mean global temperature that is consistent with the available geological data. The results are shown in table 5.19. The greenhouse model used in the computations is clearly oversimplified; the proposed course of the CO_2 pressure is therefore uncertain (see Kuhn and Kasting 1983). The figures suggest that CO_2 alone

TABLE 5.19.

The CO_2 pressure required to produce a mean global surface temperature T_s during earth history; the solar constant, S, is assumed to have increased linearly with time (Owen et al. 1979).

Time (10^9 yr BP)	S ($W\ m^{-2}$)	P_{CO_2} (bar)	T_s (K)
4.25		0.31	310
3.5		0.070	296
3.0	1133	0.033	293
2.5	1171	0.018	292
2.0	1209	0.0086	290
1.5	1247	0.0029	288
1.0	1284	0.00065	286
0.5	1322	0.00032	287
0	1360	0.00032	290

could have generated a greenhouse effect sufficient to account for the presence of liquid water on the surface of the Earth at least 3800 m.y. ago and the formation of evaporites approximately 3500 m.y. ago. This is an agreement with the conclusions reached by Henderson-Sellers, Benlow, and Meadows (1980). The results are intriguing but somewhat worrisome. They suggest that P_{CO_2} was about twice the present level 1.0 b.y. ago, and 100 times the present level 3.0 b.y. ago. It seems likely that such a drastic difference would have affected CO_2 consumption during weathering, especially since both periods predate the appearance of vascular plants. The data presented in this chapter do not show the expected difference. Perhaps the data for the Sheba greywackes can be discounted on the grounds that the sample is geographically too restricted; but Cameron and Garrels' (1980) data for the Canadian Shield are difficult to dismiss on these grounds. It could also be argued that temperatures during the Archean were barely above 0°C, that the required CO_2 pressures were therefore lower than those shown in table 5.19, and that the rates of chemical weathering were slower; however, the absence of evidence for extensive Archean glaciations and the presence of what appear to be the remains of 3.5 b.y.-old evaporites make this a rather unattractive hypothesis. The problem seems to be unresolved at present; its solution may involve a better understanding of the paleogeography of the Earth (Endal and Schatten 1982; see also Nesbitt and Young 1982) or of the greenhouse effect produced by atmospheric CO_2, or the demonstration that gases other than CO_2 contributed significantly to the greenhouse effect on the early Earth.

The whole matter of changes in P_{CO_2} with time is therefore still in a quite uncertain state. It is very likely that before the development of vascular plants P_{CO_2} was significantly higher than today; it is most likely that P_{CO_2} has been positively correlated with the intensity of CO_2 degassing, and it seems likely that a higher CO_2 content contributed to an enhance greenhouse effect on the early Earth. These qualitative statements must now be converted into believable figures for the course of P_{CO_2} during Earth history.

References

Adams, J.A.S., and Weaver, C. E. 1958. Thorium-to-uranium ratios as indictors of sedimentary processes: Example of concept of geochemical facies. *Bull. Am. Assoc. Petrol. Geologists* 42:387–430.

Allaart, J. H. 1976. The pre-3760 m.y. old supracrustal rocks of the Isua area, central West Greenland, and the associated occurrence of quartz banded ironstone. In *The Early History of the Earth*, edited by B. F. Windley, 177–189. New York: Wiley.

Allègre, C. J., and Othman, D. B. 1980. Nd-Sr isotopic relationship in granitoid rocks and continental crust development: a chemical approach to orogenesis. *Nature* 286:335–342.

Anhaeusser, C. A., Roering, C., Viljoen, M. J., and Viljoen, R. P. 1968. The Barberton Mountain Land: A model of the elements and evolution of an Archean fold belt. *Trans. Geol. Soc. South Afr.* 71 (Annexure): 225–254.

Anhaeusser, C. A., Mason, R., Viljoen, M. J., and Viljoen, R. P. 1969. A reappraisal of some aspects of Precambrian shield geology. *Bull. Geol. Soc. Amer.* 80:2175–2200.

Augustsson, T., and Ramanathan, V. 1977. A radiative-convective model study of the CO_2 climate problem. *J. Atmos. Sciences* 34:448–451.

Baadsgaard, H. 1976. Further U-Pb dates on zircons from early Precambrian rocks of the Godthaabsfjord area, West Greenland. *Earth Planet. Sci. Lett.* 33:261–267.

Boak, J. L., and Dymek, R. F. 1980. An occurrence of kyanite at Isua, West Greenland, and implications for early Archean geothermal gradients. *Abstracts with Programs* 12(7):389. Geol. Soc. Amer., Boulder, Colo.

———. 1982. Metamorphism of the ca. 3800 Ma supracrustal rocks at Isua, West Greenland: Implications for early Archean crustal evolution. *Earth Planet. Sci. Lett.* 59:155–176.

Bridgwater, D., Watson, J., and Windley, B. F. 1973. The Archean craton of the North Atlantic region. *Phil. Trans. R. Soc. Lond.* A273: 493–512.

Bridgwater, D., and McGregor, V. R. 1974. Field work on the very early Precambrian rocks of the Isua area, southern West Greenland. *Geological Survey of Greenland Report* 65:49–54.

Bridgwater, D., Collerson, K. D., and Myers, J. S. 1978. The development of the Archean gneiss complex of the North Atlantic region. In *Evolution of the Earth's Crust*, edited by D. H. Tarling, 19–69. New York: Academic Press.

Calvert, S. E. 1976. The mineralogy and geochemistry of near-shore sediments. In *Chemical Oceanography*, 2d ed., vol. 6, edited by J. P. Riley and R. Chester, chap. 33. New York: Academic Press.

Cameron, E. M., and Baumann, A. 1972. Carbonate sedimentation during the Archean. *Chem. Geol.* 10:17–30.

Cameron, E. M., and Jonasson, I. R. 1972. Mercury in Precambrian shales of the Canadian Shield. *Geochim. Cosmochim. Acta* 36:895–1005.

Cameron, E. M., and Garrels, R. M. 1980. Geochemical compositions of some Precambrian shales from the Canadian Shield. *Chem. Geol.* 28:181–197.

Cawley, J. L., Burruss, R. C., and Holland, H. D. 1969. Chemical weathering in central Iceland: An analog of pre-Silurian weathering. *Science* 165:391–392.

Chester, R., and Aston, S. R. 1976. The geochemistry of deep-sea sediments. In *Chemical Oceanography*, 2d ed., vol. 6, edited by J. P. Riley and R. Chester, chap. 34. New York: Academic Press.

Clarke, F. W. 1924. *The Data of Geochemistry*. 5th ed. U.S. Geol. Surv. Bull. 770.

Cooper, J. A., James, P. R., and Rutland, R.W.R. 1982. Isotopic dating and structural relationships of granitoids and greenstones in the East Pilbara, Western Australia. *Precamb. Res.* 18:199–236.

Daly, R. A. 1933. *Igneous Rocks and the Depths of the Earth*. New York: McGraw-Hill.

Davis, R., Jr., and Evans, J. C., Jr. 1976. Report on the Brookhaven solar neutrino experiment. *Bull. Amer. Phys. Soc.* 21:683.

DePaolo, D. J. 1980. Crustal growth and mantle evolution: Inferences from models of element transport and Nd and Sr isotopes. *Geochim. Cosmochim. Acta* 44:1185–1196.

Eade, K. E., and Fahrig, W. F. 1971. Geochemical evolutionary trends of continental plates: A preliminary study of the Canadian Shield. *Geol. Surv. Can. Bull.* 179.

Endal, A. S., and Schatten, K. H. 1982. The faint young Sun-climate paradox: Continental influences. *J. Geophys. Res.* 87:7295–7302.

Eriksson, K. A. 1977. Tidal deposits from the Archean Moodies Group, Barberton Mountain Land, South Africa. *Sediment. Geol.* 18:257–281.

Frakes, L. A. 1979. *Climates Throughout Geologic Time.* New York: Elsevier.

Garrels, R. M., and Mackenzie, F. T. 1971. *Evolution of Sedimentary Rocks.* New York: W. W. Norton.

Gill, R.C.O., Bridgwater, D., and Allaart, J. H. 1983. The geochemistry of the earliest known basic igneous rocks, Isua, West Greenland (in press).

Green, J., and Poldervaart, A. 1955. Some basaltic provinces. *Geochim. Cosmochim. Acta* 7:177–188.

Hamilton, P. J., O'Nions, R. K., Evensen, N. M., Bridgwater, D., and Allaart, J. H. 1978. Sm-Nd isotopic investigations of Isua supracrustals and implications for mantle evolution. *Nature* 272:41–43.

Hargraves, R. B. 1976. Precambrian geologic history. *Science* 193:363–371.

Harrison, C.G.A. 1980. Spreading rates and heat flow. *Geophys. Res. Letters* 7:1041–1044.

Hart, M. H. 1978. The evolution of the atmosphere of the Earth. *Icarus* 33:23–39.

Henderson-Sellers, A., Benlow, A., and Meadows, A. J. 1980. The early atmospheres of the terrestrial planets. *Quart. J. Roy. Astr. Soc.* 21:74–81.

Hill, T. P., Werner, M. A., and Horton, M. J. 1967. Chemical composition of sedimentary rocks in Colorado, Kansas, Montana, Nebraska, North Dakota, South Dakota, and Wyoming. U.S. Geol. Surv. Prof. Paper 561.

Holland, H. D. 1978. *The Chemistry of the Atmosphere and Oceans.* New York: Wiley.

Holser, W. T., and Kaplan, I. R. 1966. Isotope geochemistry of sedimentary sulfates. *Chem. Geol.* 1:93–135.

Hower, J., Eslinger, E. V., Hower, M. E., and Perry, E. A. 1976. Mechanism of burial metamorphism of argillaceous sediments: 1. Mineralogical and chemical evidence. *Bull. Geol. Soc. Amer.* 87:725–737.

Hunt, J. M. 1972. Distribution of carbon in crust of Earth. *Bull. Am. Assoc. Pet. Geol.* 56:2273–2277.

Knopf, A. 1916. The composition of the average igneous rock. *Jour. Geol.* 24:620–622.

Kuhn, W. R., and Kasting, J. F. 1983. Effects of increased CO_2 concentrations on surface temperature of the early Earth. *Nature* 301:53–55.

Lowe, D. R. 1980. Archean sedimentation. *Ann. Rev. Earth Planet. Sci.* 8:145–167.

Lowe, D. R., and Knauth, L. P. 1977. Sedimentology of the Onverwacht

Group (3.4 billion years), Transvaal, South Africa, and its bearing on the characteristics and evolution of the early Earth. *Jour. Geol.* 85:699–723.

Margolis, S. V., Kroopnick, P. M., and Showers, W. J. 1982. Paleoceanography: The history of the ocean's changing environment. In *Rubey Memorial Vol. 2.* Newark, N.J.: Prentice-Hall.

Moorbath, S., O'Nions, R. K., and Pankhurst, R. J. 1975. The evolution of early Precambrian crustal rocks at Isua, West Greenland: Geochemical and isotopic evidence. *Earth Planet. Sci. Lett.* 27:229–239.

Moore, C. B., Lewis, C. F., and Kvenvolden, K. A. 1974. Carbon and sulfur in the Swaziland Sequence. *Precamb. Res.* 1:49–54.

Nanz, R. J., Jr. 1953. Chemical composition of Precambrian slates with notes on the geochemical evolution of lutites. *Jour. Geol.* 61:51–64.

Nesbitt, H. W., and Young, G. M. 1982. Early Proterozoic climates and plate motions inferred from major element chemistry of lutites. *Nature* 299:715–717.

Newman, M. J., and Rood, R. T. 1977. Implications of solar evolution for the Earth's early atmosphere. *Science* 198:1035–1037.

Nockolds, S. R. 1954. Average chemical compositions of some igneous rocks. *Bull. Geol. Soc. Amer.* 65:1007–1032.

Nockolds, S. R., and Allen, R. 1953. The geochemistry of some igenous rock series. *Geochim. Cosmochim. Acta* 4:105–142.

O'Nions, R. F., Carter, S. R., Evensen, N. M., and Hamilton, P. J. 1979. Geochemical and cosmochemical application of Nd isotope analysis, *Ann. Rev. Earth Planet. Sci.* 7:1–38.

Owen, T., Cess, R. D., and Ramanathan, V. 1979. Enhanced CO_2 greenhouse to compensate for reduced solar luminosity on early Earth. *Nature* 277:640–642.

Perry, E., and Hower, J. 1970. Burial diagenesis in Gulf Coast pelitic sediments. *Clays and Clay Minerals* 18:165–177.

Pettijohn, F. J. 1975. *Sedimentary Rocks.* 3d ed. New York: Harper and Row.

Poldervaart, A. 1955. Chemistry of the Earth's Crust. In *Crust of the Earth*, edited by A. Poldervaart, 119–144. *Geol. Soc. Amer. Special Paper 62.*

Pretorius, D. A. 1975. The depositional environment of the Witwatersrand goldfields: A chronological review of speculations and observations. *Inf. Circ. Econ. Geol. Res. Unit Univ. Witwatersrand 95.*

Reimer, T. O. 1975. Untersuchungen über Abtragung, Sedimentation und Diagenese im frühen Präkambrium am Beispiel der Sheba-Formation (Südafrika). *Geol. Jahrb.* 17:3–108.

———. 1982. Sulfur isotopes and the derivation of detrital barytes of the

Archean Fig Tree Group (South Africa). In *Sedimentary Geology of Highly Metamorphosed Precambrian Complexes*, edited by A.V. Sidorenko, 63–74. Moscow: Nauka Publishing House.

Ronov, A. B. 1956. Chemical composition and conditions of formation of Paleozoic carbonate beds of the Russian Platform (with lithologic-geochemical maps). *Trudy. Geol. Inst. Akad. Nauk S.S.S.R.*, no. 4, 256–343 (in Russian).

Ronov, A. B., and Yaroshevsky, A. A. 1967. Chemical structure of the Earth's crust. *Geokhimiya*, 1285–1309.

Ronov, A. B., Migdisov, A. A., and Barskaya, N. V. 1969. Tectonic cycles and regularities in the development of sedimentary rocks and paleogeographic environments of sedimentation of the Russian Platform (an approach to a quantitative study). *Sedimentology* 13:179–212.

Ronov, A. B., and Migdisov, A. A. 1971. Geochemical history of the crystalline basement and the sedimentary cover of the Russian and North American Platforms. *Sedimentology* 16:137–185.

Ronov, A. B., and Yaroshevsky, A. A. 1976. A new model for the chemical structure of the Earth's crust. *Geochemistry International* 13 (6): 89–121.

Ronov, A. B., Khain, V. E., Balukovsky, A. N., and Seslavinsky, K. B. 1980. Quantitative analysis of Phanerozoic sedimentation. *Sediment. Geol.* 25:311–325.

Sagan, C., and Mullen, G. 1972. Earth and Mars: Evolution of atmospheres and surface temperatures. *Science* 177:52–56.

Stanley D. J., Editor. 1972. *The Mediterranean Sea: A Natural Sedimentation Laboratory*. Assisted by G. Kelling and Y. Weiler. Stroudsburg, Pa.: Dowden, Hutchinson and Ross.

Trask, P. D., and Patnode, H. W. 1942. *Source Beds of Petroleum*. Tulsa, Okla.: American Association of Petroleum Geologists.

Veizer, J. 1983. Geologic evolution of the Archean: Early Proterozoic Earth. In *Origin and Evolution of Earth's Earliest Biosphere: An Interdisciplinary Study*, edited by J. W. Schopf, chap. 10. Princeton, N.J.: Princeton University Press.

Viljoen, M. J., and Viljoen, R. P. 1969a. An introduction to the geology of the Barberton granite-greenstone terrain. *Geol. Soc. South Afr. Spec. Publ.* 2:9–28.

———. 1969b. The geological and geochemical significance of the upper formations of the Onverwacht Group. *Geol. Soc. South Afr. Spec. Publ.* 2:113–152.

Vinogradov, A. P., and Ronov, A. B. 1956a. Evolution of the chemical composition of clays of the Russian Platform. *Geochemistry*, no. 2, 123–139.

————. 1956b. Composition of the sedimentary rocks of the Russian Platform in relation to the history of its tectonic movements. *Geochemistry*, no. 6, 533–559.

Visser, D.J.L. 1956. The geology of the Barberton Area. *Geol. Soc. South Afr. Spec. Publ.*, no. 15.

Vitrac, A. M., Lancelot, J., Allègre, C. J., and Moorbath, S. 1977. U-Pb ages on single zircons from the early Precambrian rocks of West Greenland and the Minnesota River Valley. *Earth Planet. Sci. Lett.* 35:449–453.

Vogt, J.H.L. 1931. The average composition of the Earth's crust, with particular reference to the contents of phosphoric and titanic acids. *Norske Videnskaps-Akad. Oslo, Skrifter I. Mat. naturv. Kl.*, no. 7, 1–48.

Von Brunn, V., and Hobday, D. K. 1976. Early Precambrian tidal sedimentation in the Pongola Supergroup of South Africa. *Jour. Sed. Petrol.* 46:670–679.

Vos, R. G. 1975. An alluvial plain and lacustrine model for the Precambrian Witwatersrand deposits of South Africa. *Jour. Sed. Petrol.* 45:480–493.

Weaver, C. E. 1967. Potassium, illite, and the ocean. *Geochim. Consmochim. Acta* 31:2181–2196.

White, W. A. 1959. Chemical and spectrographic analyses of Illinois clay materials. *Illinois State Geol. Survey Circ. 282.*

Windley, B. F. 1977. *The Evolving Continents.* New York: John Wiley.

6

Carbonates, Clays, and
Exchange Reactions

1. Introduction

The previous chapter explored the balance between the acid supply by degassing and the acid demand for weathering during the past 3.8 b.y. The available data are incomplete, but they show very clearly that carbonation reactions were important in shaping the chemistry of the earliest known sedimentary rocks, and that their intensity has been roughly comparable during much of Earth history. Two observations made in the previous chapter are somewhat puzzling: the abnormally small proportion of dolomite in Mesozoic and Cenozoic carbonates, and the pattern of changes in the Na_2O/K_2O ratio of shales since the Archean. This chapter explores some possible explanations for these observations.

Until quite recently the conversion of igneous into sedimentary rocks was thought to involve only the titration of continental igneous rocks with acid volatiles, the deposition of the solid residues as sediments, and the accumulation of salts in the oceans (see, for instance, Garrels and Mackenzie 1971, pp. 247–249; Horn and Adams 1966). Although mass-balance calculations made in this fashion are adequate for many elements, they fail rather badly for others. The effects of submarine weathering and of seawater cycling through hot volcanic rocks in mid-ocean ridges seem to be capable of explaining at least some of these failures.

2. The Chemistry of Carbonate Sediments

The Phanerozoic record of carbonate sediments is very extensive. Paleozoic carbonates contain a good deal of dolomite; Mesozoic and Cenozoic carbonates contain progressively less dolomite. This trend has been documented thoroughly for the sediments of the Russian and North American platforms (see chapter 5) and seems to be a worldwide phenomenon; the proportion of dolomite in recent carbonate sediments the world over is extremely small compared to their proportion in Paleozoic carbonates (see, for instance, Holland 1978, chap. 5). Several interpretations of this change have been offered. One of these is based on the notion that most dolomites are formed by the replacement of limestones long after deposition; Mesozoic and Cenozoic limestones are thought to be too young to have been dolomitized extensively, and the increase in the proportion of dolomites with increasing age is ascribed to a progressive increase in the likelihood of replacement of limestone by dolomite.

This view has been challenged during the past two decades by the discovery of modern dolomite associated with evaporites (see, for instance, Holland 1978, chap. 5) and in zones where seawater is diluted by meteoric waters (Land, Salem, and Morrow 1975; Folk and Land 1975). Similar environments on a large scale have been identified in the geologic record (see, for instance, Friedman 1980a,b); it now seems likely that most dolomites were formed during or shortly after primary carbonate sedimentation and that most of the magnesium in dolomites was derived from concentrated or diluted seawater during the formation of these sediments (see, for instance, Zenger, Dunham, and Ethington 1980). If this view turns out to be correct, then the observed decrease in the dolomite/limestone ratio is at least in part a reflection of changes in the operation of the atmosphere-biosphere-ocean-crust system during the course of the Phanerozoic Era.

The pre-Phanerozoic record of carbonate sediments is rather incomplete. The Precambrian carbonates that have been preserved must be a small proportion of those that once existed; the $(MgO + CaO)_{sil.}$ deficit of Precambrian shales is impossible to explain in any other way (see chapter 5). It would be fortuitous if the dolomite/limestone ratio in preserved Precambrian carbonates were the same as that of all their contemporary carbonates, but the composition of the remaining carbonates and the composition of contemporaneous shales and clastic sediments are probably fair indicators of the evolution of the dolomite/limestone ratio during the Precambrian.

Limestones and dolomites have both been found in Archean terrains. Veizer (personal communication, 1981) has surveyed the available data, and has found that Archean limestone occurrences outnumber Archean dolomite occurrences. However, the total number of Archean carbonate occurrences is small, and little quantitative significance can be attached to the apparent preponderance of limestones in Archean carbonates. In the Proterozoic Era, dolomites are very well represented among carbonates, but limestones are by no means absent.

One of the oldest, most extensive, and best-described early Proterozoic carbonate sequence is that of the Transvaal Dolomite, an epeiric shelf carbonate sequence in South Africa (see figure 6.1). This carbonate sequence is more than 2200 m.y. old, and was deposited in supratidal, intertidal, and subtidal settings (Eriksson, McCarthy, and Truswell 1975; Eriksson and Truswell 1974; Visser and Grobler 1973; Eriksson 1972; Toens 1966). A generalized model of its depositional environment is

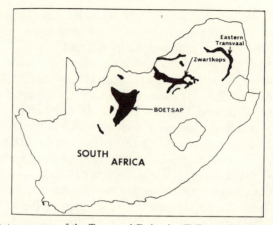

FIGURE 6.1. Outcrop areas of the Transvaal Dolomite (Eriksson, McCarthy, and Truswell 1975). (Reproduced by permission of the Society of Economic Paleontologists and Mineralogists.)

shown in figure 6.2. Deeper, subtidal zones contain large-scale elongate stromatolitic mounds with a delicate "crinkled" minor structure. The rocks in this zone are pure dolomite. The shallow subtidal zone is characterized by elongate domes that may contain delicate columnar stromatolites. Where wave base impinges on the sea floor to produce turbulent conditions, oolites, oncolites, ripple marks, and scour breccias form in the agitated zone. Shoreward, columnar stromatolites and some flat domes dominate the intertidal area. Material in the shallow subtidal, agitated, and intertidal zones is composed of limestones and dolomitic limestones. The setting of the Transvaal Dolomite does not appear to be unusual. Many ancient dolomites are apparently not supratidal, and a sabkha

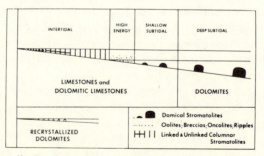

FIGURE 6.2. Generalized environmental model for the deposition of the Transvaal Dolomite (Eriksson, McCarthy, and Truswell 1975). (Reproduced by permission of the Society of Economic Paleontologists and Mineralogists.)

model does not seem to be appropriate to explain the intense dolomitization of such carbonates.

Stromatolites are very widespread in Proterozoic sediments. They have been recorded in the most ancient Archean sedimentary rocks of Australia (Dunlop et al. 1978; Walter, Buick, and Dunlop 1980; Lowe 1980) and of Africa (Muir and Grant 1976; Orpen and Wilson 1981), but they reached their most impressive development during the Proterozoic Era (see, for instance, Schopf 1975, 1977; Schopf et al. 1977; Schopf and Prasad 1978; Awramik 1982). During the Phanerozoic Era the abundance of stromatolites declined; their decline was probably a consequence of the evolution and diversification of grazing animals that feed on surface algal mats, and of burrowing animals that destroy sedimentary laminations (Garrett 1970), as well as in response to substrate competition and changing sedimentologic conditions caused by the evolution of skeletal metazoans (Pratt 1982).

The role of algae in the formation of carbonate sediments is still not entirely clear. Some algae, notably the coralline red algae, secrete $CaCO_3$ within their cell walls. Photosynthesis by algae may also produce environmental changes that are sufficient to initiate inorganic precipitation. A third important contribution is provided by the blue-green algae, which readily trap fine sediment within the framework of their mucilaginous thalli; this process seems to be the most important mode of origin for algal structures such as those of figure 6.3 (Donaldson 1963).

An additional active role for algae has been proposed by Gebelein and Hoffman (1971) (see also Davies, Ferguson, and Bubela 1975; C. B. Jones 1981) for the formation of the many ancient carbonate rocks that are composed of dolomite interlaminated with calcite on a centimeter and millimeter scale. Most of these interlaminated carbonates are stromatolitic or cryptalgal. The dolomite laminae tend to be uniform in thickness, to have bituminous residues, and to lack epiclastic textures. Calcite laminae are variable in thickness, lack bituminous residues, and have an epiclastic texture. The laminated internal structure of analogous recent stromatolites shows an alternation of algal-rich layers and sediment-rich layers. The characteristics of algal-rich and sediment-rich laminae are identical to those of the dolomite and calcite laminae, respectively, in ancient sediments. Dolomite laminae in ancient interlaminated sediments may therefore have been derived from algal-rich laminae, and calcite laminae from sediment-rich layers. Laboratory experiments on modern cyanobacteria tend to support this interpretation. The Mg/Ca ratio in the sheath material of *Schizotrix calciola*, a major stromatolite-forming cyanobacterium, is three to four times greater than that in surrounding seawater. The release of excess magnesium after the death of the algae may

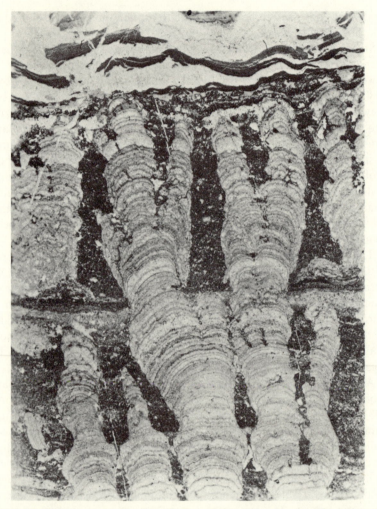

FIGURE 6.3. Digitate stromatolites from the Denault Formation, Newfoundland; plane polarized light; 4.5x (Donaldson 1963). (Reproduced by permission of the Geological Survey of Canada.)

be responsible for the precipitation of dolomite in laminae previously occupied by such algae. This process might produce a highly selective distribution of dolomite controlled by the original distribution of organic layers enriched in magnesium.

Most of the Proterozoic carbonate sequences contain significant quantities of chert. In some sedimentary sequences these host superbly preserved microfossils (see, for instance, Barghoorn and Tyler 1965; Schopf

1968). Such cherts have yielded much of our present store of knowledge regarding the nature of life in the Precambrian. Much of the chert must have been deposited contemporaneously or penecontemporaneously with sedimentation. There seem to be few quantitative studies of the proportion of chert in Precambrian carbonate sections, but the available data suggest that chert frequently accounts for 5 to 10% of carbonate sections, and that it is usually distributed in a rather heterogeneous manner (see, for instance, Jolliffe 1966; Toens 1966; Eriksson 1972). These observations are consistent with a concentration of dissolved silica in the Precambrian oceans well in excess of its concentration in the oceans today (see, for instance, Maynard 1976) and with the precipitation of chert in mildly evaporitic carbonate settings.

The proportion of dolomite in Precambrian carbonate sequences is somewhat greater than that in Paleozoic carbonates and very much greater than that of Mesozoic and Cenozoic carbonates. This is shown quite well by the data in figure 5.22, in tables 6.1 and 5.15, and in the summary table 6.2. The latter table lists the number of moles of MgO, CaO, and CO_2 per kg in the several groups of carbonates. These cannot

TABLE 6.1.

Composition of Archean and Proterozoic carbonates from the Russian Platform and Canadian Shield (Ronov and Migdisov 1970) (in percentages).

	Russian Platform		Canadian Shield; Archean and Proterozoic
	Archean	Proterozoic (Pt 1-2)	
No. of Analyses	7	25	11
SiO_2	13.76	15.62	18.27
TiO_2	0.02	0.12	0.18
Al_2O_3	2.16	2.14	3.00
Fe_2O_3	1.99	1.29	0.29
FeO	2.73	1.73	1.62
MnO	0.504	0.051	0.04
MgO	15.65	13.61	10.75
CaO	28.51	30.14	30.73
K_2O	0.28	0.28	0.67
Na_2O	0.31	0.11	0.32
P_2O_5	0.026	0.078	0.10
C_{org}	—	0.01	0.47
CO_2	31.36	30.46	32.17
SO_3 (total)	0.06	0.56	0.57
H_2O	3.61	1.20	0.82
Total	100.97	98.30	100.00

TABLE 6.2.

Dolomite and limestone in carbonates of the Russian and North American platforms; data from Ronov and Migdisov (1970).

	$\dfrac{MgO}{(mol/kg)}$	$\dfrac{CaO}{mol/kg}$	$\dfrac{CO_2}{mol/kg}$	$\left(\dfrac{MgO}{MgO + CaO}\right)\dfrac{mol}{mol}$	$\left(\dfrac{MgO^*}{MgO^* + CaO}\right)\dfrac{mol}{mol}$
Russian Platform					
Archean	3.88	5.09	7.13	0.43	0.29
Proterozoic (Pt 1-2)	3.37	5.38	6.92	0.39	0.22
Late Proterozoic	1.62	4.64	5.28	0.26	0.12
Paleozoic	2.22	7.06	8.35	0.24	0.17
Mesozoic and Cenozoic	0.25	7.36	7.17	0.03	0
Canadian Shield					
and North American Platform					
Archean and Proterozoic	2.67	5.49	7.31	0.33	0.25
Paleozoic	2.12	7.57	9.58	0.22	0.21
Mesozoic and Cenozoic	0.71	7.93	8.30	0.08	0.04

be converted directly into dolomite/limestone ratios; the sum of the number of moles of (MgO + CaO)/kg is always greater than the number of moles of CO_2/kg in these carbonates, because the carbonates are impure and some Mg and Ca is present in silicate minerals. A strong maximum value for the fraction of magnesium carbonate in the carbonate fraction of these average samples is almost certainly given by the ratio (MgO/MgO + CaO). A reasonable minimum value can be obtained if one assumes that all of the CaO in these sedimentary rocks is present as a constituent of carbonates, and that MgO*, the number of moles of MgO present in carbonate phases, is given by the difference,

$$MgO^*(mol/kg) = CO_2(mol/kg) - CaO(mol/kg). \qquad (6.1)$$

The fraction of magnesium in carbonate phases is then given by the expression

$$\frac{MgO^*}{MgO^* + CaO} = \frac{MgO^*}{CO_2}. \qquad (6.2)$$

It is likely that the actual value of the Mg/Ca ratio in Precambrian carbonates is closer to the minimum than to the maximum values in table 6.2. The average value of the ratio $MgO^*/(MgO^* + CaO)$ of Precambrian carbonates is approximately 0.22. The mole ratio of $Mg^{+2}/(Ca^{+2} + Mg^{+2})$ in average modern river water is approximately 0.30. This implies that if all of the Mg^{+2} and Ca^{+2} in modern rivers were precipitated as constituents of dolomite and limestone, the dolomite/limestone ratio would be somewhat greater than the dolomite/limestone ratio in Precambrian carbonates.

In average igneous rocks the ratio MgO/CaO is approximately unity. This is well illustrated by figures 6.4 and 6.5, in which the MgO and CaO content of the igneous rocks listed in table 5.6 are plotted against their iron content. The composition of very magnesian rocks such as komatiites has not been included; the quantity of such rocks on a continental scale is probably sufficiently small, so that their deletion does not affect the following argument.

The observation that the Mg/Ca ratio in Precambrian carbonates is considerably less than unity indicates either that the release of magnesium during weathering in the Precambrian Era was unequal to that of calcium, or that the Mg/Ca ratio of carbonates from this era differs considerably from the release ratio of these elements; this could have been true if Mg was sequestered and/or Ca was added prior to carbonate sedimentation. The removal of MgO from igneous rocks during weathering has apparently always been less complete than the removal of calcium. The composition of the Archean and Proterozoic shales of the Canadian Shield in

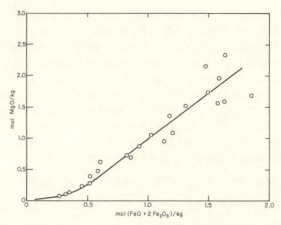

FIGURE 6.4. The relationship between the number of moles of MgO and the number of moles of $(FeO + 2Fe_2O_3)$ in igneous rocks; data from table 5.6.

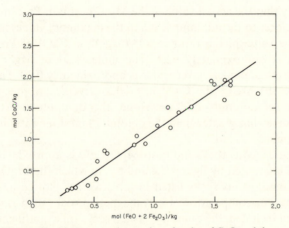

FIGURE 6.5. The relationship between the number of moles of CaO and the number of moles of $(FeO + 2Fe_2O_3)$ in igneous rocks; data from table 5.6.

table 5.9 illustrates this point quite well. The shales are quite poor in carbonates. The value of $(MgO)_{sil.}$ can therefore be calculated with some confidence. The moles of CaO/kg in the composites is greater than the number of moles of CO_2/kg. It seems likely that most of the CO_2 is present as a constituent of $CaCO_3$, that the remaining CaO is present in one or several silicate phases, and that all or nearly all of the MgO is a constituent of one or more silicate phases. If the CO_2 in these composites is present partly or entirely as a constituent of dolomite, the calculated values of $(MgO)_{sil.}$ are too high, but only marginally so, because the quantity of CO_2 in most of these shales is much smaller than that of MgO. The calculated values of $(MgO)_{sil.}$ were converted to values of $(MgO)_{sil.\,corr.}$ by the technique described in chapter 5:

$$(MgO)_{sil.\,corr.} \equiv (MgO)_{sil.} \times \frac{15.6}{wt\,\%\,Al_2O_3}. \tag{6.3}$$

The value of $(FeO + 2Fe_2O_3)$ for the composites was computed from the data in table 5.9, and $(FeO + 2Fe_2O_3)_{corr.}$ was calculated as before by normalizing to an Al_2O_3 content of 15.6 wt %,

$$(FeO + 2Fe_2O_3)_{corr.} \equiv (FeO + 2Fe_2O_3) \times \frac{15.6}{wt\,\%\,Al_2O_3}. \tag{6.4}$$

The values of $(MgO)_{orig.}$, the MgO content of igneous rocks that were presumably the parent rocks of these shales, were then obtained from figure 6.4. Figure 6.6 is a plot of the value of $(MgO)_{sil.\,corr.}$ for the shale composites versus $(MgO)_{orig.}$. The results are somewhat surprising. All but three of the composite shale analyses plot close to the line along which $(MgO)_{sil.\,corr.}$ is equal to $(MgO)_{orig.}$. The three composites whose composition does not plot close to this line have values of $(MgO)_{sil.\,corr.}$ that are within the range of the other composites, and values of $(MgO)_{orig.}$ that are considerably greater. It seems likely that iron has been added to these shales from an axtraneous source, and that the calculated values of

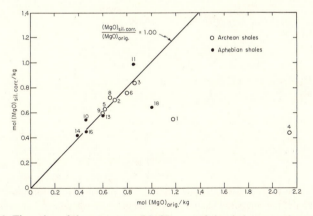

FIGURE 6.6. The value of the parameter $(MgO)_{sil.\,corr.}$ of the shale composites of table 5.9 plotted against their $(MgO)_{orig.}$ content; see text for a description of the method by which these functions were calculated. The numbers are those of the several composites in table 5.9.

$(MgO)_{orig.}$ are too high for this reason. If this interpretation is correct, then the data in figure 6.6 as a whole indicate that the calculation scheme is valid, and that very little MgO was lost during the conversion of igneous rocks into these Archean and Proterozoic shales.

If shales such as these were representative of Archean and Proterozoic shales as a whole, the abundance of dolomite in carbonates from these eons demands a supply of magnesium other than from the weathering of igneous rocks. At present the only other known supply of magnesium is the submarine alteration of basalts under rather special conditions. On the whole, the alteration of oceanic basalts seems to be a sink rather than a source of magnesium (see below), and it is unlikely that this process supplied the magnesium required for the formation of Archean and Proterozoic dolomites. It seems more likely that the Canadian shales of table 5.9 are products of abnormally mild weathering. This interpretation is supported by the data in figure 6.7 for the Precambrian lutite composite due to Nanz (1953) (see table 4.8 in Holland 1978) and for the late Precambrian (Sn) shales of the Russian Platform.

The pattern of points in figure 6.7 is quite different from that in figure 6.6. There is no alignment of data along the line of slope 1.0. Most of the composite shales fall well below this line, and there is a conspicuous clustering of the analyses for average Cambrian, Carboniferous, Jurassic, and Cretaceous shales. The Devonian and Silurian shales plot close to

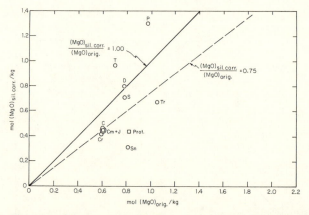

FIGURE 6.7. The value of the parameter $(MgO)_{sil.\,corr.}$ of shales of the Russian Platform in table 5.12 and of the composite analyses of Precambrian lutites (table 4.8 in Holland 1978) plotted against their $(MgO)_{orig.}$ content; see text for the method by which these functions were calculated. The symbols are those in table 5.12.

the line of slope 1.0; the analyses of Triassic and Permian shales plot well above this line.

Some of the scatter in figure 6.7 is due to variations in the carbonate content of the analyzed shale composites. The increase in $(MgO)_{sil. corr.}$ with increasing CO_2 content in Phanerozoic shales was pointed out in the previous chapter; the Permian shales of the Russian Platform are particularly carbonate-rich, and their position in figure 6.7 is probably related to this. On the other hand, the Silurian shales, which fall slightly below the line of slope 1.0, are almost as carbonate-rich as the Permian shales. Clearly, differences in the carbonate content of these shales are not the only reason for the scatter of data in figure 6.7. Differences in provenance and in the intensity of weathering of the parent rocks must also be important.

A precise measure of the worldwide release of magnesium to carbonates during weathering demands the availability of the results of worldwide sampling. In the absence of such sampling, one can only dash in a very rough, mean line through the data in figure 6.7; the line of slope 0.75 in this figure is heavily biased by the cluster of points close to $(MgO)_{orig.} = 0.60$ mol/kg, $(MgO)_{sil. corr.} = 0.45$ mol/kg. The virtue of this line is that it is consistent with the composition of Proterozoic and Paleozoic carbonates. Since $(MgO)_{orig.} \approx (CaO)_{orig.}$ in parent igneous rocks, and since nearly all CaO in igneous rocks is released during weathering, data points falling along a line of slope 0.75 in figure 6.7 imply that the ratio $[Mg^{+2}/(Mg^{+2} + Ca^{+2})]$ in the cations that are released during weathering is

$$\frac{Mg^{+2}}{Mg^{+2} + Ca^{+2}} \approx \frac{0.25}{0.25 + 1.00} = 0.20.$$

This value is close to the ratio $[MgO*/(MgO* + CaO)]$ in pre-Mesozoic carbonates. One can therefore entertain the notion that much of the magnesium and calcium released during weathering was deposited as constituents of carbonates during all but the last few hundred million years of Earth history. It is not possible to make a quantitative estimate of the difference between the supply of Mg^{+2} and Ca^{+2} by weathering and its use in the deposition of limestones and dolomites during the Precambrian and Paleozoic eras because neither the supply nor the use terms are well known; but it seems likely that the Mg/Ca ratio in pre-Mesozoic carbonates differed by no more than a factor of two from the Mg/Ca ratio in the cations that were released during weathering. This is certainly not true of Mesozoic and Cenozoic carbonates, and the difference stands in need of explanation.

3. The Magnesium Metabolism of the Oceans

The minor role that the deposition of magnesium as a constituent of carbonates plays in the output of magnesium from the oceans today has been appreciated for some time. During the 1960s it was generally believed (see, for instance, Mackenzie and Garrels 1966) that the formation of authigenic clays in marine sediments was the major sink of Mg^{+2} in the oceans. This belief was severely shaken by Drever's (1974) lack of success in finding the authigenic magnesian clays that should have been produced by this removal mechanism. Since then it has been discovered that Mg^{+2} is removed very efficiently from seawater by reaction with basalt at elevated temperatures and in many instances at low temperatures as well. It now seems likely that much of the present-day river input of Mg^{+2} is removed from the oceans by the reaction of seawater with oceanic basalts.

The laboratory evidence for the efficacy of basalt-seawater interaction as a mechanism for removing Mg^{+2} from the oceans was reviewed by Holland (1978, pp. 190–200). Since then the analyses of water from four warm springs on the crest of the Galapagos spreading ridge (Edmond et al. 1979) and from hot springs at the crest of the East Pacific Rise at latitude 21°N (Edmond, Von Damm, and McDuff 1981) have confirmed the inference that Mg^{+2} is removed quantitatively from seawater during reaction with mid-ocean ridge basalts at elevated temperatures. However, there is still considerable uncertainty regarding the fraction of the river input of Mg^{+2} that is removed from the oceans by this process. Livingstone (1963) proposed that the average Mg^{+2} content of river water is 4.1 ppm. Meybeck (1979) has suggested a mean Mg^{+2} content of 3.3 ppm corrected for anthropogenic inputs. The best current estimate for the total annual flux of river water to the oceans is probably 4.6×10^{19} cc/yr (for a review, see Holland 1978, chap. 3).

The nonanthropogenic input of Mg^{+2} to the oceans is therefore approximately

$$\frac{3.3 \times 10^{-6} \dfrac{\text{gm } Mg^{+2}}{\text{gm river water}} \times 4.6 \times 10^{19} \dfrac{\text{gm river water}}{\text{yr}}}{24.33 \dfrac{\text{gm } Mg^{+2}}{\text{mol}}}$$

$$= 0.6 \times 10^{13} \frac{\text{mol}}{\text{yr}}.$$

If the reaction of seawater with basalt at elevated temperatures is the major mechanism by which the river flux of Mg^{+2} is removed from the oceans, then the flux of seawater through hot basalt must be

$$\left(\frac{dV_0}{dt}\right)_{\text{calc.}} \cong \frac{0.6 \times 10^{13} \text{ mol/yr}}{53 \times 10^{-3} \text{ mol Mg}^{+2}/\text{kg seawater}}$$

$$\cong 1.1 \times 10^{14} \text{ kg seawater/year.}$$

The uncertainty in this number is determined mainly by the present uncertainty of ca. 30% in the river flux of Mg^{+2}.

The uncertainty in the actual quantity of seawater that reacts annually with hot basalt is considerably larger. The total quantity of heat transported from mid-ocean ridges by the circulation of seawater through hot basalt has been estimated to be $(4.9 \pm 1.2) \times 10^{19}$ cal/yr (Sleep and Wolery 1978; Edmond et al. 1979; Edmond, Von Damm, and McDuff 1981). This figure is based on the magnitude of the observed heat flow anomalies in the vicinity of the spreading axis on MORs and on the relationship between the ^{3}He content of hot spring waters at MORs, and the ^{3}He budget of the atmosphere (see, for instance, Lupton et al. 1980; Craig and Lupton 1981). If all of this heat is released at or close to the spreading axis of MORs by the flow of hydrothermal solutions that exist through the ocean floor at temperatures of ca. 300°C, then the flux of seawater through MORs is approximately

$$\frac{(4.9 \pm 1.2) \times 10^{19} \text{ cal/yr}}{3.3 \times 10^5 \text{ cal/kg}} \approx (1.5 \pm 0.4) \times 10^{14} \text{ kg/yr.}$$

All of the present river flux of Mg^{+2} to the oceans could therefore be removed by the reaction of seawater with hot basalt. It has, however, become quite clear that much of the heat extracted from MORs by circulating seawater does not emerge with high temperature solutions at spreading axes. The quantity of heat released at lower temperatures at some distance from the spreading axes is not well known but is probably between 65 and 85% of the total heat that is extracted hydrothermally from MORs (Hart and Staudigel 1982). Mg^{+2} is stripped rapidly out of seawater by reaction with basalt at temperatures above 150°C and is apparently removed at a significant rate at temperatures as low as 70°C. The volume of heated seawater required to remove a given quantity of heat from MORs increases as the exit temperature of the heated seawater decreases. The total quantity of Mg^{+2} removed from the oceans by seawater cycling through MORs therefore depends on the distribution of maximum and exit temperatures in seawater cycled through MORs. It can be shown that the removal rate of Mg^{+2} could well equal the river input of this element, but the actual removal rate is still somewhat uncertain.

There are independent indications that not all of the river flux of Mg^{+2} is removed from the oceans at or close to MORs. The reaction of seawater with basalt far from MORs at temperatures close to 0°C also seems to be removing Mg^{+2} from the oceans. In the interstitial waters of marine sediments the concentration of Ca^{+2} typically increases with depth, while the concentration of Mg^{+2} typically decreases with depth. Two examples of such profiles are shown in figures 6.8a and 6.8b. The data are for interstitial waters from DSDP Holes 417 and 418 drilled into 108 m.y.-old oceanic basement in the area shown in figure 6.9. The observed calcium and magnesium gradients are probably produced by the alteration of basalts underlying the sediments in this area and by the transport of the two ions through the sediment column above the basalts. This interpretation is supported by the isotopic composition of oxygen in the interstitial waters (Lawrence, Gieskes, and Broecker 1975; McDuff and Gieskes 1976).

Rather curiously, the altered basalts from DSDP Drill Holes 417 and 418 are depleted in both calcium and magnesium (Donnelly, Thompson, and Salisbury 1980). Calcium is very much more depleted than magnesium. These data, together with the interstitial water profiles, suggest that calcium and magnesium were both removed from the basalts in the past, and that small amounts of magnesium are now being added to the basalts at the expense of the remaining calcium. Clearly, the flux of magnesium from and into basalts depends critically on the conditions of the alteration process (Thompson and Humphris 1977). Until we know more about the alteration of basalts below the sea floor, little can be said with assurance

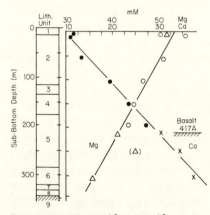

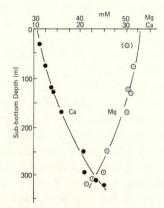

FIGURE 6.8a. The Mg^{+2} and Ca^{+2} content of interstitial water from DSDP Holes 417A and 417D (Gieskes and Reese 1980). (Reproduced by permission of the authors.)

FIGURE 6.8b. The Mg^{+2} and Ca^{+2} content of interstitial water from DSDP Hole 418 (Gieskes and Reese 1980). (Reproduced by permission of the authors.)

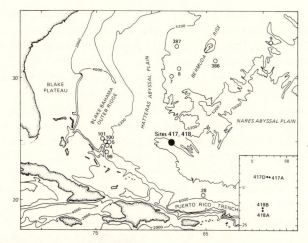

FIGURE 6.9. The location of DSDP Holes 417 and 418 (Scheidegger and Stakes 1980). (Reproduced by permission of the authors.)

about the overall movement of magnesium between basalt and seawater at low temperatures, but it will be argued below that the effects of reactions between basalt and seawater at temperatures above 70°C are quantitatively more important for the magnesium metabolism of the oceans (see also Gieskes and Lawrence 1981).

At present the mass balance equation for magnesium in the oceans can be written in the form,

$$\text{R.W.}m_{Mg^{+2}} \frac{dV_{R.W.}}{dt} \cong {}_0m_{Mg^{+2}} \cdot \frac{dV_0}{dt} + \Phi_{Dol} + \Phi_{BLT}, \qquad (6.5)$$

where

$\text{R.W.}m_{Mg^{+2}}$ = concentration of Mg^{+2} in average river water (mol/kg),

$\dfrac{dV_{R.W.}}{dt}$ = the annual flux of river water in kg/year,

${}_0m_{Mg^{+2}}$ = the concentration of Mg^{+2} in seawater (mol/kg),

$\dfrac{dV_0}{dt}$ = the annual flux of seawater through hot basalts (kg/yr),

Φ_{Dol} = the rate of removal of Mg^{+2} as a constituent of dolomite (mol/yr),

Φ_{BLT} = the rate of transfer of Mg^{+2} from seawater to basalt due to reactions at low temperatures.

Equation 6.5 is clearly incomplete. Cation exchange reactions on clays, the formation of authigenic clays, and the deposition of magnesian calcites have been neglected; it has been argued previously (Holland 1978, chap. 5) that these processes only dispose of small fractions of the river input today, and nothing has occurred in the meantime to require a revision of this assessment. Φ_{Dol} is, of course, also small today, but the term has been retained because it was important in the past.

It was suggested above that the formation of dolomite was the dominant process of magnesium removal in pre-Mesozoic time, and it was shown in chapter 5 that the return of Mg^{+2} to shales via the formation of magnesian clays in the oceans cannot explain the small amount of dolomite in Mesozoic and Cenozoic carbonates. The shift in the mechanism of Mg^{+2} output therefore seems to have involved a decrease in the rate of dolomite formation at the expense of an increase in Mg^{+2} loss to submarine basalts. Since this loss is accompanied by a gain in Ca^{+2} (see table 6.3), the total rate of carbonate deposition in the oceans need not have been affected by the increased loss of Mg^{+2} to submarine basalts.

TABLE 6.3.

Comparison of composition of seawater with composition of high-temperature vent solutions at 21°N, and estimated composition of the 350°C hydrothermal end member in the Galapagos warm spring water (Edmond et al. 1982).

	Galapagos	21°N	Seawater
Li (μmol kg^{-1})	1142–689	820	28
K (mmol kg^{-1})	18.8	25.0	10.1
Rb (μmol kg^{-1})	20.3–13.4	26.0	1.32
Mg (mmol kg^{-1})	0	0	52.7
Ca (mmol kg^{-1})	40.2–24.6	21.5	10.3
Sr (μmol kg^{-1})	87	90	87
Ba (μmol kg^{-1})	42.6–17.2	95–35	0.145
Mn (μmol kg^{-1})	1140–360	610	0.002
Fe (μmol kg^{-1})	+	1800	—
Si (mmol kg^{-1})	21.9	21.5	0.160
SO$_4$ (mmol kg^{-1})	0	0	28.6
H$_2$S (mmol kg^{-1})	+	6.5	0

Note: The ranges in the Galapagos results derive from the different trends for composition versus heat observed between individual vent fields. +, Non-conservative to subsurface mixing; −, seawater concentration not accurately known.

The relative importance of the three terms on the right side of equation 6.5 depends on a variety of factors. Only those that control the first term are obvious. Since the removal of Mg^{+2} from seawater that reacts with basalts at temperatures in excess of ca. 150°C is nearly complete, the rate of Mg^{+2} removal by this process is simply equal to the product of the Mg^{+2} content of seawater and the rate of seawater cycling through hot basalts, dV_0/dt. The greater importance of Mg^{+2} removal by this mechanism today than in pre-Mesozoic time is presumably due either to a greater Mg^{+2} concentration in seawater today, or to a more rapid rate of cycling of seawater through mid-ocean ridges, or to both. The next section will explore the evidence bearing on the rate at which seawater has cycled through mid-ocean ridges in the past.

4. The Cycling of Seawater through Mid-Ocean Ridges, Past and Present

EVIDENCE FROM THE LITHIUM CONTENT OF SEDIMENTARY ROCKS

Several physical, chemical, and isotopic properties of marine sediments and sedimentary rocks are potential indicators of the rate of cycling of seawater through mid-ocean ridges. Among these indicators, the lithium content of sedimentary rocks is one of the most interesting and potentially most useful. In his paper on chemical fractionation in sedimentary environments Wedepohl (1968) pointed out that the concentration of a number of elements in sedimentary rocks is significantly greater than their concentration in igneous rocks. His data are shown in figures 6.10 and 6.11. The excess of most of these elements in sedimentary rocks is explained quite readily by degassing of the interior of the Earth. This is certainly true for the excess of carbon, nitrogen, sulfur, selenium, chlorine, bromine, and iodine. It is probably also true for arsenic, antimony, and mercury. Whether it is true for molybdenum, indium, and tin is debatable. The small but real excess of lithium in sedimentary rocks is almost certainly not due to this process. At the time of writing, Wedepohl (1968) ventured that the behavior of this element "cannot yet be explained." There is now convincing evidence that the lithium excess in sedimentary rocks is due to the cycling of seawater through mid-ocean ridges (Holland 1978, p. 200).

Wedepohl's (1968) figures for the lithium content of magmatic rocks of the upper continental crust (30 ppm), for greywackes (51 ppm), and for shales (66 ppm) are similar to those in earlier and in more recent compilations. Table 6.4 summarizes a good deal of the available data for the

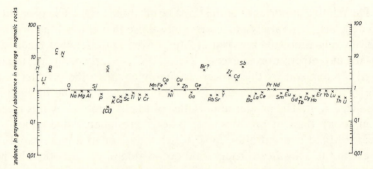

FIGURE 6.10. The ratio of the abundance of elements in greywackes to their abundance in average magmatic rocks (Wedepohl 1968). (Reproduced by permission of Pergamon Press, Ltd.; copyright by Pergamon Press, Ltd.)

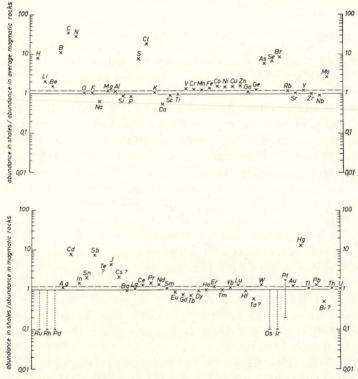

FIGURE 6.11. The ratio of the abundance of elements in shales to their abundance in average magmatic rocks (Wedepohl 1968). (Reproduced by permission of Pergamon Press, Ltd.; copyright by Pergamon Press, Ltd.)

TABLE 6.4.

Lithium content of igneous rocks (Styrt 1977).

Rock Type	Russian Samples		Non-Russian Samples		All Samples		
	No. of samples	Mean Li content (ppm)	No. of samples	Mean Li content (ppm)	No. of samples	Mean Li content (ppm)	$3\sigma/\sqrt{n}$
Granitic rocks	281	41	149	68	430	50	4.8
Intermediate igneous rocks	439	30	92	22	531	28	2.1
Basaltic rocks	103	29	163	13	266	19	3.3

lithium content of igneous rocks (Styrt 1977). The analyses of granites, rhyolites, granophyres, and igneous rocks of similar composition have been grouped as granitic rocks; diorites, andesites, trachytes, granodiorites, dacites, syenites, and monzonites are grouped as intermediate rocks; basalts, dolerites, and gabbros are grouped as basaltic rocks. The data in each of the three groups have been divided into Russian and non-Russian subsets. The means of the lithium content in the two subsets are distinctly different. This is particularly true for the granitic rocks; this difference is probably due in part to differences in rock classification and in part to the relative abundance of high-lithium granites in the two sample sets.

The analytical data for each of the three rock types were plotted on probability paper, the concentration of lithium at the 84% and 16% cumulative frequencies was determined, and the standard deviation, σ, was calculated by the usual expression,

$$\sigma = \frac{f_{84} - f_{16}}{2}. \tag{6.6}$$

The last column in table 6.4 shows the values of three times the standard error for each of the three sets. The precise meaning of this function is not entirely clear; many of the analyses used in the computations are themselves averages, and nothing is known about the distribution of individual concentrations in these populations. Nevertheless, it is likely that the best value of the lithium content of average igneous rocks is close to 30 ppm; the uncertainty in this figure is probably ca. 5 ppm.

Table 6.5 is a compilation of data for the lithium content of sedimentary rocks. The data are again divided into subsets, and the standard error

TABLE 6.5.

Lithium content of sedimentary rocks (Styrt 1977).

Rock Type	Russian Samples		Non-Russian Samples		All Samples		
	No. of samples	Mean Li content (ppm)	No. of samples	Mean Li content (ppm)	No. of samples	Mean Li content (ppm)	$3\sigma/\sqrt{n}$
Shales and clays	417	74	122	62	539	71	5.1
Sandstones	467	31	18	24	485	31	1.8
Carbonates	324	14	28	22	352	15	1.2

was computed as described above. The Russian data are dominated by those of Ronov et al. (1970), who analyzed composites of the several types of sedimentary rocks from the late Precambrian to the Quaternary of the Russian Platform. The origin of the composites is somewhat obscure; they are not identical to those of Vinogradov and Ronov (1956a,b). Ronov et al. (1970) point out that there is a rather good correlation between the Li and the Al content of the composite samples. This correlation is demonstrated by the relatively small scatter of the Li/Al ratios in table 6.6 and in the plot of figure 6.12. The average Li/Al ratios of shales and clays, sands and silts, and carbonates are very similar. The solid line in figure 6.12 is the line of best fit drawn through the average Li and Al content of the three rock types. Along this line the Li/Al ratio is equal to 0.95×10^{-3} gm/gm. This is considerably greater than the Li/Al ratio in average igneous rocks. If the average Li content of igneous rocks is taken to be (30 ± 5) ppm and their average Al_2O_3 content 15.6% (Al = 8.3%), then

$$\left(\frac{Li}{Al}\right)_{ign.} \approx \frac{(30 \pm 5)}{8.3} \times 10^{-4} = (0.36 \pm 0.06) \times 10^{-3} \text{ gm/gm}.$$

The difference between this ratio and that of the sedimentary rocks of the Russian Platform is nearly a factor of three. Not a single data point in figure 6.12 corresponds to a rock analysis in which the Li/Al ratio is as low as 0.36×10^{-3}. Clearly, sizable quantities of Li have been added during the conversion of the parent igneous rocks to the sedimentary rocks of the Russian Platform. Data of equivalent quality for sedimentary rocks from other parts of the world are lacking. However, the similarity in table 6.5 of the average Li content of sedimentary rocks from non-

Russian areas to those of the Russian Platform suggests that the data in table 6.6 are reasonably representative.

We can use these data to calculate the mean rate of supply of lithium to igneous rocks during their conversion to sedimentary rocks. The Li increment is approximately

$$\Delta\left(\frac{Li}{Al}\right) \times \frac{wt\%\,Al}{100}$$

$$\approx \{[0.95 \pm 0.05] \times 10^{-3} - [0.36 \pm 0.06] \times 10^{-3}\} \times \frac{8.3}{100}$$

$$\approx (49 \pm 8) \times 10^{-6} \text{ gm Li/gm igneous rock.}$$

Most of the sediments that are deposited today are mixtures of old sedimentary rocks that have been eroded and are being redeposited together with the weathering products of igneous and high-grade metamorphic rocks. The recycled sedimentary rocks and some of the high-grade metamorphic rocks have already acquired a lithium excess; the igneous component has not. The weathering products of igneous and high-grade metamorphic rocks account for approximately 25% of the mass of modern sediments. The rate of transport of sediments to the oceans is ca. 2×10^{16} gm/yr. If Li is being supplied at an average rate, ΔLi, the uptake of lithium today is therefore approximately

$$\Delta Li \le (49 \pm 8) \times 10^{-6} \times (0.25 \pm 0.05) \times (2.0 \pm 0.2) \times 10^{16} \text{ gm/yr}$$
$$\le (2.5 \pm 0.8) \times 10^{11} \text{ gm/yr.}$$

The hydrothermal input of Li due to the cycling of seawater through MORs can be calculated roughly using the data in table 6.6. The Li content of high-temperature vent waters at 21°N is ca. 790 μmol/kg greater than that of seawater. If one-third of the heat that is lost from MORs is lost by the circulation of seawater at ridge axes, the quantity of Li added to the oceans at spreading axes via high temperature solutions is ca.

$$790 \times 10^{-6} \frac{\text{mol Li}}{\text{kg seawater}} \times 6.9 \frac{\text{gm Li}}{\text{mol Li}} \times 0.5 \times 10^{14} \frac{\text{kg seawater}}{\text{yr}}$$

$$= 2.7 \times 10^{11} \text{ gm Li/yr.}$$

The agreement between this figure and that for the rate of Li uptake is excellent. It seems likely (see above) that the fraction of heat lost at ridge

TABLE 6.6.

Lithium content of composites of sedimentary rocks from the Russian Platform (Ronov et al., 1970).

Stratigraphic Range	Clays and Clay Shales					Sands and Silts					Carbonates				
	No. of av. spp.	No. of spec. in sp.	Li(ppm)	Al(%)	$\left(\dfrac{Li}{Al}\right) \times 10^3$	No. of av. spp.	No. of spec. in sp.	Li(ppm)	Al(%)	$\left(\dfrac{Li}{Al}\right) \times 10^3$	No. of av. spp.	No. of spec. in sp.	Li(ppm)	Al(%)	$\left(\dfrac{Li}{Al}\right) \times 10^3$
Riphean, Serdob Series (Rf_3^{serd})	3	44	37	7.74	0.48	5	291	21	3.62	0.58	3	106	15	3.46	0.43
Vendian, Valdai Series (V)	11	478	74	9.14	0.81	12	390	46	5.95	0.77	—	—	—	—	—
Cambrian (Cm)	10	217	78	9.18	0.85	5	129	28	3.99	0.70	8	473	16	2.26	0.71
Ordovician (O)	10	187	47	7.08	0.66	14	214	25	4.24	0.59	5	379	18	2.16	0.83
Silurian (S)	6	531	50	7.55	0.66	—	—	—	—	—	45	780	20	2.58	0.77
Devonian (D_2)	20	264	107	8.32	1.28	29	783	43	3.40	1.26	79	2762	10	1.18	0.85
Frasnian stage (D_3^{fr})	23	638	71	7.72	0.92	32	745	36	3.58	1.00	37	2213	25	1.13	2.21
Famennian stage (D_3^{fm})	5	94	169	6.17	2.74	9	117	40	4.39	0.91	22	717	4	0.28	1.43
Lower Carboniferous (C_1)	25	206	137	10.55	1.30	14	121	55	3.45	1.59	24	1229	14	0.24	5.83
Middle Carboniferous (C_2)	8	111	88	8.32	1.06	6	31	38	5.96	0.64	14	520	12	0	—
Upper Carboniferous (C_3)	—	—	—	—	—	—	—	—	—	—	15	715	12	0	—
Lower Permian (P_1)	—	—	—	—	—	—	—	—	—	—	—	—	—	—	—
Ufa stage (P_2^{uf})	2	20	60	7.28	0.82	4	82	40	5.33	0.75	—	—	—	—	—

Kazan stage (P_2^{kaz})	3	63	73	6.03	1.21	3	50	31	4.33	0.71	4	66	27(?)	0.56	4.80
Tatar stage (P_2^{tat})	4	290	55	5.68	0.97	5	318	32	4.30	0.74	—	—	—	—	—
Triassic (T)	3	111	89	7.46	1.19	2	44	44	5.09	0.86	—	—	—	—	—
Lower Jurassic (J_1)	2	10	68	10.21	0.67	2	18	44	2.67	1.65	—	—	—	—	—
Middle Jurassic (J_2)	36	351	110	11.36	0.97	31	216	55	5.16	1.06	17	72	20	2.84	0.70
Upper Jurassic (J_3)	40	598	70	9.66	0.72	6	45	29	3.89	0.74	—	—	—	—	—
Lower Cretaceous (C_1)	33	363	70	9.72	0.72	6	91	35	4.08	0.86	32	425	10	1.23	0.81
Cretaceous (C_2)	3	68	54	4.99	1.08	4	33	22	2.38	0.92	—	—	—	—	—
Paleogene (Pg_1)	5	54	37	5.32	0.69	12	97	44	3.11	1.41	—	—	—	—	—
Buchak stage (Pg_2^{b})	31	171	55	10.60	0.52	49	308	15	3.12	0.48	—	—	—	—	—
Kiev stage (Pg_2^{kv})	—	—	—	—	—	28	148	22	3.22	0.68	—	—	—	—	—
Kharkov stage (Pg_2^{ch})	45	482	50	7.62	0.66	70	464	20	3.15	0.63	—	—	—	—	—
Poltara stage (Pg_3^{pl})	—	—	—	—	—	24	124	8	1.68	0.48	—	—	—	—	—
Lower Neogene (N_1)	9	86	55	7.78	0.71	19	100	14	2.26	0.62	12	95	9	1.08	0.83
Upper Neogene (N_2)	—	—	—	—	—	3	21	30	4.34	0.69	7	15	5	0.52	0.96
Quaternary	4	34	40	6.10	0.65	3	34	18	3.46	0.52	—	—	—	—	—
Sum or mean	341	5471	80	8.44	0.95	397	5014	36	4.15	0.87	324	10567	13	0.93	1.40

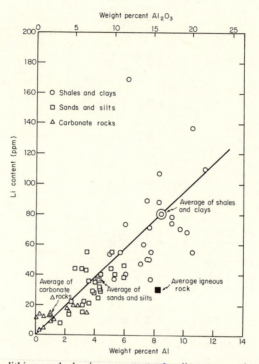

FIGURE 6.12. The lithium and aluminum content of sedimentary rocks of the Russian Platform (Ronov et al. 1970); averages for the several rock types are shown by the position of the double circle, double square, and double triangle.

axes is somewhat less than one-third. However, it is also likely that Li is released from MOR basalts down to temperatures at least as low as 150°C. The cycling of seawater through MORs is therefore the most likely source of the Li excess in sediments and in sedimentary rocks.

Other possible explanations do, however, come to mind. It is possible that the oceans are currently receiving more than their normal input of lithium, but neither the lithium content of the most recent sediments of the Russian Platform nor that of sediments in the Gulf of Paria suggest that this is so. Table 6.7 shows that the average Li/Al ratio in sediments from the Gulf of Paria is 0.77×10^{-3} gm/gm, a figure that is lower than the average Li/Al ratio of sediments on the Russian Platform, and somewhat higher than the Li/Al ratio of Quaternary sediments of the Russian Platform. The Li content of Upper Tertiary sediments from the vicinity of the Japan Trench suggests the same conclusion. Figure 6.13 is a plot of the Li content of sediments from DSDP Drill Holes 434, 434B,

TABLE 6.7.

Average Li, Rb, and Cs content and the Li/Al, Li/Mg, K/Rb, K/Cs, and Rb/Cs ratio of sediments from the Gulf of Paria (Hirst 1962).

	Li	Rb	Cs	Li/Al[‡]	Li/Mg[‡]	K/Rb	K/Cs	Rb/Cs
3 Delta sands	22	41	1.7	0.76	0.92	230	4500	26
12 Platform sands	20	47	1.9	0.62	0.37	250	7700	36
Open platform	10	41	1.1	0.32	0.20	230	7700	36
Protected platform	23	50	2.5	0.71	0.43	260	7700	36
6 Greenish muds	89	160	10.0	1.08	0.60	140	2000	16
4 Bluish clays	87	151	7.3	1.12	0.66	150	3200	21
Delta clay	61	146	7.9	0.80	0.88	140	2500	19
Fe corrections	7.2	36	12.0	0.15	0.08	200	4500	26

Li, Rb, and Cs contents in ppm.
[‡] Li/Al, $\times$ 10^3. Li/Mg, $\times$ 10^3.

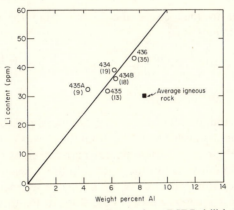

FIGURE 6.13. The Li and Al content of sediments from DSDP drill holes near the Japan Trench (Nohara 1980).

435, 435A, and 436. The location of these holes is shown in figure 6.14. The chemical characteristics of the inner trench sediments of Sites 434 and 435 indicate that they are of island arc origin, and that they contain a much larger fraction of first-generation sediments than the sedimentary rocks of the Russian Platform. This probably explains the difference between the Li content of the Japan Trench sediments and those of the Russian Platform. If all first-generation sediments behave like those at Sites 434, 435, and 436, and if these comprise 25% of the present-day

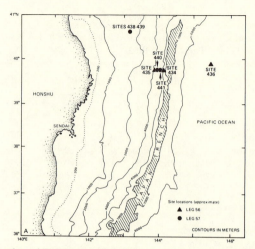

FIGURE 6.14. Location of DSDP Sites 434–441 near the Japan Trench (Barron 1980).

sediment input to the oceans, the Li loss from the oceans to first-generation sediments is ca.

$$(50 - 30) \times 10^{-6} \frac{\text{gm Li}}{\text{gm sediment}} \times 0.25 \times 2 \times 10^{16} \frac{\text{gm sediment}}{\text{yr}}$$

$$= 1 \times 10^{11} \text{ gm Li/yr.}$$

On the other hand, if sediments of all generations pick up ca. 20 ppm Li on each sedimentary cycle, the Li flux from the oceans to sediments would be ca. 4×10^{11} gm Li/yr. These figures may represent reasonable minimum and maximum values of the current Li flux to sediments. The same range spans the estimated value of the Li flux to sediments based on the composition of sedimentary rocks of the Russian Platform. This is somewhat reassuring, but many more analyses of marine sediments are needed before these conclusions can be accepted with confidence.

Submarine weathering of basalts is responsible for a significant flux of Li out of the oceans. Lithium, potassium, rubidium, and cesium are strongly concentrated in submarine basalts that react with seawater at low temperatures. A careful study of the uptake of the alkali metals during this process has been reported by Donnelly, Thompson, and Salisbury (1980), and by Staudigel et al. (1981) and Staudigel, Hart, and Richardson (1981) for basalts in DSDP cores. The relationship between the uptake of lithium and potassium in these basalts is shown in figures 6.15 and 6.16. The basalts in Hole 417A are considerably more altered than those in Hole 417D. In both holes the lithium and potassium content of the basalts

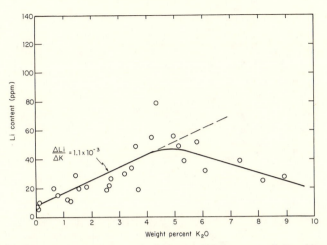

FIGURE 6.15. The relationship between the lithium and K_2O content of altered basalts from DSDP Hole 417A (Donnelly, Thompson, and Salisbury 1980).

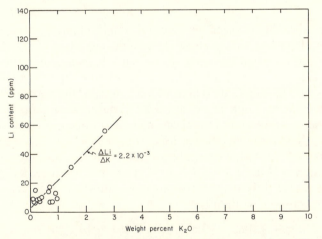

FIGURE 6.16. The relationship between the lithium and K_2O content of altered basalts from DSDP Hole 417D (Donnelly, Thompson, and Salisbury 1980).

increases initially with increasing alteration. In the most highly altered samples of Hole 417A the lithium content decreases with increasing potassium content. This pattern is probably due to the positive correlation between Li and K in samples where clays are the dominant alteration products and an inverse correlation between the two elements where zeolites and K-feldspar are the dominant alteration minerals.

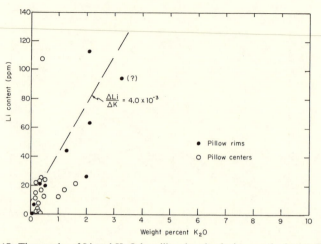

FIGURE 6.17. The uptake of Li and K_2O by pillow basalts during weathering on the ocean floor (Rivers 1976).

The relationship between the Li and K uptake of pillows dredged from the floor of the Atlantic Ocean has been studied by Rivers (1976). His data are not very plentiful, but they do indicate that the glassy rims of his pillows are more altered than pillow interiors, and they suggest that the Li/K_2O ratio in his pillows is higher than in basalts undergoing alteration well below the sea floor. The ratio ($\Delta Li/\Delta K$) during alkali uptake in these pillows seems to be quite variable; its minimum value is at least as low as 1×10^{-3}, its maximum value at least as high as 4×10^{-3} gm Li/gm K.

The flux of K into the oceanic crust due to low-temperature alteration is estimated to be ca. 1.5×10^{13} gm/yr (Hart and Staudigel 1982). If this estimate proves to be correct, the flux of Li into the oceanic crust due to low-temperature alteration is

$$\leq 4 \times 10^{-3} \times 1.3 \times 10^{13} = 0.5 \times 10^{11} \text{ gm/yr.}$$

This is only 20% of the estimated Li flux into marine sediments. Li transfer from the oceans into the oceanic crust does not, therefore, appear to be a major mechanism by which this element is lost from the oceans.

Since our knowledge of the present-day fluxes of lithium is still limited, it should not come as a surprise that only rough indications of the past magnitude of these fluxes can be extracted from the geologic record. Figure 6.18 is a plot of variations in the Li/Al ratio in shales, sands, and silts of the Russian Platform since late Precambrian time. The variations in the Li/Al ratio are not large. However, the fluctuations in the Li/Al ratio

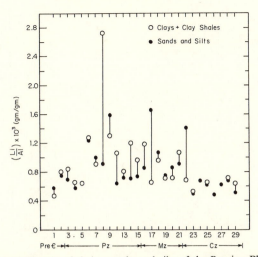

FIGURE 6.18. The Li/Al ratio of shales, sands, and silts of the Russian Platform; the data are from table 6.6; the numbers along the time axis refer to the periods as numbered in this table.

of shales are reasonably well correlated with those of sands and silts, and may well be real. Both sets of ratios are higher in the Paleozoic and Mesozoic than in the late Precambrian and Cenozoic rocks of the Russian Platform. Ronov et al. (1970) suggested that these differences are related to changes in the environment of deposition of sediments on the Russian Platform during the past billion years. Differences in source rocks and weathering regimes may also be involved. It is also possible that the fluctuations in the Li/Al ratio reflect changes in the rate of lithium input to the oceans due to variations in the rate of seawater cycling through mid-ocean ridges. A test of these several interpretations requires sets of data equivalent to those of Ronov et al. (1970) for sedimentary rocks in other parts of the world. Unfortunately, there are few such data.

Figure 6.19 is a plot of the Li content of sediments from DSDP Drill Holes 415, 415A, and 416 from the continental margin off Morocco (see figure 6.20). The sediments consist of pelagic carbonates, and terrigenous and terrigenous-carbonate turbidites. The sections include sporadic parts of the Cenozoic, Cretaceous, and Upper Jurassic. The chemical analyses of these sediments give a general idea of the changes in chemical composition of the sedimentary rocks in the Moroccan Basin from the Tithonian to the Pleistocene (Migdisov et al. 1980). Core 415 sampled Pleistocene to Miocene sediments, Core 415A sampled Paleocene to Albian sediments, and Core 416 sampled Miocene to Upper Jurassic (Berr.-Tith.)

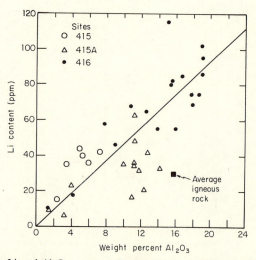

FIGURE 6.19. The Li and Al_2O_3 content of sediments from DSDP Holes 415, 415A, and 416 on the continental margin off Morocco (Migdisov et al. 1980).

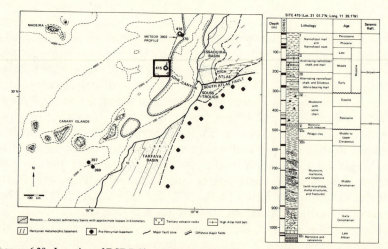

FIGURE 6.20. Location of DSDP Sites 415 and 416 and the stratigraphy at Site 415 (Lancelot et al. 1980). (Reproduced by permission of the authors.)

sediments. With the exception of some samples in Core 415A, the Li and Al data scatter about a Li/Al ratio of 0.9×10^{-3} gm/gm. There is no obvious difference between the value of this ratio in Mesozoic and Cenozoic sediments. The mean Li/Al ratio agrees well with that of Mesozoic sediments on the Russian Platform.

In spite of the present uncertainties in the data, the Li content of shales affords a rough test of the hypothesis advanced by Fryer, Fyfe, and Kerrich (1979) that hydrothermal fluxes dominated the chemistry of the early oceans, and that the early oceans were buffered by large-scale interaction of seawater with ocean floor basalts (Veizer et al. 1982). If this was indeed the case, we might expect to see the effect of intense hydrothermal activity both on the chemistry of those elements that are added to and on those that are removed from the oceans by hydrothermal processes. One might expect to find, for instance, that the Li content of Archean shales is significantly greater and the Mg content of Archean carbonates significantly smaller than during the Phanerozoic Eon. Figure 6.21 is a plot of the Li and Al content of Cameron and Garrels's (1980) Archean and Proterozoic (Aphebian) shale composites from the Canadian Shield. The line of best fit through the data for sedimentary rocks of the Russian Platform has been drawn in this figure for purposes of comparison. The weighted average of the Li/Al ratio in the Aphebian shales is significantly lower than that of the sediments of the Russian Platform; the weighted average of the Li/Al ratio of the Archean shale composites is lower still. These data therefore do not support the notion that the oceans were "volcanogenic" during the deposition of the Canadian shales in the latter part of the Archean. The same conclusion can be drawn from the data in figure 6.22.

Additional reinforcement for this conclusion is provided by the data of Lonka (1967), who studied the concentration of a number of trace elements, including Li, in 174 phyllites and mica shists from eighteen areas in Finland. These metasediments range in age from ca. 1.8 to 2.7 b.y.b.p. The average Li and Ga content of metasediments in the eighteen areas is shown in figure 6.23 together with the Li and Ga content of average igneous rock. The phyllites and mica schists are clearly enriched in Li; their average Li/Ga ratio is ca. 2.2 times that of igneous rocks. The Al content of the metasediments has not been determined. However, Ga and Al are so well correlated in sedimentary and in igneous rocks that the Li/Al ratio in the Finnish phyllites and mica schists is almost certainly also ca. twice that of average igneous rocks. They are therefore as enriched in Li as the Proterozoic shales of figure 6.21 and slightly less enriched in Li than the sedimentary rocks of the Russian Platform.

It is tempting to use the data in figure 6.21 to derive an approximate upper bound on the rate of cycling of seawater through hot basalt during the Archean. At steady state the mass balance for lithium in the oceans requires that

$$_1m_{\text{Li}}\frac{dW_1}{dt} + {_2m_{\text{Li}}}\frac{dW_2}{dt} + \Delta_3 m_{\text{Li}}\frac{dV_0}{dt} \cong {_4m_{\text{Li}}}\frac{dW_4}{dt} + R_{\text{Li}}, \qquad (6.7)$$

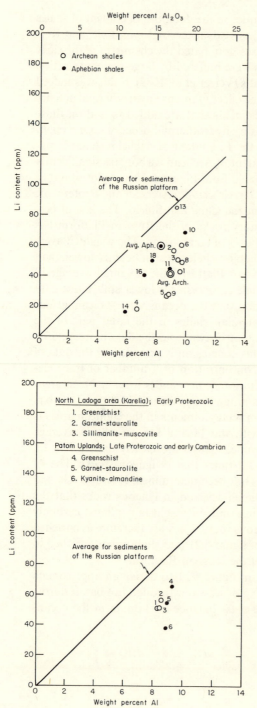

FIGURE 6.21. The Li and Al content of Archean and Proterozoic (Aphebian) shales of the Canadian Shield; for data see tables 5.9 and 5.10.

FIGURE 6.22. The Li and Al content of metamorphosed Proterozoic shales from the USSR. (Ronov, Migdisov, and Lobach-Zhuchenko et al. 1977).

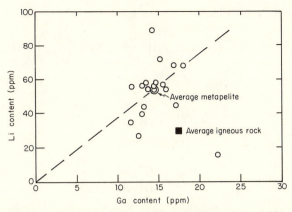

FIGURE 6.23. The Li and Ga content of Precambrian phyllites and mica schists from Finland (Lonka 1967). These sedimentary rocks were deposited between ca. 1.8 and 2.7 b.y.b.p. The figure for the Ga content of average igneous rock is that of Burton and Culkin (1970).

where

$_1m_{Li}$ = lithium content of igneous rocks undergoing weathering (gm Li/kg rock),

$\dfrac{dW_1}{dt}$ = rate of erosion of igneous rocks (kg/yr),

$_2m_{Li}$ = lithium content of sedimentary and metamorphic rocks undergoing weathering (gm Li/kg rock),

$\dfrac{dW_2}{dt}$ = rate of erosion of sedimentary and metamorphic rocks (kg/yr),

$\Delta_3 m_{Li}$ = mean increase in the concentration of Li in seawater during reaction with hot basalt (gm Li/kg seawater),

$\dfrac{dV_0}{dt}$ = mass of seawater which reacts annually with hot basalt (kg/yr),

$_4m_{Li}$ = average lithium content of newly deposited sediments (gm Li/kg),

$\dfrac{dW_4}{dt}$ = rate of deposition of sediments (kg/yr),

R_{Li} = removal of lithium from the oceans by low-temperature submarine weathering of basalts (gm/yr).

Equation 6.7 can be recast into the form

$$\left[_4m_{\text{Li}} - (_1m_{\text{Li}}f_1 + _2m_{\text{Li}}f_2)\right] \frac{dW_4}{dt} = \Delta_3 m_{\text{Li}} \left(\frac{dV_0}{dt}\right) - R_{\text{Li}}, \quad (6.8)$$

where

$$f_1 = \frac{dW_1/dt}{dW_4/dt}. \quad (6.9)$$

and

$$f_2 = \frac{dW_2/dt}{dW_4/dt}. \quad (6.10)$$

The sum of the terms on the left-hand side of equation 6.8 represents the quantity of lithium that is added annually to igneous rocks and old sedimentary rocks during their conversion to new sediments. The terms on the right side of the equation show that the quantity of the added lithium is the difference between the supply of Li to the oceans by hydrothermal processes and the removal of Li by low-temperature submarine weathering of basalt.

The value of the term $\left[_4m_{\text{Li}} - (_1m_{\text{Li}}f_1 + _2m_{\text{Li}}f_2)\right]$ for Archean shales depends on the Li content of material in the source areas of the shales. If present-day values for $_1m_{\text{Li}}$, f_1, $_2m_{\text{Li}}$, and f_2 were used, the left-hand side of equation 6.8 would be negative. A maximum positive value of the term $\left[_4m_{\text{Li}} - (_1m_{\text{Li}}f_1 + _2m_{\text{Li}}f_2)\right]$ for the Archean can be obtained if f_2, the fraction of recycled sediments, is set equal to zero, and if $_1m_{\text{Li}}$, the Li content of igneous rocks in the source areas of the shales, is assumed to be abnormally low. If $_1m_{\text{Li}}$ is taken to be 20×10^{-3} gm/kg rock, then for the Archean shales

$$\left[_4m_{\text{Li}} - (_1m_{\text{Li}}f_1 + _2m_{\text{Li}}f_2)\right] \le [42 \times 10^{-3} - 20 \times 10^{-3}]$$
$$= 22 \times 10^{-3} \text{ gm/kg}.$$

This is comparable to the estimated present-day Li increment.

The value of the term dW_4/dt during the Archean is not known, but it may well have been greater than its present-day value. If this was indeed so, the river flux of solutes was also probably higher than it is at present, because the mean intensity of weathering of sedimentary rocks has remained roughly constant (see chapter 5). For purposes of estimating the relative flux of solutes from rivers and hydrothermal reactions, differences in dW_4/dt can therefore probably be neglected.

Nothing is known about $\Delta_3 m_{\text{Li}}$, the increment of Li during the reaction of seawater with hot basalt during the Archean Era; it seems likely, however, that this increment has changed little with time, because the composition of basalts has remained nearly invariant.

The rate of submarine basalt weathering during the Archean is also not known. However, rather extraordinary rates would have to be invoked to make the Li data fit the hypothesis of a "volcanogenic" Archean ocean. The presence of significant quantities of dolomite in Archean carbonates also makes such a hypothesis unattractive.

THE ISOTOPIC COMPOSITION OF STRONTIUM IN SEAWATER

The data for the geochemistry of lithium have given a rather limited insight into the past rate of seawater cycling through mid-ocean ridges. The evidence derived from the evolution of several isotopic ratios in seawater is potentially more useful. Most isotope ratios in seawater integrate the effects of inputs and outputs to the oceans quite well, and are influenced much less than elemental abundances in sediments by local variations in environments of deposition. Three isotope ratios have been linked to seawater cycling through mid-ocean ridges: $^{87}Sr/^{86}Sr$, $^{143}Nd/^{144}Nd$, and $^{18}O/^{16}O$. The discovery that the $^{87}Sr/^{86}Sr$ ratio of marine carbonates is an excellent indicator of the Sr^{87}/Sr^{86} ratio in seawater at the time of deposition, and the discovery that the $^{87}Sr/^{86}Sr$ ratio of limestones, and hence of seawater, has varied significantly with time must be counted among the most interesting contributions of isotope geochemistry to paleo-oceanography. The first thorough study of the isotopic composition of strontium in marine carbonates during the Phanerozoic Era was carried out by Peterman, Hedge, and Tourtelot (1970); this study has been followed by a number of others (Veizer and Compston 1974; Brass 1976; Faure, Assereto, and Tremba 1978; Starinsky et al. 1980; and Burke et al. 1982); these have confirmed the essential correctness of the pattern described by Peterman, Hedge, and Tourtelot, and have refined many of its details. The summary plot of Burke et al. (1982) for the variations of the $^{87}Sr/^{86}Sr$ ratio in marine carbonates during the Phanerozoic Era is reproduced in figure 6.24.

The data for Precambrian carbonates are still quite sparse. Veizer and Compston (1974) have shown that the $^{87}Sr/^{86}Sr$ ratio decreases toward values close to 0.700 in very ancient carbonates. The simplicity of the proposed pattern in figure 6.25 is probably misleading. The coverage of the great length of Precambrian time is still quite spotty, and fluctuations in the $^{87}Sr/^{86}Sr$ ratio in seawater similar to those that occurred during the Phanerozoic Eon almost certainly occurred during the Precambrian as well (Veizer, personal communication, 1982).

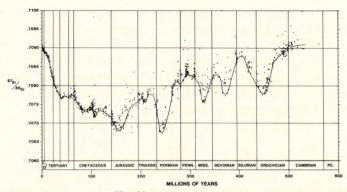

FIGURE 6.24. Variation of the $^{87}Sr/^{86}Sr$ ratio of seawater during the Phanerozoic Eon (Burke et al. 1982).

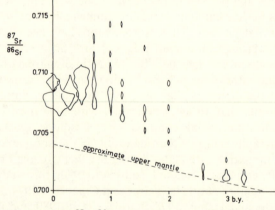

FIGURE 6.25. Variation of the $^{87}Sr/^{86}Sr$ ratio during the past 3.3 b.y. (Veizer 1976). (Reproduced by permission of John Wiley & Sons, Ltd.)

The $^{87}Sr/^{86}Sr$ ratio in seawater is influenced by the cycling of seawater through mid-ocean ridges. During the reaction of seawater with basalt, both chemical and isotopic exchange takes place. The Sr content of MOR basalts is sufficiently large compared to that of seawater, so that during reaction at and above 300°C the isotopic composition of Sr in seawater rapidly approaches that of MOR basalts except in settings where the sea-water/basalt ratio is very large. Albarède et al. (1981) have shown that seawater that has cycled through the East Pacific Rise at 21°N latitude (see Edmond et al. 1979) has a $^{87}Sr/^{86}Sr$ ratio of 0.7030, only slightly higher than that of unaltered basalts from this area. Since strontium is one of the few elements in seawater whose concentration in solution is affected only slightly by reaction with MOR basalts at temperatures between ca. 300° and 400°C (see table 6.3), the effect of seawater cycling on this element is essentially restricted to the alteration of its isotopic composition. Reaction between seawater and basalt at temperatures below 100°C has nearly the same effect as reaction at high temperatures.

The isotopic composition of Sr in rivers is the other major influence on the isotopic composition of Sr in seawater. Variations in the $^{87}Sr/^{86}Sr$ ratio of rivers and in the relative magnitude of the river input of Sr to the oceans compared to the effects of the reaction of seawater with oceanic basalts have probably dominated the evolution of the isotopic composition of strontium in the oceans.

The isotopic composition of Sr in river and lake waters is quite variable, because it is determined by the isotopic composition of Sr in each drainage basin. Table 6.8 shows that in river and lake waters in Precambrian shield areas the $^{87}Sr/^{86}Sr$ ratio ranges from ca. 0.712 to 0.730; rivers and lakes whose drainage basin is dominated by limestones tend to have $^{87}Sr/^{86}Sr$ ratios between 0.706 and 0.709; those in areas of young volcanics tend to have $^{87}Sr/^{86}Sr$ ratios between 0.704 and 0.706. Large rivers integrate the effect of variations in the $^{87}Sr/^{86}Sr$ ratio within their drainage areas, and it is surely no accident that the $^{87}Sr/^{86}Sr$ ratio of the Amazon (0.7110) is similar to that of the Mississippi (0.7101). Table 6.8 does not include data for enough of the major rivers of the world to permit the computation of a firm figure for the average isotopic composition of strontium entering the oceans today, but it is likely that the world average is close to the $^{87}Sr/^{86}Sr$ ratio of the Amazon (Wadleigh, Veizer, and Brooks 1981).

The isotopic composition of Sr in world-average river water could have varied significantly with time. Precambrian shield areas seem to be exposed particularly prominently today, and the high $^{87}Sr/^{86}Sr$ ratio in such areas contributes significantly to the value of the $^{87}Sr/^{86}Sr$ ratio in Mississippi and Amazon river water. If there were times in the past when most shields were covered by more recent sedimentary rocks, the $^{87}Sr/^{86}Sr$ ratio in world-average river water during these periods was significantly lower. A simple calculation illustrates this point. If the strontium that is currently entering the oceans is derived from several sources i, then the flux of ^{87}Sr is

$$\frac{d_7 M_{\text{R.W.}}}{dt} = \sum_i {}_7\phi_i \frac{dM_i}{dt}, \qquad (6.11)$$

where

$$_7\phi_i = \text{the mole fraction of } {}^{87}Sr \text{ in source } i,$$

TABLE 6.8.

The $^{87}Sr/^{86}Sr$ ratio of rivers and lakes.

Location	Ratio	Source of Data
United States		
Mississippi R., Ark.	0.7101	Brass (1976)
Russian R., Calif.	0.7053	"
Eel R., Calif.	0.7063	"
Mad R., Calif.	0.7066	"
Colorado R., Texas	0.7085	"
Brazos R., Texas	0.7086	"
Red R., La.	0.7090	"
Wateree R., S.C.	0.7096	"
Susquehanna R., Pa.	0.7137	"
Lake Superior, Wisc.	0.7146	"
Lake Superior, Wisc.	0.7176*	Faure et al. (1963)
Lake Superior, Wisc.	0.7164	Hart and Tilton (1966)
Lake Huron, Mich.	0.7086*	Faure et al. (1963)
Lake Erie, Mich.	0.7119*	Faure et al. (1963)
Great Salt Lake, Utah	0.7174	L.M. Jones and Faure (1972)
Bear R., Utah	0.7219	"
Jordan R., Utah	0.7206	"
Weber R., Utah	0.7129	"
Locomotive Springs, Utah	0.7110	"
Wailuku R., Hawaii	0.7036	Brass (1976)
Canada		
Kenogami L., Ont.	0.7153*	Faure et al. (1963)
Remi L., Ont.	0.7199*	"
Kabinakagami R., Ont.	0.7263*	"
Klotz L., Ont.	0.7191*	"
Little Longlac., Ont.	0.7251*	"
Fairbank L., Ont.	0.7209*	"
Cassel L., Ont.	0.7182*.	"
St. Maurice R., Que.	0.7157*	Faure et al. (1963)
Muskrat L., Ont.	0.7121*	"
Ottawa R., Ont.	0.7161*	"
Mattawa R., Ont.	0.7233*	"
Lake Nipissing, Ont.	0.7194*	"
Marten R., Ont.,	0.7192*	"
St. Lawrence R., L'Islet, Que.	0.7134*	"
St. Lawrence R., Trois Pistoles, Que.	0.7141*	"
St. Francis R., Que.	0.7108*	"
Lake George, Ont.	0.7184	Jones and Faure (1972)
Lake Manitouwadge, Ont.	0.7193	Pushkar in Jones and Faure (1972)

TABLE 6.8. (*Continued*)

Location	Ratio	Source of Data
Egypt		
Nile R.	0.7060	Brass (1976)
Europe		
Rhone R., France	0.7089	Brass (1976)
Po R., Italy	0.7097	Brass (1976)
South America		
Amazon R., Brazil	0.7110	Brass (1976)
Asia		
Cauveri R., India	0.7130	Brass (1976)
Antarctica		
Lake Vanda	0.7149	Jones and Faure (1967)
Lake Bonney	0.7136	"

* These values are reported for an E. and A. standard value of 0.7108, about 0.0029 higher than the value used by Brass (1976).

and

$$\frac{dM_i}{dt} = \text{the number of moles of Sr entering the oceans each year from source } i \text{ (mol/yr).}$$

By analogy, the flux of ^{86}Sr to the oceans is

$$\frac{d_6 M_{\text{R.W.}}}{dt} = \sum_i {}_6\phi_i \frac{dM_i}{dt}, \qquad (6.12)$$

where ${}_6\phi_i$ is the fraction of ^{86}Sr in source i. The $^{87}Sr/^{86}Sr$ ratio in the river flux of Sr to the oceans is therefore

$$r_{\text{R.W.}} = \frac{d_7 M_{\text{R.W.}}/dt}{d_6 M_{\text{R.W.}}/dt} = \frac{\sum_i {}_7\phi_i \frac{dM_i}{dt}}{\sum_i {}_6\phi_i \frac{dM_i}{dt}}. \qquad (6.13)$$

Since the values of ${}_6\phi_i$ vary only very slightly, equation 6.13 can be recast without significant loss of accuracy into the more useful form,

$$r_{\text{R.W.}} \cong \frac{\sum_i \frac{{}_7\phi_i}{{}_6\phi_i} \frac{dM_i}{dt}}{\sum_i \frac{dM_i}{dt}} \cong \sum_i r_i f_i, \qquad (6.14)$$

where

$$r_i \equiv \frac{7\phi_i}{6\phi_i}, \qquad (6.15)$$

and

$$f_i \equiv \frac{dM_i/dt}{\sum_i dM_i/dt}. \qquad (6.16)$$

Equation 6.14 can be used easily to determine the effect of changes in the source regions of river Sr on the $^{87}Sr/^{86}Sr$ ratio of river Sr entering the oceans. Roughly 20% of the total Sr flux today is probably coming from areas in which the $^{87}Sr/^{86}Sr$ ratio is approximately 0.715. If the Sr flux from this source were replaced by an equal Sr flux with a $^{87}Sr/^{86}Sr$ ratio of 0.706, then the value of $r_{R.W.}$ would decrease by 0.0018. This is about 80% of the difference between the present-day value of the $^{87}Sr/^{86}Sr$ in ocean water and its Phanerozoic minimum value during the Jurassic and Permian. Changes in the sources of river Sr can therefore have a significant effect on the isotopic composition of Sr in the oceans.

The average concentration of Sr in river waters is ca. 70 μg/kg (Turekian 1969). The total river flux of Sr to the oceans is therefore approximately

$$\frac{dM_{R.W.}}{dt} \approx \frac{70 \times 10^{-6} \text{ gm Sr/kg R.W.} \times 4.6 \times 10^{16} \text{ kg R.W./yr}}{87.62 \text{ gm Sr/mol}}$$

$$\approx 3.7 \times 10^{10} \text{ mol/yr.}$$

Within the oceans the major influence on the isotopic composition of Sr appears to be the reaction of seawater with hot basalt at and near mid-ocean ridges and with cold basalt during submarine weathering. This is shown schematically in figure 6.26. There seems to be little change in the concentration of Sr in seawater during reaction with hot basalt. This was first observed in laboratory experiments by Mottl and Holland (1978) and has been confirmed by analyses of the Galapagos warm springs and of the hot solutions at 21°N (see table 6.3). The total flux of Sr through MORs is by no means negligible. If the flux of heat released via high-temperature (ca. 300°C) vents close to the axis of MORs is between 15% and 35% of the total flux released hydrothermally at MORs (see above), then the flow of Sr through hot basalt is at least as large as

$$87 \times 10^{-6} \frac{\text{mol Sr}}{\text{kg seawater}} (0.4 \pm 0.2) \times 10^{14} \frac{\text{kg seawater}}{\text{yr}}$$

$$= (0.35 \pm 0.17) \times 10^{10} \frac{\text{mol Sr}}{\text{yr}}.$$

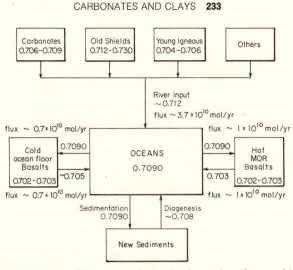

FIGURE 6.26. Some aspects of the geochemical cycle of strontium; figures within the boxes denote the mean $^{87}Sr/^{86}Sr$ ratio of the contained Sr.

To this we must add the flux of Sr through MORs at temperatures less than 300°C but at temperatures high enough so that isotopic exchange of Sr with the ocean crust can take place. The total Sr flux through MORs at elevated temperatures, $dM_{H.B.}/dT$, is probably ca. 1.0×10^{10} mol/yr.

The $^{87}Sr/^{86}Sr$ ratio of seawater after reacting with hot basalt is close to that of unaltered basalt (Albarède et al. 1981). Thus the annual change in the number of moles of ^{87}Sr in the oceans due to such reactions is

$$\left(\frac{d_7 M_0}{dt}\right) \approx (_7\phi_B - {_7}\phi_0)\frac{dM_{H.B.}}{dt}, \tag{6.17}$$

and the change in the number of moles of ^{86}Sr is

$$\left(\frac{d_6 M_0}{dt}\right)_{H.B.} \approx (_6\phi_B - {_6}\phi_0)\frac{dM_{H.B.}}{dt}. \tag{6.18}$$

The two changes are related, because the sum of the fraction of the strontium isotopes in any given sample is equal to unity:

$$_4\phi + {_6}\phi + {_7}\phi + {_8}\phi \equiv 1.00. \tag{6.19}$$

Since the relative proportions of ^{84}Sr, ^{86}Sr, and ^{88}Sr are essentially invariant,

$$\frac{_4\phi}{_6\phi} = \alpha = \frac{0.56}{9.86} = 0.0568 \tag{6.20}$$

and

$$\frac{_8\phi}{_6\phi} = \beta = \frac{82.56}{9.86} = 8.373. \tag{6.21}$$

Thus,

$$_6\phi(1 + \alpha + \beta) + _7\phi = 1.00, \tag{6.22}$$

which is equivalent to

$$_6\phi = \frac{1.00 - _7\phi}{9.430} \tag{6.23}$$

and

$$\frac{d_6\phi}{d_7\phi} = -0.106. \tag{6.24}$$

The $^{87}Sr/^{86}Sr$ ratio, r, is therefore given by the expression,

$$r \equiv \frac{_7\phi}{_6\phi} = \frac{9.430_7\phi}{(1.00 - _7\phi)}. \tag{6.25}$$

From this it follows that

$$_7\phi = \frac{r}{9.430 + r}, \tag{6.26}$$

and

$$_6\phi = \frac{1}{9.430 + r}. \tag{6.27}$$

If these expressions are substituted into equations 6.18 and 6.19, one obtains

$$\left(\frac{d_7M_0}{dt}\right)_{H.B.} = \left[\frac{r_B}{9.430 + r_B} - \frac{r_0}{9.430 + r_0}\right]\frac{dM_{H.B.}}{dt}, \tag{6.28}$$

and

$$\left(\frac{d_6M_0}{dt}\right)_{H.B.} = \left[\frac{1}{9.430 + r_B} - \frac{1}{9.430 + r_0}\right]\frac{dM_{H.B.}}{dt}. \tag{6.29}$$

Since the difference between r_B and r_0 is very small compared to 9.430, these expressions can be simplified and evaluated directly:

$$\left(\frac{d_7 M_0}{dt}\right)_{\text{H.B.}} \approx \left(\frac{r_B - r_0}{9.430 + r_0}\right)\frac{dM_{\text{H.B.}}}{dt} \tag{6.30}$$

$$\approx \left(\frac{0.7030 - 0.7090}{9.430 + 0.7090}\right) \times 1.0 \times 10^{10} \text{ mol/yr}$$

$$\approx -6 \times 10^6 \text{ mol/yr}$$

$$\left(\frac{d_6 M_0}{dt}\right)_{\text{H.B.}} \approx \frac{(r_0 - r_B)}{(9.430 + r_0)^2} \cdot \frac{dM_{\text{H.B.}}}{dt} \tag{6.31}$$

$$\approx \frac{(0.7090 - 0.7030)}{(9.430 + 0.7090)^2} \cdot \frac{dM_{\text{H.B.}}}{dt}$$

$$\approx +0.6 \times 10^6 \text{ mol/yr.}$$

The behavior of Sr during the reaction of seawater with cold basalts during sea-floor weathering seems to be similar to its behavior at temperatures of several hundred degrees. The available data, which are rather scant, indicate that the Sr content of basalt changes little during halmyrolysis. This is illustrated by the data in figures 6.27 and 6.28. During alteration of these basalts potassium is added in large amounts. The K_2O content of altered basalts is a fair measure of their degree of alteration; the Sr content of these basalts is changed little, even after the addition of as much as 9% K_2O.

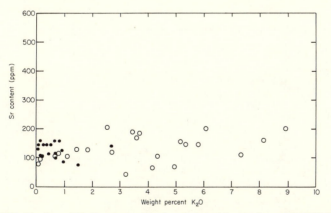

FIGURE 6.27. The Sr and K_2O content of fresh and altered basalts from DSDP Holes 417A and 417D (Donnelly, Thompson, and Salisbury 1980); open circles, Hole 417; closed circles, Hole 418.

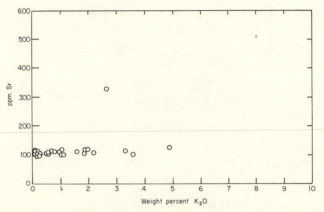

FIGURE 6.28. The Sr and K_2O content of fresh and altered basalts from DSDP Holes 417A and 418A (Humphris, Thompson, and Marriner 1980).

However, the change in the isotopic composition during alteration is quite significant. Figure 6.29 shows the relationship between the isotopic composition of Sr and the potassium content of basalts from DSDP Holes 417A, 417D, and 418A (Staudigel, Hart, and Richardson 1981). The data are very scattered. The $^{87}Sr/^{86}Sr$ ratio of basalts from Hole 418A have clearly approached the $^{87}Sr/^{86}Sr$ ratio of contemporary seawater at a much lower increment of K than the basalts of Hole 417A.

The conversion of these data into an estimate of the flux of ^{87}Sr and ^{86}Sr between seawater and submarine basalt is obviously risky. If we

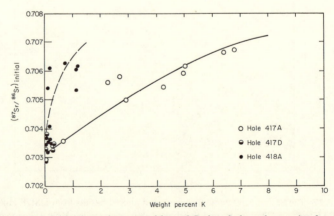

FIGURE 6.29. The initial isotopic composition of Sr in whole-rock samples from DSDP Holes 417A, 417D, and 418A (Staudigel, Hart, and Richardson 1981).

accept Hart and Staudigel's (1982) figure of 1.5×10^{13} gm/yr for the flux of K into the upper 600 m of oceanic crust and take the mean rate of ocean floor generation to be 3 km^2/yr, then the mean addition of K to oceanic crust is approximately

$$\frac{1.5 \times 10^{13} \text{ gm K/yr}}{3 \times 10^{10} \dfrac{\text{cm}^2}{\text{yr}} \times 600 \times 10^2 \text{ cm} \times 3.0 \text{ gm/cc}} = 0.28 \times 10^{-2} \frac{\text{gm K}}{\text{gm crust}}$$

Figure 6.29 indicates that a K uptake of this magnitude is apt to increase the $^{87}\text{Sr}/^{86}\text{Sr}$ ratio of the basalt roughly to 0.7055 if it behaves on average like basalts in Hole 418A and to ca. 0.7035 if it behaves on average like basalts in Hole 417A. The Sr content of unaltered basalts is ca. 110 ppm (see figure 6.22). Such a change in isotopic composition therefore would affect ca.

$$110 \times 10^{-6} \frac{\text{gm Sr}}{\text{gm basalt}} \times \frac{3 \times 10^{10} \text{ cm}^2 \times 0.6 \times 10^5 \text{ cm} \times 3.0 \text{ gm/cm}^3}{87.62 \text{ gm Sr/mol}}$$

$$\approx 0.7 \times 10^{10} \text{ mol Sr/yr}.$$

The loss of ^{87}Sr to ocean-floor basalts with properties intermediate between those of Holes 417A and 418A would be roughly

$$\left(\frac{d_7 M_0}{dt}\right)_{\text{C.B.}} \approx -\left[\frac{(0.7046 \pm 0.0012) - 0.7029}{9.430 + 0.703}\right] \times 0.7 \times 10^{10} \text{ mol/yr}$$

$$\approx -(1.1 \pm 0.9) \times 10^6 \text{ mol/yr},$$

and

$$\left(\frac{d_6 M_0}{dt}\right)_{\text{C.B.}} \approx +\left[\frac{(0.7046 \pm 0.0012) - 0.7029}{(9.430 + 0.703)^2}\right] \times 0.7 \times 10^{10} \text{ mol/yr}$$

$$\approx +(0.11 \pm 0.08) \times 10^6 \text{ mol/yr}.$$

Staudigel et al. (1981) showed that the alteration of the basalts in these holes probably took place some 100 m.y. ago; at that time the $^{87}\text{Sr}/^{86}\text{Sr}$ ratio in seawater was only ca. 0.7070. Basaltic alteration with present-day seawater therefore probably involves a flux of ^{87}Sr into and a flux of ^{86}Sr out of the oceanic crust that is ca. 50% greater than the rates calculated above. Even so, the effect of basalt alteration at low temperatures on the isotopic composition of Sr in seawater is probably considerably smaller than the effect of the alteration of oceanic basalts at temperatures in excess of 70°C.

The residence time of Sr in the oceans is approximately 4 m.y. (Holland 1978, p. 154). This is geologically short, and it seems likely that the oceans are roughly at steady state with respect to Sr. If so, the influx of ^{87}Sr and ^{86}Sr is roughly in balance with their outflow, and the ratio of the influx of the two isotopes is equal to 0.7090, their ratio in the outflow. Thus,

$$0.7090 \cong \frac{_7\phi_R \times 3.7 \times 10^{10} - 7 \times 10^6}{_6\phi_R \times 3.7 \times 10^{10} + 0.7 \times 10^6}. \tag{6.32}$$

This equation can be solved for $r_{R.W.}$ by using equations 6.26 and 6.27. The computed value of $r_{R.W.}$, 0.711, is equal to that of the Amazon River (0.711) and to the best observational estimate of $r_{R.W.}$ today. This agreement is obviously reassuring.

There is, however, one uncertainty in the above calculation that should be mentioned. The Sr content of carbonate sediments at and below the ocean floor tends to decrease progressively with age. This trend is illustrated by the data in figure 6.30 (see also Graham et al. 1982). The decrease in Sr content is diagenetically controlled, and the return flow of Sr from these sediments to the oceans has continued for at least 120 m.y. The observed decrease is consistent with the recrystallization of marine

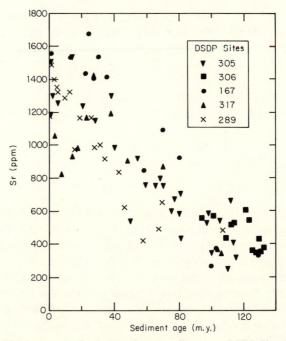

FIGURE 6.30. The Sr content of pelagic carbonate sediments at DSDP Sites 289, 167, 305, 317, and 306, plotted as a function of their age (Manghnani, Schlanger, and Milholland 1980). (Reproduced by permission of Plenum Publishing Corp.)

carbonates under conditions in which the distribution coefficient k_{Sr}^C is close to 0.04 (Baker, Gieskes, and Elderfield 1982). Since there has been a significant increase in the $^{87}Sr/^{86}Sr$ ratio of seawater during the past 120 m.y., the $^{87}Sr/^{86}Sr$ ratio of the diagenetic flux of Sr is significantly less than 0.7090. This implies that the computed current value for $r_{R.W.}$ is somewhat but not significantly larger than 0.711 (Elderfield and Gieskes 1982).

The reasonable agreement between the calculated fluxes and the requirements of steady state for Sr in the present-day oceans is encouraging. It shows that either the important terms in the isotopic mass-balance equations have been evaluated properly or that the errors in these terms compensate each other. If the important terms have been evaluated properly, then it is likely that the changes in the isotopic composition of Sr in seawater during the Phanerozoic Eon are the result of changes in the isotopic composition of Sr in world-average river water and/or of variations in the exchange rate of Sr with oceanic basalts. If the increase in the $^{87}Sr/^{86}Sr$ ratio of seawater since the Jurassic has been due mainly to an increase in the value of $r_{R.W.}$ related to a significant increase in the contribution of Sr from old shields to the Sr in world river water, then there should be an accompanying increase in the mean $^{87}Sr/^{86}Sr$ ratio of the detrital fraction of marine sediments. It would be useful but difficult to determine whether such an increase exists.

If the change in the $^{87}Sr/^{86}Sr$ ratio in seawater during the Phanerozoic Eon is entirely due to changes in the intensity of the reaction of seawater with oceanic basalts, the available data can be used to place a rough maximum on the required changes in the rate of this reaction. From equations 6.18, 6.19, 6.30, 6.31, and 6.32, it follows that r_0, the $^{87}Sr/^{86}Sr$ ratio of seawater in the past, has been determined largely by the expression,

$$r_0 \approx \frac{{}_7\phi_{R.W.}\dfrac{dM_{R.W.}}{dt} + (1+\gamma)({}_7\phi_B - {}_7\phi_0)\dfrac{dM_{H.B.}}{dt}}{{}_6\phi_{R.W.}\dfrac{dM_{R.W.}}{dt} + (1+\gamma)({}_6\phi_B - {}_6\phi_0)\dfrac{dM_{H.B.}}{dt}}, \qquad (6.33)$$

where γ is the fraction of the ^{87}Sr and ^{86}Sr exchange of seawater with basalt that takes place during submarine weathering. The present value of γ is ca. 0.2 (see above). The second term in the denominator of equation 6.33 is very small compared to the magnitude of the first term. If the smaller term is neglected, and if ${}_6\phi_{R.W.}$ is set equal to ${}_6\phi_B$ and ${}_6\phi_0$ as suggested above, equation 6.33 becomes

$$r_0 \approx r_{R.W.} + (1+\gamma)(r_B - r_0)\left(\frac{dM_{H.B.}/dt}{dM_{R.W.}/dt}\right). \qquad (6.34)$$

This is easily rearranged to the form

$$r_0 \approx \frac{r_{\text{R.W.}} + (1 + \gamma)r_B\left(\dfrac{dM_{\text{H.B.}}/dt}{dM_{\text{R.W.}}/dt}\right)}{1 + (1 + \gamma)\left(\dfrac{dM_{\text{H.B.}}/dt}{dM_{\text{R.W.}}/dt}\right)}. \tag{6.35}$$

When the ratio of the magnitude of the flux $dM_{\text{H.B.}}/dt$ to that of the flux $dM_{\text{R.W.}}/dt$ is very small, $r_0 \approx r_{\text{R.W.}}$; when the ratio of the two fluxes is large, $r_0 \approx r_B$. The lowest value of r_0 during the Phanerozoic was ca. 0.7068 (see figure 6.24). r_B must have been close to 0.7030 during the entire Phanerozoic; if $r_{\text{R.W.}}$ was always close to 0.711 during the Phanerozoic Eon, the value of the term,

$$(1 + \gamma)\frac{dM_{\text{H.B.}}/dt}{dM_{\text{R.W.}}/dt},$$

was never greater than ca. 1.1, and

$$(1 + \gamma)\frac{m_0 \, dV_0/dt}{dM_{\text{R.W.}}/dt} \leq 1.1. \tag{6.36}$$

It is likely that the concentration of Sr in seawater has been correlated roughly with the rate of river input of this element to the oceans. Thus, very roughly, the term $(1 + \gamma)\,dV_0/dt$ has not been more than about three times its present value during the course of the Phanerozoic Eon if the isotopic composition of river Sr has remained constant and if the rather unsatisfactory computations outlined above are essentially correct.

If changes in the rate of seawater cycling through MORs have been largely responsible for the fluctuations in the $^{87}\text{Sr}/^{86}\text{Sr}$ ratio of seawater during the Phanerozoic Eon, the present and the Cambrian are periods of minimum seawater cycling through MORs; seawater cycling through MORs increased from the Cambrian to the Permian and has decreased since the Jurassic. Large fluctuations are superimposed on this long-term trend. The long-term trend coincides with the progressive assembly of Pangaea between the Cambrian and the Permian and with the breakup of this supercontinent during the last 150 m.y. (see, for instance, Ziegler et al. 1979). The large fluctuations superimposed on the trend of the $^{87}\text{Sr}/^{86}\text{Sr}$ curve can be correlated with several periods of Phanerozoic orogeny. The Ordovician minimum coincides with the Taconic orogeny; the mid-Devonian minimum coincides with the Acadian orogeny in North America. The Carboniferous minimum coincides with the Hercynian orogeny, the minimum in the Permian marks the assembly of Pangaea, and the minimum in the Jurassic coincides with the Timmerian orogeny. The collision of India with Eurasia during the early Tertiary appears to have left a relatively minor imprint on the history of the $^{87}Sr/^{86}Sr$ ratio in seawater. Although it seems unlikely that these time coincidences are fortuitous, the causal connections between the orogenies and the $^{87}Sr/^{86}Sr$ ratio of seawater are still unclear.

Excursions in the $^{87}Sr/^{86}Sr$ ratio of seawater similar to those in the Phanerozoic were almost certainly superimposed on the gradual decrease of this ratio in Precambrian time (see figure 6.25). The definition of these excursions will be an important contribution to our understanding of the evolution of the Earth, and should help to define the intensity of sea-water cycling through the oceanic crust during the Precambrian, especially if work in progress on the variation of the $^{143}Nd/^{144}Nd$ ratio in seawater with time resolves the present uncertainties regarding the relative importance of changes in the isotopic composition of Sr in world average river water and the reaction of Sr in seawater with the oceanic crust.

THE ISOTOPIC COMPOSITION OF OXYGEN IN SEAWATER

The decade of the 1970s saw a major shift in the interpretation of controls on the $\delta^{18}O$ value of seawater. In 1972 Li showed that the mean $\delta^{18}O$ value of sediments and sedimentary rocks is $+17 \pm 1‰$, some 10‰ more positive than the average $\delta^{18}O$ value of igneous rocks. He proposed that this ^{18}O excess was balanced by a decrease in the $\delta^{18}O$ value of water that had been released at high temperatures in isotopic equilibrium with silicate melts and that is now present on the Earth as seawater. The $\delta^{18}O$ value of oxygen in water equilibrated with average igneous rock melts is approximately $+7.5‰$. The isotopic composition of oxygen in present-day standard mean ocean water is, by definition, 0‰. If ocean water was originally released from silicate melts, the isotopic composition of its oxygen has become ca. 7.5‰ more negative. A simple mass balance calculation shows that this decrease could be balanced by the conversion of $(2.4 \pm 0.4) \times 10^{24}$ gm of average igneous rocks into sedimentary rocks (Li 1972, p. 122). This figure agrees admirably with the current best estimate for the mass of sedimentary rocks.

Unfortunately, the above calculation does not account for the $\delta^{18}O$ of metamorphic rocks; these are also enriched in $\delta^{18}O$, and comprise a large fraction of the continental crust. As Perry and Tan (1972) have pointed out, the ^{18}O excess in metamorphic rocks is large compared to that in sedimentary rocks; Li's (1972) ^{18}O balance calculation is therefore seriously in error. A slightly modified form of Perry and Tan's (1972) inventory of the crustal ^{18}O excess, is shown in table 6.9. The figures for the

TABLE 6.9.

Crustal reservoirs of $\delta^{18}O$ (modified from Perry and Tan 1972).

	Moles of "O_2" ($\times 10^{22}$)				
Rock Types	Ronov and Yaroshevsky (1969)	Ronov and Yaroshevsky (1976)	$\delta^{18}O$ (‰)	$\Delta\delta^{18}O$ (‰)	Oxygen gain $\equiv$ $\Delta\delta^{18}O \times$ moles oxygen ($\times 10^{22}$)
Granites, granitoids, and related rocks	4.49	} 8.9	+9	+3	} 27
Granodiorite, diorite, and related rocks	4.65		+9	+3	
Gabbros, basalts, and metamorphic equivalents		23.0	+6	0	0
Crystalline schists	2.09	} 8.1	+13	+7	} 50
Gneisses	9.01		+12	+6	
Marble	0.41	0.3	+15	+9	3
Sedimentary rocks	3.12	3.6	+17	+11	40

number of moles of oxygen in the first column of figures were taken from Perry and Tan (1972). The figures in the second column are based on a revised version of Ronov and Yaroshevsky's 1969 model of the Earth's crust (Ronov and Yaroshevsky 1976). The figures for the average $\delta^{18}O$ values of granites and granodiorites were taken from Taylor (1968), those for gneisses and schists from Garlick and Epstein (1967), and those for sedimentary rocks from Perry and Tan (1972). The values of $\Delta\delta^{18}O$ are the differences between the average $\delta^{18}O$ values of the several rock types and +6‰, the approximate $\delta^{18}O$ value of mantle material. The final column is the ^{18}O excess in each of the rock types. The sum of these excesses, some 120×10^{22} mil-mol, is three times the excess of ^{18}O in sedimentary rocks alone. Although all of the figures in the last column are rather uncertain, there is little doubt about the existence of a large imbalance between the crustal ^{18}O excess and the apparent ^{18}O deficit of the H_2O that is now ocean water.

Muehlenbachs and Clayton (1976) pointed out that this imbalance is probably due to the reaction of seawater with the oceanic crust at high and low temperatures. Under hydrothermal conditions the $\delta^{18}O$ value of submarine basalts decreases by approximately (1.3 ± 0.5)‰. The supporting data for this figure are shown in table 6.10. At low temperatures the flow of ^{18}O is in the opposite direction: weathered basalts on and below the ocean floor are enriched in ^{18}O with respect to seawater. This

TABLE 6.10.

The $\delta^{18}O$ value of hydrothermally altered submarine rocks (Muehlenbachs and Clayton 1976).

	$\delta^{18}O$	Number of samples
Greenstones	+5.3 ± 1.1	14
Gabbros	+4.6 ± 0.6	3
Serpentinites	+3.2 ± 1.2	20
Intrusives	+4.4 ± 1.4	4

effect is well documented by the data in figures 6.31 and 6.32. There is a great deal of scatter in the latter figure, but the results do indicate that the $\delta^{18}O$ value of altered basalts in the oceanic crust increases by approximately 4‰ during the first tens of millions of years of exposure to cold seawater at and below the ocean floor.

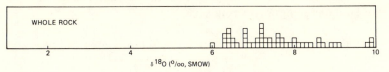

FIGURE 6.31. The isotopic composition of whole-rock samples from DSDP Legs 51–53 (Friedrichsen and Hoernes 1980). (Reproduced by permission of the authors.)

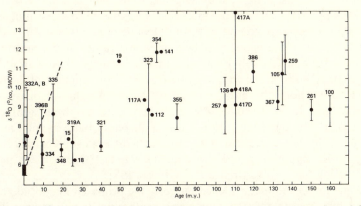

FIGURE 6.32. $\delta^{18}O$ value of basalts from DSDP sites versus their age as estimated from information given in the Initial Reports. Length of the bars represents total range of $\delta^{18}O$ found; the dot is the average value. Box at zero age delimits the $\delta^{18}O$ of fresh basalt. The dashed line is the weathering trend found in dredged basalts (Muehlenbachs 1980). (Reproduced by permission of the author.)

It is very difficult to convert such data into a believable set of rates for the several processes that are affecting the $\delta^{18}O$ value of seawater. A first attempt in this direction was made by Muehlenbachs and Clayton (1976). Table 6.11 reproduces many of their figures, together with results that seem more probable at the time of writing in 1982. During weathering on land and in the oceans the $\delta^{18}O$ value of solids increases and that of the weathering medium decreases. Muehlenbachs and Clayton (1976) arrived at the rate of weathering on land by dividing the mass of the continents by the age of the Earth. Their figure for the rate of ^{18}O increase due to weathering on land was then computed by assuming that the change in $\delta^{18}O$ of material undergoing weathering is +10‰. This procedure neglects the cycling of sediments through the mantle. The best figure for the actual rate of weathering is 20×10^{15} gm/yr (Holeman 1968). Approximately 75% of the material undergoing weathering consists of sedimentary rocks; the remaining 25% consists of a mixture of igneous and high-grade metamorphic rocks. The relative proportion of igneous and high-grade metamorphic rocks is not well known. If the mixture consists of approximately equal quantities of the two, then the average change of $\delta^{18}O$ in material undergoing weathering is ca.

$$(0.75 \times 0‰) + (0.125 \times 10‰) + (0.125 \times 5‰) \approx 1.8‰.$$

The uncertainty in this figure is large, probably on the order $\pm 0.4‰$. The ^{18}O use during weathering on land computed in this fashion $(5.6 \pm 2.0) \times 10^{14}$ mil-mol/yr, is approximately seven times as large as the rate of ^{18}O use computed by Muehlenbachs and Clayton (1976). These authors proposed that the oceanic crust is weathered to a depth of 600 m and that the average change in the $\delta^{18}O$ value of basalt is $+2.0 \pm 0.5‰$. It now appears that 600 m may be a minimum rather than a mean value (but see Ito and Clayton 1983), and that the average change in $\delta^{18}O$ is larger than +2‰. If the data in figure 6.32 are representative of the ocean floor as a whole, the mean change in the $\delta^{18}O$ value of ocean floor basalts may be closer to +4‰, and this figure has been adopted in table 6.11. The preferred ^{18}O use rate is therefore twice as high as that proposed by Muehlenbachs and Clayton (1976); even so, it is only half as large as the figure proposed above for the rate of ^{18}O use in continental weathering (see also Lawrence and Gieskes 1981).

Some rather indirect evidence supports the proposed magnitude of the $\delta^{18}O$ flux associated with submarine weathering. It was shown in the previous section that ca. 0.3% K are probably added to the upper 600 m of oceanic crust during low-temperature alteration by seawater. Muehlenbachs's (1980) data for the relationship between the K content of

TABLE 6.11.

Estimated ^{18}O budget of the oceans.

	Mass of material involved ($\times 10^{15}$ gm/yr)		Oxygen weight fraction	Moles of "O_2" exchanged ($\times 10^{13}$ mol/yr)		Change in $\delta^{18}O$		Oceanic ^{18}O flux in 10^{14} mil-mol/yr	
	M + C (1976)*	this volume		M + C (1976)*	this volume	M + C (1976)*	this volume	M + C (1976)*	this volume
Processes that decrease the $\delta^{18}O$ of sea water									
Continental weathering	0.5 ± 0.2	20 ± 3	0.50	0.8 ± 0.3	31 ± 5	10 ± 1	1.8 ± 0.4	−0.8 ± 0.3	−5.6 ± 2.0
Submarine weathering	4.9 ± 0.8	4.9 ± 0.8	0.45	6.9 ± 0.9	6.9 ± 0.9	2.0 ± 0.5	4 ± 1	−1.4 ± 0.6	−2.8 ± 1.2
Processes that increase the $\delta^{18}O$ of sea water									
High-temperature hydrothermal alteration	8.1 ± 2.7	50 ± 10	0.45	11 ± 3	70 ± 14	−1.3 ± 0.5	−1.6 ± 0.5	+1.4 ± 1.1	+17 ± 8
Dehydration of oceanic crust	0.9 ± 0.3	0.9 ± 0.3	0.89	2.5 ± 1.0	2.5 ± 1.0	−2.5 ± 1	−2.5 ± 1	+0.6 ± 0.6	+0.6 ± 0.6
Total								−0.2 ± 2.6	+9 ± 12

* M + C (1976) = Muehlenbachs and Clayton (1976)

altered basalts from DSDP Hole 417A and their $\delta^{18}O$ value are shown in figure 6.33. Additional samples analyzed since then also fall on or close to this line. A basalt to which 0.3% K has been added should have a $\delta^{18}O$ value ca. 3‰ more positive than that of fresh MOR basalt. This is encouraging, but the assumptions underlying the calculation and the dependence of the average value of $\delta^{18}O$ on the nature of the distribution of K in altered basalts should temper undue enthusiasm for the result of this computation.

The figure for the quantity of ocean crust affected by hydrothermal circulation in the first column of table 6.11 is based on a spreading rate of 2.94 km²/yr and a penetration depth of 1.0 km for hydrothermally cycled seawater. This depth now seems to be too small for the mean penetration depth of seawater in or near MORs. Christensen and Salisbury (1975) have pointed out that there are perhaps as much as 5 km of metaintrusives in the ocean crust.

H. Craig has shown that seawater that has reacted with MOR basalts at 21°N on the East Pacific Rise at temperatures of ca. 350°C reemerges with a $\delta^{18}O$ value of +1.6‰. If we accept the value of $(0.4 \pm 0.2) \times 10^{14}$ kg for the mass of seawater that cycles annually through MORs and emerges at temperatures of 300–350°C, the gain of ^{18}O by the oceans

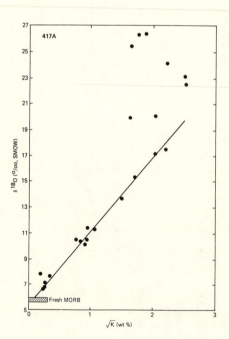

FIGURE 6.33. $\delta^{18}O$ value of basalts from Hole 417A versus the square root of their potassium content (Muehlenbachs 1980). (Reproduced by permission of the author.)

due to this process is approximately

$$+1.6\text{‰} \times (0.4 \pm 0.2) \times 10^{14}\, \frac{\text{kg seawater}}{\text{yr}} \times 27.5\, \frac{\text{mol ``O}_2\text{''}}{\text{kg seawater}}$$

$$= (17 \pm 8) \times 10^{14}\, \frac{\text{mil-mol}}{\text{yr}}.$$

The effect of seawater cycling through MORs at lower temperatures is uncertain, since the sign of the $\delta^{18}O$ exchange changes toward lower temperatures. It is clear, however, that the reaction of seawater with MOR basalts at and very close to ridge axes plays an important role in the $\delta^{18}O$ balance of the oceans.

Submarine weathering and hydrothermal alteration both introduce water into the oceanic crust. When oceanic crust is subducted, this chemically bound water is released into the mantle, its ^{18}O equilibrates with hot mantle rocks, and it is returned to the surface in isotopic equilibrium with the associated magmas at a $\delta^{18}O$ value between ca. $+6$‰ and $+9$‰. A good deal of the water of hydration in the oceanic crust is almost certainly hydrothermally cycled seawater. Muehlenbachs and Clayton (1976) suggested that the water content of the oceanic crust is ca. 2%, and that the change in $\delta^{18}O$ of this water on release from the oceanic crust is ca. 2.5‰. This figure is very uncertain. Fortunately, the calculated $\delta^{18}O$ flux due to the dehydration process is probably small compared to that of the other three fluxes in table 6.11.

The estimated fluxes in the last two columns of table 6.11 differ very significantly, but both sets yield a $\delta^{18}O$ balance for the oceans that is within the large uncertainties of the sums. Both sets of fluxes indicate that there is a rapid exchange of ^{18}O between the hydrosphere and the lithosphere, and that reactions between seawater and the oceanic crust are important for the geochemical cycle of the oxygen isotopes.

A rough notion of the buffering effect of the interaction of seawater with the oceanic crust can be extracted from the data in table 6.11. If we accept the figures for the flux of continental and submarine weathering, neglect the effect of water cycling, and adjust the $\delta^{18}O$ flux due to hydrothermal alteration so that the $\delta^{18}O$ fluxes balance, then the mass balance equation for $\delta^{18}O$ in the oceans can be written in the form,

$$\frac{d_{18}\delta_0}{dt} \times 3.9 \times 10^{22} = -0.56 \times 10^{14}[_{18}\delta_0 + 10]$$

$$- 0.14 \times 10^{14}[_{18}\delta_0 + 20]$$

$$- 5.2 \times 10^{14}[_{18}\delta_0 - 1.6], \qquad (6.37)$$

where $_{18}\delta_0$ is the $\delta^{18}O$ value of ocean water in per mil relative to SMOW. The coefficient of $d_{18}\delta_0/dt$ is the number of moles of "O_2" in ocean water. The first term on the right side of equation 6.37 maintains that the $\delta^{18}O$ flux due to continental weathering is proportional to $[_{18}\delta_0 + 10]‰$, and that of submarine weathering is proportional to $[_{18}\delta_0 + 20]‰$. The latter figure may be in need of explanation. Clays in altered ocean-floor basalts tend to have $\delta^{18}O$ values close to $+26‰$, i.e., ca. $20‰$ more positive than unaltered basalt. The figure of $+4‰$ for the change in the $\delta^{18}O$ value of basalts during weathering in the seventh column of figures of table 6.11 is a measure both of the $\delta^{18}O$ enrichment during complete alteration and the average degree of alteration of basalts in the upper 600 m of the ocean crust.

When $_{18}\delta_0 = 0$, the system is in balance, and $d_{18}\delta_0/dt$ is equal to zero. If the rate of sea-floor spreading changes by a factor ψ, it is likely that the rate of $\delta^{18}O$ use in sea-floor basalt weathering and the rate of $\delta^{18}O$ release by hydrothermal alteration will also change approximately by a factor ψ. If the rate of continental weathering remains unchanged, equation 6.37 becomes

$$\frac{d_{18}\delta_0}{dt} \times 3.9 \times 10^{22} = -0.56 \times 10^{14}[_{18}\delta_0 + 10]$$

$$- 0.14 \times 10^{14}\psi[_{18}\delta_0 + 20]$$

$$- 5.2 \times 10^{14}\psi[_{18}\delta_0 - 1.6]. \qquad (6.38)$$

After a new steady state has been reached, $d_{18}\delta_0/dt = 0$, and

$$_{18}\delta_0 = \left(\frac{\psi - 1}{0.1 + 0.95\psi}\right). \qquad (6.39)$$

Figure 6.34 is a plot of this function. When $\psi = 0$, i.e., when no sea-floor spreading is taking place, $_{18}\delta_0 = -10‰$. As ψ approaches 1.0, $_{18}\delta_0$ rapidly approaches $0‰$. When sea-floor spreading is more rapid than today, $\psi > 1$. The $\delta^{18}O$ value of seawater increases, but only very slightly with increasing ψ, and as ψ approaches ∞, $_{18}\delta_0$ approaches $+1.1‰$. For most likely values of ψ, the $\delta^{18}O$ value of ocean water therefore lies between $-2‰$ and $+1.0‰$. It follows that the cycling of seawater through MORs buffers the $\delta^{18}O$ value of seawater very effectively, unless the assumptions and the data on which these calculations are based turn out to be seriously in error. Muehlenbachs and Clayton (1976) arrived at the same conclusion. It could be argued that the rate of weathering on land is probably linked to the rate of sea-floor spreading through the rate of degassing of CO_2 and other acid volatiles (see chapter 5). If this is cor-

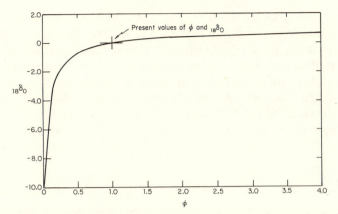

FIGURE 6.34. The probable effect of changes in the sea-floor spreading rate on the isotopic composition of oxygen in ocean water.

rect, then variations in the rate of sea-floor spreading will have an even smaller effect on $_{18}\delta_0$ than indicated by equation 6.39.

The rate of approach of the $\delta^{18}O$ value of seawater to a new steady state can be estimated, at least roughly, from equation 6.37. This equation can be recast into the form,

$$\frac{d_{18}\delta_0}{dt} = -\frac{5.9 \times 10^{14}}{3.9 \times 10^{22}}\,_{18}\delta_0 = -1.5 \times 10^{-8}\,_{18}\delta_0. \qquad (6.40)$$

Integration of equation 6.40 yields

$$_{18}\delta_0 = \,_{18}\delta_0{}'e^{-1.5 \times 10^{-8}t}, \qquad (6.41)$$

where $_{18}\delta_0'$ is the initial value of $\delta^{18}O$ at $t = 0$. The half life for the readjustment of steady state is therefore approximately

$$\tau_{1/2} \sim \frac{0.693}{1.5 \times 10^{-8}} \sim 46 \times 10^6 \text{ yrs.} \qquad (6.42)$$

This half-life is somewhat shorter than that suggested by Muehlenbachs and Clayton (1976).

The evolution of the isotopic composition of seawater during the last 4 b.y. has been the subject of a lively debate since the publication of Perry's (1967) paper on the isotopic composition of oxygen in ancient cherts. Oxygen in these cherts was found to be isotopically lighter than in recent cherts. Numerous additional measurements of the $\delta^{18}O$ value of cherts of all ages have confirmed the reality of Perry's trend, but have narrowed the difference between the $\delta^{18}O$ value of young and old cherts. Figure 6.35 summarizes a good deal of the available data. The highest values of $\delta^{18}O$ in cherts from the Onverwacht Group are distinctly lower than those of Phanerozoic cherts. Although there is general agreement on this point, there is little consensus concerning the interpretation of the difference. Degens and Epstein (1962) favored a greater influence of low-^{18}O meteoric waters on the $\delta^{18}O$ value of older cherts than on the $\delta^{18}O$ value of younger cherts. Perry (1967) and Perry et al. (1978) have favored the ef-

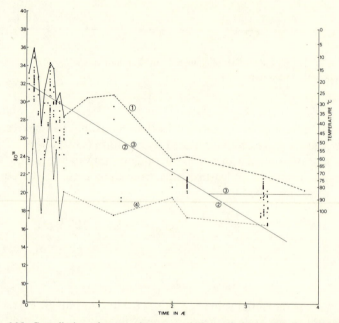

FIGURE 6.35. Compilation of oxygen isotope variations of cherts with time (Knauth and Lowe 1978). Curve 1 connects the largest δ-values for cherts at any given time. This curve is the current best estimate of the oxygen isotopic composition of cherts formed in marine or near-marine environments. Curve 2 represents the initial estimate of this secular change by Perry (1967). Curve 3 is the recent estimate by Perry, Ahmad, and Swulius (1978). Curve 4 connects the lowest δ-values reported for cherts at any given time. Over the time intervals where the data are more numerous, there is a suggestion of sympathetic changes of curves 1 and 4. The difference in δ-values for curves 1 and 4 at any given time is, in most cases, a measure of the isotopic variation of meteoric waters involved in early chert diagenesis. Although the Precambrian data are few, there is some suggestion that this isotopic difference increases with time. The dashed portions of curves 1 and 4 are for time intervals where the data are minimal. New Onverwacht data are shown at 3.4 b.y. Other data are from references 1, 2, 3, 4, 5, and 26 in Knauth and Lowe 1978. Data from deep-sea cherts and magadii-type cherts are excluded. (Reproduced by permission of Elsevier Scientific Publishing Company.)

fect of secular changes in the $\delta^{18}O$ value of seawater; Knauth and Epstein (1976) have championed the effects of changes in climate. The issue is currently unresolved. Hein and Yeh (1981) have pointed out that the $\delta^{18}O$ value of cherts in marine sediments from DSDP Sites 463, 464, 465, and 466 ranges from $+36.8‰$ to $+28.4‰$, that there is a general decrease in $\delta^{18}O$ with time, and that this decrease depends in a complicated manner on the mineralogy of the cherts. Knauth and Lowe (1978) have acknowledged the great variability in the origin of the cherts of the Onverwacht Group. It is still unclear to what extent the $\delta^{18}O$ values of the Onverwacht cherts reflect the origin of these cherts, the $\delta^{18}O$ value and temperature of the waters in which they formed, and the effects of diagenesis. If the lower $\delta^{18}O$ values of these ancient cherts are entirely due to temperature effects, the temperature in their area of deposition may have been as high as 70°C (Knauth and Lowe 1978).

The buffering effect of seawater cycling through hot basalts appears to be so strong, that large variations in the $\delta^{18}O$ value of seawater since Archean times are unlikely. The Li content of Cameron and Garrels's (1980) Archean shales is consistent with an intensity of Archean seawater cycling roughly equivalent to that during more recent periods of Earth history (see above). If this turns out to have been the case, the $\delta^{18}O$ value of chemical sediments can tell us something about conditions during deposition and/or diagenesis, but very little about variations in the rate of cycling of seawater through mid-ocean ridges during the past 3.5 b.y.

SEA-FLOOR SPREADING RATES AND CHANGES IN SEA LEVEL

It seems likely that the rate of seawater cycling through MORs and the quantity of weathering at and below the ocean floor are roughly proportional to the global rate of sea-floor spreading. One could, however, imagine that the functional relationship is complicated. At present, high-temperature hydrothermal activity at the axis of fast-spreading ridges such as the East Pacific Rise is much more intense than at slow-spreading ridges such as the Mid-Atlantic Ridge. The total amount of seawater cycling through MORs at a given time is therefore apt to depend not only on the mean rate of production of new ocean floor but on the range and distribution of spreading rates in the world ocean (for a summary of pertinent data, see Anderson and Skilbeck 1981).

The identification of periods of particularly rapid sea-floor spreading is still a somewhat uncertain undertaking. Hays and Pitman (1973) published estimates of spreading rates of most of the world's ridge segments between 110 m.y. and 10 m.y. ago, and proposed that the mean spreading rate between 110 and 85 m.y.b.p. was more than twice the mean spreading rate between 85 m.y.b.p. and 10 m.y.b.p. Their conclusion was severely criticized by Berggren et al. (1975), who pointed out that the existing data do not require any change at all in the average spreading rate between 110 m.y.b.p. and 85 m.y.b.p.; they suggested that the period of rapid sea-floor spreading proposed by Hays and Pitman (1973) could well be an artifact of small errors in the timing of magnetic reversals during the Jurassic and Cretaceous periods. In their reply to this critique Larson and Pitman (1975) expressed doubt that their proposed increase in spreading rates could be dismissed on the grounds given by Berggren et al. (1975), but they agreed that additional geochronologic and biostratigraphic work on the Early Cretaceous and Late Jurassic is required to settle the matter. The more recent paper by Pitman (1978) improved on the earlier computations of spreading rates during the Cretaceous; the history of the Pacific ridges was better known when these new computations were made, and a reasonably complete picture of the evolution of the Indian Ocean ridges had been developed. However, most of the uncertainties in the magnetic polarity time scale and in the correlation of this time scale with the paleontologic time scale remained, and the validity of the quantitative aspects of Pitman's (1978) conclusions is still uncertain. Nevertheless, the basic thrust of Pitman's (1978) argument surely has merit; changes in spreading rate and in the length of spreading ridges have probably been among the important causes of those worldwide eustatic changes in sea level that are not related to changes in the volume of continental ice sheets. The reality of worldwide changes in sea level and the comings and goings of the sea have been known and discussed for more than a century. If changes in the volume of spreading ridges turn out to be the major cause for these marine transgressions and regressions, then the stand of sea level in the past can be used to define, at least roughly, changes in the rate of basalt-seawater interaction at low and high temperatures.

Schlanger, Jenkyns, and Premoli-Silva (1981) have reviewed the recent literature dealing with sea level changes during the Cretaceous marine transgression. The differences of opinion regarding the magnitude of these changes are somewhat alarming. The minimum figure given recently is that of Watts and Steckler (1979), who estimated that the rise in sea level from a datum at 110 m.y.b.p. was some 100 m at the height of the Cretaceous transgression. The maximum figure proposed for this rise is that of Hancock and Kauffman (1979), who suggested an early Albian to mid-Maastrichtian sea level rise of 650 m. Figure 6.36 shows the relative changes in sea level during the Phanerozoic as proposed by Vail and Mitchum (1979); these authors estimate that sea level reached a high point near the end of the Campanian (late Cretaceous) some 350 m above present sea level, and low points during the Early Jurassic, middle

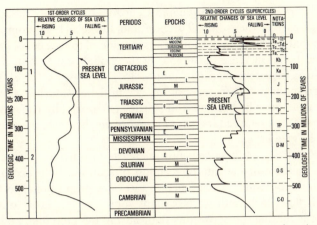

FIGURE 6.36. First- and second-order global cycles of relative changes of sea level during Phanerozoic time (Vail and Mitchum 1979). (Reproduced by permission of the American Association of Petroleum Geologists.)

Oligocene, and late Miocene about 150, 250, and 200 m, respectively, below present sea level.

Several attempts have been made to use these data and information regarding changes in the degree of continental flooding in order to estimate the fluctuations in heat flow from the Earth's interior during the Phanerozoic. Figures 6.37a and 6.37b show the data used by Turcotte and Burke (1978); figure 6.37c is their proposed curve for the variation of heat flow from the interior of the Earth during the Phanerozoic Eon. The

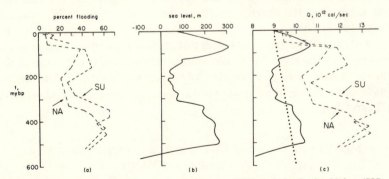

FIGURE 6.37. (a) Percent flooding of North America (NA) and the Soviet Union (SU) as given by Hallam (1977) on the basis of paleogeographic maps. (b) Relative paleo-sea level as determined by Vail, Mitchum, and Thompson (1977). (c) Paleo-heat flow curves calculated by Turcotte and Burke (1978). (Reproduced by permission of Elsevier Scientific Publishing Company.)

sea level curve for the Mesozoic and Cenozoic is probably the least controversial part of their data. Turcotte and Burke's (1978) computations indicate that the large Cretaceous transgression of the sea could have been produced by an increase of ca. 15% in the global heat flow; if the continental heat flow remained constant, a 15% increase in the global heat flow corresponds to an increase of ca. 25% in the oceanic heat flow. More recent calculations by Sprague and Pollack (1980) and by Harrison (1980) indicate that the required increase in the oceanic heat flux may have been somewhat greater, ca. 35%.

The conversion of these heat-flow increases into figures for the accompanying increase in the rate of reaction of seawater with ocean basalts is not straightforward. If the reaction rates are proportional to the oceanic heat flow, then both were perhaps some 25–35% faster during the Cretaceous marine transgressions than during the preceding and following periods.

Such changes would have had a rather small effect on the $^{87}Sr/^{86}Sr$ ratio of seawater. It could be argued that this is consistent with the absence of a large excursion in the $^{87}Sr/^{86}Sr$ ratio of seawater during the Cretaceous (see figure 6.24). However, the time variation of the $^{87}Sr/^{86}Sr$ ratio in seawater during the past 250 m.y. is clearly not correlated with the time variation of sea level; this suggests that one or both of these parameters are essentially unrelated to the cycling of seawater through the oceanic crust.

The extensive flooding of the continents during Cambro-Ordovician time is well documented (see figure 6.36), but the accompanying changes in sea level are even less well known than those for the Cretaceous marine transgression. Hallam's (1977) data in figure 6.37 could imply a considerable increase in the oceanic heat flow during this period (Turcotte and Burke 1978). The course of the $^{87}Sr/^{86}Sr$ ratio in limestones during the lower Paleozoic in figure 6.24 does not show a pronounced minimum at the height of the lower Paleozoic marine transgression. The cause of this transgression had an effect on the isotopic composition of marine Sr that was apparently no more pronounced than the cause of the Cretaceous transgression.

Summary of Data for Changes in the Rate of Seawater Cycling through MORs

The data in this section are somewhat unsatisfactory. The available analyses of the lithium content of sedimentary rocks suggest very strongly that seawater cycling through MORs has not varied greatly since the

Archean. However, the scatter in the analyses is large, and the data do little to define the course of the intensity of basalt-seawater interaction during the Phanerozoic Eon. The $\delta^{18}O$ value of seawater turns out to be too insensitive an indicator as well. The evidence from sea level changes disagrees strongly with that from the history of the $^{87}Sr/^{86}Sr$ ratio in seawater. It will be shown in chapter 9 that the time variation of the $^{87}Sr/^{86}Sr$ ratio in seawater is rather well correlated with the time variation of the $\delta^{34}S$ value of seawater, and that the variation of the composition of both isotopes may well have been determined largely by changes in the cycling rate of seawater through the oceanic crust. If this turns out to be correct, the curve in figure 6.24 is the best available indicator of changes in the rate of seawater cycling through the oceanic crust during the Phanerozoic Eon.

5. A Hypothesis for the Variation in the Abundance of Dolomite during the Phanerozoic Eon

The proposal that the time variation of the $^{87}Sr/^{86}Sr$ ratio in seawater is a fair indicator of variations in the rate of seawater cycling through the oceanic crust during the Phanerozoic Eon has interesting implications for the interpretation of the history of carbonate sedimentation during the past 600 m.y. The general increase in the rate of seawater cycling from the Cambrian to the Jurassic that is indicated by the data in figure 6.24 implies that the proportion of the river input of Mg to the oceans that was removed by transfer to the oceanic crust increased in a general way during the first 400 m.y. of the Phanerozoic Eon. We can make only a very rough estimate of the magnitude of the effects of this increase on dolomite formation. As a rough first order approximation, let the rate of dolomite formation in equation 6.5 be set equal to

$$\Phi_{Dol.} \approx \alpha_0 m_{Mg^{+2}}, \tag{6.43}$$

where $_0 m_{Mg^{+2}}$ is the concentration of Mg^{+2} in seawater and where α is a parameter that depends on a host of chemical, physical, and biological factors. If we neglect the term Φ_{BLT} in the mass balance equation for Mg^{+2}, equation 6.5 can be rearranged into the form,

$$_0 m_{Mg^{+2}} \approx \frac{_{R.W.} m_{Mg^{+2}} \dfrac{dV_{R.W.}}{dt}}{\dfrac{dV_0}{dt} + \alpha}. \tag{6.44}$$

The rate of dolomite formation is therefore given very approximately by the equation

$$\Phi_{\text{Dol.}} \approx \frac{\text{R.W.}\, m_{\text{Mg}+2}\, \dfrac{dV_{\text{R.W.}}}{dt}}{\dfrac{1}{\alpha}\dfrac{dV_0}{dt} + 1}.\tag{6.45}$$

As dV_0/dt increases, $\Phi_{\text{Dol.}}$ decreases and approaches zero when dV_0/dt becomes sufficiently large so that essentially all of the river input of Mg^{+2} is transferred into the oceanic crust. The decrease in the dolomite content of carbonate rocks on the Russian Platform during the Paleozoic Eon (see figure 5.22) is consistent with this very simplified analysis. The similarity of the Mg variation with time in carbonates of the North American Platform suggests that this pattern may have been worldwide. The curve in figure 5.22 for the Mg content of carbonate sediments is not particularly detailed, and it is possible that its generality obscures several of the rather major fluctuations that are visible in figure 6.24 for the $^{87}Sr/^{86}Sr$ ratio in seawater during the Paleozoic Eon.

Unfortunately the relationship between figure 6.24 and figure 5.22 disappears completely during the last 150 m.y. of Earth history. The expected rise in the fraction of dolomite in carbonate rocks accompanying the well-documented rise in the $^{87}Sr/^{86}Sr$ ratio of seawater during this period did not occur. Several possible explanations for the divergence come to mind. It is possible, for instance, that the rise in the $^{87}Sr/^{86}Sr$ ratio of seawater since the Jurassic is due to an increase in the mean $^{87}Sr/^{86}Sr$ ratio of the river input of Sr to the oceans rather than due to a decrease in the cycling of seawater through the oceanic crust. It is also possible, however, that the ease with which dolomite was formed in carbonate sediments decreased during this period. This is equivalent to a decrease in the value of the coefficient α in equation 6.43. If seawater cycling through the oceanic crust has decreased by a factor of ca. 3 since the Jurassic, then the value of α must have decreased by at least the same factor to account for the paucity of dolomite in Jurassic and more recent carbonate sediments.

One can imagine several reasons for changes in the coefficient α. Baker and Kastner (1981) have proposed that the effect may be due to changes in the SO_4^{-2} content of seawater. It will be shown in chapter 9 that the SO_4^{-2} concentration of seawater has probably varied significantly during the past 900 m.y., but that it has probably never been as low as seems to be required to explain the inferred time variation in α. There are almost certainly other dissolved constituents in seawater that also act as catalysts for or inhibitors of dolomite nucleation and crystal growth. Unfortunately, too little is known about the kinetics of dolomite formation (see, for instance, Gaines 1980) to evaluate these potential effects.

Biological effects look somewhat more promising. The sedimentology of Precambrian dolomites indicates that a good many of them were deposited in non-sabkha environments, and that dolomite deposition was frequently linked to the presence of algal mats in intertidal and shallow subtidal settings (see above). These algal mats have become progressively rarer during the Phanerozoic Eon. It is therefore possible that the formation of dolomite was considerably easier in pre-Mesozoic times than it is at present, that a smaller degree of supersaturation of the oceans with respect to dolomite was required for dolomite formation in the pre-Mesozoic oceans, and that the Precambrian oceans were able to eliminate the river input of Mg^{+2} without requiring the high Mg^{+2}/Ca^{+2} ratio of present-day seawater.

Other biological changes during the past 150 m.y. could also have been involved. Foraminifera evolved a planktonic mode of life in the Middle Jurassic (see, for instance, Vincent and Berger 1981). Their first morphologic proliferation began in Albian times. Calcareous nannofossils developed roughly contemporaneously. The earliest unequivocal "coccolith" is found in the earliest Jurassic, and the group evolved and radiated rapidly shortly thereafter (see, for instance, Gartner 1981). The development of planktonic calcareous organisms during the Jurassic has had a profound effect on environments of carbonate deposition. A large fraction of post-Jurassic carbonate sediments has accumulated in deep waters. These sediments consist largely of low magnesium calcite, which is rather resistant to dolomitization, especially at the low temperatures characteristic of the deep ocean floor. These observations probably go far toward explaining the lack of dolomitization in deep sea carbonate sediments. They do not, however, explain why aragonitic carbonate sediments deposited in shallow tropical settings during the past 150 m.y. have not been more thoroughly dolomitized.

The hypothesis that the formation of dolomite has been particularly difficult during the last 150 m.y., that the rate of seawater cycling through MORs has decreased, and that the Mg content of seawater has increased significantly since the Jurassic seems to fit all of the available data and to contradict none (see, for instance, chapter 9). The hypothesis is not, however, firmly based, and the explanations for the proposed difficulty of dolomite formation since the Jurassic have a distinctly ad hoc flavor. It is likely that a definition of the $^{143}Nd/^{144}Nd$ ratio variations in marine sediments and of the Mg^{+2}/Cl^- ratio in seawater during the Phanerozoic Eon will do much to test the validity of the hypothesis. Even if the hypothesis turns out to be correct, it may be very difficult to establish with certainty the reasons for the severe inhibition of dolomite formation during the past 150 m.y.

6. A Hypothesis for Time Variations in the Potassium Content of Shales

The decrease in the potassium content of the shales of the Russian Platform during the Phanerozoic Era is one of the best documented of the time variations in the composition of sedimentary rocks (see figure 5.22). A similar decrease has been found in the potassium content of shales of the North American Platform, and this similarity has suggested to some that the decrease may be worldwide (see, for instance, Garrels and Mackenzie 1971, pp. 230–237). It will be shown that the change in the K_2O content of shales was probably not worldwide, and that, if it does turn out to have been worldwide, the change was probably not related to changes in the nature of weathering or to changes in the intensity of basalt-seawater interaction.

It is useful to set the observed changes in the K_2O content of shales within a framework similar to that developed earlier in this chapter for the analysis of changes in the $MgO_{sil.}$ content of sedimentary rocks. The K_2O content of sedimentary rocks depends on the K_2O content of their parent rocks and on the multitude of additions and subtractions that take place during the conversion of igneous rocks into sedimentary rocks. It is usually impossible to identify all of the reasons for differences between the composition of any two shales, but plots similar to those developed earlier can at least identify those differences that are unlikely to be due to differences in the composition of their igneous source rocks.

The variation of the K_2O content of igneous rocks with total iron content was described in chapter 5; the igneous rock trend line for K_2O in figure 5.9 has been reproduced in figure 6.38. Since the average igneous rock is a mixture of igneous rocks along this trend line, the K_2O content of average igneous rock must lie within the crescent-shaped area between the igneous rock trend line and the line connecting its end points. The composition of Clarke's average igneous rock (see table 5.5, analysis no.11) falls, as expected, within the crescent-shaped area.

The analyses of shales of the Russian Platform in table 5.12 have been used to calculate the function $(FeO + 2Fe_2O_3)_{corr.}$ for these shales by multiplying the total number of moles of iron per kg of shale by the ratio 15.6/wt. percent Al_2O_3; the function $K_2O_{corr.}$ for the shales was calculated by multiplying the number of moles of K_2O per kg of shale by the ratio 15.6/wt. percent Al_2O_3. These Al_2O_3-normalized K_2O and total iron concentrations have been plotted in figure 6.38. With the exception

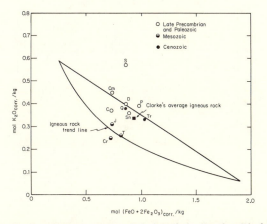

FIGURE 6.38. The K_2O and iron content of shales of the Russian Platform; data from Vinogradov and Ronov (1956b).

of the Silurian shales, the compositional data for the Russian shales plot within or close to the crescent-shaped area. This implies that the observed variations in the K_2O content of shales on the Russian Platform require little or no addition or subtraction of K_2O during their derivation from a variety of parent igneous rocks. The K_2O content of the late Precambrian, Paleozoic, and Cenozoic shales can be explained in terms of an ultimate derivation from mixtures of granitic and basaltic rocks; the K_2O content of the Mesozoic rocks can be explained in terms of an ultimate derivation from intermediate igneous rocks. Only the production of the Silurian shales requires the addition of large quantities of potassium from another source.

These observations do not prove that additions and subtractions of potassium have not been important during the formation of the Russian shales. The relatively low K_2O content of the Mesozoic shales of the Russian Platform could, for example, be explained equally well by the loss of potassium and iron from igneous rocks of a composition similar to that of Clarke's average igneous rock. This ambiguity attaches to the composition of most shales. Ronov, Migdisov, and Lobach-Zhuchenko (1977) have shown that the concentration of all but the volatile elements in shales remains nearly invariant during low- and medium-grade metamorphism. It is therefore of interest to compare their analyses of Russian metapelites, particularly Precambrian metapelites, with those of unmetamorphosed Russian shales. Their data for metapelites in figure 6.39 fall within or close to the K_2O-total iron crescent. There seems to be little change with time in the position of the analyses within this crescent; it

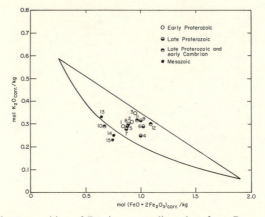

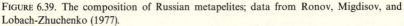

FIGURE 6.39. The composition of Russian metapelites; data from Ronov, Migdisov, and Lobach-Zhuchenko (1977).

Anal. No.	Area	Age
1, 2, 3	North Ladoga	Early Proterozoic·
4, 5, 6	North Baykal Uplands	Late Proterozoic
7, 8, 9	Vitim-Patom Uplands	Late Proterozoic
10, 11, 12	Patom Uplands	Late Proterozoic and early Cambrian
13, 14, 15	Central Range in Kamchatka	Mesozoic

is interesting and perhaps significant that analyses 13, 14, and 15 of Mesozoic metapelites from the Central Range in Kamchatka fall in the same low-K_2O part of the crescent as the composition of unmetamorphosed Mesozoic shales of the Russian Platform.

Figure 6.40 shows that Cameron and Garrels' (1980) Archean shales from the Canadian Shield fall in much the same part of the crescent as well; the analyses of their Aphebian shales, however, scatter widely. Their shale composites from Albanene (14), Attikamagen (16), and Menihek (18) are greatly enriched in potassium; in this they are similar to some of the North American Devonian black shales. Without the three K-rich composites, the average $K_2O_{corr.}$ content of their Aphebian shales would be similar to that of their average Archean shale and to the average of the Proterozoic Russian metapelites in figure 6.39. Figure 6.41 is a summary plot of a great deal of DSDP data for sediments, largely Cenozoic, from the Caribbean, the western Atlantic Ocean, and the Pacific Ocean. The sediment analyses from many of the DSDP sites represented in this plot fall within the K_2O-total iron crescent, but a sufficient number of analyses fall below the crescent, so that the mean $K_2O_{corr.}$ content of all sites falls in the lowermost part of the crescent. A rather

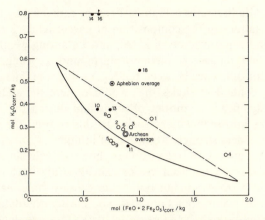

FIGURE 6.40. The K_2O and total iron content of Archean and Aphebian shales from the Canadian Shield; data from Cameron and Garrels (1980); see table 5.9.

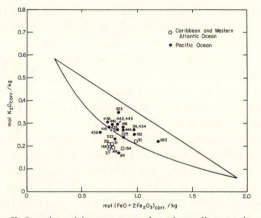

FIGURE 6.41. The K_2O and total iron content of marine sediments; the numbers in this figure refer to DSDP sites. Sources of data: Sites 27, 29, 30, 31, 148, 149, 154, 34, 84, 166, 192, 283, 322, 323, 325, Donnelly and Wallace 1976; Sites 434, 435, 436, 438, 439, 440, 441, Sugisaki 1980a; Sites 442, 443, 444, Sugisaki 1980b.

more complete compilation and a more thorough analysis of the DSDP data are needed to relate these analyses to those of platform sediments. In particular, the effect of basal sediments in DSDP holes has to be assessed; these sediments are frequently enriched in iron due to hydrothermal processes, and some are strongly enriched in potassium due to palagonitization (see, for instance, Murdmaa et al. 1979).

All of the data in figures 6.38 to 6.41 can be explained by the combined effects of variations in the composition of parent igneous rocks and the dispersion and concentration of K, Al, and Fe during weathering, transport, deposition, diagenesis, metamorphism, and recycling. There seems to be no major difference between the K_2O content of igneous rocks and the $K_2O_{corr.}$ content of shales that would require the addition or subtraction of significant quantities of potassium during the conversion of igneous rocks into sedimentary rocks. In this respect, the geochemistry of potassium differs strongly from that of lithium.

The data in figures 6.38 to 6.41 do not, however, rule out the possibility that some potassium has been lost or gained during the formation of all shales, or that periods of major potassium gain by shales have alternated with periods of major potassium loss. Specifically, the data do not rule out the possibility that the decrease in the $K_2O_{corr.}$ content of shales of the Russian Platform during the Phanerozoic Era were part of a worldwide pattern of K_2O loss. Two mechanisms look potentially feasible: changes in the behavior of potassium during terrestrial weathering due to the development of higher land plants, and changes in the reaction of seawater with oceanic basalts; it will be shown that both mechanisms seem to be inadequate to explain the observed effects.

Basu (1981) has suggested that in the absence of land plants, i.e., before the Silurian period, K-feldspars were stable during terrestrial weathering. As evidence for this he cites the presence of abundant fresh K-feldspar and the absence of plagioclase in the Cambrian-Ordovician arenites of Ohio, Wisconsin, Minnesota, Missouri, Texas, Wyoming, and Montana. If his hypothesis is valid, the development of higher land plants exerted a considerable influence over the distribution of potassium in sedimentary rocks. The available evidence does not, however, support this proposition. There is no significant change in the K_2O content of sandstones of the Russian Platform during the Phanerozoic Eon (see table 5.13). The average K_2O content of sandstones deposited between the Upper Proterozoic Era and the end of the Silurian period is ca. 2.3%; the average K_2O content of more recent sandstones on the Russian Platform is ca. 2.6%. The abundance of illite and illitic clays in pre-Silurian sedimentary rocks is also difficult to explain without the destruction of potash feldspars during weathering and the subsequent incorporation of released potassium into clay minerals. Whatever the correct explanation of the fresh Cambrian-Ordovician potassium feldspars discussed by Basu (1981) turns out to be, it probably does not involve the development of the cover of higher land plants during the latter part of the Paleozoic Era.

The effect of the interaction of seawater with oceanic basalts is also insufficient to account for the observed differences in the K_2O content

TABLE 6.12.

Marine fluxes of potassium in units of 10^{13} gm/yr.

Flux	Source
River Flux into the Oceans	
7.4	Hart and Staudigel (1982)
	Holland (1978)
Hydrothermal Flux into Oceans	
0.4	Hart and Staudigel (1982) "Best value"
2.5	Hart and Staudigel (1982) Maximum value
2.3 $\pm$ 1.2	This chapter
Weathering Flux into Oceanic Crust	
1.5	Hart and Staudigel (1982)

of shales on the Russian and North American platforms. Table 6.12 summarizes the available data for the present-day marine fluxes of potassium. All of the figures are uncertain, but the uncertainties are not large enough to invalidate the proposed qualitative conclusion. The river flux of K is considerably larger than the hydrothermal flux from MORs and the flux into the cold oceanic crust. The difference between the hydrothermal input of K to the oceans and the output via the weathering of the oceanic crust probably falls between ca. $+1 \times 10^{13}$ gm/yr and -1×10^{13} gm/yr. If this difference is added uniformly to or subtracted from marine sediments that enter the oceans at the rate of ca. 2×10^{16} gm/yr, the K content of these sediments will be changed by

$$\Delta K \approx \frac{\pm 1 \times 10^{13} \text{ gm K/yr}}{2 \times 10^{16} \text{ gm sediment/yr}} \times 100 = \pm 0.05\%.$$

This is a very small fraction of the changes in the $K_2O_{corr.}$ content of shales of the Russian and North American platforms during the Phanerozoic Eon (see table 5.11).

A nearly independent cross-check of this calculation can be obtained by another route. The ratio of the excess K to that of Li in the high-temperature vent waters at 21°N (see table 6.3) is

$$\frac{\Delta K}{\Delta Li} = \frac{(25.0 - 10.1) \text{ mmol}}{(0.820 - 0.028) \text{ mmol}} = 106 \frac{\text{gm K}}{\text{gm Li}}.$$

If all of the K and Li from this source were added to marine sediments and if the resulting Li increment is 20 ppm, the K increment would be ca. 0.2%. Since K is lost more readily than Li to the oceanic crust by submarine weathering (see above), the actual maximum increase must be less than 0.2%. This is consistent with the maximum value of 0.05% for ΔK calculated above.

If the interpretation proposed earlier for the variation of the $^{87}Sr/^{86}Sr$ ratio in seawater is correct, then seawater cycling through MORs has been significantly more intense during several parts of the Phanerozoic Eon than it is presently. However, it is easy to show that even at these more rapid rates of seawater cycling, the sinks and sources of K within the oceanic crust would have been much too small to account for the changes in the $K_2O_{corr.}$ content of sediments on the Russian and North American platforms. It is likely, therefore, that the large observed changes in the composition of the Russian and North American shales do not reflect global changes in the composition of shales; they are probably due to changes in the composition of their particular source rocks with time and/or to changes in the losses and gains of potassium by shales on these two platforms during conversion into new sedimentary rocks.

7. Summation

The reaction of seawater with ocean basalts at temperatures between $0°$ and ca. $400°C$ is almost certainly of importance for the chemistry of the oceans today. Mid-ocean ridge basalts are probably the major sink of the river input of Mg as well as an important source of Ca and K for the oceans. Lithium is among the elements that are also released in geochemically significant quantities to seawater at MORs; a part of this lithium is returned to the ocean crust during reaction with ocean-floor basalts at low temperatures. The remainder is added to marine sediments and almost certainly accounts for the observed difference between the concentration of Li in sedimentary and igneous rocks.

The reaction of seawater with ocean basalts affects not only elemental abundances in the oceans and in sedimentary rocks, but isotopic abundances as well. The reaction of seawater Sr with ocean basalts reduces the $^{87}Sr/^{86}Sr$ ratio in the reacting seawater to values close to those of MOR basalts. The effect of this process on the $^{87}Sr/^{86}Sr$ ratio of the oceans as a whole is significant. In its absence the $^{87}Sr/^{86}Sr$ ratio in seawater would probably be close to 0.7110.

The reaction of seawater with basalts affects the isotopic composition of oxygen in seawater even more strongly than the isotopic composition of strontium. The reaction of seawater with MOR basalts at hydrothermal temperatures decreases the $\delta^{18}O$ value of the basalts with which it reacts by 1 to 2‰, and this process almost certainly acts as an effective long-term buffer for the isotopic composition of oxygen in the entire body of ocean water.

Variations in the intensity of the reaction of seawater with ocean basalts are still poorly defined, but all of the presently available criteria suggest that the rate of seawater cycling through MORs has not varied by more than approximately a factor of three during the Phanerozoic Eon. The concentration of Li in Archean and Proterozoic shales indicates that seawater cycling through basalts was active during these periods. However, the process does not seem to have dominated marine chemistry 2 or 3 b.y. ago, and it is probably a misnomer to call the oceans of that period "volcanogenic."

The fluctuations in the $^{87}Sr/^{86}Sr$ ratio of seawater during the Phanerozoic Era have probably been produced in large part by changes in the rate of seawater cycling through MORs. The single most compelling observation supporting this inference is the similarity between the time variation of the $^{87}Sr/^{86}Sr$ ratio (figure 4.24) and that of the $\delta^{34}S$ value (figure 9.12) of seawater during the Phanerozoic Era. Both parameters decreased in a general way from the Cambrian to the Permian and increased from the Jurassic to the present.

The decrease in the dolomite/limestone ratio between the Cambrian and the end of the Jurassic is to be expected if the course of the isotopic composition of Sr and S during this interval signals a rather unsteady increase in the cycling of seawater through MORs. The relative scarcity of dolomite in carbonate sediments of the past 150 m.y. is inconsistent with the proposed interpretation of the isotopic data. The scarcity of dolomite during this period seems to demand that dolomite formation has become progressively more difficult since mid-Mesozoic time. Several biological factors could be invoked to explain this change.

The present flux of K into the oceans due to high-temperature reactions of seawater with MORs is probably much smaller than the river flux of K to the oceans. The magnitude of the K flux out of the oceans due to submarine weathering of basalts is comparable to that of the high-temperature flux into the oceans; the sign of the net flux of K due to reactions of seawater with oceanic crust may be either positive or negative. The absolute value of the net flux is too small to account for more than a small fraction of the observed changes in the K content of the shales of the Russian and North American platforms if these changes reflect worldwide changes in the K content of shales. No other mechanisms are currently available to account for such large changes on a worldwide scale, and it seems likely that the observed changes are not worldwide.

Although the length of this chapter testifies to the size of the body of knowledge that has accumulated during the past decade regarding the nature and intensity of basalt-seawater interaction at low and high temperatures, the number of qualifying adjectives in the chapter is a measure of the inadequacy of the presently available data to define the intensity of basalt-seawater interaction in the past. If the next decade turns out to be as fruitful as the last, it may be possible to test many of the inferences and hypotheses that have been ventured in this chapter; we may even be able to illuminate issues that are presently beyond the limits of decent speculation.

References

Albarède, F., Michard, A., Minster, J. F., and Michard, G. 1981. $^{87}Sr/^{86}Sr$ ratios in hydrothermal waters and deposits from the East Pacific Rise at 21°N. *Earth Planet. Sci. Lett.* 55:229–236.

Anderson, R. N., and Skilbeck, J. N. 1981. Oceanic heat flow. In *The Oceanic Lithosphere*, vol. 7 of *The Sea*, edited by C. Emiliani, chap. 7. New York: Wiley-Interscience.

Awramik, S. M. 1982. The pre-Phanerozoic fossil record. In *Mineral Deposits and the Evolution of the Biosphere*, edited by H. D. Holland and M. Schidlowski, 67–81. New York: Springer-Verlag.

Baker, P. A., and Kastner, M. 1981. Constraints on the formation of sedimentary dolomite. *Science* 213:214–216.

Baker, P. A., Gieskes, J. M., and Elderfield, H. 1982. Diagenesis of carbonates in deep-sea sediments: Evidence from Sr/Ca ratios and interstitial dissolved Sr^{2+} data. *Jour. Sed. Petrol.* 52:71–82.

Barghoorn, E. S., and Tyler, S. A. 1965. Microorganisms from the Gunflint chert. *Science* 147:563–577.

Barron, J. A. 1980. Lower Miocene to Quaternary diatom biostratigraphy of Leg 57, off northeastern Japan, Deep Sea Drilling Project. In *Initial Reports of the Deep Sea Drilling Project*, vol. 56, 57, part 2, chap. 17. Washington, D.C.: U.S. Govt. Printing Office.

Basu, A. 1981. Weathering before the advent of land plants: Evidence from unaltered detrital K-feldspars in Cambrian-Ordovician arenites. *Geology* 9:132–133.

Berggren, W. A., McKenzie, D. P., Sclater, J. G., and Van Hinte, J. E. 1975. World-wide correlation of Mesozoic magnetic anomalies and its implications: Discussion and Reply. *Bull. Geol. Soc. Amer.* 86:267–269.

Brass, G. W. 1976. The variation of the marine Sr^{87}/Sr^{86} ratio during Phanerozoic time: Interpretation using a flux model. *Geochim. Cosmochim. Acta* 40:721–730.

Burke, W. H., Denison, R. E., Hetherington, E. A., Koepnick, R. B., Nelson, H. F., and Otto, J. B. 1982. Variation of seawater $^{87}Sr/^{86}Sr$ throughout Phanerozoic time. *Geology* 10:516–519.

Burton, J. D., and Culkin, F. 1970. Gallium. In *Handbook of Geochemistry*, edited by K. H. Wedepohl, vol. 2.3, secs. 31-B to 31-0. New York: Springer-Verlag.

Cameron, E. M., and Garrels, R. M. 1980. Geochemical compositions of some Precambrian shales from the Canadian Shield. *Chem. Geol.* 28:181–197.

Christensen, N. I., and Salisbury, M. H. 1975. Structure and constitution of the lower oceanic crust. *Rev. Geophys. Space Phys.* 13:57–86.

Craig, H., and Lupton, J. E. 1981. Helium-3 and mantle volatiles in the ocean and the oceanic crust. In *The Oceanic Lithosphere*, vol. 7 of *The Sea*, edited by C. Emiliani, chap. 11. New York: Wiley Interscience.

Davies, P. J., Ferguson, J., and Bubela, B. 1975. Dolomite and organic material. *Nature* 255:472–474.

Degens, E. T., and Epstein, S. 1962. Relationship between $^{18}O/^{16}O$ ratios in coexisting carbonate, cherts, and diatomites. *Bull. Amer. Assoc. Petrol. Geologists* 46:534–542.

Donaldson, J. A. 1963. Stromatolites in the Denault Formation, Marion Lake, Coast of Labrador, Newfoundland. Bulletin 102, Geological Survey of Canada.

Donnelly, T. W., and Wallace, J. L. 1976. Major and minor element chemistry of Antarctic clay-rich sediments: Sites 322, 323, and 325, DSDP Leg 35. In *Initial Reports of the Deep Sea Drilling Project*, vol. 35, chap. 23. Washington, D.C.: U.S. Govt. Printing Office.

Donnelly, T. W., Thompson, G., and Salisbury, M. H. 1980. The chemistry of altered basalts at Site 417, Deep Sea Drilling Project, Leg 51. In *Initial Reports of the Deep Sea Drilling Project*, vol. 51, 52, 53, part 2, chap. 54. Washington, D.C.: U.S. Govt. Printing Office.

Drever, J. I. 1974. The magnesium problem. In *The Sea*, edited by E. D. Goldberg, vol. 5, chap. 10. New York: Wiley-Interscience.

Dunlop, J.S.R., Muir, M. D., Milne, V. A., and Groves, D. I. 1978. A new microfossil assemblage from the Archean of western Australia. *Nature* 274:676–678.

Edmond, J. M., Measures, C., McDuff, R. E., Chan, L. H., Collier, R., Grant, B., Gordon, L. I., and Corliss, J. B. 1979. Ridge crest hydrothermal activity and the balances of the major and minor elements in the ocean: The Galapagos data. *Earth Planet. Sci. Lett.* 46:1–18.

Edmond, J. M., Von Damm, K. L., McDuff R. E., and Measures, C. I. 1982. Chemistry of hot springs on the East Pacific Rise and their effluent dispersal. *Nature* 297:187–191.

Elderfield, H., and Gieskes, J. M. 1982. Sr isotopes in interstitial waters of marine sediments from Deep Sea Drilling Project cores. *Nature* 300:493–497.

Epstein, S. 1967. The stable isotopes. *Eng. and Sci.* 31:47–51.

Eriksson, K. A. 1972. Cyclic sedimentation in the Malmani dolomite, Potchefstroom Synclinorium. *Trans. Geol. Soc. South Afr.* 75:86–95.

Eriksson, K. A., and Truswell, J. F. 1974. Tidal flat associations from a Lower Proterozoic carbonate sequence in South Africa. *Sedimentology* 21:293–309.

Eriksson, K. A., McCarthy, T. S., and Truswell, J. F. 1975. Limestone formation and dolomitization in a lower Proterozoic succession from South Africa. *Jour. Sed. Petrol.* 45:604–614.

Faure, G., Hurley, P. M., and Fairbairn, H. W. 1963. An estimate of the isotopic composition of strontium in rocks of the Precambrian Shield of North America. *J. Geophys. Res.* 68:2323–2329.

Faure, G., Assereto, R., and Tremba, E. L. 1978. Strontium isotope composition of marine carbonates of Middle Triassic to Early Jurassic age, Lombardic Alps, Italy. *Sedimentology* 25:523–543.

Folk, R. L., and Land, L. S. 1975. Mg/Ca ratio and salinity: Two controls over crystallization of dolomite. *Bull. Amer. Assoc. Petrol. Geologists* 59:60–68.

Friedman, G. M. 1980a. Review of depositional environments in evaporite deposits and the role of evaporites in hydrocarbon accumulation. *Bull. Cent. Rech. Explor.-Prod. Elf-Aquitaine* 4:589–608.

———. 1980b. Dolomite is an evaporite mineral: Evidence from the rock record and from sea-marginal ponds of the Red Sea. In *Concepts and Models of Dolomitization*, edited by D. H. Zenger, J. R. Dunham, and R. L. Ethington, 69–80. Soc. Econ. Paleon. Mineral., Spec. Publ. no. 28.

Friedrichsen, H., and Hoernes, S. 1980. Oxygen and hydrogen isotope exchange reactions between seawater and oceanic basalts from Legs 51 through 53. In *Initial Reports of the Deep Sea Drilling Project*, vol. 51, 52, 53, part 2, chap. 44. Washington, D.C.: U.S. Govt. Printing Office.

Fryer, B. J., Fyfe, W. S., and Kerrich, R. 1979. Archean volcanogenic oceans. *Chem. Geol.* 24:25–33.

Gaines, A. M. 1980. Dolomitization kinetics: Recent experimental studies. In *Concepts and Models of Dolomitization*, edited by D. H. Zenger, J. B. Dunham, and R. L. Ethington, 87–110. Soc. Econ. Paleon. Mineral., Spec. Publ. no. 28.

Garlick, G. D., and Epstein, S. 1967. Oxygen isotope ratios in coexisting minerals of regionally metamorphosed rocks. *Geochim. Cosmochim. Acta* 31:181–214.

Garrels, R. M., and Mackenize, F. T. 1971. *The Evolution of Sedimentary Rocks*. New York: W. W. Norton.

Garrett, P. 1970. Phanerozoic stromatolites: Noncompetitive ecologic restriction by grazing and burrowing animals. *Science* 169:171–173.

Gartner, S. 1981. Calcareous nannofossils in marine sediments. In *The Oceanic Lithosphere*, vol. 7 of *The Sea*, edited by C. Emiliani, chap. 7. New York: Wiley-Interscience.

Gebelein, C. D., and Hoffman, P. 1971. Algal origin of dolomite in interlaminated limestone-dolomite sedimentary rocks. In *Carbonate Cements*, edited by O. P. Bricker, 319–326. Baltimore: Johns Hopkins University Press.

Gieskes, J. M., and Reese, H. 1980. Interstitial water studies, Legs 51–53. In *Initial Reports of the Deep Sea Drilling Project*, vol. 51, 52, 53, part 2, chap. 11. Washington, D.C.: U.S. Govt. Printing Office.

Gieskes, J. M., and Lawrence, J. R. 1981. Geochemical significance of diagenetic reactions in ocean sediments: An evaluation of interstitial water data. In *Geology of Oceans*, vol. C4 of *Proc. XXVI Int'l Congress of Geology*, Paris.

Graham, D. W., Bender, M. L., Williams, D. F., and Keigwin, L. D., Jr. 1982. Strontium-calcium ratios in Cenozoic planktonic foraminifera. *Geochim. Cosmochim. Acta* 46:1281–1292.

Hallam, A. 1977. Secular changes in marine inundation of USSR and North America through the Phanerozoic. *Nature* 269:769–772.

Hancock, J. M., and Kauffman, E. G. 1979. The great transgressions of the Late Cretaceous. *J. Geol. Soc. London* 136:175–186.

Harrison, G.C.A. 1980. Spreading rates and heat flow. *Geophys. Res. Letters* 7:1041–1044.

Hart, S. R., and Tilton, G. R. 1966. The isotope geochemistry of strontium and lead in Lake Superior sediments and water. In *The Earth Beneath the Continents*, edited by J. S. Steinhart and T. J. Smith, 127–137. Geophysical Monograph 10, American Geophysical Union.

Hart, S. R., and Staudigel, H. 1982. The control of alkalis and uranium in seawater by ocean crust alteration. *Earth Planet Sci. Lett.* 58: 202–212.

Hays, J. D., and Pitman, W. C. III. 1973. Lithospheric plate motion, sea level changes and climatic and ecological consequences. *Nature* 246: 18–22.

Hein, J. R., and Yeh, H.-W. 1981. Oxygen-isotope composition of chert, from the Mid-Pacific Mountains and Hess Rise, Deep Sea Drilling Project Leg 62. In *Initial Reports of the Deep Sea Drilling Project*, vol. 62, chap. 30. Washington, D.C.: U.S. Govt. Printing Office.

Hirst, D. M. 1962. The geochemistry of modern sediments from the Gulf of Paria: I. The relationship between the mineralogy and the distribution of major elements. *Geochim. Cosmochim. Acta* 26:309–334.

Holeman, J. N. 1968. The sediment yield of major rivers of the world. *Water Resource Res.* 4:737–747.

Holland, H. D. 1978. *The Chemistry of the Atmosphere and Oceans.* New York: Wiley.

Horn, M. K., and Adams, J.A.S. 1966. Computer-derived geochemical balances and element abundances. *Geochim. Cosmochim. Acta* 30: 279–297.

Humphris, S. E., Thompson, R. N., and Marriner, G. F. 1980. The mineralogy and geochemistry of basalt weathering, Holes 417A and 418A. In *Initial Reports of the Deep Sea Drilling Project*, vol. 51, 52, 53, part 2, chap. 47. Washington, D.C.: U.S. Govt. Printing Office.

Ito, E., and Clayton, R. N. 1983. Submarine metamorphism of gabbros from the mid-Cayman Rise: an oxygen isotopic study. *Geochim. Cosmochim. Acta* 47:535–546.

Jolliffe, A. W. 1966. Stratigraphy of the Steeprock Group, Steep Rock Lake, Ontario. *Special Paper No. 3*, 75–98. The Geol. Association of Canada.

Jones, C. B. 1981. Periodicities in stromatolite lamination from the Early Proterozoic Hearne Formation, Great Slave Lake, Canada. *Palaeontology* 24:231–250.

Jones, L. M., and Faure, G. 1967. Origin of the salts in Lake Vanda, Wright Valley, Southern Victoria Land, Antarctica. *Earth Planet. Sci. Lett.* 3:101–106.

———. 1972. Strontium isotope geochemistry of Great Salt Lake, Utah. *Bull. Geol. Soc. Amer.* 83:1875–1880.

Knauth, L. P., and Epstein, S. 1976. Hydrogen and oxygen isotope ratios in nodular and bedded cherts. *Geochim. Cosmochim. Acta* 40:1095–1108.

Knauth, L. P., and Lowe, D. R. 1978. Oxygen isotope geochemistry of cherts from the Onverwacht Group (3.4 billion years), Transvaal,

South Africa, with implications for secular variations in the isotopic composition of cherts. *Earth Planet. Sci. Lett.* 41:209–222.

Lancelot, Y., Winterer, E. L., Basellini, A., Boutefeu, G. A., Boyce, R. E., Čepek, P., Fritz, D., Galimov, E. M., Melguen, M., Price, I., Schlager, W., Sliter, W., Taguchi, K., Vincent, E., and Westberg, J. 1980. Site 415, Agadir Canyon, Deep Sea Drilling Project Leg 50. In *Initial Reports of the Deep Sea Drilling Project*, vol. 50, chap. 3. Washington, D.C.: U.S. Govt. Printing Office.

Land, L. S., Salem, M.R.I., and Morrow, D. W. 1975. Paleohydrology of ancient dolomites: Geochemical evidence. *Bull. Amer. Assoc. Petrol. Geologists* 59:1602–1625.

Larson. R. L., and Pitman, W. C., III. 1972. World-wide correlation of Mesozoic magnetic anomalies and its implications. *Geol. Soc. of Amer. Bull.* 83:3645–3662.

Lawrence, J. R., Gieskes, J. M., and Broecker, W. S. 1975. Oxygen isotope and cation composition of DSDP pore waters and the alteration of Layer II basalts. *Earth Planet. Sci. Lett.* 27:1–10.

Lawrence, J. R., and Gieskes, J. M. 1981. Constraints on water transport and alteration in the oceanic crust from the isotopic composition of pore water. *J. Geophys. Res.* 86:7924–7934.

Li, Y-H. 1972. Geochemical mass balance among lithosphere, hydrosphere and atmosphere. *Amer. Jour. Sci.* 272:119–137.

Livingstone, D. A. 1963. Chemical composition of rivers and lakes. In *Data of Geochemistry*, 6th ed., edited by M. Fleischer. U.S. Geol. Survey Prof. Paper 440G.

Lonka, A. 1967. Trace-elements in the Finnish Precambrian phyllites as indicators of salinity at the time of sedimentation. Bull. 228 of the Comm. Géol. de Finlande.

Lowe, D. R. 1980. Stromatolites 3400 Myr old from the Archean of western Australia. *Nature* 284:441–443.

Lupton, J. E., Klinkhammer, G. P., Normark, W. R., Haymon, R., MacDonald, K. C., Weiss, R. F., and Craig, H. 1980. Helium-3 and manganese at the 21°N East Pacific Rise hydrothermal site. *Earth Planet. Sci. Lett.* 50:115–127.

McDuff, R. E., and Gieskes, J. M. 1976. Calcium and magnesium profiles in DSDP interstitial waters: Diffusion or reaction? *Earth Planet. Sci. Lett.* 33:1–10.

Mackenzie, F. T., and Garrels, R. M. 1966. Chemical mass balance between rivers and oceans. *Amer. Jour. Sci.* 264:507–525.

Manghnani, M. H., Schlanger, S. O., and Milholland, P. D. 1980. Elastic properties related to depth of burial, strontium content and age, and diagenetic stage in pelagic carbonate sediments. In *Bottom-Interacting*

Ocean Acoustics, edited by W. A. Kupferman and F. B. Jensen. New York: Plenum.

Maynard, J. B. 1976. The long-term buffering of the oceans. *Geochim. Cosmochim. Acta* 40:1523–1532.

Meybeck, M. 1979. Concentration des eaux fluviales en éléments majeurs et apports en solution aux océans. *Rev. de Géol. Dynam. et de Géogr. Phys.* 21:215–246.

Migdisov, A. A., Girin, Yu. P., Galimov, E. M., Grinenko, V. A., Baraskaya, N. V., Krivitsky, V. A., Sobornov, O. P., and Cherkovsky, S. L. 1980. Major and minor elements and sulfur isotopes of the Mesozoic and Cenozoic sediments at Sites 415 and 416, Leg 50, Deep Sea Drilling Project. In *Initial Reports of the Deep Sea Drilling Project*, vol. 50, chap. 32. Washington, D.C.: U.S. Govt. Printing Office.

Mottl, M. J., and Holland, H. D. 1978. Chemical exchange during hydrothermal alteration of basalt by seawater: I. Experimental results for major and minor components of seawater. *Geochim. Cosmochim. Acta* 42:1103–1115.

Muehlenbachs, K. 1980. The alteration and aging of the basaltic layer of the sea floor: Oxygen isotope evidence from DSDP/IPOD Legs 51, 52 and 53. In *Initial Reports of the Deep Sea Drilling Project*, vol. 51, 52, 53, part 2, chap. 42. Washington, D.C.: U.S. Govt. Printing Office.

Muehlenbachs, K., and Clayton, R. N. 1976. Oxygen isotope composition of the oceanic crust and its bearing on seawater. *J. Geophys. Res.* 81:4365–4369.

Muir, M. D., and Grant, P. R. 1976. Micropaleontological evidence from the Onverwacht Group, South Africa. In *The Early History of the Earth*, edited by B. F. Windley, 595–604. New York: Wiley.

Murdmaa, I. O., Gordeev, V. V., Emelyanov, E. M., and Bazilevskaya, E. S. 1979. Inorganic geochemistry of Leg 43 sediments. In *Initial Reports of the Deep Sea Drilling Project*, vol. 43, chap. 30. Washington, D.C.: U.S. Govt. Printing Office.

Nanz. R. H., Jr. 1953. Chemical composition of Precambrian slates with notes on the geochemical evolution of lutites. *Jour. Geol.* 61:51–64.

Nohara, M. 1980. Geochemical history of Japan Trench sediments sampled during Leg 56, Deep Sea Drilling Project. In *Initial Reports of the Deep Sea Drilling Project*, vol. 56, 57, part 2, 1251–1257. Washington, D.C.: U.S. Govt. Printing Office.

Orpen, J. L., and Wilson, J. F. 1981. Stromatolites at ~ 3500 Myr and a greenstone-granite unconformity in the Zimbabwean Archean. *Nature* 291:218–220.

Perry, E. C., Jr. 1967. The oxygen isotopic chemistry of ancient cherts. *Earth Planet. Sci. Lett.* 3:62–66.

Perry, E. C., Jr., and Tan, F. C. 1972. Significance of oxygen and carbon isotope variations in Early Precambrian cherts and carbonate rocks of Southern Africa. *Bull. Geol. Soc. Amer.* 83:647–664.

Perry, E. C., Jr., Ahmad, S. N., and Swulius, T. M. 1978. The oxygen isotopic composition of 3800 m.y. old metamorphosed chert and iron formation from Isukasia, West Greenland. *Jour. Geol.* 86:223–239.

Peterman, Z. E., Hedge, C. E., and Tourtelot, H. A. 1970. Isotopic composition of strontium in seawater throughout Phanerozoic time. *Geochim. Cosmochim. Acta* 34:105–120.

Pitman, W. C. III. 1978. Relationship between eustacy and stratigraphic sequences of passive margins. *Bull. Geol. Soc. Amer.* 89:1389–1403.

Pratt, B. R. 1982. Stromatolite decline: A reconsideration. *Geology* 10: 512–515.

Rivers, M. L. 1976. The chemical effects of low-temperature alteration of ocean floor basalt. Bachelor's thesis, Harvard University.

Ronov, A. B., and Yaroshevsky, A. A. 1969. Chemical composition of the Earth's crust. In *The Earth's Crust and Upper Mantle*, edited by P. J. Hart, 37–57. Geophysical Monograph 13, Amer. Geophys. Union.

Ronov, A. B., and Migdisov, A. A. 1970. Evolution of the chemical composition of the rocks in the shields and sediment cover of the Russian and North American Platform. *Geochemistry International* 7:294–324.

Ronov, A. B., Migidsov, A. A., Voskresenskaya, N. T., and Korzina, G. A. 1970. Geochemistry of lithium in the sedimentary cycle. *Geochemistry International* 7:75–102.

Ronov, A. B., and Yaroshevsky, A. A. 1976. A new model for the chemical structure of the Earth's crust. *Geochemistry International* 13(6):89–121.

Ronov, A. B., Migdisov, A. A., and Lobach-Zhuchenko, S. B. 1977. Regional metamorphism and sediment composition evolution. *Geochemistry International* 14(1):90–112.

Scheidegger, K. F., and Stakes, D. S. 1980. X-ray diffraction and chemical study of secondary minerals from Deep Sea Drilling Project Leg 51, Holes 417A and 471D. In *Initial Reports of the Deep Sea Drilling Project*, vol. 51, 52, 53, part 2, chap. 50. Washington, D.C.: U.S. Govt. Printing Office.

Schlanger, S. O., Jenkyns, H. C., and Premoli-Silva, I. 1981. Volcanism and vertical tectonics in the Pacific Basin related to global Cretaceous transgressions. *Earth Planet. Sci. Lett.* 52:435–449.

Schopf, J. W. 1968. Microflora of the Bitter Springs Formation, late Precambrian, Central Australia. *Jour. Paleont.* 42:651–688.

——. 1975. Precambrian paleobiology: Problems and perspectives. *Ann. Rev. Earth Planet. Sci.* 3:213–249.

————. 1977. Biostratigraphic usefulness of stromatolitic Precambrian microbiotas: A preliminary analysis. *Precamb, Res.* 5:143–173.

Schopf, J. W., Dolnik, T. A., Krylov, I. N., Mendelson, C. V., Nazarov, B. B., Nyberg, A. V., Sovietov, Yu. K., and Yakshin, M. S. 1977. Six new stromatolitic microbiotas from the Proterozoic of the Soviet Union. *Precamb. Res.* 4:269–284.

Schopf, J. W., and Prasad, K. N. 1978. Microfossils in collenia-like stromatolites from the Proterozoic Vempalle Formation of the Cuddapah Basin, India. *Precamb. Res.* 6:347–366.

Sleep, N. H., and Wolery, T. J. 1978. Egress of hot water from midocean ridge hydrothermal systems: Some thermal constraints. *J. Geophys. Res.* 83:5913–5922.

Sprague, D., and Pollack, H. N. 1980. Heat flow in the Mesozoic and Cenozoic. *Nature* 285:393–395.

Starinsky, A., Bielski, M., Lazar, B., Wakshal, E., and Steinitz, G. 1980. Marine $^{87}Sr/^{86}Sr$ ratios from the Jurassic to Pleistocene: Evidence from groundwaters in Israel. *Earth Planet. Sci. Lett.* 47:75–80.

Staudigel, H., Hart, S. R., and Richardson, S. H. 1981. Alteration of the oceanic crust: Processes and timing. *Earth Planet. Sci. Lett.* 52:311–327.

Staudigel, H., Muehlenbachs, K., Richardson, S. H., and Hart, S. R. 1981. Agents of low temperature ocean crust alteration. *Contrib. Mineral. Petrol.* 77:150–157.

Staudigel. H., and Hart, S. R., 1983. Alteration of basaltic glass: Mechanisms and significance for the oceanic crust-seawater budget. *Geochim. Cosmochim. Acta* 47:337–350.

Styrt, M. M. 1977. Variations in lithium content between igneous and sedimentary rocks. Bachelor's thesis, Department of Geological Sciences, Harvard University, Cambridge, Mass.

Sugisaki, R. 1980a. Major element chemistry of the Japan Trench sediments, Legs 56 and 57 Deep Sea Drilling Project. In *Initial Reports of the Deep Sea Drilling Project*, vol. 56, 57, part 2, chap. 55. Washington, D.C.: U.S. Govt. Printing Office.

————. 1980b. Major-element chemistry of argillaceous sediments at Deep Sea Drilling Project Sites 442, 443 and 444, Shikoku Basin. In *Initial Reports of the Deep Sea Drilling Project*, vol. 58, chap. 23. Washington, D.C.: U.S. Govt. Printing Office.

Taylor, H. P., Jr. 1968. The oxygen isotope geochemistry of igneous rocks. *Contr. Mineral. and Petrol.* 19:1–71.

Thompson, G., and Humphris, S. 1977. Seawater-rock interactions in the oceanic basement. In *Second International Symp. on Water-Rock Interaction*, Proc. vol. 3, edited by H. Pacquet and Y. Tardy, 3–8. Science Géologique, Univ. Louis Pasteur, Strasbourg, France.

Toens, P. D. 1966. Precambrian dolomite and limestone of the Northern Cape Province. Geol. Survey of the Republic of South Africa, Memoir 57.

Turcotte, D. L., and Burke, K. 1978. Global sea-level changes and the thermal structure of the Earth. *Earth Planet. Sci. Lett.* 41:341–346.

Turekian, K. K. 1969. The oceans, streams and atmosphere. In *Handbook of Geochemistry*, vol. 1, edited by K. H. Wedepohl. New York: Springer-Verlag.

Vail, P. R., Mitchum, R. M., Jr., and Thompson, S. III. 1977. Seismic stratigraphy and global changes of sea level: Part 4, Global cycles of relative changes of sea level. In *Seismic Stratigraphy: Application to Hydrocarbon Exploration*, edtied by C. E. Payton, 83–87. Amer. Assoc. Petr. Geol., Memoir 26.

Vail, P. R., and Mitchum, R. M., Jr. 1979. Global cycles of relative changes of sea level from seismic stratigraphy. In *Geological and Geophysical Investigations of Continental Margins*, edited by J. S. Watkins, L. Montadert, and P. W. Dickerson, 469–472. Amer. Assoc. Petr. Geol., Memoir 29.

Veizer, J. 1976. $^{87}Sr/^{86}Sr$ evolution of seawater during geologic history and its significance as an index of crustal evolution. In *The Early History of the Earth*, edited by B. F. Windley, 569–578. New York: Wiley-Interscience.

Veizer, J., and Compston, W. 1974. $^{87}Sr/^{86}Sr$ composition of seawater during the Phanerozoic. *Geochim. Cosmochim. Acta* 38:1461–1484.

Veizer, J., Compston, W., Hoefs, J., and Nielsen, H. 1982. Mantle buffering of the early oceans. *Naturwissenschaften* 69:173–180.

Vincent, E., and Berger, W. H. 1981. Planktonic foraminifera and their use in paleoceanography. In *The Oceanic Lithosphere*, vol. 7, chap. 25 of *The Sea*, edited by C. Emiliani. New York: Wiley-Interscience.

Vinogradov, A. P., and Ronov, A. B. 1956a. Composition of the sedimentary rocks of the Russian Platform in relation to the history of its tectonic movements. *Geochemistry*, no. 6, 533–559.

———. 1956b. Evolution of the chemical composition of clays of the Russian Platform. *Geochemistry*, no. 2, 123–139.

Visser, J.N.J., and Grobler, N. J. 1973. The transition beds at the base of the dolomite series in the northern Cape Province. *Trans. Geol. Soc. South Afr.* 75:265–274.

Wadleigh, M. A., Veizer, J., and Brooks, C. 1981. Strontium and its isotopes in Canadian rivers: Consequences for the exogenic cycle. *Abstracts with Programs* 13(7):574. Geol. Soc. Amer.

Walter, M. R., Buick, R., and Dunlop, J.S.R. 1980. Stromatolites 3400–3500 Myr old from the North Pole area, Western Australia. *Nature* 284:443–445.

Watts, A. B., and Steckler, M. S. 1979. Subsidence and eustacy at the continental margin of eastern North America. In *Deep Drilling Results in the Atlantic Ocean: Continental Margins and Paleoenvironment*, edited by M. Talwani, W. Hay, and W.B.F. Ryan, 218–234. Amer. Geophys, Union, Maurice Ewing Series 3.

Wedepohl, K. H. 1968. Chemical fractionation in the sedimentary environment. In *Origin and Distribution of the Elements*, edited by L. H. Ahrens, 999–1016. New York: Pergamon.

Zenger, D. H., Dunham, J. B., and Ethington, R. L., eds. 1980. *Concepts and Models of Dolomitization*. Soc. of Econ. Paleont. and Mineral., Spec. Publ. 28.

Ziegler, A. M., Scotese, C. R., McKerrow, W. S., Johnson, M. E., and Bambach, R. K. 1979. Paleozoic paleogeograhy. *Ann. Rev. Earth Planet. Sci.* 7:473–502.

7

Oxygen in the Precambrian Atmosphere: Evidence from Terrestrial Environments

1. Introduction

During the first half billion years of Earth history the atmosphere was probably mildly reducing (see chapter 4). It is highly oxidizing today, and the evolution of the oxidation state of the atmosphere is a fascinating chapter of historical geochemistry. A great deal has been written on the subject. Much of this literature is now of little interest, because many of the proposed indicators of the oxygen pressure in the past have turned out to be of doubtful validity. In this chapter and in chapter 8 I have assembled the data that I consider to be most reliable as indicators of the oxidation state of the atmosphere in the past. This chapter deals with the evidence for the evolution of atmospheric oxygen that can be gleaned from the history of terrestrial environments; chapter 8 is concerned with the evidence that can be derived from the history of marine environments. Fortunately, the several lines of evidence converge.

2. The Evidence from Paleosols

Ancient soils (paleosols) and their lithified equivalents are potentially among the most useful indicators of past levels of atmospheric oxygen. They developed in contact or in communication with the atmosphere, and their original mineralogy and chemistry were shaped in large part by the contemporary atmosphere. The use of the present-day mineralogy and chemical composition of ancient soil horizons to infer past levels of atmospheric oxygen is complicated, however, because soils are altered significantly during diagenesis and metamorphism. The usefulness of paleosols as indicators of atmospheric evolution is therefore constrained not only by their rarity and by the difficulty of identifying paleosols with certainty, but by the necessity of peering through the dark glass of intervening events (see, for instance, Retallack 1981). This section attempts to show that it is nevertheless well worth making the effort.

PALEOSOLS IN THE ELLIOTT LAKE AREA, ONTARIO

Gay and Grandstaff's (1979) paper on the chemistry and mineralogy of two paleosols near Elliot Lake, Ontario, Canada, contains one of the most complete descriptions of Precambrian paleosols. The paleosols are located some 30 km north of Lake Huron, approximately half way between Sudbury and Sault Ste. Marie (see figure 7.1). The geology of the area has been studied extensively (see, for instance, Derry 1960; Pienaar 1963; Roscoe 1969; and Robertson 1978), in part because the area harbors the Huronian uraniferous conglomerates that are the largest single source of uranium ore in Ontario. The stratigraphy of the area is summarized in table 7.1. Archean granites and greenstones of the Whiskey Lake Greenstone Belt are overlain unconformably by Huronian sedimentary rocks, which were intruded ca. 2.2 b.y.b.p. by the Nipissing diabase. The rocks in the Blind River-Elliot Lake region have been folded into the broad, westward plunging Quirke Syncline and Chiblow Anticline.

The Archean Whiskey Lake Greenstone Belt consists largely of basic and acidic volcanics and metasediments, greywacke, conglomerates, iron formation, and various intrusives. These rocks underwent greenschist facies metamorphism, and were later intruded by Archean granites and quartz diorites that yield ages of ca. 2.4 to 2.7 b.y. (Van Schmus 1965).

The Huronian Supergroup consists of suites of subarkose, greywacke, argillite, and limestone, together with oligomictic and polymictic conglomerates. The maximum total stratigraphic thickness of the Huronian sequence is approximately 8.5 km (Robertson 1978). Paleosols developed between the top of the Archean and the base of the Huronian section. Their thickness ranges from a few cm to 20 m; this variability is probably due in part to localized soil erosion. The paleosol horizon is not

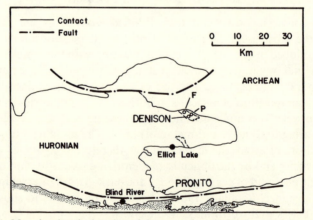

FIGURE 7.1. Map showing the location of the two paleosols (Denison and Pronto) studied by Gay and Grandstaff (1979), and those studied by Pienaar (P) (1963) and by Fryer (F) (1977). Major ore zones are shown by dashed lines. (Reproduced by permission of Elsevier Scientific Company.)

TABLE 7.1.

Stratigraphy of the Elliot Lake area. (Gay and Grandstaff 1979)

Stratigraphy		Source
Nipissing Diabase	2.2 b.y.	Van Schmus 1965, and Fairbairn et al. 1969
Huronian Supergroup		
Cobalt Group		
Bar River		
Gordon Lake		
Lorraine		
Gowganda	2.29 b.y.	Fairbairn et al. 1969
Quirke Lake Group		
Serpent		
Espanola		
Bruce		
Hough Lake Group		
Mississagi		
Pecors		
Ramsay Lake		
Elliot Lake Group		
McKim		
Matinenda		
Copper Cliff		
Archean		
Paleosols		
Granite	2.5–2.7 b.y.	Van Schmus 1965
Greenstone		

particularly well dated, but it must be younger than the Archean intrusives and older than the Nipissing diabase. The best estimate at present places the age of the paleosols at ca. 2.3 to 2.4 b.y.b.p. (Gay and Grandstaff 1979).

The two paleosol sections sampled by Gay and Grandstaff (1979) differ considerably. The first, developed on greenstone, was exposed by a recent tunnel in the Consolidated Denison Mine; the second, developed on granite, is located west of the Pronto Mine. Both paleosols are overlain by the Huronian Matinenda Formation. The mineralogy of the Denison paleosol is shown in figure 7.2. The underlying greenstone is a fine-grained, dark-green, massive to slightly foliated metaigneous rock, which has undergone greenschist facies metamorphism. Its major minerals are actinolite, chlorite, quartz, and sphene. Minor and accessory minerals include epidote, microcline, rutile, leucoxene, hematite, biotite, muscovite, and pyrite. The composition of the unweathered parent greenstone is unusual

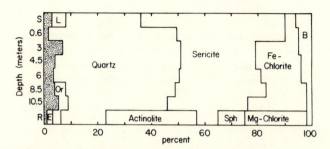

FIGURE 7.2. The mineralogy of the Denison greenstone and of its paleosol (Gay and Grandstaff 1979). Or = orthoclase; E = epidote; Sph = sphene; B = biotite; L = leucoxene. Other accessory minerals include rutile, pyrite, and magnetite. (Reproduced by permission of Elsevier Scientific Publishing Company.)

in that the sum of the weight percentage of $(Na_2O + K_2O + CaO)$ totals slightly less than 2% (see table 7.2).

The depth variation in the concentration of the major and of some of the minor elements in the paleosol is plotted in figure 7.3. The paleosol in this area has a thickness of 10.5 m. Erosion may have removed the upper portions of the soil horizon prior to the deposition of the Huronian sequence. For this reason and because the soil has been thoroughly compacted, its original thickness must have been considerably greater. The mineralogy of the paleosol differs markedly from that of the underlying greenstone. Apparently the parent minerals were destroyed completely during weathering; the mineralogy of the paleosol was clearly altered subsequent to weathering, probably during diagenesis and during the mild metamorphism to which the area was subjected some 1.7 b.y. ago (see section 3). The texture of the paleosol is quite different from that of the unweathered greenstone; the differences are readily interpreted in terms of the effects of weathering and lithification (Gay and Grandstaff 1979).

The changes in the chemical composition of the paleosol with depth have many features that are common to modern soils (see, for instance, Holland 1978, chap. 2), but also some that are rather unusual. The concentration of CaO and MgO decreases upward, but Na_2O is nearly invariant; curiously K_2O increases dramatically toward the top of the profile. This indicates that K_2O was added to the paleosol during or after diagenesis, and that aluminosilicates that were developed during weathering were converted to sericite by "reverse weathering reactions" during K_2O addition. Similar changes have been observed in many other Precambrian paleosols.

The total iron content of the Denison paleosol decreases rapidly upward, largely because of a decrease in FeO content. The Fe_2O_3 concentration changes rather little. The Fe_2O_3/FeO ratio increases markedly

TABLE 7.2.

Chemical composition of the Denison Greenstone and its paleosol (Gay and Grandstaff 1979).

Depth (M)	0	0.3	0.6	2	3	4.5	6	8.5	10.5	14.5	20	30
SiO_2	63.94	71.72	73.06	72.42	69.70	67.78	60.19	73.30	55.90	51.64	51.39	62.55
TiO_2	0.93	0.20	0.27	0.42	0.61	0.86	0.58	0.96	1.04	2.25	2.01	1.27
Al_2O_3	21.03	13.75	12.49	12.94	12.75	12.60	12.97	12.25	14.33	9.77	10.24	8.30
FeO	1.87	1.44	1.72	3.16	5.97	6.49	10.46	3.38	13.06	13.23	12.51	9.47
Fe_2O_3	2.60	2.30	1.85	1.78	0.34	2.20	4.39	0.22	2.92	4.48	3.30	2.54
MgO	0.96	1.34	1.85	1.17	1.58	1.52	2.61	1.42	3.87	13.68	12.84	11.25
CaO	0.07	0.07	0.16	0.17	0.19	0.16	0.17	0.49	0.83	0.53	1.07	1.15
Na_2O	0.45	0.43	0.30	0.27	0.35	0.26	0.29	0.31	0.37	0.12	0.67	0.32
K_2O	5.92	5.02	4.59	4.38	3.16	3.17	3.19	3.70	2.90	0.12	0.73	0.44
P_2O_5	0.27	0.39	0.39	0.10	0.41	0.42	0.44	0.32	0.28	0.48	0.41	0.44
MnO	0.022	0.012	0.015	0.019	0.048	0.070	0.078	0.039	0.147	0.245	0.222	0.16
H_2O^-	0.11	0.33	0.15	0.27	0.35	0.18	0.23	0.38	0.20	0.10	0.21	0.36
H_2O^+	2.25	3.24	2.47	2.72	4.11	4.12	3.98	3.20	4.36	3.22	4.68	2.54
Total	100.42	100.24	99.32	99.82	99.57	99.83	100.12	100.47	100.21	99.87	99.48	100.79
						(ppm)						
Mo	4.6	5.0	3.6	9.1	12.2	11.0	12.0	18.4	17.2	21.0	27.0	25.8
Ni	32	35	32	29	22	32	48	32	52	331	630	297
Cr	56	38	28	35	53	35	38	45	133	132	560	408
Cu	33	—	17	16	65	153	—	110	—	—	—	160
C^0	140	1200	230	2500	—	—	—	—	—	—	—	—

upward without benefit of the oxidation of FeO to Fe_2O_3 (see figure 7.3b). Not unexpectedly, the concentration of Mo, Ni, Cr, and Cu decreases upward. Small quantities (0.02 to 0.25%) of elemental carbon are present in the upper 2 m of the paleosol. The TiO_2 content of the paleosol decreases rapidly upward. This behavior is most unusual, and may be related to the mineralogy of titanium in the Denison greenstone. Titanium in the parent greenstone is present largely as a constituent of sphene, a mineral that was completely destroyed during weathering. Much of the titanium released during this destruction was apparently lost from the soil. The peculiar behavior of titanium may, however, be due to an unusual history for the paleosol.

The observed loss of iron is typical of modern gley soils developed under oxygen-poor or anoxic conditions. In these environments iron is maintained in the divalent state during weathering and tends to be flushed out of soils in solution. By contrast, iron is oxidized and retained, largely as a constituent of ferric hydroxide, in weathering horizons that are developed under oxidizing conditions (see, for instance, Holland 1978, chap. 2). Fe^{+3} in the Denison greenstone was not significantly reduced to Fe^{+2} during weathering and was not subsequently removed in solution. Considerable care must be exercised in interpreting the Fe_2O_3/FeO ratio of this paleosol, because it may well have been altered during diagenesis.

The parent rock from which the Pronto paleosol was formed is quite different from the parent rock of the Denison paleosol. The bedrock in the Pronto locality is a medium-grained, leucocratic, subaluminous, pink alkali granite with an average grain size of 5 mm. Quartz, microcline, and partially sericitized plagioclase are the major minerals; biotite that has been largely altered to chlorite is the most important minor component. The accessory minerals include apatite and zircon. The chemical composition of the granite and of the associated paleosol is shown in table 7.3; the associated mineralogical variations are shown in figure 7.4, and the depth variation in the concentration of the various major and of some minor components is shown in figure 7.5. The base of the paleosol can be recognized by a change of texture and by an abrupt change in the abundance of plagioclase. Oligoclase comprises about 16% of the parent rock; its abundance is much less in the paleosol, and the mineral is absent close to the upper contact of the paleosol. Microcline was altered much less completely than oligoclase. The difference in the susceptibility of the two feldspars to chemical weathering during the development of this soil is the same today.

Relatively few opaque or heavy mineral grains occur in samples of the paleosol. Leucoxene, magnetite, and some rutile are generally present. Curiously, pyrite comprises approximately 3% of the uppermost paleosol

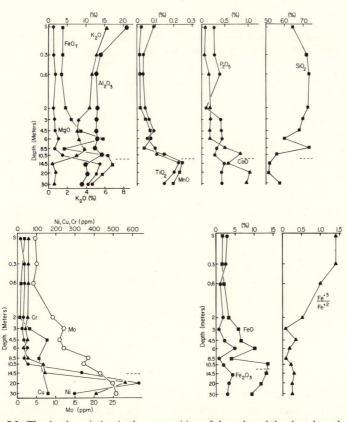

FIGURE 7.3. The depth variation in the composition of the paleosol developed on the Dension greenstone (Gay and Grandstaff 1979). The dashed lines in these figures at a depth of 10.5 m indicate the base of the paleosol as determined by mineralogical and textural criteria. (a) Major elements; (b) FeO and Fe_2O_3; (c) minor elements. (Reproduced by permission of Elsevier Scientific Publishing Company.)

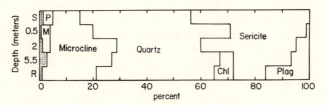

FIGURE 7.4. The mineralogy of the parent alkali granite and of the Pronto paleosol. P = pyrite; M = magnetite; Chl = chlorite. Minor minerals represented by the stippled area include leucoxene, rutile, apatite, and zircon. Figure from Gay and Grandstaff (1979). (Reproduced by permission of Elsevier Scientific Publishing Company.)

TABLE 7.3.

Composition of the Pronto Paleosol and of the parent alkali granite. The base of the paleosol is at a depth close to 5.5 m (Gay and Grandstaff 1979).

Depth (M)	0	0.5	2.0	3.5	5.5	7.0	9.0
SiO_2	66.92	77.50	72.20	72.76	73.66	72.53	73.76
Al_2O_3	15.11	10.93	14.54	13.61	12.70	12.04	11.54
TiO_2	0.54	0.37	0.21	0.21	0.09	0.13	0.14
MgO	0.61	0.39	0.66	0.90	0.64	0.31	0.42
FeO	0.57	0.28	0.29	0.43	0.72	0.48	0.57
Fe_2O_3	6.37	2.48	1.30	1.02	0.99	0.03	0.21
Na_2O	0.29	0.17	0.20	0.19	2.92	4.36	4.18
CaO	0.25	0.13	0.15	0.18	0.48	0.26	0.31
K_2O	5.01	4.54	7.26	7.70	5.72	5.42	5.43
MnO	0.013	0.004	0.005	0.006	0.005	0.011	0.013
P_2O_5	0.21	0.28	0.40	0.31	0.39	0.33	0.63
H_2O^-	0.29	0.13	0.15	0.12	0.14	0.14	0.11
H_2O^+	4.60	2.34	2.22	2.76	1.86	3.82	3.10
Total	100.78	99.54	99.59	100.20	100.32	99.86	100.41
				(ppm)			
Ni	49	28	22	25	26	25	25
Cu	67	9.2	—	—	6.8	5.6	—

sample. This mineral was almost certainly introduced during or after diagenesis. Its presence and textures indicate that later solutions containing reduced sulfur species penetrated the upper part of the paleosol. Some chlorite is present in the lowest paleosol horizon, but none was found at higher levels. It seems likely that some 40–55% of clay minerals were present at one time in the paleosol. These have since been altered to sericite, presumably at the time when K_2O was added to the Denison paleosol.

The concentration of P, Ca, Na, and Fe^{+2} is less in the Pronto paleosol than in the parent granite; the concentration of Fe^{+3}, Al, K, Ti, Si, Mg, and total iron is greater in the paleosol, and the Fe_2O_3/FeO ratio is higher in the paleosol than in the parent rock. The concentration of Cu, Mn, and Ni increases toward the erosion surface. TiO_2 increases relative to Al_2O_3 by a factor of about 3, and total iron increases by a factor of about 7.

The Pronto profile is similar to those of modern soils developed under oxidizing conditions. The increase in TiO_2 with respect to Al_2O_3 reflects the higher mobility of Al than that of Ti during weathering. The increase of Fe_2O_3 and total iron in the Pronto profile is undoubtedly due in part

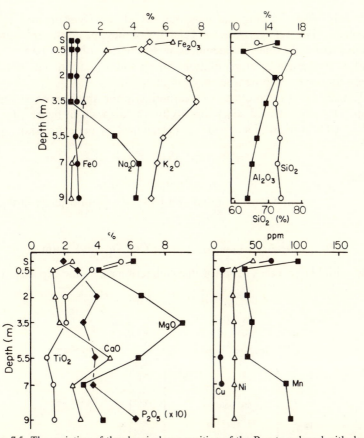

FIGURE 7.5. The variation of the chemical composition of the Pronto paleosol with depth (Gay and Grandstaff 1979). (Reproduced by permission of Elsevier Scientific Publishing Company.)

to the excellent retention of iron hydroxide in oxidized soil horizons, but it is probably also due to the addition of $Fe_2O_3 \cdot nH_2O$ from neighboring areas. The pronounced difference between the Denison and Pronto paleosols is rather surprising and, at first sight, puzzling. Similar differences exist today, and are explained quite readily in terms of differences in the nature of the parent rocks, the degree of soil ventilation, and the vegetative cover. During the Precambrian Era the vegetative cover on land was almost certainly restricted to algal mats. Like their modern counterparts these were probably no more than a few millimeters thick. The intense amplification of the CO_2 content and the concomitant reduction of the O_2 content of soil air that are common features of soils in heavily vegetated areas today (see Holland 1978, p. 23) were therefore almost certainly absent during the development of the Denison and Pronto paleosols. The differences in the behavior of these paleosols are therefore almost certainly due to differences in the nature of their parent rocks and/or in the degree of ventilation during soil formation. The effects of these differences can be treated semiquantitatively.

Consider first the progress of weathering in the absence of soil ventilation. This is illustrated schematically by the diagram in figure 7.6a. The initial CO_2 content of soil water is determined by the expression,

$$m_{H_2CO_3}^i = B_{CO_2} \cdot P_{CO_2}, \tag{7.1}$$

where

$$P_{CO_2} = \text{the partial pressure of } CO_2 \text{ in the atmosphere (atm)},$$

and

$$B_{CO_2} = \text{Henry's Law constant of } CO_2 \text{ in mol/kg atm.}$$

The initial O_2 content of soil water is determined by the expression,

$$m_{O_2}^i = B_{O_2} \cdot P_{O_2}, \tag{7.2}$$

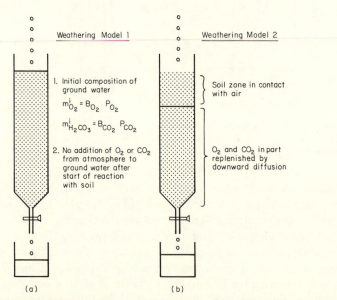

FIGURE 7.6. Schematic diagram to illustrate the progress of chemical weathering without (Model 1) and with (Model 2) soil ventilation. In both models the initial composition of ground water is determined by equilibrium with the ambient atmosphere. In Model 1, no CO_2 or oxidants can be added after reaction within the soil horizon has begun. In Model 2, CO_2 and oxidants used during weathering can be replenished from the atmosphere.

where the significance of the symbols is equivalent to that in equation 7.1. The first weathering reactions in soils are carbonation reactions, in which cations in the parent rock react with H_2CO_3 to form a dilute bicarbonate solution (see, for instance, Schott, Berner, and Sjoberg et al. 1981; Berner and Schott 1982). The attack of H_2CO_3 on silicates containing FeO releases Fe^{+2}. In the presence of dissolved O_2 and/or other oxidants, Fe^{+2} is oxidized to Fe^{+3} and is usually precipitated as a constituent of ferric hydroxide. If the initial supply of H_2CO_3 is exhausted by acid titration reactions before O_2 is exhausted by oxidation reactions, all of the Fe^{+2} released during acid attack is apt to be oxidized, precipitated out, and hence retained in the soil horizon. Highly oxidized soil profiles develop under these circumstances. On the other hand, if the initial supply of O_2 is exhausted before the supply of H_2CO_3, Fe^{+2} released after O_2 depletion is not oxidized to Fe^{+3}, and therefore tends to be flushed out of the soil. Such soils therefore become depleted in iron, and paleosols are produced that contain a significantly smaller concentration of iron than their parent rocks. The relationship between the ratio $m_{O_2}^i/m_{H_2CO_3}^i$ and the ratio, R, of the O_2 demand to the H_2CO_3 demand of the parent rock during weathering determines whether a given soil becomes depleted in iron. Let us define the parameter, R, such that when

$$\frac{m_{O_2}^i}{m_{H_2CO_3}^i} < R. \tag{7.3}$$

Model 1 soils are depleted in iron. The oxygen demand of igneous rocks is determined almost entirely by the quantity of oxygen required to oxidize their contained "FeO" via the reaction,

$$4\text{"FeO"} + O_2 \longrightarrow 2Fe_2O_3. \tag{7.4}$$

The concentration of "MnO" is usually sufficiently small, so that the quantity of oxygen consumed during the oxidation of "MnO" to MnO_2 via the reaction,

$$2\text{"MnO"} + O_2 \longrightarrow 2MnO_2, \tag{7.5}$$

is negligible compared to the quantity of oxygen consumed during the oxidation of "FeO". In some sedimentary and metamorphic rocks the concentration of organic carbon and sulfide minerals is sufficiently large, so that oxygen use via the reactions,

$$C^0 + O_2 \longrightarrow CO_2 \tag{7.6}$$

and

$$FeS_2 + \tfrac{15}{4}O_2 + 2H_2O \longrightarrow \tfrac{1}{2}Fe_2O_3 + 2H_2SO_4, \tag{7.7}$$

is equal to or greater than oxygen use via equation 7.4 (see, for instance, chapter 9). A host of other oxidation reactions accompany weathering under oxidizing conditions, but few of these account for a significant fraction of the oxygen consumed during weathering. In virtually all soil horizons the total oxygen demand, D_{O_2}, in mol O_2/kg of parent rock is therefore essentially equal to

$$D_{O_2} \cong 0.25 m_{\text{"FeO"}} + 0.5 m_{\text{"MnO"}} + m_{C^0} + 3.75\, m_{FeS_2}. \qquad (7.8)$$

The CO_2 demand of igneous rocks depends in large part on the concentration of the cations that can appear balanced by HCO_3^- in soil and ground water (see chapter 5). The reaction of calcium silicates with H_2CO_3 can be written in the form,

$$\text{"CaO"} + 2H_2CO_3 \longrightarrow Ca^{+2} + 2HCO_3^- + H_2O; \qquad (7.9)$$

that of magnesium, sodium, and potassium silicates in the form,

$$\text{"MgO"} + 2H_2CO_3 \longrightarrow Mg^{+2} + 2HCO_3^- + H_2O, \qquad (7.10)$$

$$\text{"Na}_2\text{O"} + 2H_2CO_3 \longrightarrow 2Na^+ + 2HCO_3^- + H_2O, \qquad (7.11)$$

$$\text{"K}_2\text{O"} + 2H_2CO_3 \longrightarrow 2K^+ + 2HCO_3^- + H_2O. \qquad (7.12)$$

It could be argued that reactions with "FeO"-silicates should be included in the CO_2 demand. This would, however, be incorrect. When enough O_2 is present in a soil water so that all of the Fe^{+2} released into solution is oxidized and reprecipitated, there is no net loss of CO_2 to the soil water due to reactions with "FeO". CO_2 loss due to reactions involving iron occurs only when Fe^{+2} is lost from soil horizons. Since the parameter R is defined so as to set the limit to conditions under which no iron is lost from soil horizons, reactions involving "FeO"-containing silicates do not contribute to the CO_2 demand of parent rocks as here defined.

In sedimentary rocks containing significant quantities of C^0 and FeS_2, the generation of CO_2 by the oxidation of C^0 may not be negligible, and the use of CO_2 in the dissolution of carbonate minerals as well as the production of sulfuric acid by the oxidation of FeS_2 cannot be neglected. Thus D_{CO_2}, the CO_2 demand as defined here, is given approximately by the expression,

$$D_{CO_2} \cong 2[m_{\text{"CaO"}} + m_{\text{"MgO"}} + m_{\text{"Na}_2\text{O"}} + m_{\text{"K}_2\text{O"}}]$$
$$+ m_{CaCO_3} + 2m_{CaMg(CO_3)_2} - m_{C^0} - 4m_{FeS_2}. \qquad (7.13)$$

The definition of the parameter R in equation 7.3 is equivalent to the expression,

$$R = \frac{D_{O_2}}{D_{CO_2}}. \qquad (7.14)$$

However, the value of R calculated in this manner can be used in equation 7.3 only if oxidation reactions are preceded by carbonation reactions. This seems to be generally true. Fe^{+2} must be released into solution before significant oxidation of "FeO" can take place, because the rate of diffusion of O_2 into silicate minerals at temperatures below $100°C$ is extremely slow compared to the residence time of mineral grains in most soil horizons. A second assumption underlying the use of the parameter R is less easily justified. As defined in equation 7.14, the parameter R involves all of the potentially important oxidation and carbonation reactions. If weathering of the minerals in a given rock is sequential, the effective value of R at any given time may differ significantly from the mean value of R. The implications of this complication are worth exploring in greater detail.

The systematics of soil development in Model 1 systems, where neither CO_2 nor O_2 is added to rainwater after passing into a soil horizon, are somewhat different from those in Model 2 systems, which are open to atmospheric gases. When

$$\frac{m^i_{O_2}}{m^i_{H_2CO_3}} > R$$

in such an open soil-water system, all Fe^{+2} released into solution is apt to be oxidized and reprecipitated as in Model 1 systems. H_2CO_3 will be exhausted before O_2 and hence will tend to be resupplied more rapidly than O_2 by downward mixing from the atmosphere-soil water interface. It is unlikely, however, that this resupply will be sufficiently fast to produce a loss of iron from the soil profile. If $m^i_{O_2}/m^i_{H_2CO_3} < R$, O_2 is exhausted before H_2CO_3. If O_2 is resupplied rapidly, the ratio of O_2 to H_2CO_3 supplied to soil water can be much greater than $m^i_{O_2}/m^i_{H_2CO_3}$. Hence soil horizons that are depleted considerably in iron during closed-system (Model 1) weathering may suffer little or no loss of iron during open-system (Model 2) weathering. It follows that the loss of iron from a soil profile is a good indication that $m^i_{O_2}/m^i_{H_2CO_3} < R$, whereas the retention of iron in a soil can tell us little about the relationship between $m^i_{O_2}/m^i_{H_2CO_3}$ and R, unless it can be shown that soil development took place in a system that was closed or nearly closed with respect to the resupply of atmospheric gases.

The Pronto and the Denison paleosols developed at essentially the same time. If the proposed analysis of the behavior of iron during the formation of soils is correct, then the difference in the behavior of iron in the two soils should be related to their R value. This is indeed the case. The R value of the Denison greenstone is ca. 6.2×10^{-2}, that of the Pronto granite ca. 0.6×10^{-2}. The loss of ca. 80% of iron in the Denison paleosol and the loss of very little, if any, iron from the Pronto paleosol is readily explained if the value of the ratio $m^i_{O_2}/m^i_{H_2CO_3}$ in rainwater during the formation of these soils was between 0.6×10^{-2} and 6.2×10^{-2}. If this is correct, then between ca. 2.3 and 2.4 b.y.b.p.,

$$0.6 \times 10^{-2} < \frac{B_{O_2}}{B_{CO_2}} \cdot \frac{P_{O_2}}{P_{CO_2}} < 6.2 \times 10^{-2}.$$

Table 7.4 lists the values of B_{O_2} and B_{CO_2} between $0°$ and $30°C$. The value of both parameters decreases significantly with increasing temperature, but their ratio changes only moderately in this temperature interval. If we take a value of 3.3×10^{-2} for the ratio B_{O_2}/B_{CO_2} during the development of the Denison and Pronto soils, it follows that

$$0.2 < \frac{P_{O_2}}{P_{CO_2}} < 2$$

in the atmosphere at that time. Today the P_{O_2}/P_{CO_2} ratio is close to 600. Pronto and Denison paleosol data therefore suggest that the ratio of the pressure of the two gases in the atmosphere has changed dramatically during the past 2.3 b.y.

The above calculations are uncertain, in part because little is known about the influence of the decay of organic matter on the O_2 and CO_2 content of ground waters in Precambrian soils (see, for instance, Campbell 1979). The data in tables 7.2 and 7.3 do include a few figures for the concentration of reduced carbon in the upper 2 m of the two paleosols. The observed concentrations are small. The maximum recorded value occurs at a depth of 2 m in the Denison paleosol. If the 2500 ppm of reduced carbon at this level are typical of the original $C_{org.}$ content of the Denison soil, then the oxidation of organic matter demanded some

TABLE 7.4.

Values of Henry's Law constants B_{O_2} and B_{CO_2} between $0°C$ and $30°C$.

$T(°C)$	$B^{*}_{O_2}$	$B^{\dagger}_{CO_2}$	B_{O_2}/B_{CO_2}
0	2.1×10^{-3}	7.8×10^{-2}	2.8×10^{-2}
10	1.7	5.4	3.1
20	1.3	3.9	3.5
30	1.1	3.0	3.7

* Based on Brewer 1975, pp. 561–562.
† From Holland 1978, table 2.1.

0.2 mol O_2/kg soil. This is approximately half the O_2 demand of FeO in the parent greenstone. Inclusion of the oxidation of this quantity of $C_{org.}$ in the above calculations would raise the maximum permitted value of the P_{O_2}/P_{CO_2} ratio in the contemporary atmosphere during the formation of these soils, but the increase would not be large.

The proposed interpretation of the Denison and Pronto paleosols is rather startling, and corroborative data from several contemporaneous paleosols are obviously needed. To date only one other paleosol from the same period of Earth history has been studied thoroughly. That is the soil horizon developed on the Hekpoort basalt in South Africa.

PALEOSOLS ON THE OLD HEKPOORT BASALT, SOUTH AFRICA

The Hekpoort basalt is part of the Pretoria Group of South Africa, and covers some 100,000 km^2 in the Transvaal; the Ongeluk Lava, its equivalent in the northwestern Cape, covers a similar area. The Hekpoort basalt is up to 500 m thick and consists of subaerial flows that are massive at their base and amygdaloidal at their top. The uppermost few meters of the Hekpoort are altered throughout the subbasin in the Transvaal. This altered zone is thought to be a paleoweathering profile. Rb/Sr dates of the Hekpoort Basalt place its age at 2224 $\pm$ 21 m.y. (Button 1979). The overlying sediments are intruded by the Bushveld complex whose age is 2095 $\pm$ 24 m.y. (Hamilton 1976). The age of the Hekpoort paleosol horizon must therefore be close to 2200 m.y., slightly less than the age of the Denison and Pronto paleosols.

The metamorphic grade of the paleosol at the top of the Hekpoort Basalt varies considerably in the Transvaal Basin. An exceptionally good, essentially unmetamorphosed section is exposed in a cut along the road between Waterval Onder and Waterval Boven in the Eastern Transvaal (see figure 7.7). This is the section that has been studied by Button (1979), and Button and Retallack (1981). The Pretoria Group sediments there dip approximately 5 degrees to the west. A weakly developed, nonpenetrative cleavage dips in the same direction at about 25 degrees.

A stratigraphic section of the Hekpoort paleosol is shown in figure 7.8, its mineralogy in table 7.5. The upper, particularly sericite-rich phase of the paleosol ranges in thickness from less than 1 m to more than 5 m. The lower, more chlorite-rich zone may be up to twice as thick. The contact between the paleosol and the overlying clastic sediments is sharp and slightly undulating. The contact of the sericite zone with the underlying chloritic zone is fairly sharp. In some exposures, veinlike downward projections of sericitic material extend into the chloritic zone. In other exposures sericitic material can be found well within the chloritic zone.

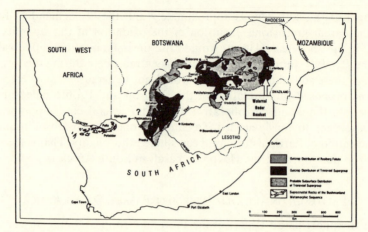

FIGURE 7.7. Location map for the Hekpoort paleosol described by Button (1979).

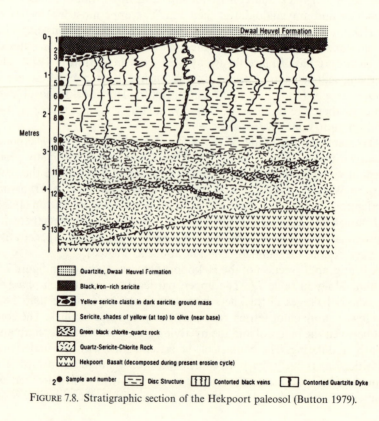

FIGURE 7.8. Stratigraphic section of the Hekpoort paleosol (Button 1979).

TABLE 7.5.

Mineralogy of the Hekpoort Paleosol (Button 1979) (expressed as percentages).

Sample No.	Sericite	Chlorite	Quartz	Stilp-nomelane	Epidote	Opaques	Number of Points Counted
1	81.6	7.8	0	0	0	10.7	385
2	77.4	0.3	0	0	0	22.3	726
3	84.0	0	0	0	0	16.0	362
4	87.9	0	0	0	0	12.1	420
5	87.0	1.2	0	0	0	11.9	345
6	80.4	5.1	0.3	0	0	14.2	393
7	77.8	5.9	1.0	0	0	15.2	388
8	79.4	6.3	4.2	0	0	10.1	378
9	0	51.5	37.3	5.1	0	6.0	332
10	77.6	10.1	4.5	0	0	7.8	515
11	51.5	13.8	29.6	0.7	0	4.4	297
12	33.9	5.9	52.3	5.4	0	2.6	354
13	58.6	11.9	15.2	0	3.8	10.5*	638

* At least 4.1% showed cubic outlines.

These are presumably joint-controlled regions where weathering has penetrated to greater than normal depth. In most exposures the top of the underlying basalt has been weathered during the contemporary erosion cycle.

Table 7.6 shows that the unweathered Hekpoort Basalt has a rather normal chemical composition. The composition of the paleosol samples in table 7.7 and figure 7.9 is similar in many respects to that of the Denison paleosol. MgO and CaO have been largely removed; the Na_2O concentration has also decreased during weathering; K_2O has increased greatly, perhaps during an episode of K-addition much later than the weathering of the Hekpoort Basalt. The total iron content has decreased. As in the Denison profile, much of the loss of iron has been at the expense of FeO, but contrary to experience at Denison, the Fe_2O_3 content of the Hekpoort paleosol is considerably larger than that of the parent basalt. Iron was apparently leached downward to what may have been the water table. This indicates that the ratio of available O_2 to CO_2 in the weathering medium was less than the ratio of O_2 demand for the oxidation of FeO to Fe_2O_3 and of MnO to MnO_2, to the CO_2 demand for the conversion of MgO, CaO, Na_2O, and K_2O to bicarbonates. Analyses 1 and 2 in table 7.6 indicate that the ratio, R, for the Hekpoort Basalt is approximately 4.5×10^{-2}; this value is somewhat lower than the value of R for the parent greenstone of the Denison paleosol (6.2×10^{-2}). The

TABLE 7.6.

Chemical analyses of the Hekpoort Basalt and the chlorite-quartz zone of the Hekpoort paleosol (Button 1979).

Sample No.	1* AB1236	2* AB1235	3[†] Z19	4[†] Z21	5[†] Z29
SiO_2	57.28	55.06	55.6	55.2	59.0
TiO_2	0.67	0.67	0.74	0.84	0.42
Al_2O_3	15.41	14.43	15.9	15.1	9.1
Fe_2O_3	1.00	0.85	11.1[‡]	12.3[‡]	25.4[‡]
FeO	8.00	7.96			
MnO	0.15	0.18	0.16	0.20	0.20
MgO	5.65	7.45	6.7	5.5	3.1
CaO	6.46	7.74	7.7	7.94	0.05
Na_2O	1.43	1.79	1.2	1.0	<0.2
K_2O	1.39	0.91	0.31	0.25	0.02
H_2O^+	2.76	2.72	0.06[#]	1.35[#]	3.29[#]
H_2O^-	0.29	0.17			
P_2O_5	0.08	0.07	0.12	0.14	<0.02
CO_2	0.03	0.03	nd	nd	nd
Total	100.60	100.03	99.59	99.82	100.58

Analysts:
* National Institue for Metallurgy, Johannesburg.
[†] Bergström and Bakker, Johannesburg.
[‡] Total iron as Fe_2O_3.
[#] Loss on ignition.
nd = not determined
1. Hekpoort Basalt, from Kindergoed 332, Waterval Boven district (Button, 1973).
2. Hekpoort Basalt, from Sterkspruit 296. Nelspruit district (Button 1973).
3. Hekpoort Basalt, from Waterval 386, Swartruggens district (Button, unpublished data).
4. Hekpoort Basalt, from Otterfontein 438, Koster district (Button unpublished data).
5. Chlorite-quartz rock from basal zone of the Hekpoort palaeosaprolite from southern approach cutting to Daspoort Tunnel, Pretoria (Button, unpublished data).

somewhat more complete conversion of FeO to Fe_2O_3 in the Hekpoort than in the Denison paleosol agrees rather well with the somewhat lower ratio of the O_2 to the CO_2 demand in the Hekpoort Basalt than in the greenstone at the Denison locality. However, the history of the two paleosols may have been rather different, and the higher Fe_2O_3 content

TABLE 7.7.

Chemical composition of samples from the Hekpoort paleosol (Button 1979).

Sample No.	1	2	3	4	5	6	7	8	9	10	11	12	13
SiO_2	44.2	45.7	46.1	46.6	47.3	47.9	48.4	46.4	53.4	45.9	59.4	67.3	51.8
TiO_2	1.79	1.64	1.74	1.80	1.66	1.70	1.65	1.67	0.76	1.49	1.15	0.74	1.06
Al_2O_3	33.5	34.8	34.8	35.1	34.9	33.6	33.0	32.5	13.9	29.7	22.4	14.7	21.3
Fe_2O_3	3.66	2.60	1.77	0.49	0.85	1.50	2.12	3.89	8.80	4.39	3.65	3.45	10.80
FeO	2.30	0.25	0.20	0.35	0.30	0.55	0.40	1.00	16.20	5.04	2.75	5.87	3.60
MnO	0.117	0.05	0.05	<0.01	<0.01	0.01	0.02	0.07	0.24	0.13	0.12	0.15	0.20
MgO	0.5	0.5	0.3	0.3	0.3	0.5	0.6	0.6	1.8	0.9	0.8	1.2	1.4
CaO	0.13	0.09	0.09	0.02	0.11	0.08	0.04	0.05	0.03	0.13	0.17	0.06	0.37
Na_2O	0.5	0.6	0.6	0.6	0.5	0.6	0.6	0.5	0.1	0.4	0.4	0.2	0.4
K_2O	8.96	9.42	9.58	9.92	9.76	9.22	8.95	9.00	0.41	7.53	5.86	3.04	4.85
H_2O^+	5.2	4.8	4.8	4.7	4.8	4.6	5.3	4.99	6.45	5.28	3.88	3.99	5.20
H_2O^-	0.4	0.4	0.45	0.3	0.4	0.3	0.2	0.4	0.6	0.3	0.3	0.3	0.4
P_2O_5	0.11	0.05	0.04	0.02	0.07	0.05	0.04	0.05	0.02	0.06	0.14	0.06	0.06
CO_2	<0.1	<0.1	<0.1	0.1	0.1	<0.1	0.2	<0.1	0.1	<0.1	<0.1	0.2	<0.1
Total	100.37	100.90	100.52	100.30	101.05	100.61	101.52	101.12	102.81	101.25	101.02	101.26	101.44
Zr*	290	340	340	320	290	300	290	270	120	250	210	130	175
Cr*	250	200	200	160	155	145	135	180	82	135	125	110	92
Rb*	330	340	330	360	360	350	350	370	49	340	280	170	260
Sr*	76	67	60	57	56	54	56	71	7	62	58	38	75
Ba*	1160	780	590	460	340	270	220	250	98	220	210	180	460
Ni*	137	104	95	65	89	124	147	153	226	132	190	264	258
Cu*	254	28	27	14	22	86	117	74	136	7	102	102	186
Zn*	11	<5	<5	<5	<5	<5	<5	<5	72	22	27	45	77

Note: See figure 7.7 for location of samples in profile.
Analysts: Bergström and Bakker, Johannesburg.
* ppm

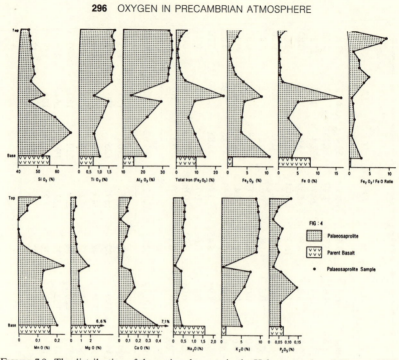

FIGURE 7.9. The distribution of the major elements in the Hekpoort paleosol (Button 1979).

of the Hekpoort sample may reflect the influence of these differences rather than that of the rather minor difference in the ratio of the O_2 demand to the CO_2 demand.

The data for the three paleosols discussed above are plotted in figure 7.10. They are consistent with the interpretation that the P_{O_2}/P_{CO_2} ratio in the atmosphere 2.2–2.3 b.y.b.p. was such that during the weathering of rocks with R values as low as the Pronto granite, excess O_2 was present in rain water, whereas during the weathering of rocks with R values as high as the Denison greenstone and the Hekpoort basalt, O_2 was depleted before CO_2 by weathering reactions in soil zones. The inclusion of the data for the Hekpoort paleosol narrows the likely range for the P_{O_2}/P_{CO_2} ratio in the atmosphere during the early part of the Proterozoic Era.

ARCHEAN PALEOSOLS IN SOUTH AFRICA

Preliminary data for two Archean paleosols from South Africa have kindly been made available by Prof. Grandstaff and Mr. Edelman of

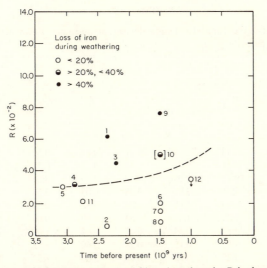

FIGURE 7.10. Summary plot of the behavior of iron in paleosols; R is the ratio of the O_2 demand to the CO_2 demand during the weathering of the parent rocks. (1) Denison paleosol, Ontario, Canada; (2) Pronto paleosol, Ontario, Canada; (3) Hekpoort paleosol, Transvaal, South Africa. (4) Dominion Reef paleosol, South Africa; (5) Amsterdam paleosol, South Africa; (6) Athabaska meta-arkose paleosol, Hole 20, Saskatchewan, Canada; (7) Athabaska meta-arkose paleosol, Holes 23, 32, 51, Saskatchewan, Canada; (8) Athabaska meta-arkose paleosol, Hole 50, Saskatchewan, Canada; (9) Athabaska metapelite paleosol, Hole 81, Saskatchewan, Canada; (10) Athabaska metasemipelite paleosol, Hole 65, Saskatchewan, Canada; (11) Abitibi paleosol, Canada; (12) Cape Wrath paleosol, Northwest Scotland.

Temple University. Chemical and mineralogical data for their ca. 2.8–2.9 b.y.-old Dominion Reef paleosol are summarized in table 7.8. Data for the 3.0–3.1 b.y.-old Amsterdam paleosol are summarized in table 7.9. Both paleosols developed on granitic rocks. The Al_2O_3 content of the upper level of both paleosols is much higher than that of the unweathered parent rocks. Both paleosols are therefore products of intense weathering. The total iron content of the Dominion Reef paleosol changes very little from base to top. If the parent rock was chemically homogeneous, the observed decrease in the ratio of total iron to Al_2O_3 toward the top of the profile implies the loss of approximately 30% of the initial iron during the formation of the paleosol. The $(FeO)_T/Fe_2O_3$ ratio of the Amsterdam paleosol is rather irregular. Apparently, rather little, if any, iron was lost during the formation of this soil.

The FeO/Fe_2O_3 ratio in both paleosols is surprisingly high. It seems likely that this is due to the reduction of some Fe_2O_3 in the original

TABLE 7.8.

Preliminary data for the chemical and mineralogical composition of Edelman and Grandstaff's Dominion Reef paleosol.

				Depth in Meters			
	0	*1*	*3*	*5*	*9*	*13*	*29*
Chemistry							
(wt. % oxides)							
SiO_2	70.45	72.58	70.01	70.81	70.51	75.70	72.33
TiO_2	0.41	0.45	0.38	0.25	0.35	0.26	0.29
Al_2O_3	19.12	17.09	17.32	14.17	14.55	13.04	13.42
FeO_T	3.21	2.92	3.33	4.05	3.33	2.74	3.46
FeO	2.06	1.67	1.94	2.80	2.74	2.02	2.56
Fe_2O_3	1.28	1.39	1.54	1.39	0.66	0.80	1.00
MgO	0.37	0.34	0.47	0.64	1.08	0.74	0.68
CaO	0.10	0.15	0.28	0.66	1.52	1.52	1.60
Na_2O	0.25	0.06	0.42	3.66	4.32	3.51	3.81
K_2O	6.51	5.32	6.73	3.84	3.24	3.79	3.57
P_2O_5	0.03	0.05	0.10	0.06	0.09	0.07	0.07
MnO	0.02	0.02	0.03	0.06	0.06	0.04	0.04
LOI	2.32	2.28	2.22	1.36	1.57	1.68	1.14
Total	102.93	101.39	102.75	99.71	100.89	102.57	100.05
Mineralogy (%)							
Qtz	43.6	36.6	42.1	21.3	33.5	42.6	28.6
Plag	(−)	(−)	1.5	43.2	37.0	30.0	37.7
Mic	5.7	1.2	13.1	23.2	19.6	18.2	25.6
Chl	(−)	(−)	(+)	6.3	4.09	5.0	5.2
Epi	(−)	(−)	(−)	1.3	0.9	1.6	1.1
Sph	(−)	0.5	0.3	(−)	0.2	0.6	0.7
Carb	(−)	(−)	(−)	(−)	1.1	0.6	0.4
Rut	(−)	(−)	0.25	(−)	(−)	0.2	0.09
Zir	0.2	0.5	(−)	0.6	0.7	0.6	0.4
Alan	(−)	(−)	(−)	(−)	(−)	0.2	0.1
Pyr	(−)	(−)	(−)	0.4	0.4	0.4	0.2
Ser	50.5	60.9	42.1	3.1	2.3	cf	cf
Leu	(+)	0.5	0.8	0.6	0.4	(−)	(−)

soil to "FeO" during diagenesis and/or metamorphism. It is possible that iron was gained or lost from the paleosols during these events. At least to that extent the position of the points representing these paleosols in figure 7.10 is uncertain. It is heartening, however, that there is no obvious difference between the inferred behavior of iron in these soils and in the three early Proterozoic paleosols discussed above.

TABLE. 7.9.

Preliminary data for the chemical and mineralogical composition of Edelman and Grandstaff's Amsterdam paleosol.

	Depth in Meters					
	0	1.7	2.9	3.2	6.2	14.9
Chemistry						
(wt. % oxides)						
SiO_2	68.04	66.80	65.99	68.01	71.58	68.52
TiO_2	0.63	0.54	0.61	0.45	0.40	0.40
Al_2O_3	22.40	21.96	18.07	15.98	15.23	14.17
FeO_T	4.72	4.85	6.40	7.09	4.59	3.29
FeO	4.07	4.41	5.98	6.14	4.59	2.60
Fe_2O_3	0.78	0.49	0.47	1.06	0.01	0.77
CaO	0.42	0.40	0.25	0.16	0.31	1.97
MgO	0.04	0.26	1.17	1.27	1.41	0.86
K_2O	2.46	2.27	5.23	4.46	3.55	2.43
Na_2O	0.18	0.18	0.24	0.22	0.29	4.29
P_2O_5	0.37	0.32	0.21	0.35	0.17	0.42
MnO	0.01	0.02	0.03		0.04	0.04
Total	99.40	97.65	98.25	92.10	97.58	96.47
Mineralogy (%)						
Qtz	42.2	41.3	43.6	35.2	41.6	28.9
Plag	(−)	(−)	(−)	(−)	(+)	39.5
Mic	(−)	(−)	(−)	(+)	1.0	13.0
Chl	0.4	0.2	7.7	8.8	6.7	11.8
Ser	32.0	40.4	45.0	52.1	47.0	1.6
Epi	(−)	(−)	(−)	(−)	(−)	2.0
Sph	(−)	(−)	(−)	(−)	(−)	0.6
Biot	(−)	(−)	(−)	(−)	(−)	0.4
Apat	2.0	0.6	0.4	(+)	(+)	0.6
Brook	(−)	(−)	(−)	(−)	0.2	(−)
Zir	2.2	0.2	0.4	0.9	0.5	(+)
Ilmen	(−)	(−)	(−)	(−)	0.5	(−)
Leu	2.6	2.0	0.6	0.9	(−)	(−)
Fe stain	(+)	(+)	0.2	(−)	(−)	(−)
Carb	(−)	(−)	0.2	(−)	(−)	(+)
Anda	17.2	13.5	(−)	(−)	(−)	(−)
Pyr	1.3	1.8	2.0	2.2	2.3	2.5

MID-PROTEROZOIC PALEOSOLS IN THE ATHABASKA BASIN

In the Athabaska Basin of northern Saskatchewan the Athabaska sandstone overlies an extensive paleosol horizon developed on the Archean crystalline basement. Intensive exploration for uranium ore in this area has made available numerous drill core sections through this paleosol. The results of a detailed mineralogical and chemical study of a number of these cores have been reported by Macdonald (1980) (see figure 7.11). The crystalline basement in this area belongs to the Wollaston domain. Drill cores intersected meta-arkoses and pelitic and semipelitic gneisses and schists, which have been deformed by two major phases of almost parallel, northeast-trending folding. The Athabaska formation has an age of 1513 ± 24 m.y. (Bell and Blenkinsop 1980), and consists of quartz sandstones with minor shale and conglomerate. The sequence is essentially undeformed and unmetamorphosed, and has a maximum preserved thickness of 1600 m.

The meta-arkoses in the basement are quite homogeneous. Soils developed on this unit were therefore studied more thoroughly by Macdonald than soils developed on the rather heterogeneous metasemipelites and metapelites. The analytical data for all of the analyzed cores are summarized in table 7.10.

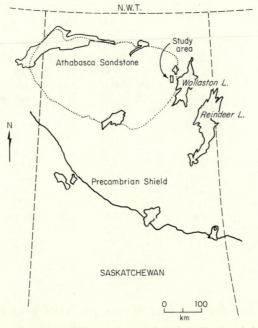

FIGURE 7.11. Location map for drill cores through the paleosol horizon below the Athabaska sandstone (Macdonald 1980).

TABLE 7.10.

Analyses of drill core samples through paleosols at the base of the Athabaska Sandstone, west of Wollaston Lake, Saskatchewan (Macdonald 1980).

Hole No.	Depth	SiO_2	TiO_2	Al_2O_3	Fe_2O_3	FeO	MnO	MgO	CaO	Na_2O	K_2O	H_2O	C
20	0.3	75.0	0.46	18.2	0.57	0.26	0.01	0.05	0.05	0.38	0.19	5.99	0.10
20	1.4	71.9	0.46	18.1	2.25	0.42	0.01	0.24	0.05	0.13	0.69	5.46	0.08
20	3.4	71.3	0.27	14.5	2.28	0.74	0.00	0.79	0.17	2.51	1.40	4.68	0.08
20	4.9	74.4	0.34	15.7	1.97	0.30	0.00	0.63	0.05	0.04	1.78	4.17	0.09
20	6.4	67.2	0.55	19.3	0.57	1.05	0.02	1.68	0.02	0.05	2.71	5.40	0.09
20	7.9	64.6	0.45	18.9	3.24	0.55	0.00	2.38	0.05	0.11	5.37	3.18	0.11
20	9.4	72.3	0.49	15.8	1.79	0.36	0.00	3.24	0.05	0.05	3.74	2.80	0.61
20	11.0	67.4	0.34	16.0	1.75	0.21	0.00	4.30	0.31	1.90	3.44	3.70	0.19
20	12.4	70.2	0.51	14.1	2.33	0.44	0.00	3.08	0.42	0.05	3.86	3.36	0.04
20	14.0	68.7	0.51	16.1	1.00	0.52	0.00	7.04	0.28	0.05	2.14	4.65	0.04
20	15.5	64.7	0.30	15.3	1.94	1.22	0.01	3.31	0.29	1.75	7.01	2.60	0.04
20	17.1	64.9	0.41	15.6	1.46	1.29	0.00	3.36	0.31	1.65	6.63	2.35	0.04
20	18.6	64.3	0.34	15.9	2.20	0.71	0.00	5.31	0.24	2.19	3.79	3.69	0.02
20	20.1	66.7	0.38	15.3	1.40	0.92	0.01	5.08	0.44	1.69	3.33	3.53	0.26
20	21.6	67.4	0.40	13.7	1.77	1.77	0.03	3.53	0.36	0.81	6.95	1.47	0.11
20	23.2	63.8	0.63	15.0	1.72	2.74	0.04	4.24	0.21	1.84	6.66	1.86	0.03
20	24.7	68.8	0.24	15.2	0.90	1.56	0.02	2.12	1.28	5.33	4.27	0.71	0.07
20	26.2	69.8	0.03	15.3	0.50	1.75	0.03	1.73	1.65	6.10	3.26	0.63	0.13
20	27.7	69.5	0.24	15.3	0.96	1.34	0.02	1.32	2.88	6.55	1.98	0.74	0.18
20	29.3	66.9	0.28	15.6	0.64	1.81	0.03	1.80	3.58	6.15	2.82	0.63	0.17
20	32.3	70.1	0.43	15.9	0.25	2.05	0.02	1.70	1.91	6.93	2.56	0.11	0.13
20	35.3	70.5	0.53	15.8	0.20	1.88	0.00	1.52	1.71	6.59	2.28	0.31	0.09
20	36.9	69.9	0.38	15.8	0.04	1.79	0.02	1.36	2.41	6.47	2.25	0.23	0.13
20	38.4	70.3	0.42	16.0	0.03	2.27	0.01	1.57	1.17	6.12	2.98	0.60	0.17
20	39.9	71.6	0.22	15.7	0.03	1.75	0.02	1.09	1.96	7.35	1.82	0.12	0.12
20	41.4	70.5	0.37	16.0	0.15	1.85	0.02	1.67	2.11	6.20	2.32	0.71	0.04

(continued)

TABLE 7.10. *(continued)*

Analyses of drill core samples through paleosols at the base of the Athabaska Sandstone, west of Wollaston Lake, Saskatchewan (Macdonald 1980).

Hole No.	Depth	SiO_2	TiO_2	Al_2O_3	Fe_2O_3	FeO	MnO	MgO	CaO	Na_2O	K_2O	H_2O	C
23	0.6	72.5	0.46	16.0	1.89	0.25	0.00	0.13	0.15	3.63	0.37	5.45	0.04
23	2.1	70.8	0.59	18.7	2.14	0.29	0.00	0.03	0.18	2.43	0.55	6.74	0.06
23	3.7	71.0	0.51	18.3	2.25	0.11	0.02	0.63	0.17	1.81	1.98	4.90	0.19
23	5.2	70.0	0.42	20.0	1.86	0.26	0.00	0.13	0.41	1.52	0.53	6.61	0.09
23	6.7	70.5	0.47	17.2	1.01	0.71	0.02	1.15	0.53	2.27	4.18	2.98	0.12
23	8.2	70.2	0.43	16.7	0.29	0.93	0.01	2.85	0.65	0.68	3.38	3.36	0.02
23	17.4	68.3	0.48	16.3	0.65	1.79	0.03	1.95	3.25	4.91	2.28	0.85	0.09
23	32.6	67.9	0.42	16.6	0.65	1.99	0.03	1.20	3.45	5.05	1.86	0.32	0.08
32	0.0	65.6	0.39	23.7	0.02	0.16	0.00	0.08	0.32	0.29	0.35	8.28	0.02
32	1.5	70.3	0.22	20.9	0.75	0.25	0.00	0.06	0.05	0.27	0.67	6.85	0.01
32	3.0	71.3	0.23	20.4	0.67	0.14	0.00	0.02	0.05	0.41	0.43	6.89	0.04
32	4.6	73.5	0.19	20.2	0.30	0.14	0.00	0.08	0.15	0.29	0.36	6.47	0.03
32	6.1	75.9	0.12	16.1	0.54	0.27	0.00	0.06	0.11	0.53	0.23	5.44	0.02
32	7.6	72.7	0.18	17.4	0.87	0.27	0.00	0.27	2.60	0.84	0.40	5.65	0.03
32	9.1	74.0	0.15	17.2	0.40	0.34	0.00	0.21	0.65	2.18	0.67	5.29	0.03
32	10.7	77.9	0.12	15.8	0.16	0.53	0.00	0.50	0.05	0.40	0.84	4.89	0.02
32	12.2	72.2	0.27	16.4	0.74	0.63	0.02	2.68	0.05	0.26	2.96	2.99	0.04
32	14.0	70.9	0.20	16.4	0.39	0.75	0.02	2.02	0.69	0.40	4.91	2.85	0.02
32	15.2	71.4	0.24	15.8	0.40	0.94	0.03	1.79	1.35	1.64	5.62	1.51	0.04
32	18.3	73.3	0.26	16.2	0.32	0.99	0.03	0.77	2.00	4.33	2.52	0.40	0.02
32	22.9	72.0	0.25	15.8	0.03	1.22	0.03	0.61	2.20	4.66	1.89	0.36	0.02
50	0.0	73.5	0.20	19.3	0.21	0.13	0.00	0.03	0.40	0.50	0.13	6.38	0.06
50	1.5	72.9	0.22	18.7	1.35	0.27	0.00	0.09	0.05	0.41	0.25	5.87	0.02
50	3.0	70.1	0.20	19.6	2.49	0.13	0.00	0.07	0.11	0.27	0.50	6.18	0.03
50	4.6	74.9	0.16	17.1	1.35	0.28	0.00	0.22	0.45	0.41	0.78	5.04	0.13
50	6.1	73.2	0.16	16.5	3.31	0.32	0.00	0.38	0.57	0.34	1.19	4.50	0.05

(continued)

50	7.6	73.5	0.19	17.3	0.58	0.34	0.00	1.23	0.19	0.24	1.33	4.34	0.09
50	9.1	73.4	0.19	17.2	0.71	0.34	0.01	1.97	0.55	0.22	1.62	4.00	0.11
50	10.7	80.5	0.17	12.2	0.05	0.43	0.00	1.50	0.13	0.19	1.15	2.80	0.10
50	12.2	76.2	0.16	14.3	0.95	0.41	0.01	2.08	0.04	0.17	1.15	3.51	0.08
50	13.7	75.2	0.28	16.7	0.75	0.49	0.01	1.81	0.10	0.16	0.97	4.32	0.06
50	15.2	71.0	0.05	17.0	0.77	0.30	0.01	3.03	0.64	0.22	2.74	2.41	0.08
50	16.8	74.8	0.17	14.8	1.40	0.40	0.01	2.53	0.11	0.19	2.58	2.01	0.08
50	18.3	75.8	0.09	13.7	0.08	0.39	0.01	1.06	0.11	0.49	6.54	0.80	0.05
50	19.8	72.8	0.21	15.2	0.05	0.41	0.01	1.08	0.29	0.42	7.23	0.70	0.06
50	21.3	74.4	0.19	14.6	0.69	0.41	0.01	1.53	0.27	0.24	5.99	1.87	0.02
50	32.0	72.7	0.17	14.3	0.05	0.54	0.01	1.01	0.57	1.79	6.99	0.50	0.02
50	38.1	75.9	0.10	13.2	0.17	0.61	0.01	0.44	0.15	2.81	6.73	0.13	0.02
51	0.0	72.8	0.52	12.7	6.73	0.41	0.01	0.18	0.24	0.22	0.63	2.97	0.08
51	1.2	76.5	0.33	14.9	1.57	0.27	0.01	0.12	0.04	0.17	0.49	4.90	0.02
51	3.0	75.0	0.22	15.8	1.74	0.27	0.01	0.09	0.06	0.05	0.86	4.89	0.02
51	4.6	73.4	0.17	15.0	2.00	0.47	0.03	0.66	0.05	0.05	1.06	4.55	0.02
51	6.1	80.5	0.24	11.4	1.62	0.47	0.03	1.50	0.04	0.08	1.45	3.23	0.02
51	7.6	74.6	0.16	14.2	1.81	0.47	0.02	3.00	0.04	0.05	1.58	3.88	0.02
51	9.1	68.0	0.38	17.1	2.35	0.41	0.01	2.58	0.13	0.05	7.09	3.61	0.02
51	10.7	65.2	0.44	18.5	1.55	0.41	0.03	3.69	0.50	0.24	4.17	4.15	0.02
51	12.2	69.4	0.20	14.7	0.38	0.61	0.01	1.48	0.18	0.37	7.78	1.69	0.02
51	13.7	69.9	0.30	15.0	0.75	1.02	0.03	1.95	0.56	1.42	5.75	2.00	0.02
51	16.8	70.0	0.18	14.6	0.75	1.30	0.03	1.22	1.67	4.49	3.49	0.25	0.02
51	21.3	60.9	0.31	19.7	0.63	2.26	0.03	3.12	1.65	5.86	4.68	0.34	0.14
65	0.4	44.6	1.03	29.5	4.95	0.13	0.01	1.66	0.04	0.07	8.23	4.18	0.09
65	2.7	65.0	0.69	22.4	1.18	0.20	0.01	0.66	0.04	0.05	2.60	6.46	0.04
65	4.1	67.4	0.80	20.1	0.83	0.27	0.01	1.35	0.04	0.05	4.12	4.15	0.02
65	5.3	61.2	0.85	21.1	4.63	0.13	0.01	1.84	0.04	0.05	5.17	3.92	0.02
65	6.8	67.8	0.91	19.5	2.56	0.27	0.02	0.95	0.09	0.05	5.28	3.14	0.05
65	7.9	66.8	0.81	18.6	3.54	0.27	0.01	0.81	0.04	0.05	3.72	3.98	0.07

(continued)

TABLE 7.10. *(continued)*

Analyses of drill core samples through paleosols at the base of the Athabaska Sandstone, west of Wollaston Lake, Saskatchewan (Macdonald 1980).

Hole No.	Depth	SiO_2	TiO_2	Al_2O_3	Fe_2O_3	FeO	MnO	MgO	CaO	Na_2O	K_2O	H_2O	C
65	9.3	66.7	0.77	18.0	3.88	0.27	0.01	1.09	0.06	0.17	5.42	2.41	0.03
65	10.5	67.5	0.78	19.8	0.97	0.20	0.00	1.02	0.04	0.10	5.51	2.74	0.04
65	11.8	65.0	0.75	18.7	4.33	0.27	0.01	1.32	0.07	0.11	5.08	2.93	0.10
65	13.0	71.6	0.26	15.6	4.80	0.34	0.01	2.45	0.04	0.05	3.17	3.30	0.02
65	14.3	63.5	0.84	18.0	7.30	0.54	0.02	2.74	0.07	0.16	4.38	3.70	0.07
65	15.6	66.7	0.71	17.0	4.21	0.82	0.02	3.40	0.04	0.05	3.75	3.58	0.18
65	16.9	66.1	0.67	17.3	1.52	1.43	0.01	4.36	0.04	0.05	3.25	3.95	0.15
65	18.3	66.6	0.70	17.0	2.62	0.91	0.00	5.20	0.04	1.22	2.21	4.08	0.21
65	19.6	68.3	0.62	17.6	2.72	0.09	0.02	3.61	0.04	2.25	2.96	3.22	0.29
65	21.0	73.6	0.61	15.3	0.46	1.82	0.00	3.90	0.21	0.05	2.39	3.17	0.17
65	22.4	68.9	0.69	16.3	0.25	0.64	0.00	4.81	0.03	0.08	2.10	4.03	0.51
65	24.8	73.4	0.61	14.4	0.12	0.63	0.01	4.44	0.08	0.05	2.04	3.21	0.66
65	26.2	71.6	0.61	15.7	0.05	0.68	0.01	3.94	0.16	0.08	3.56	3.74	0.19
65	27.5	64.9	0.65	16.6	0.58	1.37	0.02	5.86	0.15	0.05	2.84	4.81	0.73
65	30.1	78.7	0.46	11.4	0.82	0.82	0.01	4.17	0.05	0.05	2.07	2.15	0.73
65	31.2	57.0	0.63	16.2	6.02	0.82	0.05	11.8	0.07	4.21	2.04	4.43	0.57
65	32.5	57.0	0.65	15.2	3.20	4.45	0.03	10.7	0.05	0.05	3.56	3.35	0.53
65	33.8	59.8	0.66	15.2	2.72	3.76	0.03	10.1	0.09	0.05	2.89	3.61	0.42
65	35.0	57.9	0.50	13.7	2.16	4.79	0.02	11.8	0.12	0.70	1.52	4.32	0.31
65	37.5	69.7	0.28	13.1	0.66	2.26	0.02	2.08	0.13	1.36	5.79	1.52	0.08
65	39.7	62.3	0.72	17.9	0.84	4.44	0.02	3.39	0.18	1.25	3.79	2.75	0.02
65	42.3	64.2	0.56	17.4	0.83	4.40	0.02	3.16	0.16	1.48	4.38	2.64	0.02
65	44.7	63.1	0.73	18.5	1.43	4.59	0.03	4.18	0.04	0.05	5.26	3.49	0.02
81	0.0	63.7	1.10	23.1	1.10	4.11	0.03	2.60	0.04	0.07	0.01	7.82	0.07
81	0.6	61.0	1.10	23.6	1.57	0.34	0.00	2.42	0.04	0.05	0.18	8.24	0.02
81	1.5	62.9	0.78	18.4	7.25	0.34	0.02	2.15	0.04	0.05	0.44	5.84	0.03
81	3.0	63.0	0.80	18.2	6.02	1.10	0.01	2.44	0.04	0.05	1.27	5.54	0.02

(continued)

81	4.9	56.7	0.92	18.8	5.16	2.83	0.01	3.88	2.03	0.21	3.13	4.70	0.02
81	6.1	65.5	0.80	17.9	2.00	1.82	0.01	2.88	0.04	0.05	1.95	5.23	0.02
81	7.9	62.0	0.80	18.6	3.18	1.98	0.01	4.77	0.04	3.04	3.22	4.59	0.02
81	9.1	63.9	1.15	19.7	2.00	3.01	0.01	3.95	0.04	0.10	2.79	4.69	0.02
81	10.7	65.6	1.03	18.8	0.90	2.53	0.00	4.41	0.04	0.40	2.86	4.60	0.02
81	12.2	65.3	1.14	18.3	0.86	3.15	0.01	4.44	0.04	0.05	2.56	4.70	0.08
81	13.7	63.3	1.05	18.9	1.65	3.15	0.01	4.49	0.04	1.57	2.56	4.26	0.02
81	15.5	71.1	0.77	16.8	1.70	1.30	0.01	4.07	0.04	0.05	1.79	4.19	0.03
81	16.8	64.5	0.98	18.3	2.23	1.99	0.02	4.84	0.04	2.34	1.68	4.83	0.07
81	18.3	63.8	0.84	18.0	2.88	2.26	0.02	4.77	0.10	1.97	2.02	4.41	0.07
81	19.8	67.7	0.88	16.9	3.03	2.19	0.02	4.38	0.04	0.05	2.23	3.98	0.21
81	21.3	69.0	0.91	15.1	2.09	2.74	0.01	3.05	0.04	0.05	3.70	2.72	0.08
81	22.8	59.0	1.38	20.4	1.10	3.58	0.01	3.93	0.10	0.17	4.23	4.20	0.11
81	24.3	67.1	0.93	17.1	0.54	5.62	0.04	2.49	0.08	0.86	4.74	0.83	0.17
81	25.9	65.2	0.99	17.0	0.36	7.45	0.07	2.73	0.49	0.75	2.92	0.60	0.16
81	27.4	63.0	1.04	18.8	0.27	6.78	0.06	2.68	0.85	1.06	5.06	0.81	0.17
81	30.4	63.6	0.90	17.4	0.44	7.05	0.06	2.57	0.50	0.98	3.90	0.80	0.10
81	33.5	57.0	1.10	21.9	0.17	8.70	0.11	3.06	2.53	2.22	2.90	1.06	0.10
81	36.6	58.3	1.22	20.7	1.22	7.19	0.08	3.10	2.20	1.93	2.45	1.22	0.07

Note: The samples were analyzed by the Geology Division of the Saskatchewan Research Council. The listed depths are in meters below the unconformity in each section. Drill Holes 20, 23, 32, 50, and 51 intersected meta-arkose; Drill Hole 65, metasemipelite; and Drill Hole 81, metapelite.

The values of the parameter R for the five cores that bottomed in un-weathered meta-arkose are shown in table 7.11. The computed values of R are based on the analyses of unweathered meta-arkose near the bottom of each drill hole. Since the parent rocks of the paleosols are not completely homogeneous, a fair degree of uncertainty attaches to the calculated R values. Elemental carbon was not included in the computation of the O_2 and CO_2 demand of the meta-arkoses, since this component appeared to be inert during weathering. If C^0 had been included, the calculated values of R would be somewhat higher, in some cases as much as twice the value shown in table 7.11. Fortunately, this uncertainty is of no great importance. Even if C^0 had been included in the computations, the values of R for the meta-arkoses would still have fallen in the lower part of the diagram in figure 7.10.

None of the meta-arkoses seem to have lost a significant quantity of iron during weathering. This is well illustrated by the data in figure 7.12. At the base of Drill Hole 20 nearly all of the iron is present in the divalent state. At the top of the paleosol nearly all of the iron is present in the trivalent state. The total quantity of iron changes little from the base to the top of the paleosol. The Al_2O_3 concentration increases only modestly toward the top of the paleosol. A small amount of iron may have been lost from this paleosol; but if so, the quantity is a small fraction of the initial iron content of the upper part of the paleosol. The variation

TABLE 7.11.

Values of the parameter, R, for parent rocks below the paleosols in the Athabaska Basin.

Mineral	R	Samples used in computation of R
Meta-arkoses		
Drill Hole 20	2.0×10^{-2}	Bottom 4
Drill Hole 23	1.5×10^{-2}	Bottom 2
Drill Hole 32	1.5×10^{-2}	Bottom sample
Drill Hole 50	0.8×10^{-2}	Bottom 2
Drill Hole 51	1.6×10^{-2}	Bottom 2
Semimetapelites		
Drill Hole 65	5.0×10^{-2}	Bottom 3
Metapelite		
Drill Hole 81	7.7×10^{-2}	Bottom 2

Note: Data from table 7.10.

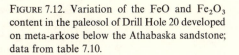

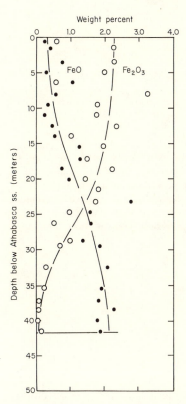

FIGURE 7.12. Variation of the FeO and Fe$_2$O$_3$ content in the paleosol of Drill Hole 20 developed on meta-arkose below the Athabaska sandstone; data from table 7.10.

in the total iron content of the paleosol developed on metapelite in Drill Hole 81 and particularly on semimetapelite in Drill Hole 65 is difficult to interpret with certainty. If the weathered rocks in the upper part of the paleosols in these drill holes initially contained as much as the fresh rock at the base of these paleosols, then a significant amount of iron was lost from both sections. It is quite clear that much more remains to be done on uniform rocks with R values in excess of 5×10^{-2} before the dashed trend line separating paleosols that have lost iron from those that have retained their iron can be accepted with confidence.

PALEOSOLS ON PRESQUE ISLE POINT, NORTHERN MICHIGAN

On Presque Isle Point near Marquette, Michigan, the late Precambrian Jacobsville sandstone overlies an irregular erosion surface that contains local concentrations of altered bedrock. These altered zones have been described as paleosols by Krimmel (1941), by Gair and Thaden (1968),

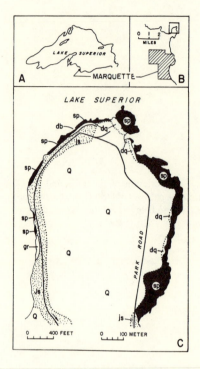

FIGURE 7.13. The geology of Presque Isle, Michigan (Kalliokoski 1975). Sp = serpentinized peridotite; gr = granodiorite; db = diabase; q = dolomite-quartz rocks; js = Jacobsville sandstone; Q = glacial overburden and beach gravel. (Reproduced by permission of the Geological Society of America.)

and by Kalliokoski (1975)*. Distinctive paleosols in this area developed on each of three rock types: granodiorite, serpentinized peridotite, and diabase (see figure 7.13). Although the age of the paleosols is uncertain, the weathering episode they represent precedes the deposition of the Jacobsville, which is now assigned to the upper Keweenawan Bayfield Group (White 1966, p. 28). The granitic rocks in the area are considered to be of Archean age. Kalliokoski (1975) has suggested that the paleosols were developed ca. 1.0 b.y. ago.

The paleosols that developed on granodiorite have been studied in greatest detail. In thin sections of the freshest granodiorite, the K-feldspar margins and interstices show a distinctive micrographic intergrowth with quartz. The feldspar was originally microperthitic, and is now altered almost completely to a clouded mass of sericite. The interstitial material is opaque and consists of a mixture of primary and secondary minerals: biotite, muscovite, chlorite, and vermiculite, heavily stained by fine-grained

* Field work on Presque Isle during the summer of 1983 convinced the author that the validity of this interpretation is by no means assured.

hematite. The interstitial alteration and pigmentation are probably weathering phenomena.

A profile of the Presque Isle paleosol and of the changes in its composition with elevation are shown in figure 7.14. The lowermost parts of the profile consist of fairly hard granodiorite. These are overlain by a slightly less coherent, TiO_2-rich saprolitic material with recognizable white feldspar pseudomorphs and tiny carbonate veinlets in a reddish matrix. Higher still in the profile, the feldspars are represented by 1×0.2 mm flecks, and the material has a weak subhorizontal foliation. The top 50 cm of the section is a red-brown, loose, soil-like material containing lumps of saprolite in a sandy matrix. The profile contains two dolomite-cemented zones, which probably represent paleocaliche horizons.

TiO_2 is enriched in the paleosol. The concentration of total iron calculated as Fe_2O_3 also increases upward, but figure 7.14b shows that Fe_2O_3 has been concentrated less than TiO_2 during weathering. There has therefore been some loss of iron during weathering of the granodiorite. Without additional chemical data it is not possible to calculate the ratio of the O_2 to the CO_2 demand during the weathering of the granodiorite. It seems likely that this ratio was intermediate between that of the Pronto granite and the Hekpoort basalt, and it is reassuring that the behavior of the Presque Isle granodiorite during weathering also seems to have been intermediate between that of these two parent rocks. There is therefore no contradiction between the data in figure 7.14 and the upper and

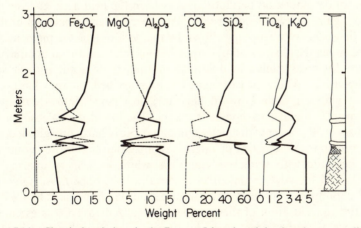

FIGURE 7.14a. Chemical variations in the Presque Isle paleosol developed on granodiorite (Kalliokoski 1975). (Reproduced by permission of the Geological Society of America.)

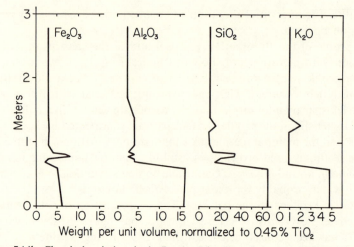

FIGURE 7.14b. Chemical variations in the Presque Isle paleosol on granodiorite, normalized to 0.45 g TiO_2 per unit volume (Kalliokoski 1975). (Reproduced by permission of the Geological Society of America.)

lower limits proposed above for the partial pressure of oxygen in the atmosphere between ca. 2.2 and 2.4 b.y.b.p.

The paleosols developed on peridotite are rather complex, in part because the parent rock is heterogeneous. The peridotite contains hydrothermal dolomite-quartz veinlets and irregularly developed blocky structures that are subparallel to the unconformity. No complete sections of paleosol developed on peridotite exist within the field area, and the profile of figure 7.15 was reconstructed from several partial paleosol profiles. The first effect of weathering shown by the peridotite is the pigmentation of the minerals; olivine and pyroxene become clouded by an opaque substance. Intricate veining by dolomite and quartz on a microscopic scale is also visible. Where the carbonate has been leached, the rock has turned into a granular sand. Upward, this horizon changes into a dark gray, brownish, or reddish carbonate-free zone. The top of the paleosol profile consists of 30–90 cm of distinctly laminated, fairly hard, fine-grained, reddish brown rock consisting of a matrix of mixed-layer chlorite-vermiculite and hematite with veinlets of purplish, white, or jaspery quartz. The Fe_2O_3 and the Al_2O_3 content of material in this uppermost horizon is considerably greater than in the parent peridotite (solid line in figure 7.15), but the concentration of these oxides normalized to TiO_2 is essentially the same as in the parent rock. All three oxides have therefore been retained in roughly the same proportion during weathering. Cr_2O_3 has been similarly concentrated in the upper part of the profile.

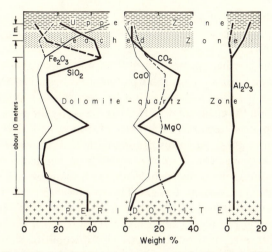

FIGURE 7.15. Chemical variation in a composite vertical paleosol profile on Presque Isle through peridotite, dolomite-quarts, caliche, leached and upper horizons (Kalliokoski 1975). (Reproduced by permission of the Geological Society of America.)

Free oxygen was obviously present in the atmosphere during the weathering of the peridotite. In the absence of more complete chemical data, the available information cannot be used to set a lower limit on P_{O_2} during the formation of the paleosols on this bedrock.

OTHER REPORTED PRECAMBRIAN WEATHERING ZONES

Data for two other Precambrian paleosols have been added to figure 7.10. Vogel (1975) has described chloritoid and kyanite-bearing acid metavolcanic rocks in the Abitibi greenstone belt. These rocks have an aluminum surplus that was probably acquired during weathering prior to metamorphism. The R value of the parent rocks is ca. 2.1×10^{-2}, and there seems to have been little, if any, loss of iron during weathering. This is consistent with the other paleosol data in figure 7.10. It is also true of the weathering profile developed on Lewisian biotite gneiss of the Cape Wrath district of northwest Scotland (Williams 1968). The soil zones in this area are overlain by late Precambrian (ca. 1.0 b.y.) Torridonian alluvial fan deposits. Unfortunately, the reported analyses of the parent rocks do not include data for their MgO content. Only the maximum value of R for these rocks is therefore known; their actual R value is probably no less than 70% of the calculated maximum value. As expected from their position in figure 7.10, there has been little or no loss of iron from the soils in this area. A similar behavior has been observed in soils developed on granites below the Proterozoic Jatulian sediments in Karelia (Koryakin 1981).

Descriptions of other occurrences of Precambrian weathering add little to the data summarized below. Sharp (1940) has described Precambrian paleosols from the Grand Canyon, but his description is qualitative and does not contain the chemical data that are necessary for extracting quantitative information regarding the level of atmospheric oxygen during the formation of these paleosols. Rankama (1955) has described a Precambrian breccia and conglomerate schists from Suodenniemi, Finland; these consist of fragments and pebbles of pre-Bothnian diorite embedded in a fine-grained granoblastic cement. The area has been sufficiently metamorphosed to cast doubt on the conclusion that the FeO/Fe_2O_3 ratio of the weathering products has been preserved. Even if this were established, the data presented by Rankama (1955) would be too incomplete to be helpful in defining the oxygen content of the atmosphere during the time of weathering. Unfortunately, the same comments also apply to the weathering products described by Serdyuchenko (1968). Schau and Henderson (1983) have described a possible paleosol below the Archean Steep Rock Group in the Western Superior Structural Province of Canada. Their data are, however, incomplete, and their interpretation is uncertain.

SUMMARY AND IMPLICATIONS OF THE PALEOSOL DATA

The summary plot of figure 7.10 shows that all of the available data for Precambrian paleosols are internally consistent. Soils developed on high-R rocks between 1.5 and 3.0 b.y.b.p. lost iron during weathering; soils developed on low-R rocks did not. The value of R for rocks at the transition from iron loss to iron retention during soil formation does not seem to have changed a great deal between 3.0 and 1.5 b.y.b.p. In late Archean times the transition seems to have been at R values close to 3×10^{-2}, and in mid-Proterozoic time at R values between ca. 2×10^{-2} and 5×10^{-2}. During the intervening 1.5 b.y. the value of the function $[(B_{O_2}/B_{CO_2})(P_{O_2}/P_{CO_2})]$ was therefore $\leq (4 \pm 1) \times 10^{-2}$, and the ratio $P_{O_2}/P_{CO_2} \leq (1.3 \pm 0.5)$. Today the ratio of P_{O_2} to P_{CO_2} is close to 600. As pointed out earlier, the composition of the Precambrian atmosphere between 3.0 and 1.5 b.y.b.p. therefore seems to have differed significantly from that of the present day. Two questions come to mind: can we separate the effect of variations in the O_2 pressure from those of the CO_2 pressure on the inferred changes in the P_{O_2}/P_{CO_2} ratio, and can anything be done to set a lower limit on the P_{O_2}/P_{CO_2} ratio during the Precambrian Era?

The first question can be answered only in the roughest manner. The CO_2 content of the Precambrian atmosphere was discussed briefly at the end of chapter 5. It was shown there that P_{CO_2} was almost certainly significantly higher then than now, but that its value between 1.5 and 3.0 b.y.b.p. is not well constrained by the acid-base balance of the Earth's crust. Other lines of evidence are obviously needed to define the course of P_{CO_2} in the atmosphere. One of these turns out to be the behavior of uraninite during chemical weathering between 2 and 3 b.y.b.p. The remainder of the chapter is devoted to this subject. The second question is also difficult to answer. Soils developed on low-R rocks tend to be relatively porous and permeable. They therefore tend to be rather open to a resupply of O_2 and CO_2 from the atmosphere. The O_2/CO_2 ratios in the Precambrian atmosphere could thus have been considerably lower than the upper bound indicated by iron loss from soils developed on high-R rocks.

It is even possible that oxygen was completely absent from the atmosphere, and that iron in Precambrian soils was oxidized by substances other than O_2. Alternative oxidants that come to mind are HNO_3, HNO_2, H_2O_2, and O_3. The first of these is the most likely. The production of nitric acid in the atmosphere by photochemical reactions and in lightning discharges has been investigated in some detail (see, for instance, Yung and McElroy 1979; Chameides and Walker 1981; Kasting and Walker 1981; and Ehhalt and Drummond 1982). Figure 7.16 shows the results of Yung and McElroy's (1979) calculations of the effect of P_{O_2} on the rate of nitrogen fixation in the atmosphere. The fixation of nitrogen by lightning in the contemporary atmosphere was taken to be 40×10^{12} gm N/yr. If 40×10^{12} gm of N are converted annually to HNO_3 and if they are removed as a constituent of 5×10^{20} gm of rain, the mean concentration of HNO_3 in rain is

$$\frac{40 \times 10^{12} \text{ gm N/yr}}{14 \frac{\text{gm N}}{\text{mol N}} \times 5 \times 10^{17} \text{ kg rain/yr}} = 6 \times 10^{-6} \frac{\text{mol}}{\text{kg}}.$$

This is only ca. 2% of the concentration of dissolved oxygen in present-day rain. The ratio of the HNO_3 to the dissolved O_2 concentration in average rainwater increases with decreasing P_{O_2} (see table 7.12). The preferred range of P_{O_2} during the period for which there are paleosol data is 1×10^{-3} to 1×10^{-2} atm (see below). In this range the calculated concentration of HNO_3 in average rainwater is less than that of m_{O_2}, but

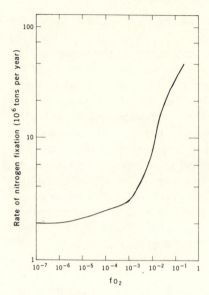

FIGURE 7.16. The rate of nitrogen fixation in the atmosphere as a function of the mixing ratio of O_2 (Yung and McElroy 1979). (Copyright 1979 by the American Association for the Advancement of Science.)

TABLE 7.12.

Concentration of O_2 and HNO_3 in average rainwater as a function of the oxygen content of the atmosphere.

P_{O_2} (atm)	m_{O_2} (mol/kg)	m_{HNO_3} (mol/kg)	m_{HNO_3}/m_{O_2}	$(m_{HNO_3} + m_{O_2})$ mol/kg
0.2	2.6×10^{-4}	6×10^{-6}	0.023	2.6×10^{-4}
0.1	1.3×10^{-4}	5×10^{-6}	0.038	1.35×10^{-4}
1×10^{-2}	1.3×10^{-5}	1.5×10^{-6}	0.12	1.5×10^{-5}
1×10^{-3}	1.3×10^{-6}	0.5×10^{-7}	0.38	1.8×10^{-6}
1×10^{-4}	1.3×10^{-7}	0.4×10^{-7}	3	5.2×10^{-7}
1×10^{-5}	1.3×10^{-8}	0.3×10^{-7}	23	3.1×10^{-7}

Notes: $B_{O_2} = 1.3 \times 10^{-3}$ (see table 7.4).
HNO_3 production rates from Yung and McElroy (1979).

it is not trivial. At lower values of P_{O_2} the HNO_3/O_2 ratio exceeds unity. If P_{O_2} is 1×10^{-5} atm, m_{HNO_3} is ca. 3×10^{-7} mol/kg. The minimum value of the CO_2 concentration in Precambrian rain is probably that of the present day; thus,

$$\frac{m_{HNO_3}}{m_{H_2CO_3}} \leq 3 \times 10^{-2}.$$

It is therefore barely possible that the low-R rocks in figure 7.10 were oxidized largely by HNO_3. It is likely, however, that the HNO_3 production figures in figure 7.16 are high by a factor of about 5 (Ehhalt and Drummond 1982), and that $m_{H_2CO_3}$ in the Precambrian was at least five times greater than today. The $m_{HNO_3}/m_{H_2CO_3}$ ratio at a P_{O_2} of 1×10^{-5} atm was therefore probably $\leq 0.1 \times 10^{-2}$ and hence insufficient to account for the observed oxidation of more than a small fraction of the iron in soils developed on low-R rocks. It is most likely, then, that molecular oxygen was the dominant oxidant but not the only one in Precambrian soils. If reducing compounds, such as formaldehyde, were present in Precambrian rainwater, the P_{O_2} may actually have been greater than those indicated in the interpretation of the paleosol data.

3. Detrital Minerals in Sedimentary Rocks

ORE DEPOSITS OF DETRITAL URANINITE

Many minerals that are thermodynamically unstable in the presence of free oxygen are decomposed so rapidly today that they occur very rarely, if ever, in the residues of atmospheric weathering. Some minerals, on the other hand, react so slowly with atmospheric oxygen at temperatures below 100°C that they survive exposure to oxygen during weathering, transport, and deposition, and are among the common detrital constituents of sedimentary rocks. Magnetite is a conspicuous member of this group.

Uraninite and pyrite are two of the minerals that are vulnerable to atmospheric oxidation. Both are rare as detrital minerals in recent sediments and sedimentary rocks, because they are so readily oxidized under present atmospheric conditions. Today both minerals persist only in areas of unusual aridity or of unusually rapid erosion and transport. Small amounts of uraninite have, for instance, been identified in the upper reaches of the Indus River in Kashmir (Zeschke 1960) and in the Hunza River, a tributary of the Indus (Simpson and Bowles 1977). In early to mid-Precambrian sediments there are, however, large ore deposits of uraninite in which this mineral is almost certainly detrital. The known deposits of this type are all older than 2 b.y. The rarity of detrital uraninite today has generated a good deal of resistance to the notion that large ore bodies of detrital uraninite could have accumulated in the past (see, for instance, Davidson 1960), and the origin of uraninite in the Canadian uranium deposits at Blind River and Elliot Lake, and in the South African uranium-gold ores of the Witwatersrand Basin has stirred more than a little passionate debate; Bergne (1964), for instance, was moved to write that "Professor Davidson's observations on the Witwatersrand banket deposits, together with the ripostes of other geologists, have provided me over the years with the greatest of interest not untinged with amusement and a little sorrow."

In the Witwatersrand Basin of South Africa (figure 7.17), auriferous and uraniferous quartz-pebble conglomerates occur in the Dominion Reef, Witwatersrand, Ventersdorp, and Transvaal supergroups. These weakly metamorphosed sedimentary/volcanic successions rest with marked unconformity on a highly metamorphosed basement complex comprising the Older Granite and the Swaziland sequence, which together constitute an early Precambrian granite-greenstone terrain. The three lower sedimentary/volcanic supergroups are closely related both areally and structurally. The maximum thickness of the triad has been estimated to be about 15,000 m. The main development of gold and uraninite-bearing conglomerates occurs in the Upper Witwatersrand. Uraniferous conglomerates also occur in the Basal Sedimentary Formation of the Dominion Reef Supergroup. The Ventersdorp Contact Reef at the base of the Ventersdorp Supergroup is auriferous and uraniferous where it lies unconformably on the eroded and beveled beds of the Witwatersrand rocks. The Black Reef in the overlying Transvaal Supergroup is also

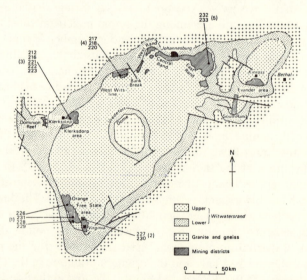

FIGURE 7.17. Simplified geologic map of the Witwatersrand Basin beneath the Ventersdorp and younger cover (Rundle and Snelling 1977). (Reproduced by permission of the Royal Society, London.)

locally auriferous and uraniferous. The gold and other heavy minerals, including uraninite in the Ventersdorp Contact Reef and in the Black Reef, are generally considered to have been derived by the erosion of underlying Witwatersrand conglomerates.

The mineralogy and petrology of the ores has been studied in considerable detail by Ramdohr (1955, 1958a,b, 1979), Liebenberg (1955, 1958, 1960), Schidlowski (1966a,b,c, 1981), Hiemstra (1968a,b), and Utter (1980). All of these authors concluded that the uraninite in the ores is largely detrital. The textural evidence for this conclusion rests in part on the size and shape of the uraninite grains (see figure 7.18), and on the relationship of the grain boundaries to zoning within the grains. The grains are similar to those of undoubtedly detrital grains of monazite and zircon, and are quite dissimilar to those of hydrothermal pitchblende.

The analyses in table 7.13 indicate that the uraninites contain significant quantities of ThO_2 (1.6–6.5%) and rare earths (1.0–3.4%). The concentration of these elements is considerably higher than in hydrothermal pitchblendes. The nature of the distribution of uraninite in the conglomerates and the high degree of correlation of uranium in the ores with the elements present in detrital minerals is also consistent with a detrital origin for much of the uraninite (Hiemstra 1968b).

It could be, and has been, argued that all these lines of evidence do not constitute proof of a detrital origin for the uraninites. Conclusive proof could be obtained by showing that the age of the uraninites is greater than the age of the enclosing sediments. This demonstration has turned out to be difficult. The lead-uranium and lead-thorium ages of material from the Witwatersrand, Ventersdorp, and Transvaal supergroups are usually discordant (Nicolaysen et al. 1962). In many of the samples there has been a major period of lead loss 2040 ± 100 m.y.b.p. as well as recent uranium loss. The total rock U-Pb data of Rundle and Snelling (1977) suggest that their samples can be regarded as having been derived from parent rocks containing 3050 ± 50 m.y.-old uranium-bearing minerals; some of their samples have experienced recent lead loss, others recent uranium loss; the remainder yielded concordant ages. Total rock U-Pb dates are less disturbed by the loss of lead during mild metamorphism than are the ages of individual minerals. However, the compensation for lead loss is usually only partial, because only part of the lead released from uraninite during metamorphism tends to be precipitated locally. The significance of total rock ages also suffers somewhat from the inclusion of uranium and lead from all radioactive minerals, not only from uraninite in a given rock sample. Monazite, whose detrital origin has never been doubted, is among the contributors of U and Pb. Nevertheless, Rundle and Snelling's (1977) proposed age of 3050 m.y. for the

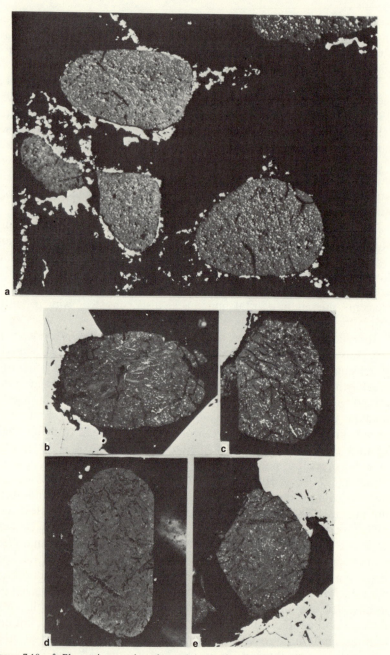

FIGURE 7.18a–f. Photomicrographs of uraninite grains from the Witwatersrand Basin (Schidlowski 1966a). (Reproduced by permission of E. Schweizerbart'sche Verlagsbuch-handlung, Stuttgart.)

a. Detrital grains of uraninite with characteristic dusting of galena; partly surrounded by PbS-overgrowths. The large grain displays a typical "muffin-shape"; Basal Reef, footwall, Loraine Gold Mines, O.F.S. Oil immersion, 315x.

b. Detrital UO_2 grain with excellent rounding, cross-cut by minor thucholite veins and partially overgrown by authigenic pyrite (left side); Elsburg A3 Reef, Loraine Gold Mines, O.F.S. Oil immersion, 375x.

c. Idiomorphic uraninite crystal with common PbS speckling; Elsburg A3 Reef. Oil immersion, 375x.

d. Idiomorphic uraninite grain (just slightly rounded) with minute veins of carbonaceous matter; Elsburg A3 Reef. Oil immersion, 315x.

e. Idiomorphic uraninite grain displaying an octahedral shape, cross-cut in different directions by thucholite stringers. Partly overgrown by pyrite (white): Elsburg A3 Reef. Oil immersion, 375x.

f. Uraninite grains, New Quirke Mine, Ontario, Canada; magnification (Meddaugh, Holland, and Shimizu 1982).

TABLE 7.13.

Chemical composition and calculated *NOC* (Non-Uranium Cation) fraction of some uraninite samples from the Witwatersrand and Elliot Lake. The *NOC* values are calculated for the uraninites at the time of formation of the uraniferous placers (Grandstaff 1980).

	Blyvoor-uitzicht (1)	Sub Nigel (1)	Dominion Reef (2)	Nordic (3)
UO_2	—	—	17.10	22.7
UO_3	—	—	46.19	42.6
U_3O_8	74.10	66.65	—	—
ThO_2	2.70	1.63	6.52	6.1
REE	1.00	2.13	3.43	4.0
$ZrO_2 + TiO_2$	1.40	1.04	0.50	0.0
SiO_2	1.78	2.07	1.43	0.0
MgO	0.16	0.30	0.11	0.0
CaO	0.81	1.54	1.30	0.7
Fe_2O_3	0.84	1.81	1.12	0.2
Al_2O_3	1.33	0.99	0.08	0.0
S	—	—	1.26	0.9
H_2O^+	--	—	2.50	—
SnO_2	—	—	1.43	—
Other	—	—	0.62	—
PbO	16.70	19.44	14.85	22.6
	100.82	97.60	98.44	99.8
NOC { *minimum*	0.13	0.17	0.21	0.15
NOC { *maximum*	0.19	0.23	0.24	0.17

References: (1) Liebenberg (1957)
(2) Hiemstra (1968)
(3) Grandstaff (1976)

uraninites is reasonable. Since the Dominion Reef and Witwatersrand sediments were probably deposited more recently than 2740 ± 19 m.y. ago (Rundle and Snelling 1977), the available data imply that the uraniferous minerals in the Dominion Reef and Witwatersrand sediments are detrital in origin.

The uranium ores of the Blind River-Elliot Lake area are similar to those of the Witwatersrand Basin. Their geologic setting was described briefly in the previous section and has been treated in detail by Pienaar (1963), Roscoe (1957, 1969), and Robertson (1978). Hand specimens of the uraniferous conglomerates from this area are virtually indistinguishable from those of the Witwatersrand. The properties of the Canadian uraninite grains are similar to those of the Dominion Reef uraninites, and they are generally considered to have a detrital origin (see, for instance, Ramdohr 1958a,b). Their age has been difficult to determine, because the area suffered a period of mild metamorphism ca. 1.7 b.y. ago. Lead loss from uraninite was extensive at that time. It has, however, been possible to reconstruct the age of the larger uraninite grains by determining U/Pb ages as a function of distance from their edges. The instrument used was an ion microprobe. The age data tend to lie along chords in a concordia plot; as shown in figure 7.19, the end points of the chords lie close to 1.7 b.y.b.p. and 2.5 b.y.b.p., respectively (Meddaugh 1983). These data demonstrate that the uraninites are at least 2.5 b.y. old, and confirm that they have suffered a major period of lead loss ca. 1.7 b.y.b.p. The isotopic composition of small blebs of galena in the vicinity of uraninite grains is highly variable (Meddaugh, Holland, and Shimizu 1982); this variability is best interpreted in terms of the effects of a more recent, rather minor period of lead loss from the uraninite grains. Lead loss during this second period probably accounts for much of the scatter in the data of figure 7.19.

The age of the sediments containing the uraninite ores must be less than that of the pre-Huronian basement and greater than the age of the Nipissing diabase that cuts the Huronian sedimentary sequence. The diabase is well dated at 2.2 b.y. (see table 7.1); the age of the basement rocks is somewhat more uncertain, but they are probably 2.5–2.7 b.y. old. The uraninites were therefore almost certainly derived from the basement and

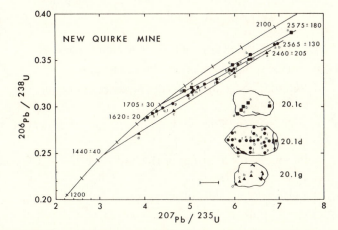

FIGURE 7.19. Pb-U age data for areas within single grains of uraninite from the New Quirke Mine, Elliot Lake, Ontario (Meddaugh 1983). (Reproduced by permission from the author.)

were included in the Huronian sediments during their deposition some 2.3–2.4 b.y. ago.

The mineralogical, petrographic, sedimentologic, and geochemical data that point to a detrital origin for much of the uraninite in the conglomerates of the Witwatersrand and in the Blind River-Elliot Lake region are weighty, and the implications of a detrital origin for uraninite in these rocks sets rather interesting limits on the composition of the atmosphere during the period between ca. 2.3 and 2.4 b.y.b.p.

THE KINETICS OF URANINITE OXIDATION AND DISSOLUTION

The survival of uraninite during weathering, transport, and deposition in a mildly oxidizing atmosphere is a matter of kinetics, not of thermodynamics. The stability field of uraninite is not well known, but the value of P_{O_2} at which UO_3 and its hydrates become stable with respect to uraninite at 25°C is much lower than the minimum value of P_{O_2} derived from the paleosol data in the previous section (see, for instance, Rich, Holland, and Petersen 1977). Whether or not uraninite accumulates as a detrital mineral under mildly oxidizing conditions depends on its rate of oxidation and dissolution in the interval of time during which it is exposed to these destructive processes. The most detailed study of the kinetics of uraninite oxidation and dissolution is that of Grandstaff (1976). His experimental data can be summarized by the equation,

$$R = 10^{20.25}(SS)(RF^{-1})(10^{-3.38-10.3NOC})(a_{\Sigma CO_2})(m_{O_2})$$
$$\cdot (a_{H+})\exp(-7045/T), \tag{7.15}$$

where R = fractional rate of dissolution of uraninite per day,

SS = specific surface area of uraninite (cm^2/gm),

RF = Retardation factor due to the presence of organic compounds,

NOC = mole fraction of the nonuranium cation component of uraninite,

$a_{\Sigma CO_2}$ = the sum of the activities of $H_2CO_3 + HCO_3^- + CO_3^=$ in solution (molar),

m_{O_2} = concentration of dissolved oxygen in solution (ppm),

T = temperature in °K.

Some of the terms in equation 7.15 are self-explanatory, others are not. The rate of dissolution of a given uraninite grain is obviously proportional to its surface area. However, it was found that organic compounds in New Jersey river and coastal waters can retard the dissolution of uraninite by a factor of 5 to 35; more detailed studies of this effect are needed, since it is possible that the dissolution of uraninite in mid-Precambrian time was also retarded by the presence of organic compounds.

The presence of cations other than uranium in uraninite decreases the rate of dissolution of this mineral. This is illustrated by the data in figure 7.20, obtained with the uraninites of table 7.14. Thorium, lead, and the rare earths are the major nonuranium cations. There are not enough data at present to separate the effects of the individual elements on the dissolution rate of uraninite, but the experiments to date indicate that the mole fraction of all the nonuranium elements can be combined, and that the sum can be treated as a single parameter without introducing serious errors in predicting the dissolution rate of uraninite.

As shown in figure 7.21, the dissolution rate of uraninite is proportional to the activity of H^+ at pH values between 4 and 6. In this pH range the oxidation and dissolution of UO_2 largely yields the complex ion $UO_2(OH)^+$ (see, for instance, Brush 1980) via the reaction,

$$UO_2 + \tfrac{1}{2}O_2 + H^+ \rightleftharpoons UO_2(OH)^+. \tag{7.16}$$

At values of the pH typical of soils and rivers, the uranyl dicarbonate

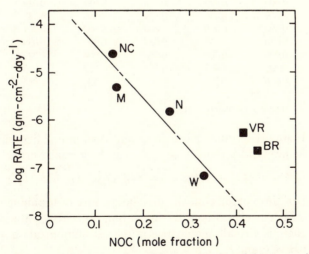

FIGURE 7.20. The rate of dissolution of uraninite as a function of the nonuranium cation fraction (*NOC*). The samples indicated by square symbols (VR and BR) are markedly heterogeneous. The measurements were performed in distilled water equilibrated with ambient air at 23°C (Grandstaff 1976). (Reproduced by permission of *Economic Geology*.)

Table 7.14.

Chemical composition and unit cell dimension of uraninites used in the measurement of uraninite dissolution rates by Grandstaff (1976). The elemental analyses were performed with an ARL electron microprobe, the UO_2 analyses by titration.

	NC	M	N	W	BR*	VR*
			Weighted percentage			
UO_2	75.5	29.3	30.8	50.0	22.7	23.2
UO_3	12.3	59.8	51.0	20.1	42.6	46.9
ThO_2	2.0	3.2	4.1	15.3	6.1	3.9
CeO_2	0.2	0.3	0.4	0.8	1.0	1.1
PbO	5.8	5.5	10.9	13.9	22.6	21.0
La_2O_3	0.0	0.0	0.0	0.1	0.1	0.6
Y_2O_3	1.5	1.4	1.6	0.0	1.9	0.2
CaO	0.2	0.2	1.9	0.1	0.7	1.1
FeO	0.1	0.1	0.1	0.1	0.1	0.2
S	0.0	0.0	0.0	0.0	0.9	1.1
	97.6	99.7	100.1	100.4	98.5	99.3
a_0	5.4398	5.4291	5.4440	5.4965	5.4865	5.4683
	±0.0008	±0.0020	±0.0002	±0.0007	±0.0010	±0.0005
			Cation mole fraction			
U (IV)	0.745	0.291	0.289	0.484	0.198	0.201
U (VI)	0.115	0.561	0.453	0.184	0.351	0.384
Th	0.020	0.032	0.039	0.152	0.055	0.034
Ce	0.003	0.005	0.006	0.012	0.014	0.015
Pb	0.069	0.066	0.124	0.163	0.239	0.221
La	0.000	0.000	0.000	0.002	0.001	0.009
Y	0.035	0.033	0.035	0.000	0.039	0.004
Ca	0.009	0.009	0.040	0.004	0.029	0.045
Fe	0.003	0.003	0.003	0.003	0.003	0.006
S	0.000	0.000	0.000	0.000	0.066	0.081

* Calculated from the average of grain analyses.

complex is almost certainly more important than uranyl hydroxide complexes. The uranyl dicarbonate complex can be formed via the reaction,

$$UO_2 + \tfrac{1}{2}O_2 + 2HCO_3^- \rightleftharpoons UO_2(CO_3)_2^{-2} + H_2O. \qquad (7.17)$$

The effect of this complex on the dissolution rate of uraninite is shown in figure 7.22. The decrease of the dissolution rate ceases with increasing pH, and there is virtually no change in the dissolution rate of uraninite between pH = 6 and 8.

The rate of uraninite dissolution is proportional to the O_2 content of the solutions and hence to the oxygen pressure of the gas phase with which the solutions are in equilibrium. This is demonstrated by the data of figure 7.23 for uraninite samples NC and M.

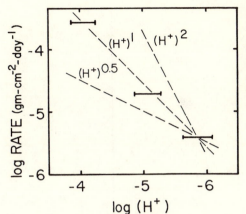

FIGURE 7.21. Variation of the dissolution rate of uraninite sample M as a function of the H^+ activity between pH 4 and 6. $T = 23°C$; the solutions were equilibrated with ambient air (Grandstaff 1976). (Reproduced by permission of *Economic Geology*.)

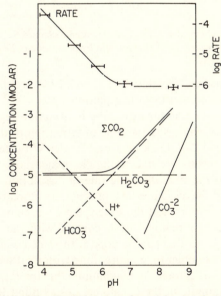

FIGURE 7.22. Variation of the dissolution rate of uraninite sample M as a function of the hydrogen ion activity together with the variation of a_{H^+}, $a_{HCO_3^-}$, $a_{CO_3^{2-}}$, and $a_{\Sigma CO_2}$. The experiments were carried out at $23°C$; the solutions were equilibrated with ambient air (Grandstaff 1976). (Reproduced by permission of *Economic Geology*.)

Dissolution rate experiments were carried out at $23°C$ and $2°C$. The measured dissolution rate at the two temperatures differed by approximately a factor of 3. Previous studies at higher temperatures and pressures (Mackay and Wadsworth 1958; Shortmann and DeSesa 1958) have shown that the temperature dependence of the dissolution rate follows

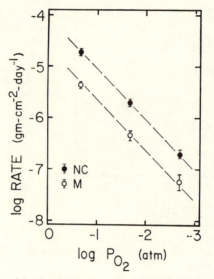

FIGURE 7.23. Variation of the dissolution rate of uraninite samples NC and M as a function of P_{O_2}; $T = 23°C$ (Grandstaff 1976). (Reproduced by permission of *Economic Geology*.)

the Arrhenius equation. The data at 23° and 2°C were therefore fitted to an Arrhenius function as shown in equation 7.15.

The use of this equation for extracting information regarding the level of atmospheric oxygen during the time of formation of the uranium ores in the Blind River-Elliot Lake area and in the Witwatersrand Basin is difficult, and the results are uncertain. The magnitude of several of the factors other than P_{O_2} that affect the rate of dissolution of uraninite are hard to gauge, and the residence time of uraninite in weathering horizons, the duration of transport, and the time spent by uraninite in the area of deposition prior to its removal from contact with the atmosphere are poorly constrained. All that can be done with some confidence is to estimate the minimum contact time of uraninite with the Precambrian atmosphere and thence to extract the maximum value for P_{O_2} consistent with the survival of uraninite in the conglomerates at Blind River and in the Witwatersrand.

An Analysis of the Survival of Detrital Uraninite in Sediments of the Indus River

A mild cross check for such calculations can be obtained by showing that the observed survival of uraninite in settings of particularly rapid erosion and transport such as the upper reaches of the Indus River is consistent with the laboratory data. This will be done first. Equation 7.15 can be recast in the form,

$$R \equiv \frac{1}{W}\frac{dW}{dt} = -k(SS), \tag{7.18}$$

where R = fractional rate of dissolution of uraninite per kg,

$\quad W$ = mass of uraninite (gm),

$\quad (SS)$ = specific surface area of uraninite (cm^2/gm),

$\quad k$ = a rate constant that includes all of the other terms in equation (7.15).

Consider a single spherical grain of uraninite of mass W, volume V, and specific gravity ρ. Its surface area, A, will be equal to

$$A = 4\pi \left(\frac{3W}{4\pi\rho}\right)^{2/3} \tag{7.19}$$

If this expression is substituted into equation 7.18, we obtain

$$\frac{dW}{dt} = -4\pi k \left(\frac{3W}{4\pi\rho}\right)^{2/3}. \tag{7.20}$$

Integration from an initial weight W_0 and volume V_0 to a weight W_1 and volume V_1 yields the expression,

$$\Delta t = \left(\frac{3}{4\pi}\right)^{1/3}\frac{\rho^{2/3}}{k}[W_0^{1/3} - W_1^{1/3}], \tag{7.21}$$

and the equivalent expression,

$$\Delta t = \left(\frac{3}{4\pi}\right)^{1/3}\frac{\rho}{k}[V_0^{1/3} - V_1^{1/3}], \tag{7.22}$$

where Δt is the time required for the weight and volume to decrease from the stated initial to the final state.

The value of the coefficient $(3/4\pi)^{1/3}$ is 0.62. If the uraninite grain had originally been cubic and remained so, the form of expressions 7.21 and 7.22 would have been the same, but the value of the numerical coefficient would have been 0.50 rather than 0.62. If the initial grain had been broken during dissolution, the time Δt would, of course, have been shorter than that calculated by means of expressions 7.21 or 7.22.

The value of the parameter k is given by the expression,

$$k = 10^{20.25}(RF^{-1})(10^{-3.38 - 10.3NOC})(a_{\Sigma CO_2})(m_{O_2})(a_{H^+})\exp(-7045/T), \tag{7.23}$$

where the parameters on the right side are the same as those in equation 7.15. Nothing is known about the retardation factor RF in the upper reaches of the Indus. A relatively low value seems probable but is by no means certain. A value of RF equal to $10^{1.0 \pm 0.5}$ covers the entire range of the New Jersey river and coastal waters tested by Grandstaff (1976) and seems a reasonable choice for the Indus. No complete analyses of the uraninites from the Indus are available, but Davidson (1960) reported a ThO_2-content of about 6%. The value of nonuranium cations (NOC) in uraninites seems to be roughly proportional to their ThO_2-content, but it is also a strong function of their age. A value of (0.3 ± 0.1) seems reasonable for the fraction of NOC of the Indus uraninites (see table 7.14). The value of $a_{\Sigma CO_2}$ for most rivers is close to $10^{-3.0}$ mol/kg, and the concentration of dissolved oxygen must be close to $10^{1.0}$ ppm. Zeschke (1960) reports the Indus waters to be nearly neutral; the activity of hydrogen has therefore been set equal to $10^{-7.0 \pm 0.5}$. The temperature varies from 5°C in winter to 24°C in the summer. The average temperature has therefore been taken to be 15 ± 5°C. Thus,

$$k \approx 10^{20.25} \times 10^{-1.0 \pm 0.5} \times 10^{-3.38 - 10.3(0.3 \pm 0.1)} \times 10^{-3.0 \pm 0.2}$$
$$\times 10^{+1.0} \times 10^{-7.0 \pm 0.5} \times \exp -24.5 \pm 0.5$$
$$\approx 10^{-6.9 \pm 2.2}.$$

The largest single component of the huge uncertainty in the estimated value of k is due to ignorance of the composition of the uraninites, a lack which should be easy to remedy. If the estimated value of k is used in equation 7.22, and if the density of the uraninite is taken to be 10 gm/cc, then

$$\Delta t \approx 0.50 \times \frac{10}{10^{-6.9 \pm 2.2}} [V_0^{1/3} - V_1^{1/3}] \text{ days}$$

$$\approx 10^{7.6 \pm 2.2} [V_0^{1/3} - V_1^{1/3}] \text{ days.} \tag{7.24}$$

The grain size of the Indus uraninites ranges from ca. 30 to 230 μm (Zeschke 1960), but approximately 70% of the grains are between 50 and 120 μm on edge. If we take 100 μm to be the mean grain size, and if we assume that on average half of the uraninite has been removed by dissolution, then,

$$\Delta t \approx 10^{7.6 \pm 2.2} [1.26 \times 10^{-2} - 1.00 \times 10^{-2}] \text{ days}$$
$$\approx 10^{2.4 \pm 2.2} \text{ years.}$$

The most likely value of Δt for such grains is therefore 250 years, the minimum value 1.6 years, and the maximum 40,000 years.

The source area of the uraninite was surely ice covered during the last glacial period; the true value of Δt is therefore almost certainly less than 10,000 years. The minimum value is obviously much too short; the proposed value is geologically reasonable. This is reassuring, but the uncertainties in the computation are so large that complacency is unwarranted. It would be useful to conduct a thorough field test of the laboratory data in the upper Indus Valley.

The Implications of the Occurrence of Ore Deposits of Detrital Uraninite for the Oxygen and Carbon Dioxide Content of the Precambrian Atmosphere

Grandstaff (1980) has applied his experimental data to derive a rough upper limit for the oxygen pressure in the atmosphere during the deposition of the uranium ores in the Blind River-Elliot Lake area and in the Witwatersrand Basin. Much the same approach that was used above in the evaluation of the Indus River data can be used to evaluate $(P_{O_2})_{max.}$ between ca. 2.7 and 2.3 b.y.b.p. Liebenberg (1957) measured the size of uraninite grains in polished sections of the Witwatersrand Basin; the average diameter of uraninite grains in the Witwatersrand sequence is about 75 μm, that of uraninite grains in the Dominion Reef and Ventersdorp Contact Reef about 100 μm. The range of grain sizes is small. Roscoe (1969) found that uraninite grains at Elliot Lake are approximately 100 μm in diameter. Grandstaff (1980) chose a diameter of 75 μm and a specific surface area of 90 cm^2/gm for uraninite grains in his calculations. In the calculations below, a mean edge length of 100 μm has been used; the difference introduced by this change in the calculated value of $(P_{O_2})_{max.}$ is not significant.

Estimates of the fractional volume lost by the uraninites due to oxidation and dissolution and by mechanical abrasion are uncertain. Grandstaff (1980) has reviewed the available data, and has taken 50% to be the maximum decrease in the volume of the uraninite grains due to chemical erosion. Thus,

$$[V_0^{1/3} - V_1^{1/3}]_{max.} \approx [(2.0 \times 10^{-6})^{1/3} - (1.0 \times 10^{-6})^{1/3}]$$
$$\approx 0.26 \times 10^{-2} \text{ cm.}$$

The uraninite grains are embedded in well-sorted quartz conglomerates. It is difficult to see how these conglomerates could have been produced without extensive chemical weathering of feldspars and other silicates. The residence time of grains in soils prior to transport away from their place of origin is generally on the order of 500–10,000 years, and it is likely that 500 years is a strong minimum for the time during which the uraninite grains were exposed to atmospheric oxygen during weathering (see, for instance, Holland 1978, chap. 2).

The length of time during which the uraninites were exposed to atmospheric oxygen during transport and in their environment of deposition is also difficult to estimate. The source regions of the uraninites are not known, and hence the transport distance can only be estimated very roughly. Furthermore, the degree of exposure of the uraninites to atmospheric oxygen during transport and deposition can only be estimated very roughly. Grandstaff (1980) proposed a minimum total elapsed atmospheric contact time of 100 years. This seems reasonable. A much shorter time is, perhaps, possible but a reduction of the 100 years estimate to zero years would only change the minimum total contact time by 16%.

If we introduce these parameters into equation 7.22, we obtain

$$k_{\text{max.}} \approx \frac{0.50 \times 10 \text{ gm/cm}^3 \times 0.26 \times 10^{-2} \text{ cm}}{600 \text{ yr} \times 365 \text{ days/yr}}$$

$$\approx 0.6 \times 10^{-7} \text{gm cm}^{-2} \text{ day}^{-1}.$$

The maximum value of the product of P_{CO_2} and P_{O_2} can now be estimated by introducing reasonable values for all of the other terms that control the value of k. The value of the organic retardation factor RF is not known, but may have been as large as that of the organic-rich Mullica River in New Jersey. A value of $10^{0.7 \pm 0.5}$ for RF seems reasonable.

The data in table 7.13 can be used to estimate the fraction of non-uranium cations, NOC, in the uraninites at the time of formation of the ores. This is not equal to the present value of NOC, because the quantity of lead generated by uranium and thorium decay since the formation of the ores is significant. Grandstaff (1980) has used a value of 0.16 for NOC in his calculations, and the same figure is used here.

In river waters equilibrated with the atmosphere (see, for instance, Holland 1978, chap. 4),

$$\frac{a_{H^+} \cdot a_{HCO_3^-}}{B_{CO_2} P_{CO_2}} = K_1, \tag{7.25}$$

Since in all normal rivers,

$$a_{HCO_3^-} \approx a_{\Sigma CO_2}, \tag{7.26}$$

$$a_{H^+} \cdot a_{\Sigma CO_2} \approx B_{CO_2} \cdot K_1 P_{CO_2}, \tag{7.27}$$

it follows that

$$a_{H^+} \cdot a_{\Sigma CO_2} \cong (10^{-1.4})(10^{-6.3}) \cdot P_{CO_3} = 10^{-7.7} P_{CO_2}. \tag{7.28}$$

The mean temperature during the formation of the placers is not well known. The mineralogy of the conglomerates indicates that temperatures were sufficiently high, so that chemical weathering was intense. This is also suggested by the nature of the Denison and Pronto paleosols. On the other hand, the presence of what have been classed as glacial sediments in the Huronian section in eastern Canada indicates that temperatures were not abnormally high. If we take $15 \pm 10°C$ as the likely temperature range (Grandstaff 1980), then the

$$k_{max.} \approx (10^{20.25})(10^{-0.7 \pm 0.5})(10^{-3.38 - 10.3 \times 0.16})$$
$$\cdot (10^{-10.8 \pm 0.3})(10^{-7.7}P_{CO_2})(10^{+1.7}P_{O_2}). \tag{7.29}$$

The coefficient of P_{O_2} differs from the value of B_{O_2}, because in Grandstaff's (1980) treatment the concentration of O_2 in ground waters was expressed in units of ppm rather than mol/kg. After simplifying and rearranging equation 7.29, we obtain

$$P_{O_2} \cdot P_{CO_2} \leq \frac{0.6 \times 10^{-7}}{10^{-2.3 \pm 0.8}} = 10^{-4.9 \pm 0.8}. \tag{7.30}$$

The uncertainty in the product $P_{O_2} \cdot P_{CO_2}$ is uncomfortably large. The relationship is nevertheless useful.

Figure 7.24 is a plot of the logarithm of the CO_2 pressure in the atmosphere as a function of the logarithm of the O_2 pressure between 2 and 3 b.y.b.p. The paleosol data yield a band of unit positive slope defined

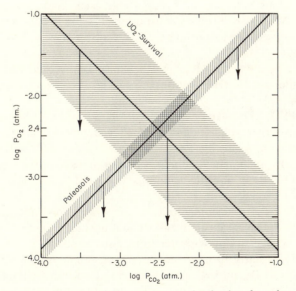

FIGURE 7.24. Limits on the range of P_{O_2} and P_{CO_2} set by the data for paleosols and the survival of uraninite as a detrital mineral between 2 and 3 b.y.b.p.

by the expression,

$$\frac{P_{O_2}}{P_{CO_2}} \leq (1.3 \pm 0.5).$$

The data for uraninite survival yield a band of unit negative slope defined by equation 7.30. At the intersection of the center line of the two bands, $P_{O_2} = 10^{-2.5} = 3 \times 10^{-3}$ atm. Since the permitted area of the diagram is limited in the high-P_{O_2} direction by the position of each band, the maximum value of P_{O_2} lies between $10^{-2.0}$ and $10^{-2.8}$ atm. This result supports the conclusion that the level of atmospheric oxygen 2 to 3 b.y.b.p. was considerably lower than its present value.

Most of the permitted area in figure 7.24 lies at P_{CO_2} values well in excess of $10^{-3.5}$ atm, the present level of CO_2 in the atmosphere. This is in agreement with the conclusion reached at the end of chapter 5. If P_{O_2} was 1×10^{-3} atm during this period, P_{CO_2} was between ca. $10^{-3.3}$ and $10^{-1.1}$ atm and most probably between $10^{-3.1}$ and $10^{-1.9}$ atm. This result suggests that CO_2 was more abundant than today, but that it was not a major component of the atmosphere between 2 and 3 b.y.b.p.

References

Bell, K., and Blenkinsop, J. 1980. Saskatchewan shield geochronology projection: Summary of investigations. Sask. Geol. Surv.

Bergne, J.A.C. 1964. Written contribution: The mode of origin of blanket ore bodies: Discussion, by C. F. Davidson, *Trans. Inst. Min. Met.* 74:504.

Berner, R. A., and Schott, J. 1982. Mechanism of pyroxene and amphibole weathering: II. Observations of soil grains. *Amer. Jour. Sci.* 282: 1214–1231.

Brewer, P. G. 1975. Minor elements in seawater. In *Chemical Oceanography*, 2d ed. vol. 1, edited by J. P. Riley and G. Skirrow, Chap. 7. New York: Academic Press.

Brush, L. H. 1980. The solubility of some phases in the system UO_3-Na_2O-H_2O in aqueous solutions at 60 and 90°C. Ph.D. diss., Harvard University.

Button, A. 1973. A regional study of the stratigraphy and development of the Transvaal Basin in eastern and northeastern Transvaal. Ph.D. diss., University of Witwatersrand, Johannesburg, South Africa.

———. 1979. Early Proterozoic weathering profile on the 2200 m.y. old Hekpoort Basalt, Pretoria Group, South Africa: Preliminary results. Inf. Circ. no. 133, Economic Geology Research Unit, University of Witwatersrand, Johannesburg, South Africa.

Button, A., and Retallack, G. J. 1981. A 2.2 billion-year-old paleosol near Waterval Onder, South Africa: Evidence of Precambrian terrestrial environments. Unpublished manuscript.

Campbell, S. E. 1979. Soil stabilization by a prokaryotic desert crust: Implications for Precambrian land biota. *Origins of Life* 9:335–348.

Chameides, W. L., and Walker, J.C.G. 1981. Rates of fixation by lightning of carbon and nitrogen in possible primitive atmospheres. *Origins of Life* 11:291–302.

Davidson, C. F. 1960. Discussion of transportation of uraninite in the Indus River, Pakistan: A paper by Zeschke. *Trans. Geol. Soc. South Afr.* 63:95–96.

Derry, D. R. 1960. Evidence of the origin of the Blind River uranium deposits. *Econ. Geol.* 55:906–927.

Ehhalt, D. H., and Drummond, J. W. 1982. The tropospheric cycle of NO_x. Paper presented at the NATO Advanced Study Institute on Chemistry of the Unpolluted and Polluted Troposphere, Corfu, Sept. 28–Oct. 10, 1981.

Fairbairn, H. W., Hurley, P. M., Card, K. D., and Knight, C. J. 1969. Correlation of radiometric ages of Nipissing diabase and Huronian metasediments with Proterozoic orogenic events in Ontario. *Can. J. Earth Sci.* 6:489–497.

Fryer, B. J. 1977. Geochemistry of early Proterozoic paleosols north of Lake Huron, Ontario. Geol. Abstracts and Proc. of the 25th Annual Meeting of the Lake Superior Conference.

Gair, J. E., and Thaden, R. E. 1968. Geology of the Marquette and Sands quadrangles, Marquette County, Michigan. U.S. Geol. Surv. Prof. Paper 397, 56–57.

Gay, A. L., and Grandstaff, D. E. 1979. Chemistry and mineralogy of Precambrian paleosols at Elliot Lake, Ontario, Canada. *Precamb. Res.* 12:349–373.

Grandstaff, D. E. 1976. A kinetic study of the dissolution of uraninite. *Econ. Geol.* 71:1493–1506.

———. 1980. Origin of uraniferous conglomerates at Elliott Lake, Canada, and Witwatersrand, South Africa: Implications for oxygen in the Precambrian atmosphere. *Precamb. Res.* 13:1–26.

Hamilton, P. J. 1976. Isotope and trace element studies of the Great Dyke and Bushveld mafic phase and the relation to early Proterozoic magma genesis in Southern Africa. *Jour. Petrol.* 12:24–52.

Hiemstra, S. A. 1968a. The mineralogy and petrology of the uraniferous conglomerate of the Dominion Reefs Mine, Klerkdorp area. *Trans. Geol. Soc. South Afr.* 71:1–65.

———. 1968b. The geochemistry of the uraniferous conglomerate of the Dominion Reefs Mine, Klerkdorp area. *Trans. Geol. Soc. South Afr.* 71:67–100.

Holland, H. D. 1978. *The Chemistry of the Atmosphere and Oceans.* New York: Wiley.

Kalliokoski, J. 1975. Chemistry and mineralogy of Precambrian paleosols in northern Michigan. *Bull. Geol. Soc. Amer.* 86:371–376.

Kasting, J. F., and Walker, J.C.G. 1981. Limits on oxygen concentration in the prebiological atmosphere and the rate of abiotic fixation of nitrogen. *J. Geophys. Res.* 86:1147–1158.

Koryakin, A.S. 1981. Results of a study of Proterozoic weathering crusts in Karelia. *Internat. Geol. Rev.* 13:973–980.

Krimmel, P. 1941. The serpentinite of Presque Isle. Master's thesis, Northwestern University.

Liebenberg, W. R. 1955. The occurrence and origin of gold and radioactive minerals in the Witwatersrand System, the Dominion Reef, the Ventersdorp Contact Reef and the Black Reef. *Trans. Geol. Soc. South Afr.* 58:101–254.

———. 1957. The occurrence and origin of gold and radioactive minerals in the Witwatersrand System, the Dominion Reef, the Ventersdorp Contact Reef and the Black Reef. In *Uranium in South Africa, 1945–1956,* vol. 1. 20–218. The Associated Scientific and Technical Societies of South Africa, Johannesburg, South Africa.

———. 1958. Mode of occurrence and theory of origin of the uranium minerals and gold in the Witwatersrand ores. *Proc. 2nd U.N. Intern. Conf. Peaceful Uses of Atomic Energy, Geneva* 2:379–387.

———. 1960. On the origin of uranium, gold and osmiridium in the conglomerates of the Witwatersrand Goldfields. *Neues Jb. Mineral.* 94:831–867.

Macdonald, C. C. 1980. Mineralogy and geochemistry of a Precambrian regolith in the Athabaska Basin. Master's thesis, University of Saskatchewan, Saskatoon, Saskatchewan, Canada.

Mackay, T. L., and Wadsworth, M. E. 1958. A kinetic study of the dissolution of uranium dioxide in sulfuric acid. *Amer. Inst. Mining Metall. Petr. Eng. Trans.* 212:597–603.

Meddaugh, W. S. 1983. Age and origin of uraninite in the Elliot Lake, Ontario uranium ores. Ph.D. diss., Harvard University.

Meddaugh, W. S., Holland, H. D., and Shimizu, N. 1982. The isotopic composition of lead in galenas in the uranium ores at Elliot Lake, Ontario, Canada. In *Ore Genesis: The State of the Art,* edited by G. C. Amstutz et al., 25–37. Berlin: Springer-Verlag.

Nicolaysen, L. O., Burger, A. J., and Liebenberg, W. R. 1962. Evidence for the extreme age of certain minerals from the Dominion Reef conglomerates and the underlying granite in the Western Transvaal. *Geochim. Cosmochim. Acta* 26:15–23.

Pienaar, P. J. 1963. Stratigraphy, petrology, and genesis of the Elliot Group, Blind River, Ontario, including the uraniferous conglomerate. Geol. Surv. of Canada Bull. 83.

Ramdohr, P. 1955. Neue Beobachtungen an Erzen des Witwatersrandes in Südafrika und ihre genetische Bedeutung. Abh. Akad. Wiss., Kl. Chem. Geol., no. 5.

———. 1958a. New observations on the ores of the Witwatersrand in South Africa and their genetic significance. *Trans. Geol. Soc. South Afr.* 61 (Annexure).

———. 1958b. Die Uran- und Goldlagerstätten Witwatersrand, Blind River District, Dominion Reef, Serra de Jacobina: Erzmikroskopische Untersuchungen und ein geologischer Vergleich. Abh. dt. Akad. Wiss. Berlin, Kl. Chem. Geol. Biol. Jg. 1958, no. 3.

———. 1979. Daten und Bilder von Mineralien des U^{+4} verschiedenen Alters und verschiedener Herkunft. *Z. dt. Geol. Ges.* 130:439–458.

Rankama, K. 1955. Geologic evidence of chemical composition of the Precambrian atmosphere. In *Crust of the Earth*, edited by A. Poldervaart, 651–664. Geol Soc. Amer. Special Paper 62.

Retallack, G. 1981. Fossil soils: Indicators of ancient terrestrial environments. In *Paleobotany, Paleoecology, and Evolution*, vol. 1, edited by K. J. Niklas, 55–102. New York: Praeger Publishers.

Rich, R. A., Holland, H. D., and Petersen, U. 1977. *Hydrothermal Uranium Deposits*. New York: Elsevier.

Robertson, J. A. 1978. Uranium deposits in Ontario. In *Uranium Deposits, Their Mineralogy and Origin*, edited by M. M. Kimberley, chap. 9. Min. Assoc. Canada. Toronto, Canada: Univ. Toronto Press.

Roscoe, S. M. 1957. Geology and uranium deposits, Quirke Lake-Elliot Lake, Blind River area, Ontario. Geol. Surv. Canada, Paper 56-7.

———. 1969. Huronian rocks and uraniferous conglomerates in the Canadian Shield. Geol. Surv. Canada, Paper 68-40.

Rundle, C. C., and Snelling, N. J. 1977. The geochronology of uraniferous minerals in the Witwatersrand Triad: An interpretation of new and existing uranium-lead age data on rocks and minerals from the Dominion Reef, Witwatersrand and Ventersdorp supergroups. *Phil. Trans. R. Soc. London* A286:567–583.

Schau, M., and Henderson, J.B. 1983. Archean chemical weathering at three localities on the Canadian Shield. *Precamb. Res.* 20:189–224.

Schidlowski, M. 1966a. Beiträge zur Kenntnis der radioaktiven Bestandteile der Witwatersrand-Konglomerate: I. Uranpecherz in den Konglomeraten des Oranje-Freistaat-Goldfeldes. *N. Jb. Mineral. Abh.* 105:183–202.

———. 1966b. Beiträge zur Kenntnis der radioaktiven Bestandteile der

Witwatersrand-Konglomerate: II. Brannerit und "Uranpecherz-geister". *N. Jb. Mineral. Abh.* 105:310–324.

———. 1966c. Beiträge zur Kenntnis der radioaktiven Bestandteile der Witwatersrand-Konglomerate: III. Kohlige Substanz ("Thucholith"). *N. Jb. Mineral. Abh.* 106:55–71.

———. 1981. Uraniferous constituents of the Witwatersrand Conglomerates: Ore-microscopic observations and implications for the Witwatersrand metallogeny. U.S. Geol. Survey Prof. Paper 1161-N.

Schott, J., Berner, R. A., and Sjöberg, E. L. 1981. Mechanisms of pyroxene and amphibole weathering: I. Experimental studies of iron-free minerals. *Geochim. Cosmochim. Acta* 45:2123–2135.

Serdyuchenko, D. P. 1968. Metamorphosed weathering crusts of the Precambrian, their metallogenic and petrographic features. In *Geology of Precambrian*, sec. 4, 37–42. Proc. of the 23rd Int'l Geological Congress, Prague.

Sharp, R. P. 1940. Ep-Archean and Ep-Algonkian erosion surfaces, Grand Canyon, Arizona. *Bull. Geol. Soc. Amer.* 51:1235–1270.

Shortmann, W. E., and DeSesa, M. A. 1958. Kinetics of the dissolution of uranium dioxide in carbonate-bicarbonate solutions. *Proc. 2nd U.N. Conf. Peaceful Uses of Atomic Energy, Geneva* 3:333–341.

Simpson, P. R., and Bowles, J.F.W. 1977. Uranium mineralization of the Witwatersrand and Dominion Reef Systems. *Phil. Trans. R. Soc. London.* A286:527–548.

Utter, T. 1980. Rounding of ore particles from the Witwatersrand gold and uranium deposit (South Africa) as an indicator of their detrital origin. *Jour. Sed. Petrol.* 50:71–76.

Van Schmus, R. 1965. The geochronology of the Blind River-Bruce Mines area, Ontario, Canada. *Jour. Geol.* 73:755–780.

Vogel, D. E. 1975. Precambrian weathering in acid metavolcanic rocks from the Superior Province, Villebon Township, southcentral Quebec. *Can. J. Earth Sci.* 12:2080–2085.

White, W. S. 1966. Geologic evidence for crustal structure in the western Lake Superior Basin. In *The Earth Beneath the Continents*, edited by J. S. Steinhart and T. J. Smith, 28–41. Amer. Geophys. Union Monograph 10.

Williams, G. E. 1968. Torridonian weathering, and its bearing on Torridonian paleoclimate and source. *Scott. J. Geol.* 4(2):164–184.

Yung, Y. L., and McElroy, M. B. 1979. Fixation of nitrogen in the prebiotic atmosphere. *Science* 203:1002–1004.

Zeschke, G. 1960. Transportation of uraninite in the Indus River, Pakistan. *Trans. Geol. Soc. South Afr.* 63:87–98.

8

Oxygen in the Precambrian Atmosphere: Evidence from Marine Environments

1. Introduction

In the last chapter it was shown that chemical processes in terrestrial environments are not particularly sensitive to changes in the oxygen content of the atmosphere, but that the available evidence for the behavior of iron in soils and the resistance of uraninite to chemical weathering indicate that oxygen was considerably less abundant in the Earth's atmosphere 2 to 3 b.y.b.p. than it is today. This chapter explores the evidence that can be extracted from the nature of marine sediments for the oxygen content of the Precambrian atmosphere.

Intuitively, this seems easier than the extraction of evidence from the history of terrestrial environments. For one, the oxygen content of seawater is quite small even today, so that any substantial reduction of atmospheric oxygen would reduce the oxygen content of surface waters to the point that anoxic conditions develop readily. For another, most of the Precambrian sedimentary rocks that have been preserved are probably marine; the potential sources of information from marine sediments are therefore extensive compared to those from Precambrian soil horizons. On the other hand, the interpretation of marine environments involves a fairly formidable set of problems. Present-day anoxic sediments, are, for instance, commonly in contact with highly oxygenated seawater. Their anoxic condition is simply the result of high organic productivity in the overlying water column and the burial of enough organic matter in sediments, so that its decomposition overwhelms the supply of oxygen and the effect of highly oxidized constituents in the sediments. The presence of widespread anoxic areas is therefore a poor indicator of a low oxygen pressure in the contemporary atmosphere. It will be shown that in spite of problems such as these, the sedimentary record of the early and middle Precambrian Era can best be understood in terms of an associated atmosphere such as that described in the last chapter. The model developed below for the state of the Precambrian oceans is still incomplete and unsatisfactory; but it does seem to do justice to the available data, and it is amenable to further testing and elaboration.

2. Evidence from Carbonate Sediments

The evolution of the dolomite/limestone ratio in carbonates was described and discussed in chapters 5 and 6. The ratio is essentially independent of the oxidation state of the atmosphere so long as CO_2 is a stable component of the atmosphere, and this condition is met even at vanishingly small oxygen pressures. Potentially important information regarding the oxidation state of the contemporary atmosphere can, however, be extracted from the minor element distribution in carbonates and from the isotopic composition of carbon in limestones and dolomites. Among the minor elements, manganese and iron are potentially the most useful, because their chemistry depends strongly on the oxidation state of their environment; fortunately, many analyses of carbonate rocks include data for the concentration of these two elements.

The number of reasonably complete analyses of carbonate rocks is somewhat limited. Table 8.1 lists all analyses of fairly pure South African carbonates from the period between 2 and 3 b.y.b.p. reported by Visser (1964). All but one of these analyses are of dolomites and limestones from the Dolomite Series of the Transvaal System, which was deposited 2.2 ± 0.1 b.y.b.p. (Button 1976). One of the analyses is of a sample of dolomite from the ca. 3.0 b.y.-old Moodies Formation.

All of the analyses of dolomites in table 8.1 contain significant quantities of FeO, Fe_2O_3, and MnO. FeO and Fe_2O_3 are also present in substantial amounts in all but one of the limestones. The MnO content of the limestones is considerably lower, on average, than that of the dolomites. For purposes of comparison, analyses of a number of Phanerozoic dolomites have been assembled in table 8.2, and a number of analyses of Phanerozoic limestones are listed in table 8.3. The number of analyses in these tables is too small to serve as a basis for far-reaching conclusions, but the figures in these tables for the concentration of Mn and Fe are not very different from those in other compilations (Veizer 1971; Veizer and Garrett 1978; and Veizer 1978) and in the much more numerous partial analyses of limestones and dolomites.

The range of the concentration of Fe_2O_3, FeO, and MnO in the analyses of the Precambrian and Phanerozoic carbonates in tables 8.1 to 8.3 has been summarized in table 8.4. In general, the ranges are comparable, but the FeO content and the MnO content of the Precambrian dolomites seem to be distinctly greater than those of the Phanerozoic dolomites.

This difference is also quite clear in the data for the MnO and FeO content of Phanerozoic carbonates in figure 8.1 and of Precambrian carbonates in figure 8.2 (Veizer 1971). The analyses are those of limestones and dolomites containing less than 0.55% Al_2O_3. Since the iron content

Composition of some Precambrian carbonates from South Africa (Visser 1964).

Sample No	Carbonate Rocks from the Dolomite Series, Transvaal System											Dolomite from the Moodies Formation
	Dolomites								Limestones			
	1068	1082	1090	1167	1169	1170	1171	1131	1161	1162	1163	1092
SiO_2	0.42	0.27	7.96	20.89	5.43	5.08	17.97	10.88	16.42	11.35	14.07	0.18
Al_2O_3	0.34	nil	0.54	1.46	0.40	0.45	2.08	2.73	2.86	2.74	2.87	tr.
Fe_2O_3	1.44	0.96	0.32	1.91	1.29	1.12	1.47	0.32	0.79	1.76	1.45	nil
FeO	0.86	0.72	1.87	0.86	0.57	1.01	2.44	0.72	0.36	0.29	0.28	0.72
MgO	20.11	21.07	19.77	19.76	18.03	18.61	17.45	1.17	1.11	1.49	1.11	21.18
CaO	30.84	30.28	28.60	27.42	32.08	31.96	28.32	48.89	48.82	49.62	50.46	30.52
Na_2O	tr.	0.05							0.17			
K_2O	tr.	nil							0.65			
H_2O^+	0.64	0.31	}0.64	5.36	}1.30	}1.22	}1.51	}0.37	0.05	0.24	0.29	
H_2O^-	0.03	0.03		0.06					0.65	0.02		
CO_2	45.53	45.82	40.03	21.71	39.82	40.63	28.90	34.93	27.56	31.67	29.42	46.89
TiO_2	0.05	nil	0.05	tr.	0.06	0.06		tr.	0.21	tr.	tr.	
P_2O_5	0.07	0.01		0.06			0.07	0.07	0.08	0.09	tr.	
F	0.01	0.015										
S	0.01	0.02	tr.		0.02	0.01	0.01		0.15	0.70		
V_2O_3	n.d.		tr.				0.01		0.01			
MnO	0.38	0.75	0.70	0.13	0.38	0.40	0.30	tr.	0.06	0.15	0.05	
Total	100.73	100.305	100.48	99.62	99.38	100.55	100.52	100.08	99.95	100.12	99.97	99.49
Moles divalent cations/kg	10.65	10.84	10.37	9.93	9.95	10.14	9.39	9.11	9.06	9.28	9.33	10.80
Moles CO_2/kg	10.35	10.42	9.10	4.93	9.05	9.23	6.44	7.93	6.26	7.19	6.69	10.66

1068 Dolomite, Dolomite Series, Transvaal System, Tweerivier 197 JQ, Brits

1082 Dolomite, Dolomite Series, Transvaal System, Droogegrond 380 JR, Pretoria

1090 Dolomite, Dolomite Series, Transvaal System, Buffelshoek 351 KQ, Rustenburg

1092 Dolomite, Moodies Formation, West of Mopani 527 MS, Soutpansberg

1131 Limestone, Transvaal System, Schoongesicht 357 JS, Belfast

1161 Limestone, altered, Transvaal System, Zeekoegat 296 JR, Pretoria

1162 Limestone, Transvaal System, Leeuw Spruit, Zeekoegat 296 JR, Pretoria

1163 Limestone, Transvaal System, Kameeldrift 298 JR, Pretoria

1167 Limestone, Transvaal System, Vissershoek 435 JQ, Pretoria

1169 Limestone, dolomitic, Dolomite Series, Transvaal System, Kwaggashoek 345 KQ. Rustenburg

1170 Limestone, dolomitic, Dolomite Series, Transvaal System, Kwaggashoek 345 KQ. Rustenburg

1171 Limestone, dolomitic, Dolomite Series, Transvaal System, Kwaggashoek 345 KQ. Rustenburg

TABLE 8.2.

Composition of some Phanerozoic dolomites.

Sample No.	Hill, Werner, and Horton (1967)							Maxwell et al. (1965)		
	15-51	15-52	15-53	15-54	15-55 + 56	15-57	28-1	555	556	571
SiO_2	7.76	0.21	0.27	0.70	0.19	0.34	0.43	13.46	18.89	5.84
Al_2O_3	0.11	0.27	0.04	0.17	0.21	0.22	0.42	1.56	0.49	0.80
Fe_2O_3	0.10	0.21	0.22	0.11	0.00	0.09	—	1.05	0.72	0.79
FeO	0.57	0.24	0.13	0.38	0.27	0.71	0.04	0.48	—	0.16
MgO	20.05	21.14	21.52	20.78	21.23	21.32	19.93	17.81	16.79	19.38
CaO	27.26	30.79	29.97	30.43	30.64	29.84	31.72	25.08	23.86	28.31
Na_2O	0.037	0.062	0.016	0.094	} 0.11	0.59	—	0.28	0.47	0.27
K_2O	0.017	0.03	0.013	0.046		tr.	0.34*	1.08	0.57	0.09
H_2O^+	} 0.05	} 0.22	} 0.07	} 0.04	—	0.32	0.02	1.23	1.57	0.63
H_2O^-					—	0.15	0.01	0.04	0.18	0.03
TiO_2	—				0.00	0.00	0.01	—	—	—
P_2O_5	0.07	tr.	0.03	0.12	tr.	tr.	tr.	—	—	—
CO_2	43.79	46.86	47.39	46.93	46.88	45.18	46.91	38.08	36.89	43.55
MnO	0.06	tr.	0.20	0.05	tr.	0.19	tr.	—	—	—
SO_3	tr.	tr.	—	—	0.03	—	—	—	—	—
FeS_2	tr.	tr.	—	tr.	0.00	0.35	—	—	—	—
SrO	tr.	—	—	—	—	—	0.09	—	—	—
Cl	0.062	0.10	0.041	0.143	0.08	tr.	0.06	—	—	—
Organic Matter	0.07	0.03	0.015	0.025	0.26	0.17	*	—	—	—
Total	100.01	100.14	99.93	100.02	100.12	99.47	100.00	100.15	100.43	99.85

* Includes organic matter.

Number	Location
15-51 to 15-54, and 15-55 + 56	Leadville limestone, Leadville district, Lake County, Colorado.
15-57	Leadville limestone (?), Lake County, Colorado.
28-1	Madison limestone, Big Horn County,
555	Magnesian limestone; lowest member of Altyn Formation, Oil City, Alta.
556	Magnesian limestone; middle member of Altyn Formation. 400 yds east of derrick at Oil City, Alta.
571	Magnesian limestone; upper member of the Altyn

Table 8.3.

Composition of some Phanerozoic limestones (Hill, Werner, and Horton, 1967; Hill and Werner 1972).

Sample No.	7-786	7-787	8-36	8-37	8-38	9-73	15-1	15-2	15-3	15-21	15-26	15-60	15-61	15-62	16-83	16-148
SiO_2	0.12	0.13	1.67	2.84	3.92	2.08	4.79	5.35	5.10	0.80	0.14	0.24	5.33	5.97	5.27	2.56
Al_2O_3	0.14	0.03	0.30	0.18	0.72	0.50	0.99	1.45	1.80	0.04	0.29	0.13	1.54	1.74	1.07	1.55
Fe_2O_3	0.00	0.07	0.16	0.13	0.04	0.34	0.14	0.45	0.03	0.07	—	0.08	0.23	0.18	0.71	0.50
FeO	0.08	0.03	0.08	0.09	0.26	tr.	0.48	0.56	0.56	0.10	0.04	—	0.48	0.64	0.32	0.47
MgO	2.02	2.54	0.48	1.03	12.72	0.16	0.56	0.57	1.01	0.25	5.71	0.19	0.56	0.75	0.56	0.06
CaO	53.75	53.09	53.71	54.14	42.14	54.05	50.54	51.29	50.27	55.10	48.72	55.87	50.69	49.82	50.36	51.98
Na_2O	0.00	0.00	—	—	—	0.17	0.14	0.03	0.07	—	—	—	0.05	0.16	0.20	—
K_2O	0.02	0.02	—	—	—	0.31	0.37	0.13	0.26	—	—	0.08	0.21	0.34	0.10	—
H_2O^+	0.06	0.08	{0.49}	{0.41}	{0.47}	—	0.29	0.24	0.49	—	0.04	—	0.51	0.07	{0.78}	1.36
H_2O^-	0.02	0.02				—	0.17	0.08	0.17	—	—	0.07	0.12	0.17		—
TiO_2	0.00	0.00	—	—	—	—	—	—	—	tr.	tr.	—	—	—	—	—
P_2O_5	0.01	0.01	—	—	—	—	—	—	—	tr.	0.07	—	—	—	0.06	—
CO_2	44.07	4.16	41.89	41.28	38.92	42.50	40.38	40.07	40.11	(43.39)*	44.68	43.47	39.90	39.84	40.34	41.13
MnO	0.01	0.01	—	—	—	0.10	—	—	—	0.13	0.01	0.05	—	tr.	—	—
SO_3	—	—	—	—	—	—	0.10	0.09	0.12	tr.	0.05	0.10	—	—	0.07	—
SrO	—	—	—	—	—	—	—	—	—	none	0.22	—	—	—	—	—
Organic Matter	—	—	—	—	—	—	1.09	0.27	0.16	—	0.36	—	0.44	0.43	—	—
Cl	—	—	—	—	—	—	—	—	—	—	0.02	—	—	—	—	—
Total	100.30	100.19	98.78	100.10	99.25	100.21	100.04	100.58	100.15	100.00	100.35	100.28	100.06	100.11	99.84	99.61

* By difference

Number	Location
7-786, 7-787	Paleozoic, upper Lynn Canal area, Skyway quadrangle, Alaska
8-36, to 8-38	Mississippian; near Mackay, Alder Creek Mining district, Custer County, Idaho.
9-73	Jurassic; Jackson Creek, Grant County, Oregon.
15-1, 15-2, 15-3	Upper Cretaceous Niobrara limestone, Boulder County, Colorado.
15-21	Pennsylvanian; Hermosa (?) Formation, Logan Mine, Rico District, Colorado.
15-26	Lower Ordovician; Manitou (?) limestone, El Paso County, Colorado.
15-60	Leadville limestome (?), La Plata County, Colorado.
15-61, 15-62	Upper Cretaceous Niobrara limestone, Larimer County, Colorado.
16-83	Fort Riley limestone Silverdale, Cowley County, Kansas.
16-148	Oread limestone, Fall River, Greenwood County, Kansas.

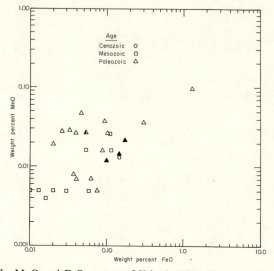

FIGURE 8.1. The MnO and FeO content of Veizer's (1971) Phanerozoic carbonates containing less than 0.55% Al_2O_3.

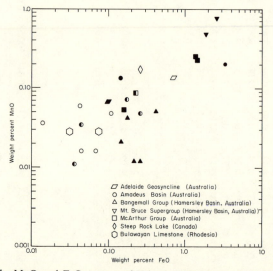

FIGURE 8.2. The MnO and FeO content of Veizer's (1971) Precambrian carbonates containing less than 0.55% Al_2O_3.

of carbonate rocks is correlated with their clay mineral content, the deletion of Veizer's high-Al_2O_3 samples removes at least some of the ambiguities due to the presence of noncarbonate components in carbonate rocks. A large fraction of Veizer's samples was taken from Australian localities. Their chemical composition has been discussed in some detail by Veizer and Garrett (1978) and by Veizer (1978).

A good indication of the manganese and iron content of Phanerozoic dolomites can be obtained from Weber's (1964) extensive data. His data for manganese are summarized in the histogram of figure 8.3. The Mn concentration of his dolostones ranges from less than 10 ppm to 3000 ppm (0.3%). The highest value corresponds to a MnO content of 0.39%. The highest MnO contents were generally found in impure dolostones. The average Mn content of Weber's (1964) "primary" dolostones (245 ppm) is nearly the same as that of his "secondary" dolostones (237 ppm). There is no conflict between these data and those in table 8.4 and figure 8.1.

The same is true for the iron content of Weber's (1964) dolostones. Figure 8.4 is a histogram of his data for this element; the range agrees with that reported in table 8.4 and figure 8.1. "Primary" dolostones contain 0.28% iron on average, "secondary" dolostones 0.20%. As expected, the total iron content of the dolostones is correlated with their content of clay impurities: average silty dolostone contains 1.76% Fe, average argillaceous dolostone 1.68% Fe.

The compilations of the U.S. Geological Survey analyses by Hill and Werner (1972) and by Hill, Werner, and Horton (1967) contain numerous semiquantitative spectrochemical analyses for Mn in carbonates of all types from the states of Kansas and Washington. Histograms of these

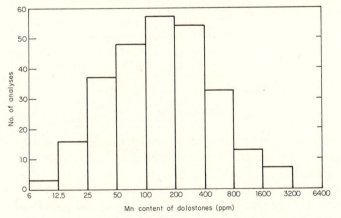

FIGURE 8.3. The manganese content of dolostones, largely Paleozoic and from the North American Platform (Weber 1964).

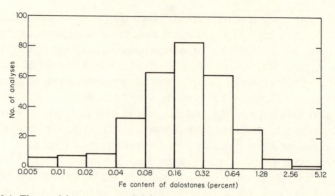

FIGURE 8.4. The total iron content of dolostones, largely Paleozoic and from the North American Platform (Weber 1964).

TABLE 8.4.

Comparison of range of concentration of Fe_2O_3, FeO, and MnO in Precambrian and Phanerozoic carbonates of tables 8.1, 8.2, and 8.3, and in figures 8.1 and 8.2 (in weight percent)

	Dolomites		
	Fe_2O_3	FeO	MnO
Precambrian	0.00–1.91	0.10–2.6	0.01–0.75
Phanerozoic	0.00–1.05	0.04–0.71	tr.–0.20

	Limestones		
	Fe_2O_3	FeO	MnO
Precambrian	0.32–1.76	0.01–0.72	0.010–0.25
Phanerozoic	0.00–0.71	0.01–1.3	0.005–7.13

* tr. = trace.

manganese analyses are similar to figure 8.3, but the peak in the Mn distribution curve for samples from both states lies between 400 and 800 ppm manganese.

Prasada Rao and Naqvi's (1977) study of a lower Ordovician carbonate sequence in Tasmania yielded values of 230 and 580 ppm for their limestones and from 1000 to 1800 ppm for their dolomites. Bencini and Turi (1974) studied the distribution of manganese in Mesozoic carbonates from Lima Valley in the Northern Apennines, and found that the Mn content of these sediments depended on the original mineralogy of the sediments, on their environment of deposition, on the supply of weathering products, and on the diagenesis of the sediments. The mean Mn content of their carbonates ranged from a low of 59 ppm for the Massiccio Limestone to a high of 1191 ppm for the Posidonia Marls.

All these figures indicate that the manganese content of carbonate rocks of all ages is highly variable, but that Phanerozoic limestones and dolomites rarely contain more than 0.3% Mn. The analyses of Precambrian carbonates suggests that this is also true of Precambrian limestones, but that early to mid-Precambrian dolomites may be considerably richer in manganese (see table 8.4, figure 8.2, and the analyses of other dolomites in Visser 1964).

The interpretation of the Mn and Fe content of carbonate rocks is complex for all the reasons cited by Bencini and Turi (1974). The concentration of these elements in the original sediments depends on the ratio of a_{Mn+2}/a_{Ca+2} and of a_{Fe+2}/a_{Ca+2} in the surrounding seawater, on the mineralogy of the carbonate sediments and on the kinetics of their precipitation. Diagenetic alteration can change the trace element content of sediments drastically. The effect of marine diagenesis on the Sr content of marine carbonate sediments was discussed briefly in chapter 6. Brand and Veizer (1980) have traced the effects of diagenesis by meteoric waters on the Sr and Mn content of carbonates in a number of areas, and have shown that the concentration of Sr normally decreases and the concentration of Mn normally increases during diagenesis. Figure 8.5 summarizes the diagenetic trends exhibited by aragonite (A), high-Mg calcite (HMC),

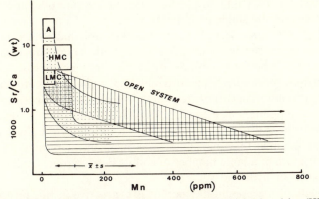

FIGURE 8.5. Summary of diagenetic trends for aragonites (A), high-Mg calcites (HMC), and low-Mg calcites (LMC) during stabilization by meteoric water. The geometric means and standard deviations are based on analyses of approximately 2100 samples from a variety of carbonate facies, localities, and ages (Brand and Veizer 1980). (Reproduced by permission of the Society of Economic Paleontologists and Mineralogists.)

and low-Mg calcite (LMC) during stabilization by meteoric waters. Diagenetic effects on the Mn content of carbonates can obviously be very large, and the reconstruction of the original composition of carbonate sediments is correspondingly difficult. Veizer, Holser, and Wilgus (1980) have attempted such reconstructions, and suggest that the Mn content of the "least altered" Archean carbonates is on the order of 200–900 ppm, well in excess of the Mn content of 30 ± 15 ppm found in many Phanerozoic calcites (Veizer, personal communication, 1982).

The observed values of the distribution coefficient of Mn^{+2} between aqueous solutions and calcite at temperatures between $0°$ and $40°$ is on the order of 5 to 18 (Bodine, Holland, and Borcsik 1965; Michard 1968; Ichikuni 1973). If the initial Mn content of Archean carbonate sediments was really ≤ 200–900 ppm, then

$$(5 - 18)\left(\frac{a_{Mn^{+2}}}{a_{Ca^{+2}}}\right)_{s.w.} \leq \frac{(200-900)/54}{400,000/40} = (0.4 - 1.7) \times 10^{-3} \quad (8.1)$$

and

$$\left(\frac{a_{Mn^{+2}}}{a_{Ca^{+2}}}\right)_{s.w.} \leq 3.4 \times 10^{-4}.$$

If, as seems likely, $a_{Ca^{+2}}$ was roughly comparable to the present value, and if $(\gamma_{Mn^{+2}})_{s.w.} \approx (\gamma_{Ca^{+2}})_{s.w.}$, then

$$(m_{Mn^{+2}})_{s.w.} \leq 3.4 \; \mu m/kg.$$

If the maximum value of the MnO content of Archean limestones had been used in this calculation, the calculated maximum Mn^{+2} concentration in seawater would have been 7.2 $\mu m/kg$.

Figure 8.6 shows that Mn^{+2} activities in the micromolar range are thermodynamically stable in an ocean of pH ca. 7.5 only at oxygen pressures of $\leq 10^{-16}$ atm. The Mn content of early and mid-Precambrian carbonates is therefore consistent with an essentially oxygen-free atmosphere at that time. Other explanations of the Mn content of these sediments are, however, more likely. The original Mn^{+2} content of primary, early and mid-Precambrian carbonate sediments may well have been somewhat less than 200 ppm; manganese may also have been present in seawater in colloidal form and may have been reduced locally to the divalent state. It is also possible that manganese was included in primary sediments as one or more of a large number of colloidal or cryptocrystalline Mn^{III}–Mn^{IV}-oxides or hydroxides.

The state of the present-day oceans is instructive in this regard. Measured concentrations of manganese in seawater are in the range of $(0.3$–40$) \times 10^{-9}$ mol/l (see, for instance, Landing and Bruland 1980; Klinkhammer and Bender 1980). Brewer (1975) proposed a value of 3.6×10^{-9} mol/l for the manganese concentration of average seawater.

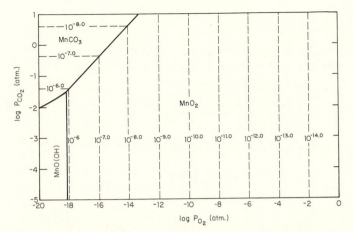

FIGURE 8.6. The activity of Mn^{+2} in mol/kg, in equilibrium with rhodochrosite, pyrolusite, and manganite at 25°C and a pH of 7.5. The ΔG_f^0 of pyrolusite was taken to be -465.1 kJ/mole, that of rhodochrosite -816.0 kJ/mole, and that of manganite -557.7 kJ/mole. The Mn^{+2} concentrations in equilibrium with rhodochrosite are in good agreement with those of Johnson (1982).

These concentrations are probably a measure of the quantity of nonionic manganese in seawater rather than a measure of the concentration of Mn^{+2} and its complexes. Figure 8.6 shows that the concentration of Mn^{+2} in seawater today is almost certainly several orders of magnitude less than 10^{-9} mol/l.

Manganese is probably included in modern carbonate sediments by one or more of the mechanisms suggested above. The increase in the Mn content of carbonates during limestone diagenesis presumably reflects the Mn^{+2} content of the mildly reducing solutions with which carbonate sediments are frequently in contact during diagenesis.

There seems to be no compelling reason, therefore, to call on a reducing atmosphere or even on an atmosphere with an O_2 content less than that of the present atmosphere to account for the observed Mn content of Precambrian carbonate rocks. If it can be proved that the Mn content of early Precambrian carbonate sediments was somewhat higher than that of carbonate sediments today, the difference could well be due to the presence of a somewhat higher concentration of colloidal Mn-oxides or hydroxides in seawater 2 to 3 b.y.b.p.

The number of analyses of Precambrian carbonates is still relatively small, and there is always the possibility that the available measurements are unrepresentative. Fortunately, the chemical composition of shales confirms that the geochemical cycle of manganese in the Precambrian was not radically different from that of the present day. The MnO content of the Precambrian shales of table 8.5 is essentially equal to that of more

TABLE 8.5.

Composition of some Precambrian shales from South Africa (Visser 1964).

Sample No.	1277	1279	1280	1281	1282	1295	1296	1297	1299	1306	1307	1310	1311	1312	1320
SiO_2	54.11	50.31	57.53	59.90	43.40	59.77	56.56	61.06	42.72	51.00	61.45	67.20	58.43	57.80	57.63
Al_2O_3	17.54	18.52	11.91	17.30	13.63	15.83	22.35	21.23	5.04	28.85	29.69	12.80	11.50	19.30	15.13
Fe_2O_3	1.43	0.65	1.27	5.00	4.32	1.52	2.71	1.04	10.43	1.68	0.32	4.07	1.43	8.50	4.87
FeO	7.45	7.50	3.31	3.00	7.04	4.81	4.89	4.67	24.36	3.74	0.22	1.08	2.59	0.70	8.41
MgO	6.97	11.76	10.66	3.90	3.95	3.65	2.03	2.51	5.73	2.05	0.09	2.42	12.51	1.20	4.18
CaO	0.22	0.70	6.05	0.50	8.41	2.20	1.68	0.88	nil	0.66	0.54	0.30	5.05	0.30	nil
Na_2O	2.14	0.33	1.58	0.50	4.03	2.00	0.50	1.05	0.29	1.11	0.38	0.22	1.21	1.00	0.60
K_2O	2.72	2.15	3.43	3.90	2.90	7.00	4.00	3.75	2.32	5.20	1.19	6.10	2.55	5.10	1.80
H_2O^+	5.14	6.72	1.15	4.80	2.07	1.00	3.33	1.93	6.48	3.76	4.90	3.70	2.38	4.70	6.12
H_2O^-	0.24	0.14	0.23	0.68	0.13	0.14	0.15	0.19	2.47	0.16	0.20	0.50	0.41	[0.40]	0.32
CO_2	0.09	0.01	1.64	n.d.	5.64	0.32	0.48	0.28	nil	—	—	nil	n.d.	—	nil
TiO_2	1.00	0.59	0.75	0.60	3.30	0.88	0.86	0.88	0.22	1.12	0.82	0.90	0.43	0.80	0.50
P_2O_5	0.14	0.11	0.05	0.14	0.54	0.66	0.33	0.10	0.04	0.09	0.07	0.09	0.06	0.02	0.16
SO_3	—	—	0.13	n.d.†	0.17	—	—	—	—	—	—	—	n.d.	—	—
S	—	nil	—	n.d.	—	—	—	—	—	—	—	nil	n.d.	—	nil
CrO	—	0.20	tr.*	—	—	—	—	—	—	—	—	—	n.d.	—	—
MnO_2	0.08	—	—	—	—	—	—	—	—	—	—	—	—	—	—
MnO	—	0.10	0.05	0.20	0.25	0.09	0.11	0.06	0.14	0.09	0.02	0.04	0.03	0.07	0.05
C	0.33	0.01	—	—	—	—	—	—	—	—	—	1.11	1.19	—	—
BaO	—	—	—	—	—	0.21	—	—	—	—	—	LOI	—	—	F = 0.05
Total	99.60	99.80	99.74	100.42	99.78	100.08	99.98	99.63	100.24	99.51	99.89	100.53	99.77	99.89	99.83

* tr. = trace;
† n.d. = not determined.

Sample No.	Location
1277	Fig. Tree Series, Swaziland System; between Zwartkoppie Bar and Uundi Bar, Barberton.
1279	Kimberley-Elsburg Series, Witwatersrand System; Withok 131 IR, Brakpan.
1280	Transvaal System; Vaalbank 511 JR Bronkhorstspruit; black.
1281	Transvaal System; Franspoort 332 JR, Bronkhorstspruit; greyish green.
1282	Transvaal System; Franspoort 332 JR, Bronkhorstspruit; altered, calcareous.
1295	Magaliesberg Stage, Pretoria Series, Transvaal System; Bothashoek 276 KT, Steelpoort River, Lydenburg; arenaceous, biotite-rich.
1296	Daspoort Stage, Pretoria Series, Transvaal System; Pretoria 264 KT, Steelpoort River, Lydenburg; slaty chiastolite, chloritoid shale.
1297	Magaliesberg Stage, Pretoria Series, Transvaal System; Naboomkoppies 263 KT, Steelpoort River, Lydenburg; andalusite shale.
1299	Transvaal System; Penge 108 KT, Lydenburg; black.
1306	Bird Reef Marker, Witwatersrand System; West Vlakfontein Gold Mining Co., Nigel.
1307	Witwatersrand System; Van Den Hevers Rust 410, Odendaalsrus; khaki shale.
1310	Dolomite Series, Transvaal System: Nooitgedacht 514, Postmasburg.
1311	Transvaal System, Kameeldrift 298 JR, Pretoria; flinty shale.
1312	Pretoria Series, Transvaal System; Pienaars River, Zeekoegat 296 JR, Pretoria.
1320	Jeppestown Series, Witwatersrand System; Vogelsruisbult gold mine, Springs.

recent shales and to the MnO content of average igneous rocks. This supports the conclusion derived from the MnO content of carbonate rocks that carbonates have never harbored more than a small fraction of the Mn which has passed through the exogenic cycle, and that shales have been the major repository of Mn in sedimentary rocks during the past 3 b.y.

Many of the same comments apply to the interpretation of the iron content of early and mid-Precambrian carbonates. The range of the FeO content of Precambrian limestones seems to be similar to that in Phanerozoic limestones (see table 8.4); the range of the FeO content in Precambrian dolomites seems to be greater than in their Phanerozoic counterparts. If the observed range of FeO concentrations in Precambrian carbonate rocks is the same as that of the carbonate sediments from which they were derived, and if the Fe^{+2} in these sediments was coprecipitated with original calcite, then the Fe^{+2} concentration in near-surface seawater was quite significant. Brand and Veizer (1980) use a figure of 1 to 20 for the distribution coefficient Fe^{2+} between aqueous solutions and calcite near 25°C. If this is roughly correct, then the Fe^{+2} concentration in early Precambrian seawater must have been approximately equal to that of Mn^{+2}. Figure 8.7 shows that an activity of Fe^{+2} in the micromolar region demands the absence of free oxygen. The value of P_{O_2} at which $a_{Fe^{+2}}$ is 10^{-6} mol/l is many orders of magnitude lower than the value of P_{O_2} at which $a_{Mn^{+2}}$ is 10^{-6} mol/l.

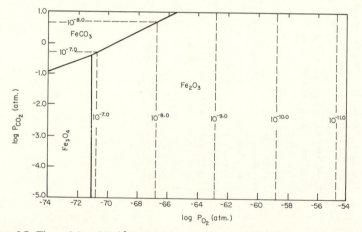

FIGURE 8.7. The activity of Fe^{+2} (mol/kg) in solutions saturated with respect to hematite or siderite in the system Fe-O-H-C at 25°C at a pH of 7.5. Free energy values used in constructing the diagram: Fe_3O_4, $-1,012.6$ kJ/mol; Fe_2O_3, -742.7 kJ/mol; $FeCO_3$, -666.7 kJ/mol.

The same doubts can, however, be raised regarding this inference that were raised regarding the interpretation of the Mn content of Precambrian carbonate rocks. The measured concentration of iron in present-day seawater falls in the range $(2-400) \times 10^{-9}$ mol/l (Brewer 1975). This is many orders greater than the concentration of Fe^{+2} in equilibrium with hematite at ambient values of P_{O_2}; the lower limit is somewhat higher than the expected concentration of Fe^{+3} and its OH^- complexes as shown in figure 8.8. The coprecipitation of Fe^{+3} with calcite and aragonite is probably very minor. The FeO content of Phanerozoic carbonates and their Precambrian analogues is therefore unlikely to be a measure of the oxidation state of contemporary surface ocean water; rather, it is apt to be a product of the admixture of impurities and the addition of FeO during diagenesis.

The MnO and the FeO content of normal Precambrian marine carbonates are therefore poor tests of the validity of the model for the Precambrian atmosphere which was developed in chapter 7. The chemistry of Precambrian carbonate sediments is simply compatible with too wide a range of oxygen and CO_2 pressures in the Precambrian atmosphere. The relatively low MnO content of Precambrian carbonates does, however, demonstrate that during the past 3 b.y. the MnO/CaO ratio of carbonates has been much less than 1.6×10^{-2}, the mol ratio of these oxides in the Earth's crust. This indicates that most of the manganese released during weathering was not removed from the hydrosphere as a constituent of carbonate sediments, as might be expected in an oxygen-free

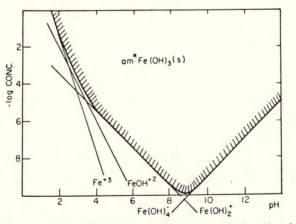

FIGURE 8.8. The concentration of Fe^{+3} (mol/l) and its OH^- complexes in solutions saturated with respect to amorphous $Fe(OH)_3$ (Stumm and Morgan 1981, p. 248). (Reproduced by permission of John Wiley & Sons, Inc.)

ocean-atmosphere system. The chemistry of marine carbonates is therefore consistent with the presence of free oxygen in the Earth's atmosphere during the past 3 b.y. The composition of the carbonate sediments associated with many Precambrian banded iron formations is quite distinct from that of the normal marine carbonates described above; the properties of these odd carbonates are used in section 4 of this chapter to define the differences between deep and shallow ocean water during much of the Precambrian Era.

3. Evidence from the Composition of Shales

The composition of Precambrian shales was discussed at some length in chapters 5 and 6. However, most of the aspects of shale chemistry considered in these chapters was not significantly influenced by the contemporary oxidation state of the ocean-atmosphere system. This section will concentrate on those aspects of the chemistry of shales which are indicators or potential indicators of the level of free oxygen in the atmosphere and oceans during sedimentation. Among these indicators, the distribution of carbon and sulfur and of their isotopes, the distribution of iron and manganese, and that of several trace elements are of particular interest.

THE DISTRIBUTION OF CARBON AND SULFUR IN SHALES

Dimroth and Kimberley (1976) have pointed out that the distribution of elemental carbon in Archean and Proterozoic shales is indistinguishable from its distribution in more recent shales and shaly sediments, and that this distribution is not at all what might be expected in sediments that are weathering products on an anoxic Earth. Extensive microscopic and field observations have confirmed the generally negative correlation between the organic carbon content and the grain size of Precambrian sedimentary rocks. Where Precambrian pelites are interbedded with arenites, organic carbon is invariably enriched in the pelitic beds. Where coarse and fine pelites are interlaminated, organic carbon is concentrated in the finer-grained pelitic laminae. Highly carbonaceous rocks are invariably mudstones. Matrix-free sandstones, cemented oolitic and stromatolitic limestones, dolostones, and oolitic and intraclastic cherts of Archean and Lower Proterozoic age are generally free of microscopically identifiable organic carbon. Sandstones, calcarenites, and coarse-grained volcaniclastic rocks containing organic carbon are rare, and when organic carbon is present in such rocks, it is nearly always concentrated in a pelitic matrix. All of these observations indicate that organic carbon in Archean and Proterozoic sediments had recently passed through the life cycle, and had been incorporated into sediments as the remains of microorganisms. The data are extremely difficult to understand if carbon in these ancient sediments was detrital in origin. It is most likely, therefore, that organic carbon in the source rocks of Archean and Proterozoic sediments was oxidized during weathering, and that the carbon dioxide produced in this manner was used then, as now, in the life cycle of the contemporary biota (see Holland 1978, pp. 270–274).

The concentration of organic carbon in Archean and Proterozoic sediments seems to be similar to that of their Phanerozoic counterparts (see, for instance, Schidlowski 1982). Cameron and Garrels's (1980) 406 Aphebian shales have a weighted C^0 content of 0.7%, their 326 lower Proterozoic shales an average of 1.6% C^0 (see table 5.9). These averages are similar to those of the C^0 content of sedimentary rocks of the Barberton area in tables 5.7 and 5.8; with those of the Russian and North American platforms in table 5.15; and with Gehman's (1962) and Trask and Patnodes's (1942) average for more recent shales. There is, however, such a wide range in the C^0 content of Precambrian shales that estimates of their average C^0 content based on the analysis of individual samples are rather uncertain. For this reason it is encouraging that the isotopic composition of carbon (see below) in ancient shales and carbonates lends strong support to the inference that the rate of C^0 burial per gram of sediment during the Archean and Proterozoic eons did not differ greatly from the rate of C^0 burial during the Phanerozoic Era.

Perhaps the strongest evidence in support of the conclusion that virtually all organic carbon in Archean and Proterozoic shales was derived from the remains of microorganisms is the correlation between the content of organic carbon and pyrite in rocks from these eras. Figure 8.9 shows the relationship between the sulfide sulfur and the elemental carbon content of lower Proterozoic black schists from Outokumpu, Finland. Figure 8.10 shows the same correlation for Cameron and Garrels' (1980) Archean and lower Proterozoic shales from Canada. The sulfur-carbon correlation in these sediments is similar to that in recent sediments and in Phanerozoic rocks (see chapter 9). It is probably due to the bacterial reduction of seawater sulfate followed by the precipitation of iron sulfides in marine sediments (see, for instance, Holland 1978, pp. 224–228, and Berner 1980). The appearance of this correlation in the Archean Eon indicates that SO_4^{-2} was already present in seawater at that time, that easily degradable organic matter was present in fresh sediments, and that sulfate-reducing bacteria were already present and active 2.7 b.y.b.p.

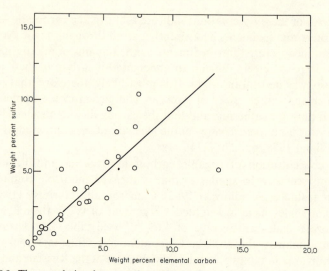

FIGURE 8.9. The correlation between the sulfide sulfur and the organic carbon content of black schists from the Outokumpu area, Finland (Peltola 1968) and from Talvivaara, Finland (Ervamaa 1981, personal communication).

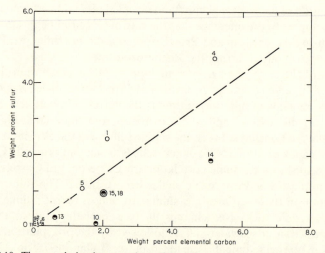

FIGURE 8.10. The correlation between the sulfide sulfur and the organic carbon content of Archean and lower Proterozoic shales from Canada (Cameron and Garrels 1980); the numbers identify the source areas of the sample in table 5.9.

(Dimroth and Kimberley 1976). These conclusions are strongly supported by the isotopic composition of sulfur in Precambrian shales (see below).

The similarity between the occurrence of pyrite in carbonaceous sediments in Archean and Proterozoic shales and in more recent sedimentary rocks is nearly matched by the similarity in the rarity of detrital pyrite in sedimentary rocks of all ages. The presence of detrital pyrite in the uraninite-bearing conglomerates of the Witwatersrand Basin and in the Blind River-Elliot Lake Region (see chapter 7) seems established beyond reasonable doubt, despite Dimroth and Kimberley's (1976) statements to the contrary; but no detrital pyrite has been found in any thin sections of the Lower Proterozoic fluvial or shallow-marine sandstones of the Labrador Trough of Quebec (Dimroth 1973), and pyrite is no more than a mineralogical rarity in the basal phase of the Archean piedmont fan conglomerates of the Rouyn-Noranda area of Quebec, as described by Rocheleau in Dimroth et al. (1975). The absence of pyrite from most Proterozoic and Archean sandstones in the Labrador Trough and in the Rouyn-Noranda area, despite the common presence of pyrite in the source areas, suggests strongly that this mineral underwent oxidative destruction during weathering, transport, and/or diagenesis.

All of this evidence indicates that the oxygen content of the atmosphere during late Archean and Proterozoic time was already sufficiently high, so that most pyrite and most elemental carbon in rocks exposed to weathering were oxidized. Pyrite was almost certainly converted to $Fe(OH)_3$ and H_2SO_4, and organic carbon to CO_2. This conclusion is consistent with the observations recorded in the previous chapter. It is not clear, however, whether an oxygen pressure as low as 0.02 PAL (4×10^{-3} atm) would have been sufficient to produce these effects. Today the oxidation of pyrite is catalyzed by the bacterial mediation of the oxidation of ferrous iron (Stumm and Morgan 1981, 471). Evidence for the presence of sulfate-reducing bacteria since the late Archean was presented above, but it is not certain that iron-oxidizing bacteria were also present at that time. The rate at which pyrite would have been oxidized in a low P_{O_2} atmosphere 2.7 b.y. ago is therefore still unknown.

Organic carbon in kerogens is almost certainly oxidized bacterially today during the weathering of carbonaceous rocks. Again, it is not known whether the cast of bacteria that existed 2.7 b.y. ago included these bacteria or a different set capable of oxidizing elemental carbon equally rapidly. For all these reasons the sedimentologic evidence presented above is consistent with the presence of an atmosphere in which the oxygen pressure was equal to or greater than 0.2 atm; it is probably also consistent with the presence of an atmosphere in which P_{O_2} was as low as the data presented in chapter 7 seem to require.

The Isotopic Composition of Carbon in Shales

The isotopic composition of organic carbon in shales is almost invariably quite different from that of carbon in associated carbonate rocks. The difference between the $^{13}C/^{12}C$ ratio of organic and carbonate carbon is due in large part to the isotopic fractionation that takes place during photosynthesis and during the conversion of organic matter to carbon residues in sedimentary rocks. Figure 8.11 summarizes the major features of these fractionation effects. Marine carbonate today has a $^{13}C/^{12}C$ ratio close to that of the PDB standard and hence, by definition, a $\delta^{13}C$ value close to zero. Plants have a lower $^{13}C/^{12}C$ ratio, and hence have negative $\delta^{13}C$ values. The range of these negative values is quite sizable; the $\delta^{13}C$ value of elemental carbon in shales is therefore also rather variable. Nevertheless, the average values of $\delta^{13}C$ in carbonates and in elemental carbon have varied surprisingly little. Figure 8.12 shows that deviations from 0‰ of the means of $\delta^{13}C$ values in limestones and dolomites have also been small. During the Phanerozoic the largest excursion apparently occurred during the Permian (see chapter 9), when the mean $\delta^{13}C$ value of carbonate rocks was close to $+2‰$ (Veizer, Holser, and Wilgus 1980).

The impression of the absence of an overall trend in the $\delta^{13}C$ value of carbonates is reinforced by the data in figure 8.13, which show that the $\delta^{13}C$ values of carbon in the oldest-known, unmetamorphosed car-

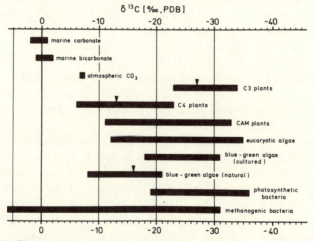

FIGURE 8.11. The isotopic composition of carbon in some carbonate and inorganic carbon reservoirs (Schidlowski 1982; see also Schidlowski 1983). (Reproduced by permission of Springer-Verlag NY, Inc.)

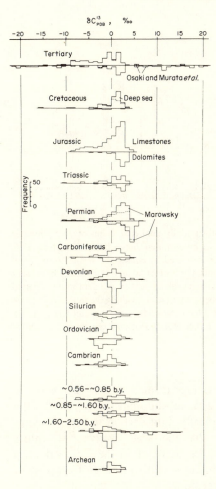
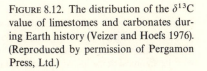

FIGURE 8.12. The distribution of the δ^{13}C value of limestomes and carbonates during Earth history (Veizer and Hoefs 1976). (Reproduced by permission of Pergamon Press, Ltd.)

bonate rocks are close to 0‰. The samples in set 2A are largely from the Bulawayan Group, and those in set 3A from sedimentary carbonates in the Barberton Mountain Land. The many samples in set 5A were collected from the Dolomite Series of the Transvaal System.

The most aberrant samples are those in set 4 from the Mid-Precambrian Lomagundi Group of Rhodesia. Their mean δ^{13}C value of $+8.2‰$ differs strikingly from the values close to 0‰ for the roughly contemporary carbonates from the Dolomite Series. The outcrop of the Lomagundi dolomites in northwestern Rhodesia extends for a distance of nearly 300 km; these are, therefore, not samples from a highly restricted, minor carbonate occurrence. Schidlowski, Eichmann, and Junge (1976) have

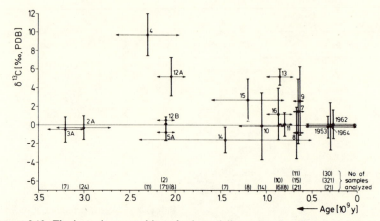

FIGURE 8.13. The isotopic composition of substantially unaltered sedimentary carbonates as a function of geologic time (Schidlowski, Eichmann, and Junge 1975). The mean δ^{13}C values of the several groups of carbonates are indicated by circles; the vertical bars represent the standard deviation of the measurements; the horizontal arrows indicate the present uncertainty in the age of the sample groups. The numbers next to the data points refer to the data tables in the original paper. The numbers in parentheses along the time axis indicate the number of samples in each data set. The Phanerozoic data are those of Craig (1953), Degens and Epstein (1962), and Keith and Weber (1964). (Reproduced by permission of Elsevier Scientific Publishing Company.)

suggested that the Lomagundi carbonates formed in a closed basin in which the δ^{13}C value of HCO_3^- was substantially increased by the removal of large quantities of isotopically light carbon in organic compounds. It is also possible, however, that the large positive δ^{13}C values of the Lomagundi carbonates record an excursion in the δ^{13}C values of marine bicarbonate as a whole. Presently neither the δ^{13}C record of Precambrian carbonates nor the Precambrian stratigraphy of southern Africa is well enough known to distinguish these two alternatives.

Figure 8.14 illustrates the great similarity between the distribution of δ^{13}C values in Phanerozoic and Precambrian samples. In both histograms the δ^{13}C value of the majority of samples falls between $-4\permil$ and $+4\permil$. The difference between the mean δ^{13}C value of the Precambrian carbonates and that of the Phanerozoic carbonates is probably not significant.

Schidlowski (1982, 1983) has recently summarized the data for the δ^{13}C value of organic carbon in Precambrian and Phanerozoic rocks. The results are comparable to those in figure 8.15. The average δ^{13}C values for each geologic period are usually close to $-25\permil$, but the standard deviations are large and could hide significant fluctuations in the mean

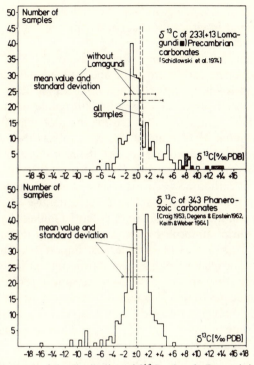

FIGURE 8.14. Histograms of the distribution of $\delta^{13}C$ values in Precambrian and Phanerozoic sedimentary carbonates (Junge et al 1975). (Copyright by the American Geophysical Union.)

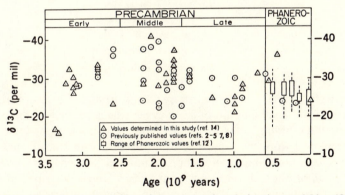

FIGURE 8.15. The isotopic composition of elemental carbon during the past 3.5 b.y. (Oehler, Schopf, and Kvenvolden 1972). (Copyright 1972 by the American Association for the Advancement of Science.)

δ^{13}C value of C^0 during the past 3 b.y. Nevertheless, the absence of a significant time trend in either δ^{13}C curve during the major part of Earth history seems well established.

The significance of the isotopic composition of the δ^{13}C value of elemental and carbonate carbon from the ancient metasediments at Isua in West Greenland is still uncertain, because the sedimentary rocks in this sequence have been subjected to amphibolite grade metamorphism. If their present δ^{13}C values are corrected for the probable effects of metamorphism, an initial δ^{13}C value of 0‰ seems likely (Schidlowski et al. 1979); if this reconstituted value is correct, the near-constancy of the δ^{13}C value of marine carbonates extends back in time at least to 3.7 b.y.b.p.

Since shifts of carbon between the reservoir of carbonate and the reservoir of elemental carbon are apt to involve the use or production of molecular oxygen, the δ^{13}C data are of potential interest for understanding the evolution of atmospheric oxygen. It is likely that the release of juvenile volatile gases was of considerable importance for the volatile budget of the atmosphere-ocean system during the earliest part of Earth history. The oxidation state of these gases was almost certainly controlled by that of the magmas from which they were released. At present, volcanic gases are mildly reducing (see table 2.6), and the apparent invariance of the oxidation state of basalts during the past 2.5 b.y. (see chapter 2) suggests that the oxidation state of volcanic gases has not varied significantly during this time interval. On average, the ratios P_{H_2O}/P_{H_2} and P_{CO_2}/P_{CO} in volcanic gases associated with basaltic magmas have probably been (see Holland 1978, p. 289):

$$105 \geq \frac{P_{H_2O}}{P_{H_2}} \geq 42 \qquad (8.2)$$

and

$$30 \geq \frac{P_{CO_2}}{P_{CO}} \geq 9, \qquad (8.3)$$

respectively. If the rate of hydrogen escape from the atmosphere was rapid, and if a small amount of free oxygen was generated by the photodissociation of water vapor, all of the carbon in volcanic gases could have been removed from the atmosphere by reaction with silicate rocks followed by the precipitation of carbonate minerals (see chapter 4). On the other hand, if photochemical reactions such as those described in chapter 4 were sufficiently rapid, or if anaerobic organisms actively extracted reduced gases from the atmosphere-ocean system, most of the volcanic hydrogen and CO could have been removed from the atmosphere as constituents of organic compounds. The calculation of the ratio of carbon in carbonates to carbon in organic compounds generated under such conditions is uncertain but nevertheless instructive. The ratio of carbon to water from the atmosphere down to the base of the crust is approximately (Holland 1978, p. 276)

$$\frac{\text{total mol C}}{\text{mol H}_2\text{O}} \approx \frac{9 \times 10^{22} \text{ gm C}/12 \text{ gm/mol C}}{2 \times 10^{24} \text{ gm H}_2\text{O}/18 \text{ gm/mol H}_2\text{O}} = 0.07.$$

If this ratio is combined with those for H_2O/H_2 and CO_2/CO above, one obtains a mean composition of volcanic gases containing ca. 7 moles $CO_2/100$ moles H_2O, (1.0–2.4) mol $H_2/100$ mol H_2O, and (0.2–0.8) mol $CO/100$ mol H_2O; this is in good agreement with the best available analyses of present-day volcanic gases (see, for instance, Holland 1978, p. 289).

If all of the H_2 and CO were removed from the atmosphere as constituents of organic compounds in which the valence of carbon is close to zero, the C^0/CO_3^{-2} ratio in sediments can be computed readily. In the reaction,

$$CO + H_2 \longrightarrow CH_2O, \tag{8.4}$$

the ratio of CO to H_2 used per mole of CH_2O produced is 1. Thus ca. 0.2 to 0.8 moles of carbon compounds would have been formed per 100 moles of H_2O. In the reaction,

$$CO_2 + 2H_2 \longrightarrow CH_2O + H_2O, \tag{8.5}$$

two moles of hydrogen are used in the reduction of CO_2. Thus, ca. (0.4–0.8) moles of organic compounds could have been generated by the H_2 remaining after reaction of the original mixture with CO. The ratio C^0/CO_3^{-2} would therefore have been ca.

$$0.30 = \frac{1.6}{5.4} \geq \frac{C^0}{CO_3^{-2}} \geq \frac{0.6}{6.4} = 0.09. \tag{8.6}$$

These limits are very wide, but they include the most probable value of this ratio in sedimentary rocks of all ages. This can be shown both on the basis of the composition of sedimentary rocks (see chapters 5 and 6) and on the basis of the isotopic composition of mean crustal carbon.

Diamonds are probably the most reliable indicators of the isotopic composition of carbon in the mantle, and it is likely that their isotopic composition is close to that of terrestrial carbon as a whole. The early work of Craig (1953) and Wickman (1956) suggested that most diamonds have $\delta^{13}C$ values close to $-5‰$ PDB. The very depleted carbonado crystals analyzed by Vinogradov et al. (1966) were generally regarded as exceptions. However, Galimov and his coworkers (see, for instance, Galimov 1978) have demonstrated that diamonds depleted in ^{13}C are more common than had been assumed. Deines (1980) has confirmed that the isotopic composition of carbon in diamonds does indeed cover a range of ca. 30‰. His frequency distribution of $\delta^{13}C$ values in diamonds in figure 8.16 shows a very pronounced mode at -5‰ to -6‰ PDB, a large negative skewness, and a sharp upper bound near 0‰. Although the reasons for the large range of $\delta^{13}C$ values in diamonds are not known with certainty, it seems likely that isotope fractionation during the formation of diamonds is very small, and that the observed range in $\delta^{13}C$ values was inherited from the carbon source of the diamonds (Deines 1980). This is consistent with the cycling of crustal carbon through the mantle.

If the $\delta^{13}C$ value of mantle carbon is taken to be ca. -5‰ PDB, and if this is equal to the mean $\delta^{13}C$ value of crustal carbon, then the crustal reservoirs of carbon in carbonate rocks ($\delta^{13}C \approx 0$‰) and of elemental carbon ($\delta^{13}C \approx -25$‰) contain respectively ca. 80% and 20% of the crustal inventory of carbon. This distribution of carbon does not require

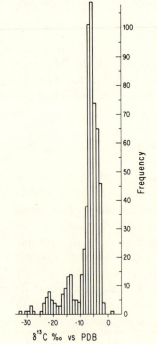

FIGURE 8.16. The $\delta^{13}C$ value of diamonds (Deines 1980). (Reproduced by permission of Pergamon Press, Ltd.)

that molecular oxygen was generated during the reduction of carbonate carbon to elemental carbon. As shown above, the relative proportion of carbonate and elemental carbon may simply reflect the preservation of the oxidation state of volcanic gases. If this is correct, the near-constancy of the isotopic composition of elemental and carbonate carbon is largely a consequence of the near-constancy of the overall redox state of the Earth's crust during the past 3 b.y.

Broecker (1970) has pointed out that the development of molecular oxygen in the atmosphere must have involved the reduction of an equivalent quantity of C^{+4}, S^{+6}, and/or Fe^{+3}. Such a reduction would have involved a shift of carbon from the carbonate to the elemental carbon reservoir; this in turn should have been accompanied by a shift in the $\delta^{13}C$ value of both reservoirs if the $\delta^{13}C$ value of crustal carbon as a whole has remained constant. No such shift is apparent in the Phanerozoic $\delta^{13}C$ record; Broecker therefore concluded that the oxygen content of the atmosphere in early Paleozoic time must have been comparable to its present value. This conclusion does not stand up to close examination. The total amount of carbon in the Earth's crust is probably close to 7.5×10^{21} moles (see above) of which ca. 1.7×10^{21} moles are present as organic carbon and ca. 5.8×10^{21} moles are present as carbonate carbon. The atmosphere contains 0.03×10^{21} moles O_2. This is such a small quantity compared to the inventory of crustal carbon that its production or its destruction would have had a negligibly small effect on the isotopic composition of carbon in the two crustal reservoirs. Rapid increases in the oxygen content of the atmosphere would be reflected in the isotopic composition of carbon in sediments deposited during the time of oxygen rise, but the identification of such periods in the sedimentary record is difficult (see chapter 9). The record of the isotopic composition of carbon in marine sediments is therefore a poor indicator of the level of atmospheric oxygen and of past changes in P_{O_2} unless these were very rapid.

The degassing of the Earth during the past 2.5 b.y. has become dominated by the recycling of crustal volatiles (see chapter 3), and the effect of juvenile volcanic gases has almost certainly become progressively less important in the carbon balance of the atmosphere-ocean-crust system. Yet the isotopic composition of carbon in the C^0 and in the carbonate reservoirs has remained roughly constant. This could be interpreted to imply that the proportion of carbon recycled from these reservoirs during metamorphism has remained essentially constant; but although this is likely, it is not an essential requirement for the maintenance of the isotopic status quo. Carbonates that are metamorphosed and lose their CO_2 contribute CO_2 in which the carbon has an isotopic composition close to

0‰. During the metamorphism of carbonaceous sediments, the mixture of released volatiles consists largely of CO_2, CH_4, CO, H_2, and H_2O. The relative proportion of these gaseous products is a function of the temperature and total pressure during metamorphism (see, for instance, Ohmoto and Kerrick 1977); but the bulk chemical composition is close to $(C^0 + H_2O)$, and the bulk isotopic composition of carbon is close to that of carbon in the starting materials. If the bulk composition of the crust is maintained, these metamorphic reactions are reversed at and near the Earth's surface. A quantity of C^0 is apt to be redeposited equal to that solubilized at depth, and at steady-state the isotopic composition of this elemental carbon will be equal to that of the solubilized carbon.

The actual situation is much more complicated, because changes in the valence of sulfur and iron are not negligible in the redox balance of the crust. Coupling between the cycle of sulfur and that of carbon has been demonstrated recently (see Veizer, Holser, and Wilgus 1980), and seems to be able to explain many of the large excursions of $\delta^{34}S$ and the much smaller excursions of $\delta^{13}C$ in Phanerozoic seawater (see chapter 9).

The Isotopic Composition of Sulfur in Shales

The isotopes of sulfur are fractionated during the reduction of SO_4^{-2} to S^{-2}, much as the isotopes of carbon are fractionated during the reduction of CO_3^{-2} to C^0. Bacterial sulfate reduction today takes place largely in anoxic sediments at the expense of organic matter and Fe_2O_3. In somewhat simplified form, the overall reaction leading to the precipitation of pyrite in anoxic sediments is

$$8SO_4^{-2} + 2Fe_2O_3 + 15CH_2O \longrightarrow$$
$$4FeS_2 + 14HCO_3^- + CO_3^{-2} + 8H_2O. \quad (8.7)$$

The fractionation of sulfur isotopes during bacterial sulfate reduction can be as large as 50‰. In extreme cases sulfides with a $\delta^{34}S$ some 50‰ more negative than the $\delta^{34}S$ value of the initial SO_4^{-2} can be produced in this fashion. Usually, however, the observed difference between the value of $\delta^{34}S$ in sedimentary sulfide minerals and the value of $\delta^{34}S$ in seawater sulfate is considerably smaller than 50‰. This is due in part to biochemical factors, and in part to variations in the openness of sediments to seawater SO_4^{-2}. If sediments are continually supplied with fresh SO_4^{-2}, the $\delta^{34}S$ value of the precipitated sulfides depends only on the $\delta^{34}S$ value of seawater SO_4^{-2} and on the fractionation of the sulfur isotopes accompanying sulfate reduction. If the system is closed after a batch of SO_4^{-2} has been added to an aliquot of sediment and if the added SO_4^{-2} is completely reduced to sulfide, the isotopic composition of the sulfide will be equal to that of the initial seawater sulfate. The openness of most sediments is intermediate in nature; the systematics of the sulfur isotope geochemistry of such partially open systems has been discussed by Ohmoto and Rye (1979).

The $\delta^{34}S$ value of sulfides formed by the reduction of seawater sulfate is variable both because $\delta^{34}S$ in seawater has varied considerably during geologic time and because local conditions of SO_4^{-2} reduction in sediments tend to be quite variable. The data of figure 8.17 indicate that the average $\delta^{34}S$ value of sulfides in Phanerozoic strata-bound marine sulfide deposits is more negative by a few to ca. 22‰ than the $\delta^{34}S$ value of contemporary seawater.

The ability to reduce SO_4^{-2} seems to be an ancient bacterial accomplishment (see, for instance, Trüper 1982), and its signature in the $\delta^{34}S$ record of sulfides in carbonaceous shales seems to extend at least as far back in time as the 2.75 b.y.-old sulfides of the Michipicoten and Woman River Iron Formations in Canada (Goodwin, Monster, and Thode 1976,

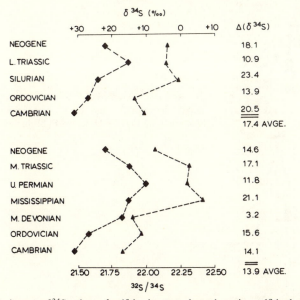

FIGURE 8.17. Average $\delta^{34}S$ values of sulfides in strata-bound marine sulfide deposits compared with the $\delta^{34}S$ value of contemporary seawater sulfate. The upper group consists of volcanic-type ores: the lower group consists of sedimentary-type ores. Diamonds represent the $\delta^{34}S$ values of seawater sulfate, triangles the $\delta^{34}S$ values of sulfides in strata-bound sulfide deposits (Sangster 1976). (Reproduced by permission of Elsevier Scientific Publishing Company.)

and Thode and Goodwin 1983). As shown in figure 8.18, the range of $\delta^{34}S$ values observed in these iron formations (sample sets 7 and 8) is in excess of 20‰; such a large range of $\delta^{34}S$ in sedimentary settings is best explained by the effects of bacterial sulfate reduction. The mean $\delta^{34}S$ value of these sulfides is close to 0%. This suggests that the $\delta^{34}S$ value of late Archean seawater was between ca. $+20‰$ and $+30‰$, if the data for Phanerozoic sulfides and sulfates are applicable to the interpretation of their Archean counterparts. Such high, positive values are not

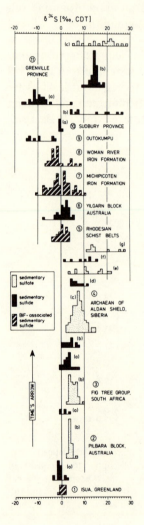

FIGURE 8.18. The record of $\delta^{34}S$ in sedimentary sulfides and sulfates between 3.7 and 1.0 b.y.b.p. (Schidlowski 1979). (Copyright © 1977 by D. Reidel Publishing Company.)

unexpected. During the Archean there were probably few cratonic settings such as those that host many of the major Phanerozoic evaporites. Most of the sulfur added to the Archean oceans may therefore have been removed as a constituent of sedimentary sulfides. If so, the mean isotopic composition of sulfide sulfur in marine settings was nearly equal to that of the river input of sulfur to the oceans (see section 3 of chapter 9). If the mean value of $\delta^{34}S$ in Archean rivers was close to 0‰, the mean isotopic composition of sulfur in Archean sedimentary sulfides would also have been close to 0‰, and the $\delta^{34}S$ value of seawater sulfate would have been such as to insure that the isotopic composition of sulfur in these outputs from the oceans equaled the isotopic composition of sulfur in the inputs to the Archean oceans.

The problem of determining the origin of the sulfur in Archean sulfides becomes progressively more difficult with increasing age of the sulfides. Ripley and Nicol (1981) have shown that the $\delta^{34}S$ values of pyrite nodules in graphitic slates from the Deer Lake greenstone sequence in northern Minnesota, which is thought to be correlated with the >2.6 b.y.-old greenstone sequences within the Vermilion District, have a range from -2.3 to $+11.1$‰, and a mean value close to $+2$‰. There is a lack of sulfide bands and occurrences of massive sulfides in the sediments, and there is no evidence for hot-spring activity in the area. Ripley and Nicol (1981) conclude that in this area, as at Woman River and at Michipicoten, bacterial sulfate reduction was the source of sulfide.

On the other hand, Fripp, Donnelly, and Lambert (1979) have found a range of 7.4‰ and a mean very close to 0‰ for sulfides from Archean (>2.8 b.y.) banded iron formations in Rhodesian greenstone belts, and propose that this constitutes evidence for a volcanic exhalative origin for the sulfur in these sulfides. Finally, at Isua, West Greenland, the spread of $\delta^{34}S$ in the few sedimentary sulfides that have been analyzed to date is only ca. 4‰, their mean is $+0.5 \pm 0.9$‰, and the sulfur in these sulfides has been assigned a magmatic origin by Monster et al. (1979).

Most of these assignments are still somewhat tentative and indefinite; firm evidence for the beginnings of bacterial sulfate reduction is still lacking. However, the data from the Woman River and Michipicoten area are consistent with the conclusion reached earlier (see figures 8.9 and 8.10) on the basis of the correlation between the C^0 and sulfide content of late Archean and lower Proterozoic shales, i.e., that considerable quantities of sulfate were present in the oceans in late Archean time, and that dissimilatory sulfate reduction and sulfide precipitation have been intense ever since (for dissenting opinions, see Cameron 1982; Skyring and Donnelly 1982).

THE CHEMICAL COMPOSITION OF CARBONACEOUS SHALES

Modern marine sediments that contain abnormally high concentrations of organic matter are enriched in a rather large suite of elements. Carbonaceous shales, the ancient analogues of these modern sediments, are similarly enriched in the same elements. The degree of enrichment of many of these elements in carbonaceous shales is roughly proportional to their content of organic carbon and to the concentration of the elements in seawater (Holland 1979). Carbonaceous sediments clearly act as excellent scavengers, and the concentration of the scavenged metals in carbonaceous shales can be used to monitor their geochemistry in the exogenic cycle. The number of analyses of Phanerozoic carbonaceous shales is respectably large (see chapter 9). Unfortunately, the number of Precambrian black shales or schists for which quantitative trace-element analyses are available is quite small. The shales analyzed by Cameron and Garrels (1980) contain too little organic carbon to act as a trustworthy guide for trace-metal enrichment during their period of sedimentation. Tyler, Barghoorn, and Barrett's (1957) study of anthracitic coal from the upper Huronian black shales of the Iron River District in northern Michigan indicates that normal trace metal enrichment occurred during the formation of these sediments, but their spectrographic analyses are too imprecise to permit more than this very general statement. Leytes, Miller, and Aleksandrova's (1977) data for the rhenium content of the Archean rocks of the Aldan Shield are too incomplete.

The most complete sets of data for the trace metal content of Proterozoic carbonaceous sediments are those of Peltola (1960, 1968) for the black schists of the Outokumpu region, and of Ervamaa (personal communication, 1981) for the black schists of the Talvivaara region, both in Finland. Marmo's (1960) less complete data for other sulfide and sulfide-graphite schists from Finland confirm the conclusions that can be drawn from Peltola's and Ervamaa's study. Figure 8.19 shows the location of the Outokumpu region in southeastern Finland and the areas where samples for Peltola's (1960) study were taken. Figure 8.20 shows the geology at Outokumpu, including the distribution of the black schist horizons and the location of the Outokumpu sulfide ore body. Major-element analyses of black schists and of schists with somewhat more modest graphite concentrations are listed in table 8.6 and 8.7. Table 8.8 shows the average trace element content of 17 black schists and 3 mica schists from the Outokumpu region as well as the concentration of the same trace elements in average Phanerozoic carbonaceous shales and in average shale.

All of the elements in table 8.8 are present in the black schists from the Outokumpu region in concentrations considerably in excess of those in

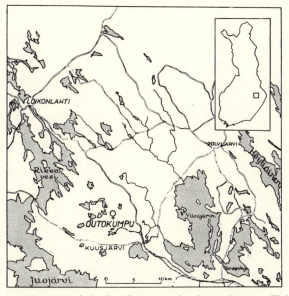

FIGURE 8.19. Location map of the Outokumpu region in southeastern Finland (Peltola 1960). (Reproduced by permission of the Geological Survey of Finland.)

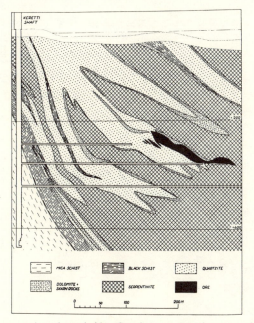

FIGURE 8.20. Cross-section through the Outokumpu complex near the Keretti Shaft (Peltola 1960). (Reproduced by permission of the Geological Survey of Finland.)

TABLE 8.6.

Analyses of argillaceous black schists from the Outokumpu Region, Finland; Peltola (1960).

	1	2	3	4	5	6	7	8	9	10	Avg.
SiO_2	33.00	34.80	40.50	49.14	48.36	58.50	60.08	52.10	62.90	61.36	50.07
TiO_2	1.45	1.03	0.98	0.92	0.62	1.08	1.06	0.47	0.93	0.59	0.91
Al_2O_3	16.75	20.19	13.01	13.22	11.28	12.92	14.78	10.73	13.98	12.24	13.91
Fe_2O_3	—	—	—	—	—	—	0.44	—	1.37	—	0.18
FeO	1.40	1.40	1.39	1.47	2.94	2.14	1.79	3.17	0.40	0.51	1.66
MnO	0.04	0.04	0.04	0.04	0.03	0.03	0.04	0.03	0.03	0.03	0.035
MgO	3.17	3.27	6.95	5.62	4.45	5.21	3.15	3.40	2.94	3.14	4.13
CaO	8.53	7.66	4.42	3.62	2.70	2.80	3.30	2.43	4.53	2.04	4.203
Na_2O	1.82	1.09	1.72	2.48	1.97	1.47	3.19	1.68	2.22	2.95	2.06
K_2O	0.48	1.70	2.52	1.89	3.14	3.34	2.13	1.62	2.11	2.79	2.172
H_2O^+	1.08	2.17	1.00	1.62	2.10	—	1.37	1.12	0.78	1.30	1.46
H_2O^-	—	—	—	—	0.14	—	—	0.19	—	0.26	—
P_2O_5	0.16	0.09	0.10	0.01	0.01	0.05	0.04	0.02	0.18	—	0.073
C	28.42	13.36	7.62	7.26	6.08	4.00	3.22	6.00	2.03	3.82	8.18
CO_2	0.04	0.07	—	—	0.10	—	0.05	0.11	0.18	0.11	0.082
F	0.10	0.13	0.22	0.06	0.08	—	0.03	0.10	0.07	0.03	0.091
S	1.45	5.32	10.48	5.26	6.10	2.90	2.83	7.80	2.04	3.90	4.81
$Fe(S)$	2.33	7.52	9.19	6.80	9.65	4.50	2.83	9.75	3.56	5.05	6.12
FeS_2	(—)	(2.20)	(19.60)	(3.75)	(—)	(—)	(4.17)	(6.68)	(—)	(2.88)	(3.93)
FeS	(3.78)	(10.50)	(—)	(7.95)	(15.54)	(7.17)	(1.30)	(10.68)	(5.54)	(5.93)	(6.84)
V_2O_5	0.04	0.23	0.20	0.10	0.09	0.09	0.11	0.11	0.11	0.06	0.114
Total	100.26	100.07	100.34	99.51	99.84	99.03	100.44	100.83	100.36	100.18	100.260

1. Graphite-rich layer of black schist. Mertala Polvijärvi.
2. Brecciated graphite- and pyrrhotite-rich schist, Outokumpu mine, Keretti shaft, 320-level.
3. Phyllitic black schist interbedded with pyritiferous "layers", Island of Kultakallio, western part of Lake Höytiäinen, Polvijärvi.
4. Massive, argillaceous, graphite- and sulphide-bearing rock. Luikonlahti, Kaavi.
5. Black schist. Sukkulansalo (drill core Oku 129, between 48.1–55.3 m) Kuusjärvi.
6. Argillaceous black schist. Outokumpu mine, Mökkivaara shaft, 320-level.
7. Argillaceous black schist. Solansaari, Polvijärvi.
8. Phyllitic argillaceous black schist with pyritiferous layers. Suopolvi, Polvijärvi.
9. Quartz-rich argillaceous black schist disseminated with pyrrhotite, Island of Kultakallio, western part of Lake Höytiäinen, Polvijärvi.
10. Quartz-rich argillaceous, phyllitic, black schist, Vitalampi, Polvijärvi.

Analyses of calcareous black schists and of intermediate sulfide and graphite-bearing schists from the Outokumpu Region, Finland; Peltola (1960).

	Calcareous Black Schists							Intermediate Sulfide and Graphite-bearing Schists				
	11	12	13	14	15	16	Avg	17	18	19	20	21
SiO_2	43.08	48.34	50.77	44.28	53.90	48.21	48.09	53.24	39.27	65.97	69.02	87.04
TiO_2	0.59	0.57	0.53	0.50	0.60	0.65	0.57	1.50	0.49	0.66	0.67	0.21
Al_2O_3	8.03	10.86	10.86	8.49	10.72	9.90	9.81	17.79	10.17	12.51	11.99	1.06
Fe_2O_3	—	—	—	—	—	—	—	—	—	—	—	0.12
FeO	0.70	1.54	1.07	1.20	1.58	1.05	1.19	1.22	2.23	3.51	4.86	0.54
MnO	0.08	0.07	0.02	0.04	0.04	0.02	0.04	0.04	0.04	0.04	0.03	0.03
MgO	10.89	8.03	8.32	7.61	7.17	4.53	7.75	3.55	2.74	2.99	2.13	1.28
CaO	14.70	9.43	9.39	5.91	6.88	5.37	8.61	6.12	4.48	1.72	2.76	2.20
Na_2O	0.74	1.82	2.03	1.38	1.94	2.14	1.67	3.27	1.01	2.94	2.32	0.23
K_2O	1.85	1.10	1.99	1.16	2.53	2.16	1.79	2.87	3.88	2.35	1.68	0.23
H_2O^+	0.98	1.42	1.11	1.10	1.42	0.85	1.16	1.10	3.20	1.37	n.d.	—
H_2O^-		0.10							0.02			0.53
P_2O_5	0.12	0.02	n.d.	0.00	0.00	n.d.	0.03	0.35	0.08	0.01	n.d.	0.02
C	4.01	1.96	3.88	7.29	5.28	5.45	4.64	0.05	2.61	1.96	1.49	0.37
CO_2	4.12	0.38	0.10			0.10	1.17	0.42	—	n.d.	0.08	1.22
F	0.15	0.04	0.06	0.07	0.04	0.07	0.07	0.18		0.06	n.d.	0.08
S	3.93	5.10	3.80	8.23	3.20	9.40	5.61	3.30	15.36	1.66	0.70	1.83
$Fe(S)$	6.23	8.75	5.68	12.72	4.36	9.86	7.93	5.34	13.38	2.67	1.12	2.25
FeS_2	(—)	(—)	(0.63)	(—)	(1.83)	(12.91)	(2.56)	(—)	(28.74)	(—)	(—)	(1.27)
FeS	(10.06)	(13.66)	(8.61)	(20.75)	(5.66)	(6.13)	(10.81)	(8.60)	(—1)	(4.33)	(1.73)	(2.66)
V_2O_5	0.14	0.13	0.18	0.06	0.06	0.08	0.10	0.05	n.d.	0.06	0.01	0.04
Total	100.34	99.66	99.79	100.04	99.72	99.84	100.23	100.39	98.96	100.48	98.86	99.28

11. Amphibole-rich black schist, Mulo, Pyhäselkä.
12. Amphibole-rich "sheal" schist, Sukkulansalo (drill core Oku 129), Kuusjärvi.
13. Calcareous black schist, Matovaara (drill core 112a, 639 m), Outokumpu.
14. Gneissose amphibole-bearing black schist, Miclonen, Itkonsalo, Kuusjärvi.
15. Amphibole-rich black schist, Luikonlahti, Kaavi.
16. Coarse-grained amphibole-bearing black schist. Ulla (drill core 115 a), Kuusjärvi.
17. Arkosic layer in phyllitic black schist, Kultakallio, Polvijärvi.
18. Sheared pyrite-rich black schist, Outokumpu mine, Keretti shaft, 320-level, Aanalyst, H.B. Viik.
19. Graphite- and sulphide-bearing quartz-rich black schist, Kykeri, Outokumpu.
20. Graphite- and suiphide-bearing quartz-rich mica schist, Kupinpuro, Polvijärvi.
21. Black quartzite, Outokumpu mine, Keretti shaft, 285-level.

TABLE 8.8.

Trace element concentrations (in ppm) in Proterozoic black schists from the Outokumpu and Talvivaara regions, Finland, in black shales, and in average shale.

	Average of 17 black schists from Outokumpu Region[a]	Average of several hundred black schist analyses from Talvivaara region[b]		95th Percentile of medians of 20 sets of black shales[c]	Average shale
		Unmineralized	Mineralized		
C(%)	4.85	6.3	7.4	5.1	0.9
S(%)	5.35	7.1	7.6	—	0.4
V	700	610	625	500	130
U	50	10–20	10–30	—	3
Cu	600	400	1400	150	50
Ni	400	400	2600	100	70
Zn	400	2300	5300	500	90
Co	200	<100	200	15	20
Pb	50	15	50	50	20
Mo	160	110	102	50	2.6
Se	20	—	—	—	0.6
Ag	2	2.3	2.7	2	0.1

[a] Peltola (1960).
[b] Ervamaa (personal communication, 1981).
[c] Vine and Tourtelot (1970).

average shale. Some elements, e.g., vanadium, lead, and zinc, are only modestly enriched. Others, such as molybdenum, uranium, and selenium (see also Koljonen 1975) are strongly enriched. The pattern of enrichment is identical to that found in modern sediments rich in organic matter and in Phanerozoic black shales. Whatever processes are responsible for these enrichments were probably in operation during the deposition of the black schists of the Outokumpu region, and their effects have not been disturbed to any major extent by the metamorphism to which the sedimentary rocks of this region have been subjected.

The black schists of the Talvivaara region north of Outokumpu are very similar both in age and chemical composition to the black schists of the Outokumpu region. Table 8.9 summarizes the average concentration of the major elements in some 700 unmineralized and mineralized samples of black schists from Talvivaara. With the exception of manganese, the concentration of the major elements was not affected significantly by a mineralization process that increased the concentration of Cu, Ni, and Zn in large volumes of these metasediments sufficiently to turn them into

TABLE 8.9.

Average composition of mineralized and unmineralized black schists of the Talvivaara Region, Finland; averages are based on approximately 700 analyses*.

	Unmineralized	Mineralized
SiO_2	51.3	47.7
TiO_2	0.7	0.7
Al_2O_3	12.4	10.0
Total Fe as Fe_2O_3	12.7	14.6
MnO	0.1	1.1
MgO	3.0	4.2
CaO	2.5	3.3
Na_2O	1.2	0.3
K_2O	3.1	3.4
H_2O^+	—	—
H_2O^-	—	—
P_2O_5	0.2	0.3
C	6.3	7.4
CO_2	—	—
S	7.1	7.6

* Ervamaa, personal communication, 1981.

a very significant low-grade resource of these metals. The data in table 8.8 show that the concentration of trace metals in the unmineralized Talvivaara black schists is quite similar to that of the Outokumpu black schists and to those in the Phanerozoic black shales studied by Vine and Tourtelot (1970).

The age of the metasediments in the Outokumpu region is not known precisely. They are clearly older than the period of metamorphism that has been well dated at 1.8 b.y.b.p.; they are younger than the basement, and hence no older than cà. 2.3. b.y. Their high trace-element content suggests that the elements listed in table 8.9 were already being released extensively during weathering at the time of formation of these black shales, that they reached the oceans in solution, and that they were removed by precipitation in sediments rich in organic matter by one or more of several possible removal mechanisms. This conclusion is still somewhat uncertain, because it is possible that a significant fraction of the trace metals concentrated in these black schists was introduced into the Precambrian oceans as a constituent of seawater that had reacted with oceanic basalts at elevated temperatures. Until the concentration of these elements in hydrothermally cycled seawater is better known, one can only claim that the pattern of trace-metal concentration in the Outokumpu schists does not require a regime of chemical release during terrestrial weathering that is significantly different from that of the present day.

It is difficult to translate this claim into a lower limit for atmospheric oxygen during the deposition of the Outokumpu sediments. The distribution of trace elements in the Precambrian paleosols described in chapter 7 is consistent with the composition of the Outokumpu schists. The rather strong concentration of uranium in the Outokumpu schists suggests that much of the uranium in rocks exposed to weathering during the deposition of these sediments was released and carried to the oceans in solution. This is consistent with the observation that a good deal of uranium in rocks is located at and close to grain boundaries and is easily taken into solution. The data are not inconsistent with the preservation of uranium in detrital minerals, including uraninite, in uranium ores such as those in the Witwatersrand Basin, and in the Blind River-Elliot Lake region.

Data such as those in Peltola's (1960, 1968) study are potentially of importance for defining the oxidation state of the ocean-atmosphere system in the past. At present, however, all that can be said is that there is no conflict between the available data and the wide limits that were set in chapter 7 on the partial pressure of oxygen 2.0–2.5 b.y. ago.

4. Evidence from Iron Formations

The origin of iron formations has been debated extensively. Many of the ideas which have been advanced to explain these unusual accumulations of iron minerals involve the oxidation state of the atmosphere-ocean system during the Precambrian Era. This section first summarizes some of the geological and chemical features of iron formations and then proposes a model for their origin that is consistent with an O_2 content of roughly 0.02 PAL (4×10^{-3} atm) in the atmosphere during the development of the major iron formations.

THE NATURE AND DISTRIBUTION OF IRON FORMATIONS

Sedimentary rocks that are abnormally rich in iron have a great range of physical and chemical characteristics. This is due in part to the great variety of iron minerals that are formed in sedimentary environments. The classification of iron-rich sediments is therefore difficult, and past usages have been diverse (see, for instance, Kimberley 1978). I shall follow the classification of James (1966), who recognized two major groups of iron-rich sedimentary rocks: (1) the "ironstones," which are noncherty and largely confined to the Phanerozoic; and (2) the "iron formations," which are typically laminated with chert and are largely but not exclusively Precambrian in age. "Ironstone" appears to have no synonyms; the rocks referred to as iron formation are also known as itabirite (Brazil), banded hematite quartzite (India), iron-bearing formation and taconite (U.S.A.), and quartz-banded ore (Scandinavia). Only in South Africa, where the term "banded iron-stone" is used for Precambrian cherty iron formations, is there overlap in the usage of the two terms.

Ironstones and iron formations both have a wide range of characteristics and associations. Table 8.10 summarizes some of the differences between the more abundant deposits of the two groups. Although some of these differences are probably related to differences in degree of metamorphism, others are not. The latter include the notably greater abundance of iron formations in the Precambrian, their greater size, the abundance of interlayered chert, and the extremely low content of the alkalis, alumina, and of many minor elements.

No complete tabulation of iron formations has been prepared. Table 8.11 does, however, include all of the world's major iron formations and nearly all those deposits for which a reasonable assessment of age and initial size can be made. The reliability of the figures in this table and in figures 8.21 and 8.22 are quite variable. For deposits such as those of the Hamersley Range in Western Australia, which have not undergone extensive postdepositional deformation and erosion, the uncertainty in the estimated tonnage may be only a factor of two. Unfortunately, the data for the thickness and distribution of many of the other deposits are scattered and incomplete; the listed tonnages for these deposits are therefore only order-of-magnitude estimates. In spite of these uncertainties the very large peak in the development of banded iron formations during the period between 2.0 and 2.5 b.y.b.p. is surely real. The deposits in all of the districts classed as "very large" in table 8.11 were formed during this period. Together, these districts contain more than 90% of all known ore in iron formations. The virtual absence of iron formations in more recent times must also be real. The preservation of sediments decreases rapidly with increasing age, and it seems most unlikely that the rarity of iron formations in Phanerozoic rocks is due to the preferential loss of these sediments.

The composition of iron formations is rather variable. However, Gole and Klein (1981) have shown that the similarities between Proterozoic

TABLE 8.10.

Comparison of typical ironstones with iron formations, after James (1966).

	Ironstones	*Iron Formations*
Age		
Minimum age	Pliocene	Cambrian or Late Precambrian
Major development during	Lower Paleozoic and Jurassic	Mid-Precambrian (2.0–2.5 b.y.b.p.)
Maximum age	Middle Precambrian (ca. 2.0 b.y.)	Early Precambrian
Thickness of Major Units	a few meters to a few tens of meters	50–600 meters
Original Areal Extent	individual basins rarely more than 100 miles in max. dimension	some basins several hundred miles long
Physical Character	massive to poorly banded; silicate and oxide-facies oolitic	thinly bedded; layers of hematite, magnetite, siderite, or silicate alternating with chert; chert ca. 50%
Mineralogy of unmetamorphosed units		
goethite	dominant	none
hematite	fairly common	common
magnetite	relatively rare	common
chamosite	dominant primary silicate	absent
glauconite	minor	absent
siderite	common	common
calcite	common	rare or absent
dolomite	common	present in some units
pelletal collophane	relatively abundant in some	absent
greenalite	none	dominant primary silicate
quartz (chert)	rare	major constituent
pyrite	common	common
Chemistry	no distinctive aspect except high iron content	remarkably low in Al, Na, K, and minor elements; phosphorus content generally much lower than in ironstones

Associated Rocks
No distinctive differences. Both ironstones and iron formations are typically interbedded with shale, sandstone, or greywacke, or their metamorphosed equivalents. Carbonate rocks: limestone and (or) dolomite in ironstone sequences not rare but subordinate to clastic rocks in immediately associated areas.

Relative abundance of types	1. oxide	1. oxide
	2. silicate (chamosite)	2. carbonate (siderite) and silicate
	3. carbonate (siderite)	3. carbonate (siderite) and silicate
	4. sulfide	4. sulfide

TABLE 8.11.

Estimated initial size and age of selected deposits of cherty banded iron formations (James and Trendall 1982).

	Ref. No.[1]	Area (may include more than one stratigraphic unit)	Class[1]	Estimated age, in m.y.[2]
Africa	AF1	Damara Belt, Namibia	Moderate	650 (590–720)
	AF2	Shushong Group, Botswana	Small	1875 (1750–2000)
	AF3	Ijil Group, Mauritania	Moderate	2100 (1700–2500)
	AF4	Transvaal-Griquatown, S. Africa	Very large	2263 (2095–2643)
	AF5	Witwatersrand, S. Africa	Small	2720 (2643–2800)
	AF6	Liberian Shield, Liberia-Sierra Leone	Large	3050 (2750–3350)
	AF7	Pongola beds, Swaziland-S. Africa	Moderate	3100 (2850–3350)
	AF8	Swaziland Supergroup, Swaziland-S. Africa	Small	3200 (3000–3400)
Australia	AU1	Nabberu Basin	Large	2150 (1700–2600)
	AU2	Middleback Range	Moderate	2200 (1780–2600)
	AU3	Hamersley Range	Very large	2500 (2350–2650)
Eurasia	EU1	Altai region, Kazakhstan-W. Siberia	Moderate	375 (350–400)
	EU2	Maly Khinghan-Uda, Far East USSR	Large	550 (500–800 (?))
	EU3	Central Finland	Moderate	2085 ± 45
	EU4	Krivoy Rog-KMA, USSR	Very large	2250 (1900–2600)
	EU5	Bihar-Orissa, India	Large	3025 (2900–3150)
	EU6	Belozyorsky-Konski, Ukraine, USSR	Moderate	3250 (3100–3400)
North America	NA1	Rapitan Group, western Canada	Moderate	700 (550–850)
	NA2	Yavapai Series, southwest USA	Small	1795 (1775–1820)
	NA3	Lake Superior, USA	Large	1975 (1850–2100)
	NA4	Labrador Trough and extensions, Canada	Very large	2175 (1850–2500)
	NA5	Michipicoten, Canada-Vermilion, USA	Moderate	2725 (2700–2750)
	NA6	Beartooth Mountains, Montana, USA	Small	2920 (2700–3140)
	NA7	Isua, Greenland	Small	>3760 ± 70
South America	SA1	Morro du Urucum-Mutun, Brazil-Bolivia	Moderate	600 (?) (450–900)
	SA2	Minas Gerais, Brazil	Very large	2350 (2000–2700)
	SA3	Imataca Complex, Venezuela	Large	3400 (3100–3700)

[1] See figure 8.21.
[2] In the absence of other data, the assigned age is the arithmetic mean of the age limits given in brackets.

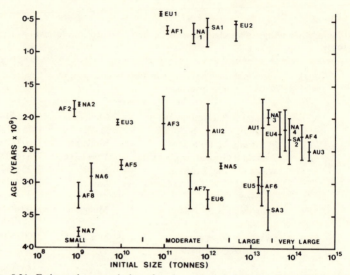

FIGURE 8.21. Estimated age and size of major, selected deposits of banded iron formations. The vertical bars indicate the uncertainty in the age of the deposits; note that the tonnage scale is logarithmic. Data from table 8.11 (in James and Trendall (1982). (Reproduced by permission of Springer-Verlag NY, Inc.)

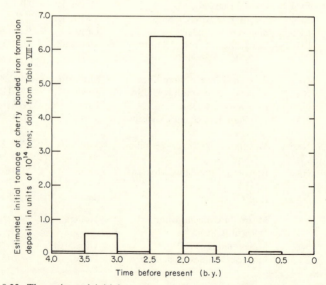

FIGURE 8.22. The estimated initial tonnage of cherty banded iron formations deposited in successive 0.5 b.y. periods of Earth history; data from table 8.11 and figure 8.21.

(1.8–2.5 b.y.) and Archean (>2.5 b.y.) banded iron formations are probably more significant than their differences. Gole and Klein's (1981) compositional data in table 8.12 and in figure 8.23 suggest that physical differences between iron formations of different ages largely reflect differences between tectonic settings during the Archean and Proterozoic eons. The similarity of the bulk composition and mineral assemblages in iron formations deposited between 3.7 and 1.8 b.y.b.p. suggests that chemical conditions during the deposition of iron formations were similar during this period of Earth history.

Within any given iron formation four types can usually be distinguished (see, for instance, James 1966). Oxide-rich banded iron formations typically consist of alternating bands of chert and hematite with or without magnetite. Where the iron oxide is dominantly magnetite, considerable quantities of siderite and of iron silicates are usually also present. The mineralogy of silicate iron formations is much more complex than that of the oxide type; this is due in large part to the effects of metamorphism on the primary mineral assemblage (see, for instance, Klein 1973, 1974; Klein and Fink 1976; Klein and Gole 1981). Initially, silicate iron formations probably consisted of complex mixtures of hydrous iron silicates, carbonates, and chert. Among the silicates, greenalite (the Fe^{+2} analogue of antigorite), minnesotaite (the Fe^{+2} analogue of talc), and stilpnomelane (a complex layer silicate) are usually the most abundant. Among the carbonates, ankerite and siderite are usually the dominant minerals. Their composition is quite variable. Manganoan and magnesiam varieties are common.

In contrast to the complexity of the mineralogy of the silicate type, that of the carbonate type is quite simple. Typically, members of carbonate iron formations consist of thinly bedded or laminated alternating layers of chert and ankerite (and/or siderite). These two types of layers are usually present in roughly equal proportions by volume. In some areas siderites are highly manganiferous.

Pyrite is the dominant iron mineral of the sulfide type. This type of BIF is present in many or most iron-rich sedimentary rocks, but it is generally subordinate in importance to the other three types. With few exceptions the highly pyritic members of iron formations are very carbonaceous black shales or their equivalents.

THE SOURCE OF IRON IN IRON FORMATIONS

The four types of iron formations clearly consist of precipitates formed under different redox conditions. Oxide iron formations were deposited

TABLE 8.12.

Averages and ranges of bulk analyses of Precambrian banded iron formations recalculated to 100% on an H_2O- and CO_2- free basis (Gole and Klein 1981).

| | ARCHEAN | | | PROTEROZOIC | | | | |
| | | | | Hamersley Basin | | Labrador Trough* | | |
	1 Yilgarn Block	2 Montana	3 Marra Mamba	4 Dales Gorge Member	5 Joffre Member	6 unmet.	7 met.	8 Biwabik
SiO_2	49.37	45.53	49.16	46.20	43.31	47.81	44.33	50.62
	21.5–68.3	33.6–59.5	26.8–67.4	24.8–60.2	7.7–59.1	16.1–87.2	15.2–63.6	46.9–53.4
TiO_2	0.04	0.06	0.18	0.04	0.07	0.04	0.10	0.06
	0.00–0.18	0.00–0.18	0.00–0.88	0.00–0.15	0.01–0.28	0.00–0.25	0.00–0.59	0.02–0.13
Al_2O_3	0.70	1.80	1.64	1.03	1.72	0.62	0.74	1.13
	0.01–3.51	0.01–5.86	0.00–6.00	0.00–5.12	0.04–7.54	0.05–1.98	0.00–4.23	0.33–2.28
Fe_2O_3	18.98	26.91	12.93	18.40	20.16	19.96	16.87	20.28
	1.7–37.6	3.4–38.3	0.0–39.5	0.1–35.9	3.1–40.1	0.0–65.0	0.0–77.9	12.0–26.3
FeO	23.65	17.51	25.49	23.88	22.53	21.69	23.68	21.43
	14.4–66.6	4.5–27.3	13.6–38.8	13.3–45.6	14.3–33.6	2.1–46.8	1.1–40.9	17.4–26.7
MnO	0.55	0.64	0.13	0.18	0.36	1.15	1.01	0.72
	0.04–2.76	0.01–3.26	0.03–0.36	0.00–0.50	0.04–2.68	0.06–2.52	0.12–2.66	0.41–1.02
MgO	3.46	3.82	5.03	3.15	4.86	4.00	6.25	3.17
	1.00–9.63	1.25–5.72	2.14–9.42	1.50–6.66	1.78–14.81	0.00–15.04	0.48–13.66	2.59–4.08
CaO	2.68	3.01	3.79	5.22	4.97	4.30	6.55	1.98
	0.32–7.51	0.51–11.48	0.00–10.47	0.74–39.6	0.05–37.4	0.00–22.3	0.00–22.4	1.02–3.17

Na₂O	0.11 0.02–0.57	0.34 0.01–0.97	0.38 0.01–0.91	0.50 0.00–4.97	0.39 0.01–2.11	0.17 0.01–3.05	0.21 0.00–1.49	0.06 0.02–0.09
K₂O	0.10 0.00–0.61	0.07 0.01–0.21	0.43 0.10–1.19	0.81 0.00–4.28	1.15 0.00–2.85	0.20 0.03–0.46	0.10 0.02–0.54	0.17 0.07–0.31
P₂O₅	0.16 0.02–0.45	0.29 <0.01–0.55	0.08 0.02–0.17	0.31 0.17–0.57	0.25 0.10–0.60	0.04 0.00–0.15	0.06 0.00–0.13	0.09 0.05–0.17
S	0.81 0.00–5.20	0.00 0.00–0.01	0.35 0.00–2.11	0.14 0.00–1.36	0.11 0.00–0.84	0.01 0.00–0.05	0.06 0.00–0.44	— —
C	0.00 0.00–0.30	0.02 0.00–0.10	0.50 0.00–1.97	0.22 0.00–1.58	0.15 0.00–0.55	0.007 0.00–0.09	0.06 0.00–0.63	0.29 0.07–0.58
—O≡S	0.31	—	0.09	0.08	0.03	—	0.02	—
Total Fe	31.65 16.2–53.7	32.43 22.6–41.0	28.85 19.6–46.5	31.43 17.7–45.7	31.61 25.7–42.1	30.82 8.6–52.4	30.21 14.3–58.6	30.84 29.1–34.2
$\dfrac{Fe^{3+}}{(Fe^{2+} + Fe^{3+})}$	0.42 0.02–0.68	0.58 0.10–0.87	0.31 0.00–0.59	0.41 0.03–0.68	0.45 0.09–0.67	0.45 0.00–0.99	0.39 0.00–0.98	0.46 0.29–0.54

Note: 0.00 ≤ 0.005%.
* MnO-rich banded iron formations not included in averages.

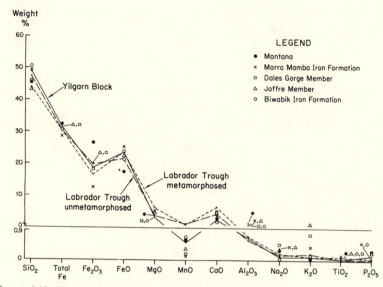

FIGURE 8.23. Plot of the averaged chemical analyses of banded iron formations in table 8.12 (Gole and Klein 1981). (Copyright by the University of Chicago.)

under the most oxidizing conditions. Those of the sulfide type were deposited under the most reducing conditions, and those of the silicate and carbonate type were deposited under intermediate redox conditions.

The settings in which the deposition of iron formations took place and the source of their ingredients have been debated extensively. Most students of iron formations have arrived at the conclusion that these sediments were deposited in relatively shallow water marine settings, probably in partly barred basins. However, Eugster and Chou (1973) have proposed that Precambrian iron formations formed in the evaporate setting of a playa-lake complex. In quite a different vein, Kimberley (1974) has proposed that iron formations are diagenetically replaced aragonite muds. Dimroth (1976) has recently leaned toward this hypothesis. However, as James and Trendall (1982) point out, it is very difficult to believe that the very large deposits of iron formations could have been formed by a postdepositional chemical transformation that shows no sign of gradational stages, that is nowhere stratigraphically discordant, and that leaves no surviving relics of unreplaced parent material.

The strongest constraint on hypotheses for the setting and origin of iron formations is surely their sheer magnitude. Figure 8.22 shows that during the period between 2.5 and 2.0 b.y.b.p., at least 6.4×10^{14} tons of iron formations were deposited. The present record of sedimentation during this period is very incomplete; the actual quantity of iron formations deposited between 2.5 and 2.0 b.y.b.p. was therefore almost certainly much greater. Iron formations contain on average approximately 30% Fe (see figure 8.23). The total amount of iron in iron formations from this period is therefore $\geq 2 \times 10^{20}$ gm, and the mean rate of iron precipitation as a constituent of iron formations during this period must have been $\geq 0.4 \times 10^{12}$ gm/yr.

There are indications that the actual rate of iron precipitation during the development of particular iron formations was considerably greater than 0.4×10^{12} gm/yr. Figure 8.24 shows the location of the Hamersley iron province of Western Australia; this area has been described by Trendall and Blockley (1970) (see also Trendall 1972 and 1973). The stratigraphic column of the Hamersley Group is shown in figure 8.25. The Dales Gorge Member of the Brockman Iron Formation has been studied particularly intensively (Trendall 1972; Ewers 1980). It contains 17 macrobands ranging in thickness from a few to about 15 m. The macrobands consist largely of chert and magnetite admixed with minor quantities of carbonates and silicates. They alternate with 16 generally thinner shaly macrobands. Much of the material in the shale macrobands is also a banded chemical sediment containing more than 15% Fe. Within the macrobands there are two finer scales of banding: the mesobands and the microbands.

FIGURE 8.24. Locality map showing the minimum area of deposition of the Brockman Iron Formation (Ewers 1980).

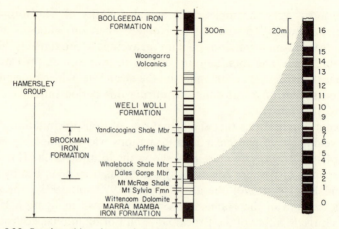

FIGURE 8.25. Stratigraphic column of the Hamersley Group showing the Brockman Iron Formation and details of the Dales Gorge Member (Ewers 1980; modified after Trendall and Blockley 1970).

Mesobands have a thickness of a few mm to a few cm. Theirs is the banding for which these rock units have been called banded iron formations. Chert mesobands consist largely of quartz together with variable quantities of one or more iron minerals. They are separated by more iron-rich mesobands consisting of magnetite and a fine-grained mixture of at least two of the following minerals: quartz, hematite, stilpnomelane, ankerite, and siderite.

Most of the mesobands have an internal lamination that has been named microbanding (Trendall 1972). It is defined by thin layers of some iron-bearing mineral or minerals within the general matrix of quartz in the chert mesobands; most commonly, the iron-bearing minerals are either a carbonate (ankerite or siderite), hematite, stilpnomelane, or a mixture of these minerals. The interval between successive microbands in any one chert mesoband is quite regular. In different mesobands the interval between successive microbands varies between about 0.2 mm and 2.0 mm. Microbands may attain a thickness of 5 mm, but wide variations of the interval between microbands never occur in a single mesoband. Most microbanded chert mesobands are between 1 and 40 mm thick and contain between 3 and 50 microbands.

Macrobands, mesobands, and microbands can be correlated over the whole of the Hamersley Basin. Trendall (1972) has made a persuasive but not conclusive case for the primary nature of banding in all three bands and for their interpretation as annual varves. If this is correct, then on average approximately 20 mg/cm^2 of iron were precipitated annually over an area of 150,000 km^2 in the Hamersley Basin (Trendall 1972) during the formation of the Dales Gorge member. This translates into a precipitation rate of 30 × 10^{12} gm Fe/yr in the basin as a whole, some 75 times the minimum mean rate of iron deposition calculated above for the period between 2.5 to 2.0 b.y.b.p. If Trendall's interpretation of the microbands is correct, even 30 × 10^{12} gm Fe/yr is a minimum figure for the rate of iron precipitation for the oceans as a whole during Dales Gorge time, since it seems unlikely that the Hamersley Basin was the only area on Earth in which iron-rich sediments were deposited during this time interval.

The rate of deposition of iron during the formation of the ores in the Hamersley Basin has both local and global implications. The rate is small but not trivial compared to the total input of iron to the oceans. At present the rate of sediment input to the oceans is ca. 2 × 10^{16} gm/yr (Holeman 1968). The concentration of iron in average sediment is close to 5%; hence the total input of iron to the oceans as a constituent of river sediments is approximately 1 × 10^{15} gm/yr. During the period between 2.5 and 2.0 b.y.b.p. the rate of sediment input may have been faster than today, but probably not by more than a factor of two or three (see chapter 6). The iron content of sediments has not changed significantly in the meantime; the sedimentary iron input to the oceans during Hamersley time was therefore roughly (1–3) × 10^{15} gm/yr; the estimated rate of deposition of iron in the Hamersley Basin is 1 to 3% of this figure.

The circulation of seawater through oceanic basalts at mid-ocean ridges could have contributed significant quantities of iron to the oceans. The analysis of the lithium content of shales in chapter 6 suggested that seawater has been cycling through basalts at a roughly constant rate. If the total quantity of seawater cycled annually during Dales Gorge time was close to 0.2 × 10^{17} cc/yr, the best estimate of the current rate, and if all this seawater emerged from hot basalts with a complement of 100 ppm dissolved iron, some 2 × 10^{12} gm Fe/yr could have been added to the oceans in this fashion. These rough calculations indicate that the iron deposited in the Hamersley Basin could have been supplied in part by hydrothermal processes or in whole by the solution of a small fraction of the land-derived supply of iron.

On a local scale, the magnitude of the estimated rate of deposition of iron in the Hamersley Basin creates problems for several of the proposed mechanisms for the formation of iron formations (Holland 1973). Lepp and Goldich (1964), Garrels, Perry, and Mackenzie (1973), and many others favored weathering as a major source of iron in these sedimentary rocks. The average content of Mg^{+2} in river water today is ca.

4 ppm. The ratio of iron to magnesium in average crustal rock is approximately 1.7. If Fe^{+2} and Mg^{+2} were released during weathering in proportion to their abundance in average rock and if all of this Fe^{+2} were carried in solution from the weathering site, then the average river today would contain ca. 7 ppm Fe^{+2}. The discussion of Precambrian sediments in chapter 5 indicated that the HCO_3^- content of rivers at that time was roughly comparable to their HCO_3^- content today, and that P_{CO_2} was probably higher then than now. An Fe^{2+} concentration of 7 ppm in contact with an anoxic atmosphere may have been possible under such circumstances.

A river containing 7 ppm iron and supplying 20×10^{12} gm Fe annually to a basin would have to have had a flow rate of ca. 4×10^{18} cc/yr. This rate is a little less than that of the Amazon River, three times that of the Congo, and seven times that of the Mississippi River (Holland 1978, p. 86). It is difficult to imagine how sediments that are as devoid of land-derived detritus as the banded iron formations of the Hamersley Basin could have been deposited near the mouth of such a large river. Proponents of weathering as a source of iron in these sediments surely face a formidable task in finding a mechanism for separating the dissolved constituents of such a river from the sediment load that almost certainly accompanied them.

It could be argued that the iron was supplied by groundwaters that reached the basin. However, the estimated flow rate of 4×10^{18} cc/yr is approximately 10% of the total present-day river runoff, and it is difficult to imagine a mechanism for channeling such a large quantity of groundwater into such a relatively small part of the Earth's surface. These figures also cast doubt on the feasibility of proposals such as Eugster and Chou's (1973) for a nonmarine origin of large banded iron formations.

Volcanic emanations have been a favorite source of iron in sedimentary iron ores. In many instances the relationship between volcanism and ore deposition is so obvious that this is an entirely reasonable proposition (James 1966; Gross 1965, 1983). On the other hand, the volcanic association in the Hamersley Basin is highly variable, and most Lake Superior-type iron formations were deposited as rather extensive units in shelf or miogeosynclinal environments (James and Sims 1973). Nevertheless, Trendall and Blockley (1970) opted for volcanism as the most likely source of iron in the Hamersley Basin. They based their contention on the inferred contemporaneous volcanic activity in adjacent areas, on the presence of material believed to represent fragmental volcanic tuff within the banded iron formation, and on analogies with the emissions of Ebeko volcano in the Kurile Islands (Zelenov 1960). Annually some 35–50 tons of iron are released in solution from this unusually productive volcano. However, even at this rate of discharge many thousands of such volcanoes would have been required to supply the quantity of iron deposited annually during Dales Gorge time. If these volcanoes were located on the periphery of the basin, they would have been spaced less than a mile apart. If volcanoes were the source of the iron, it seems necessary to call either on much more productive volcanoes or on a much longer time for the deposition of iron formation in the Hamersley Basin, as well as on rather special mechanisms to prevent the input of volcanic debris to the areas of iron ore deposition.

Submarine volcanism and the attendant circulation of seawater through mid-ocean ridges is a more feasible mechanism for producing dissolved iron; however, the calculation carried out above showed that the probable maximum rate of iron input to the oceans from this source is considerably smaller than the estimated rate of deposition of iron in the Hamersley Basin. Even if the rate of iron supply was somewhat under-estimated, it seems unreasonable to propose that virtually all of the iron introduced along the Proterozoic equivalent of mid-ocean ridges was somehow channeled into the Hamersley Basin during Dales Gorge time. Smaller iron formations, many of which are closely associated with volcanics, probably do owe their origin to processes related to volcanism, but it seems unlikely that this was true for very large Lake Superior-type iron formations, such as those of the Hamersley Basin.

We are left, then, with the oceans themselves as the most likely source of iron in Lake Superior-type iron formations unless the estimated rate of precipitation of iron in the Hamersley Basin is much too high. At first glance the oceans are a rather unattractive source for this metal. The iron content of seawater today is only ca. 2 $\mu g/kg$, a concentration that is clearly insufficient to generate large iron-ore deposits. The low concentration of iron in oxidized seawater is imposed by the minuscule solubility of $Fe(OH)_3$; in reducing marine environments it is imposed by the low solubility of pyrite. At intermediate oxidation states or under reducing conditions in the absence of sulfur species, seawater can become saturated with respect to siderite and/or ankerite. In seawater saturated with respect to both siderite and calcite, the ratio of the activity of Fe^{+2} to that of Ca^{+2} is equal to the ratio of their solubility products. Under such conditions at 25°C,

$$\frac{a_{Fe^{+2}}}{a_{Ca^{+2}}} = \frac{K_{Sid.}}{K_{Cal.}} = \frac{10^{-10.55}}{10^{-8.34}} = 6.0 \times 10^{-3}. \tag{8.8}$$

Neither Ca^{+2} nor Fe^{+2} are strongly complexed in seawater. Since an earlier discussion of this matter (Holland 1973), Davison (1979) has pointed out that the stability constant of $FeOH^+$ is almost certainly several orders of magnitude smaller than the values reported earlier, and that even at a pH of 8, less than 1% of Fe^{+2} in seawater is present as a constituent of this complex. If the calcium concentration of Proterozoic seawater was equal to that of present-day seawater, the concentration of iron at saturation with respect to siderite and calcite would therefore have been ca. 3 ppm. Seawater could, therefore, have contained rather sizable concentrations of iron in the past.

The low iron content of normal shallow water carbonate sediments deposited between 2.0 and 3.0 b.y.b.p. indicates that even during this period the iron content of normal surface seawater was extremely low. If seawater containing several ppm of iron was present during this period, it must have existed at depth and/or in abnormal near-surface environments. Borchert (1952, 1960) has suggested that some of the iron that was brought to the oceans during this period as a constituent of river sediments was reduced at depth within the oceans, was brought to the surface in regions of upwelling, was oxidized, and precipitated as a constituent of oxides and other iron minerals in near-shore, shallow-water settings. This seems to be the most promising hypothesis for the origin of iron formations (Holland 1973).

THE GENERATION OF Fe^{+2} IN MARINE SEDIMENTS

Any model for the formation of the very large Lake Superior-type iron formations must be able to explain or at least to excuse the availability of very large quantities of Fe^{+2} and SiO_2, their transport to depositional basins, and the formation of the four prominent types of iron formations. In the process of attempting this explanation, some rather severe but fortunately nonfatal conditions must apparently be imposed on the nature and on the operation of the Precambrian oceans. Recent pelagic sediments of the eastern equatorial Atlantic and Pacific oceans contain small amounts of organic matter whose decomposition gives rise to suboxic diagenesis (Froelich et al. 1979; Klinkhammer 1980; Emerson, Grundmanis, and Graham 1982; Jahnke et al. 1982). This type of diagenesis involves the reduction of O_2 and NO_3^- in interstitial waters followed by the reduction of Mn^{+3} and Mn^{+4} to Mn^{+2}, and the reduction of Fe^{+3} to Fe^{+2}. It does not extend to the reduction of SO_4^{-2} to S^{-2}; its consequences are therefore quite different from diagenesis in sediments containing large concentrations of organic matter (see, for instance, Berner 1980). The concentration profile of some important constituents of interstitial waters in suboxic sediments is shown schematically in figure 8.26. Figure 8.27 shows the actual concentration of Mn, Fe, NO_3^-, Ni, and Cu in pore waters of a core from MANOP site M in the eastern equatorial Pacific. These data are consistent with the progressive oxidation

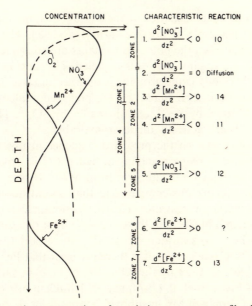

FIGURE 8.26. Schematic representation of trends in pore water profiles in suboxic sediments (Froelich et al 1979). (Reproduced by permission of Pergamon Press, Ltd.)

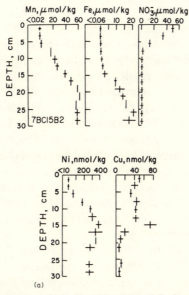

FIGURE 8.27. The concentration of Mn, Fe, NO_3^-, Ni, and Cu in pore waters as a function of depth for subcores from MANOP Site M (Klinkhammer 1980). (Reproduced by permission of Elsevier Scientific Publishing Company.)

of normal marine organic matter by the oxidant yielding the greatest free energy change per mole of organic carbon oxidized. The process continues until either all of the potential oxidants are consumed or until the store of oxidizable organic matter is depleted. The sequence of reactions predicted on thermodynamic grounds is outlined in table 8.13. From the data in this table it is clear that O_2 is the preferred oxidant, followed by NO_3^- or Mn^{+4}, then by Fe^{+3}, and thereafter by SO_4^{-2}. The disproportion reaction 6 is energetically the least favorable.

The agreement between the predicted sequence in this table and the observed sequence of redox reactions in figure 8.27 is striking. The concentration of Mn and Fe in subcores from MANOP site M at a depth of 30 cm is quite enormous compared to the concentration of these elements in the superjacent seawater. The iron concentration of 20 μmol/kg (1.1 ppm) at a depth of 30 cm in these cores is about one-third of the concentration of Fe^{+2} in seawater saturated with respect to both siderite and calcite at 25°C (see above). It is therefore clear that Fe^{+2} can be generated in sizable quantities by the reduction of Fe^{+3} in such settings.

The data of Froelich et al. (1979) and of Klinkhammer (1980) suggest that the rate of supply of Mn in suboxic settings is ca. 3×10^{-15} moles/ cm^2/sec. If the entire ocean floor were suboxic, the Mn^{+2} production rate would be ca. 20×10^{12} gm/yr. The production rate of Fe^{+2} is more uncertain, presumably comparable, but potentially larger. Aller and Benninger (1981) have studied the flux of dissolved ammonium, manganese, and silica from bottom sediments in Long Island Sound, U.S.A. The fluxes there are highly variable both in time and space. The proposed average Mn^{+2} flux of 0.42 mmoles/m^2/day (5×10^{-13} moles/cm^2/sec) is therefore rather uncertain, but their results do indicate that the estimated fluxes in suboxic settings may be somewhat lower than fluxes for the oceans as a whole. Trefry and Presley's (1982) study of the release of Mn^{+2} from sediments in the Mississippi Delta suggests that approximately 45% of the Mn in rapidly accumulating delta sediments is lost to overlying seawater. They estimate that delta sediments as a whole lose approximately $(2–3) \times 10^{10}$ gm Mn/yr. The quantity of sediment carried by the Mississippi River (1.5×10^{14} gm/yr) is about 0.7% of the total sediment load reaching the oceans today. If the behavior of Mn in the Mississippi Delta is typical of other deltas, the quantity of Mn released by reduction and reintroduction to the water column in near-shore settings is ca. $(3–4) \times 10^{12}$ gm/yr. All of these figures indicate that the reduction of Mn^{+4} to Mn^{+2} followed by transfer of Mn^{+2} from sediments to the overlying water column is a significant part of the marine geochemical cycle of this element.

Very little of the Mn and Fe that is reduced in suboxic sediments below the seawater-sediment interface today is added to the oceans proper.

TABLE 8.13.

Oxidation reactions of sedimentary organic matter (Froelich et al. 1979).

1. $(CH_2O)_{106}(NH_3)_{16}(H_3PO_4) + 138O_2 \longrightarrow 106CO_2 + 16HNO_3 + H_3PO_4 + 122H_2O$

$\Delta G^{0'} = -3190$ kJ/mole of Glucose

2. $(CH_2O)_{106}(NH_3)_{16}(H_3PO_4) + 236MnO_2 + 472H^+ \longrightarrow 236Mn^{2+} + 106CO_2 + 8N_2 + H_3PO_4 + 366H_2O$

$\Delta G^{0'} = -3090$ kJ/mole (Birnessite)
-3050 kJ/mole (Nsutite)
-2920 kJ/mole (Pyrolusite)

3a. $(CH_2O)_{106}(NH_3)_{16}(H_3PO_4) + 94.4HNO_3 \longrightarrow 106CO_2 + 55.2N_2 + H_3PO_4 + 177.2H_2O$

$\Delta G^{0'} = -3030$ kJ/mole

3b. $(CH_2O)_{106}(NH_3)_{16}(H_3PO_4) + 84.8HNO_3 \longrightarrow 106CO_2 + 42.4N_2 + 16NH_3 + H_3PO_4 + 148.4H_2O$

$\Delta G^{0'} = -2750$ kJ/mole

4. $(CH_2O)_{106}(NH_3)_{16}(H_3PO_4) + 212Fe_2O_3 \text{ (or } 424FeOOH) + 848H^+ \longrightarrow$
$424Fe^{2+} + 106CO_2 + 16NH_3 + H_3PO_4 + 530H_2O \text{ (or } 742H_2O)$

$\Delta G^{0'} = -1410$ kJ/mole (Hematite, Fe_2O_3)
-1330 kJ/mole (Limonitic Goethite, FeOOH)

5. $(CH_2O)_{106}(NH_3)_{16}(H_3PO_4) + 53SO_4^{2-} \longrightarrow 106CO_2 + 16NH_3 + 53S^{2-} + H_3PO_4 + 106H_2O$

$\Delta G^{0'} = -380$ kJ/mole

6. $(CH_2O)_{106}(NH_3)_{16}(H_3PO_4) \longrightarrow 53CO_2 + 53CH_4 + 16NH_3 + H_3PO_4$

$\Delta G^{0'} = -350$ kJ/mole

Both metals are reoxidized and reprecipitated in the upper few cm of the sediment column, where molecular oxygen is present (Klinkhammer, Heggie, and Graham, 1982). At steady-state the reduction, upward diffusion, and reprecipitation of these metals therefore gives rise to profiles of dissolved and solid Mn such as that shown in figure 8.28.

In marine areas where the concentration of molecular oxygen is very low, Mn^{+2} and Fe^{+2} produced in suboxic sediments are not reprecipitated but move into the water column. Under such circumstances the concentration of these elements in bottom waters could approach their concentration in the interstitial waters of suboxic sediments. Their actual concentration in any particular batch of bottom water would be determined by their rate of production in the sediment column, by their rate of transport into the water column, and by their removal from the vicinity of the water-sediment interface by a variety of transport mechanisms. It is clear, however, that redox reactions could be significant sources of Mn^{+2} and Fe^{+2} to the oceans in settings in which the quantity of organic matter buried with oxidized marine sediments is not sufficient to reach the SO_4^{-2} reduction stage and where bottom waters are virtually O_2-free. Such environments may well have been widespread during the Precambrian Era. If the oxygen pressure in the atmosphere was ca. $\leq 4 \times 10^{-3}$ atm (≤ 0.02 PAL), as suggested in chapter 7, then the concentration of molecular oxygen in surface seawater was ca. ≤ 5 μmol/1. An O_2 concentration of this magnitude in bottom waters could be easily

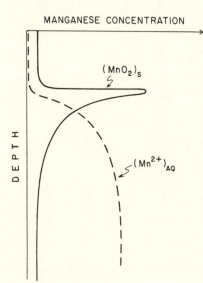

MANGANESE CONCENTRATION

$(MnO_2)_s$

$(Mn^{2+})_{AQ}$

DEPTH

FIGURE 8.28. Schematic representation of dissolved and solid phase Mn profiles in a steady-state suboxic sediment core (Froelich et al 1979). (Reproduced by permission of Pergamon Press, Ltd.)

overwhelmed by fluxes of Mn^{+2} and Fe^{+2}, such as those described by Klinkhammer (1980).

Drever (1974) has pointed out that seawater containing so little O_2 is easily transformed into a reducing solution containing sizable quantities of H_2S by the sinking of organic matter produced in the photic zone. Even in the present-day ocean, oxygen can be nearly or completely depleted by the oxidation of organic matter in the water column, and a good deal of H_2S is present in the waters of a number of enclosed or semi-enclosed basins (see, for instance, Deuser 1975). Perhaps the best known of the present-day anoxic basins is the Black Sea. Water in the upper 200 m of this basin is oxygenated. Below this depth H_2S is present, because the input of organic matter generated by near-surface photosynthesis is more rapid than the supply of oxygen and NO_3^- required for its oxidation. Bacterial oxidation in the water column therefore proceeds to the reduction of SO_4^{-2} to S^{-2}. As shown in figure 8.29, the concentration of iron under these conditions is only ca. 10 $\mu g/kg$ (0.01 ppm), i.e., much lower than the concentration of iron in the interstitial waters of suboxic sediments. The concentration of iron is somewhat higher in the transition zone close to a depth of 200 m, but it is low and variable in the upper 200 m. Manganese is present in rather high concentrations in the anoxic

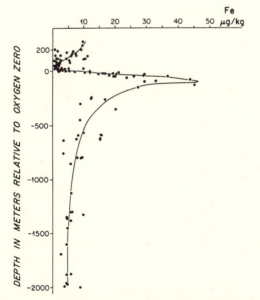

FIGURE 8.29. Profile of the concentration of dissolved iron from the Black Sea. Data from 23 stations have been plotted relative to the level at which the oxygen concentration is zero (Spencer and Brewer 1971). (Copyright by the American Geophysical Union.)

waters of the Black Sea (see fig. 8.30). Its concentration reaches a mild maximum in the transition zone and falls to very low values in the oxidizing upper 200 m of the water column. Figure 8.31 shows that there is a strong concentration of particulate manganese close to the transition zone where Mn^{+2} is oxidized and precipitated.

The data for the Black Sea are similar to those for other anoxic areas (Spencer, Bacon, and Brewer 1981; Jacobs and Emerson 1982). They show very clearly that little iron is mobilized under anoxic conditions, but that sizable quantities of manganese can be mobilized. The mobilization of equivalent quantities of iron in present-day seawater seems to require intermediate oxidation states. In the past, iron could have been mobilized under anoxic conditions if seawater was essentially sulfate free, since in the absence of S^{-2} the concentration of iron would not have been limited by the solubility of iron sulfides (Drever 1974). The chief disadvantage of models of SO_4^{-2}-free Proterozoic seawater is the difficulty they encounter in explaining the correlation between the pyrite content and the organic carbon content of Proterozoic shales. As was pointed out in section 3, the operation of the sulfur cycle between 2.0 and 3.0 b.y.b.p.

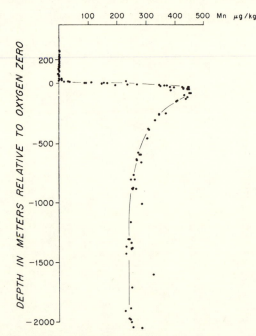

FIGURE 8.30. Profile of the concentration of dissolved manganese in the Black Sea. Data from 23 stations have been plotted relative to the level at which the oxygen concentration is zero (Spencer and Brewer 1971). (Copyright by the American Geophysical Union.)

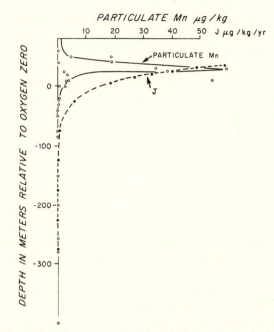

FIGURE 8.31. The profile of the concentration of particulate manganese in the Black Sea, and the calculated in situ production rate of particulate manganese required to maintain a stationary dissolved manganese profile (Spencer and Brewer 1971). (Copyright by the American Geophysical Union.)

seems to have been similar to its operation today, and it is unlikely that the oceans could have disposed of the probable influx of river SO_4^{-2} during this period at an extremely low SO_4^{-2} concentration. It therefore seems worthwhile to explore the conditions under which highly anoxic conditions would not have been widespread in the oceans while the oxygen content of the atmosphere was as low as 4×10^{-3} atm.

The chief condition is clearly that the supply of reducing organic matter was not greater than the supply of oxygen and oxidized materials in the open oceans. Calculations of the maximum permissible rate of photosynthesis in surface waters consistent with this condition are very uncertain, but it is likely that the rate of photosynthesis would have had to have been less than 1% of the average present-day rate. The open oceans would almost certainly have been nearly but not entirely lifeless. A case will be made in the final section of this chapter on other grounds for the proposition that the global rate of photosynthesis during the period between 2.0 and 3.0 b.y.b.p. was much lower than that of the present day, and that the productivity of the open oceans may indeed have been quite small.

Matters are, however, complicated by the probable introduction of H_2S into the oceans with hydrothermally cycled seawater. The H_2S concentration of such solutions today is on the order of 4 mmol/kg. At the present circulation rate of seawater through hot basalts this corresponds to a global input rate of ca. 1×10^{11} moles H_2S/yr. If surface seawater contained 5 μmol O_2/kg, if the oceans contained ca. half the present amount of water, and if the circulation time was on the order of 500 years, the supply of O_2 to the deeper oceans was ca.

$$\frac{5 \times 10^{-6} \text{ mol } O_2/\text{kg} \times 0.7 \times 10^{21} \text{ kg}}{500 \text{ yrs}} = 7 \times 10^{12} \text{ mol } O_2/\text{yr.}$$

The input of H_2S to the deeper oceans was therefore potentially significant only on a local scale compared to the rate of oxygen supply, and mildly oxic seawater could have existed over nearly the entire ocean floor. The actual O_2 content of bottom water was presumably determined in large part by the balance between the rate of supply of O_2 and of organic detritus.

All of the foregoing arguments are obviously speculative, but they do indicate that the composition of large tracts of bottom waters could have been comparable to that of present-day interstitial waters in suboxic sediments, and that the rate of Fe^{+2} supply could have been sufficient for the production of Lake Superior-type iron formations.

THE TRANSPORT OF Fe^{+2} TO NEAR-SURFACE ENVIRONMENTS

Deep waters are transported upward today at a mean rate of ca. 5 m/yr. This is consistent with an oceanic mixing time of 1000 years. In areas of strong upwelling, especially along the western coasts of continents, upward flow can be much more rapid. Walsh (1977, p. 932) has used a figure of 10^{-2} cm/sec, i.e., 3000 m/yr, for the vertical velocity of the upwelling currents off Peru, off Northwest Africa, and off Baja California. The geometry of these currents is rather irregular, but they tend to have an off-shore width of a few tens of kilometers. The circulation patterns in upwelling regions can be extremely complex, as illustrated schematically in figure 8.32. The intensity of upwelling tends to vary seasonally and can be variable on other time scales as well. The reconstruction of patterns of upwelling in areas such as the Hamersley Basin or the Labrador Trough is therefore an extremely difficult undertaking. The feasibility of supplying a sufficient quantity of iron in areas of intense upwelling of deep, iron-rich waters can, however, be tested.

The northern edge of the Hamersley Basin has a length of ca. 400 km (see figure 8.24). If an upwelling current extending 50 km off shore rose

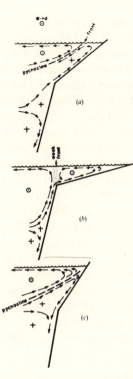

FIGURE 8.32. Some cross-stream circulation patterns in upwelling areas (Walsh 1977): (a) two-cell, possibly observed off Oregon; (b) two-cell, possibly observed of Northwest Africa; (c) one-cell, possibly observed off Peru. + indicates upward flow; ⊙ indicates flow toward the equator. (Reproduced by permission of John Wiley & Sons, Inc.)

along one-half of the basin length at a vertical rate of 3000 m/yr, and if the water in this current contained 1.0 ppm Fe^{+2}, the total annual supply of Fe^{+2} would have been approximately

$$1.0 \times 10^{-6} \frac{gm\ Fe^{+2}}{cc\ S.W.} \times 3 \times 10^{5} \frac{cm}{yr} \times 50 \times 10^{5}\ cm \times 200 \times 10^{5}\ cm$$

$$= 30 \times 10^{12} \frac{gm\ Fe^{+2}}{yr}.$$

This calculation demonstrates that it is possible to account for the estimated rate of deposition of iron in the Hamersley Basin during Dales Gorge time by means of the proposed mechanism. The calculation does not, of course, prove that the iron was actually supplied in this fashion.

If the concentration of iron in the upwelling waters was as high as suggested by Ewers (1980), 20 ppm, or by Mel'nik (1973), 100–400 ppm, the supply problem becomes even easier. However, it seems unlikely that deep ocean waters could have contained such high concentrations of Fe^{+2} for periods of time comparable to those required for the deposition of the Hamersley iron formations.

Several of the major sedimentary manganese deposits formed during the Phanerozoic Era may be analogues of Precambrian BIFs (Cannon and Force 1983). These deposits were formed during periods of warm climate accompanied by marine transgressions and the presence of widespread anoxia on the ocean floor. The recently discovered Groote Eylandt deposits in the Gulf of Carpentaria is one of the largest of these deposits. The ore horizon in this area is a member of the Albian (Lower Cretaceous) Mullaman beds. Other major manganese deposits of this type are the Oxfordian (Late Jurassic) Molango deposits in Mexico and the lower Oligocene manganese deposits of the Black Sea area; these include the deposits of the Nikopol district in the Ukraine and those of the Chiatura district in Georgia (USSR). Deposits in these areas have ore reserves on the order of 1.7 billion tons containing between 13 and 33% Mn. The ore resources at Molango, Mexico, have been estimated to consist of 15 billion tons of potential ore containing $\geq 10\%$ Mn. The geology of these deposits indicates that their contained Mn may well have been released from anoxic ocean-floor sediments and that the Mn ores were deposited from upwelling seawater in near-shore areas. It is understandable that their iron content is very low, since Fe^{+2} is precipitated nearly quantitatively as a constituent of one or more iron sulfides in highly anoxic settings, and therefore does not enter the water column in significant quantities.

Processes Leading to the Precipitation of Iron Formations

Deep, suboxic ocean waters probably had small but significant quantities of nutrients. In upwelling areas these must have become available for photosynthesizing organisms. The areas in which iron formations were formed were therefore probably areas of considerable biologic activity (Drever 1974). The residues of microorganisms that grew in the surface layers probably reacted with seawater sulfate to generate H_2S and CO_2. In such anoxic settings iron sulfides should have been precipitated, and this probably accounts for the formation of the pyrite-type iron formation in deeper, off-shore areas of iron formations. The carbon in the CO_2 generated during the oxidation of organic matter in these areas must have been isotopically light. It would have become mixed with the CO_2, HCO_3^-, and CO_3^{-2} in upwelling waters, and could have produced mixtures with decidedly negative $\delta^{13}C$ values. Such $\delta^{13}C$ values have been observed in the carbon of siderites and ankerites in carbonate iron formations. The data in figure 8.33 show that siderites and ankerites from the Dales Gorge member of the Brockman Iron Formation of the Hamersley Basin cluster around $-10\permil$ but are quite variable even within a single macroband. Baur et al. (1983) have confirmed these data, and

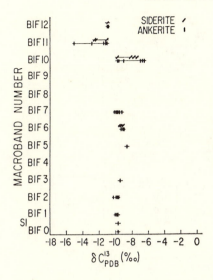

FIGURE 8.33. Distribution of $\delta^{13}C$ values in siderites and ankerites from the Dales Gorge (Becker and Clayton 1972). (Reproduced by permission of Pergamon Press, Ltd.)

have shown that there is a correlation between the isotopic composition of carbon in carbonates and the mineralogy of microbands in the Brockman Iron Formation. Isotopically very light carbon seems to be present in microbands containing little carbonate but large quantities of magnetite. The same relationship between mineralogy and $\delta^{13}C$ was found by Perry and Tan (1973) for the carbonates in the Biwabik Iron Formation of Minnesota (see figure 8.34).

Typical marine carbonates of the Hamersley Basin section have $\delta^{13}C$ values close to 0‰ (see figure 8.35). These are therefore quite normal (see section 2) and confirm the inference that the negative $\delta^{13}C$ values of the siderites and ankerites in iron formations represent local deviations rather than variations in the $\delta^{13}C$ value of the oceans as a whole.

The magnitude of the difference between the $\delta^{13}C$ value of iron carbonates and normal marine carbonates indicates that the quantity of CO_2 added to seawater by the decay of organic matter was comparable to the total dissolved carbonate content of seawater during the formation of iron formations. This is not unexpected. $\delta^{13}C$ measurements have shown that 1 mmol/1 of the 4.3 mmols/1 of total dissolved carbonate in waters of the Black Sea is derived from the oxidation of organic carbon (Deuser 1975). If the concentration of total dissolved carbonate in seawater between 2.0 and 3.0 b.y.b.p. was similar to that of present-day seawater, similar increments of CO_2 could well have been added to up-welling waters at the edge of the Hamersley Basin and in other similar settings. Since one-half a mole of S^{-2} is produced per mole of CO_2

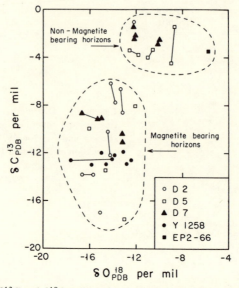

FIGURE 8.34. The δ^{13}C and δ^{18}O values of carbonates from the Biwabik Iron Formation, showing the fields of magnetite-free units (Perry and Tan 1973). Figure is from "Genesis of Precambrian Iron and Manganese Deposits", Proc. Kiev Symp. 1970, © UNESCO, 1973. (Reproduced by permission of UNESCO.)

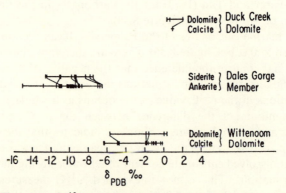

FIGURE 8.35. The values of δ^{13}C in carbonates from dolomitic limestones and iron formations from the Hamersley Basin; the lines connect the δ^{13}C value of the carbonates from the same samples (Becker and Clayton 1972). (Reproduced by permission of Pergamon Press, Ltd.)

generated during the reaction of organic matter with sulfate, the concentration of sulfide in bottom waters in the area of sulfide facies iron formation could have been on the order of 0.5 mmol/1. This is considerably greater than the concentration of ca. 0.02 mmol/1 proposed for iron in suboxic seawater. The precipitation of iron in the region of the sulfide iron formations could therefore have been essentially complete. The availability of iron for deposition in other types of BIF, which are more important quantitatively than the sulfide type, may be a consequence of the pattern of seawater movement in the region of upwelling and in the depositional basin, and due to the removal of H_2S from the water column by means other than iron sulfide precipitation.

Iron formations as a whole are poor in P_2O_5. This is consistent with the expected, relatively low concentration of nutrients in upwelling suboxic bottom waters. However, the solutions from which the metamorphosed, ca. 2.1 b.y.-old iron formations of Väyrylänkylä and some of the other iron formations in Finland were deposited (Laajoki and Saikkonen 1977) seem to have been exceptions to this rule. At Väyrylänkylä, rocks of the mixed oxide-silicate facies and of the carbonate facies contain phosphatic, noncherty, garnet-bearing iron silicate and pyrrhotite-rich black schists and phyllites. The ores are distinguished by an exceptionally high content of P_2O_5 ($\sim 2.5\%$) and CaO ($\sim 3.5\%$) due to the presence of ubiquitous marine apatite, as well as by their comparatively high S^{-2} ($\sim 2\%$) and C^0 ($\sim 1\%$) content, which are related to the frequency of iron-rich metapelite interbands.

The rise of deep upwelling seawater into shallower waters was certainly accompanied by a decrease in pressure. It may also have been accompanied by an increase in temperature, by evaporation, and by degassing (see, for instance, Drever 1974). All of these processes would have tended to lead to the precipitation of carbonate minerals. The fact that ankerite and siderite are the dominant carbonates in iron formations and that calcite is rare indicates that the ratio $m_{Fe^{+2}}/m_{Ca^{+2}}$ in the areas of carbonate precipitation was greater than ca. 6×10^{-3} (see equation 8.8). The observation that rhodochrosite is not common indicates that the Mn^{+2}/Fe^{+2} ratio was less than

$$\frac{m_{Mn^{+2}}}{m_{Fe^{+2}}} < \frac{K_{Rhod.}}{K_{Sid.}} = \frac{10^{-10.4}}{10^{-10.55}} = 1.4. \qquad (8.9)$$

On the other hand, the presence of very significant quantities of manganese is most siderites and ankerites indicates that the Mn^{+2}/Fe^{+2} ratio in the upwelling seawater was by no means negligible.

The escape of CO_2 from the upwelling waters would have led to an increase in the pH and hence toward the precipitation of iron silicates as well as of iron carbonates during the rise of deep seawater. The frequent association of silicate with carbonate iron formation may be a consequence of the coupling of the processes that led to the precipitation of minerals in both types of iron formation.

Harder (1978) has shown that at 3° and 20°C clay minerals containing iron can be synthesized in a short time only under reducing conditions. The synthesis of the silica-poor, two-layer iron clay minerals of the greenalite and chamosite types and the silica-rich three-layer iron clays of the nontronite type apparently needs reducing conditions. Unfortunately, too little is known about the stability field and the solubility product of the common iron silicates in iron formations to allow a precise definition of the conditions under which they precipitated.

Near-shore iron formations consist largely but not entirely of oxides and chert. In the Sokoman Iron Formation, for instance, a considerable part of the carbonates was deposited in peritidal environments. The iron-rich bands in oxide iron formations usually contain mixtures of magnetite and hematite. Dimroth and Chauvel (1973) have shown that magnetite and specularite formed repeatedly in the Sokoman Iron Formation of the Central Labrador Trough of Quebec. A large fraction of both minerals was formed during early diagenesis, and renewed cystallization of iron oxides took place during late diagenesis. Iron was apparently quite mobile during the diagenesis of the Sokoman Iron Formation. The metal was concentrated in the most permeable portions of the sediments; cementation materials that suffered early dehydration and cementation were affected very little by the diagenetic movement of iron.

In contrast to the high-energy environment in which the iron oxide-rich parts of the Sokoman Iron Formation were deposited, those in the Hamersley Basin were deposited in a stable depositional basin of exceptional tranquility, in which climatically controlled chemical precipitation took place in sensitive response to seasonal influences. The development of mesobanding was apparently influenced by diagenetic processes (Trendall and Blockley 1970) and cannot be used as evidence for the chemical nature of the primary depositional environment. Nevertheless, one of the requirements for any model of the formation of these sediments is that it must be able to account for the large-scale incomplete oxidation of ferrous to ferric iron within the depositional basin.

The proposed model can explain this incomplete oxidation with a fair degree of assurance. The average piston velocity of molecular oxygen over the oceans is similar to that of other gases. Peng et al. (1979) have shown that the average piston velocity of radon is close to 3 m/day. Münnich (1963) and Münnich and Roether (1967) have shown that the average piston velocity of CO_2 is ca. 5 m/day. These figures imply that on average the rate of transfer of O_2 from the atmosphere to seawater is roughly

$$\sigma_{O_2} = 0.3(_{Eq.}m_{O_2} - m_{O_2}), \tag{8.10}$$

where

σ_{O_2} = number of moles of O_2 transferred per cm^2 of ocean surface per day,

$_{Eq.}m_{O_2}$ = the concentration of O_2 in moles/liter in seawater in equilibrium with the atmosphere,

and

m_{O_2} = the actual concentration of O_2 in moles/liter in the mixed near-surface layer.

If the atmospheric oxygen pressure was 0.4×10^{-3} atm as suggested previously, its equilibrium concentration in seawater was approximately 5 μmol/l. If the waters of the Hamersley Basin were largely free of O_2, the annual rate of transfer of oxygen from the atmosphere to the waters in the Hamersley Basin was approximately

$$\frac{dM_{O_2}}{dt} \approx 0.3 \times 5 \times 10^{-6} \frac{moles}{liter} \times 365 \frac{days}{yr} \times 1.5 \times 10^{15} \, cm^2$$

$$\approx 8 \times 10^{11} \, moles \, O_2/yr. \tag{8.11}$$

Since one mole of O_2 can oxidize four moles of Fe^{+2} to Fe^{+3}, 8×10^{11} moles of O_2 can oxidize $8 \times 10^{11} \times 4 \times 56 = 180 \times 10^{12}$ gm Fe^{+2}. This is a quantity six times as large as the estimated annual rate of precipitation of iron in the Hamersley Basin during the deposition of the Dales Gorge member of the Brockman Iron Formation. This result is reassuring. It suggests that enough molecular oxygen was probably available to produce the observed amount of Fe_2O_3 in the Hamersley Basin, but the influx of O_2 was apparently sufficiently slow, so that the basin did not become particularly oxidizing. The redox balance of the basin is not known in any detail, but the presence of ferrous compounds and the scarcity of manganese in the oxide facies suggest that O_2 concentrations as high as 5 μmol/l were rarely reached in the water column of the basin.

Mechanisms other than reaction with dissolved O_2 could have played a role in the oxidation of Fe^{+2}. Cairns-Smith (1978), for instance, has suggested that UV photons could have been the oxidizing agent. It is well known that Fe^{+2} is converted to Fe^{+3} by UV light (254 nm) under fairly strongly acid conditions. H_2 is a byproduct of the reaction. Above a pH of 6.5 the production of Fe^{+3} may be related to the presence of $Fe(OH)^+$, which is sensitive to light at wavelengths up to and slightly in excess of 400 nm (Braterman, Cairns-Smith, and Sloper 1983). The quantitative significance of the photochemical oxidation of $Fe(OH)^+$ for the formation of iron formations remains to be assessed.

A large amount of manganese must have accompanied iron in the upwelling currents from suboxic settings. The major iron ores in Proterozoic banded iron formations are typically poor in manganese (see table 8.12), but manganese ores are frequently encountered within the depositional basins (see, for instance, the discussion by Button 1976, pp. 275–276). It seems likely that Mn^{+2} remained in solution until Fe^{+2} had been largely removed by oxidation and precipitation, and that for this reason the precipitation of most of the manganese oxides was separated spatially from that of the iron oxides. Photoreduction of Mn^{+3} and Mn^{+4} by dissolved organic compounds may also have been important (see, for instance, Sunda, Huntsman, and Harvey 1983).

The quantitative aspects of manganese transport and deposition are still quite uncertain, but evidence from a somewhat unexpected quarter suggests that the deposition of Fe^{+3}-Mn^{+4} oxides was extensive. Figure 8.36 shows that the concentration of the rare-earth elements in seawater

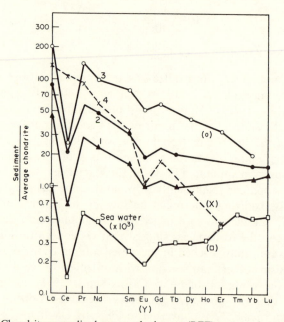

FIGURE 8.36. Chondrite-normalized rare-earth element (REE) patterns in seawater and in Fe-Mn chemical sediments (Fryer 1977.) (1) Average East Pacific Rise crest sediment; (2) Average East Pacific Rise flank sediment; (3) Cyprus ochre; (4) Silurian banded Fe-Mn chemical sediment, Maine. (Reproduced by permission of Pergamon Press, Ltd.)

today and in four sediments and sedimentary rocks rich in Fe-Mn oxides. The large depletion of Ce in seawater and in three of these sediments is probably due to the oxidation of Ce^{+3} to Ce^{+4} and the removal of Ce^{+4} from seawater with manganese nodules (Piper 1974; Elderfield et al. 1981; see also chapter 9, section 5).

Figure 8.37 shows the distribution of the rare-earth elements in modern metalliferous sediments rich in fish debris, and in apatite from the P_2O_5-rich iron formation in Väyrylänkylä. The similarity of the degree of Ce depletion in modern sediments and in the apatite-rich band of the Pääkkö iron-formation sample is striking. Similar depletions characterize the REE pattern of the Queensland phosphorites and of the apatite concentrates

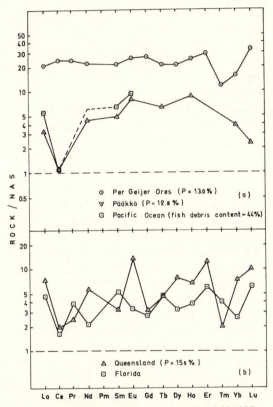

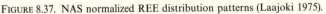

FIGURE 8.37. NAS normalized REE distribution patterns (Laajoki 1975).

a. (1) Average apatite concentrate of the Per Geijer (Kiruna) Deposit. This deposit is probably not sedimentary in origin. (2) Apatite-rich band in the Pääkkö Iron Formation. (3) Average of three metalliferous sediments rich in fish debris from the Pacific Ocean.

b. (1) Average of three pelletal phosphorites from Queensland. (2) Average of three apatite concentrations from Florida.

from Florida. It seems likely, then, that the processes that are responsible for the preferential removal of Ce from seawater today were in operation between 2.0 and 2.5 b.y.b.p. This conclusion is compatible with the deposition of at least some Fe^{+3}-Mn^{+4} oxides at the outermost oxidized fringes of iron formations.

Fryer's (1977) data for the REE content of Proterozoic banded iron formations also suggest that seawater was depleted in Ce during this period of Earth history. Graf (1978) and Appel (1983) have pointed out, however, that the REE pattern of iron formations varies strongly not only from formation to formation but within individual iron formations as well. The reasons for these variations are only partly understood, and inferences based on REE patterns in iron formations regarding the oxidation state of the atmosphere-ocean system during the Archean and during early Proterozoic time are probably not warranted. The REE data are compatible with the proposed oxidation state of the ocean-atmosphere system during late Archean and early Proterozoic time, but they cannot be used as strong support for the concept of a lower oxygen content of the atmosphere during this time period.

Quartz is the only major mineral in unmetamorphosed iron formations whose origin has not been discussed. It is surely the product of the recrystallization of one or more potential precursors; these include silica gel, silica adsorbed on compounds of iron, sodium silicate gel, and the mineral magadiite, $NaSi_7O_{13}(OH)_3$. A biological origin such as that proposed by LaBerge (1973) is usually discounted. Eugster and Chou (1973) consider that magadiite or a sodium silicate gel are the most likely precursors of the bedded cherts; they cite textural evidence and the occurrence of riebeckite in iron formations in support of this proposal. In Drever's (1974) model for iron formations, magadiite and amorphous silica gel are both considered possible precursors of chert in iron formations.

All of the likely inorganic processes of SiO_2 deposition require a SiO_2 concentration in Precambrian seawater considerably in excess of that in the present-day oceans. This is entirely reasonable. Today silica is removed from the oceans largely as the major constituent of the tests of siliceous microorganisms (see, for instance, Holland 1978, pp. 186–190). In the absence of siliceous microorganisms, the SiO_2 concentration of seawater must have been sufficiently high, so that inorganic processsses were able to remove SiO_2 from the oceans at a rate equal to the rate of input from rivers and from hydrothermal sources. This rate was almost certainly well in excess of the rate of SiO_2 deposition in the Hamersley Basin. Saturation of seawater at 25°C with respect to amorphous silica occurs at a concentration of ca. 120 ppm, and this seems a safe upper limit for the SiO_2 concentration in Precambrian seawater. As shown in figure 8.38, the precipitation of magadiite can occur at a concentration of SiO_2 considerably lower than 120 ppm at a pH between 6 and 8. The association of chert with the iron minerals of iron formations is probably best explained by the effects of mild evaporative concentration. As Drever (1974) has pointed out, the absence of $CaCO_3$ in the oxide iron formations may well be related to the reduction in the pH of seawater during the oxidation of Fe^{+2} to Fe^{+3}, followed by the precipitation of hydrous ferric oxide via the reaction,

$$4Fe^{+2} + O_2 + 4H_2O \longrightarrow 2Fe_2O_3 + 8H^+. \qquad (8.12)$$

The absence of gypsum in iron formations could be due to the initial absence of this mineral or due to its later resolution. A variety of reasons could be cited for its initial absence; but rare lozenge-shaped pseudomorphs of microcrystalline quartz in the Sokoman Iron Formation (Dimroth and Chauvel 1973) have the crystal habit of gypsum, and it may therefore be worth searching for other evidence bearing on the prior existence of this mineral in iron formations.

The discussion in this section has shown that all of the major features of banded iron formations can be fitted into a model for their deposition at a time when the oxygen content of the Earth's atmosphere was considerably lower than today. The scarcity of iron formations during the Phanerozoic may well be related to the higher concentration of atmospheric oxygen during this era and to the different nature and distribution of the biota.

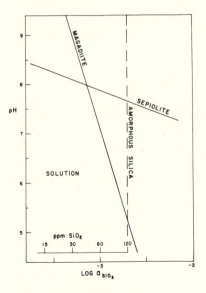

FIGURE 8.38. The equilibrium solubility of magadiite and of sepiolite in present-day seawater ($m_{Na^+} = 10^{-0.48}$ $m_{Mg^{+2}} = 10^{-1.87}$) at 25°C (Drever 1974). (Reproduced by permission of the Geological Society of America.)

5. Controls on Atmospheric Oxygen during the Precambrian Era

THE PALEONTOLOGY OF THE ARCHEAN AND PROTEROZOIC EONS

The data discussed in this and in the previous chapter demonstrate beyond reasonable doubt that free oxygen was present in the atmosphere during the latter part of the Archean and during all of the Proterozoic Era; they indicate rather strongly that P_{O_2} was on the order of 0.02 PAL during the period between 2.0 and 3.0 b.y.b.p.. P_{O_2} must have risen considerably between 2.0 b.y.b.p. and the opening of the Phanerozoic Era, 0.6 b.y.b.p., but the course of the oxygen pressure during this time period is still quite poorly defined. These observations raise a number of questions about the factors that controlled the level of atmospheric oxygen during the Precambrian, and this section will explore some possible answers. A quantitative treatment of the problem is still impossible, but this should not come as a great surpise, since there is still no truly quantitative theory to explain why the oxygen pressure in the present-day atmosphere is 0.20 atm.

In the absence of oxygen-producing photosynthetic organisms, any free oxygen in the atmosphere must have been produced by the photodissociation of water vapor followed by the loss of hydrogen into interplanetary space. The ability of this mechanism to generate an oxygenated atmosphere was discussed in section 2 of chapter 4. At present it does not look as if an atmosphere containing 4×10^{-3} atm O_2 could have been maintained by this process alone. Purely physical and chemical arguments therefore speak for the early development of organisms capable of photosynthesis involving the production of free oxygen. Unfortunately, our knowledge of life in the Archean is still very sketchy. Carbonaceous, spheroidal, probably biogenic microstructures have been reported from four Archean stratigraphic units. The simple morphology of these microstructures and their similarity to organic microstructures of known nonbiologic origin makes their unequivocal identification as biological entities virtually impossible (Knoll and Barghoorn 1977). Remains of an Archean microbiota seem to be preserved in the bedded stromatolitic cherts of the ca. 3.5. b.y.-old Warrawoona Group from the North Pole Dome of the Pilbara Block of Western Australia (Awramik, Schopf, and Walter 1983) and from cherty portions of stromatolitic limestones of the ca. 2.8 b.y.-old Fortescue Group, the lowest stratigraphic unit of the Hamersley Basin (see Schopf and Walter 1983). The Warrawoona biocenose was an entirely prokaryotic community that formed laminar microbial mats at the sediment-water interface in a shallow-water setting that was intermittently exposed. The community probably included heterotrophs and autotrophs. At least some of the microorganisms were probably anaerobic. The major mat-building taxa were filamentous, phototactic, and presumably photoautotrophic. There are no data to indicate whether these presumed photoautotrophs were oxygen-producing.

The paleontologic record of the early part of the Proterozoic Era is somewhat more extensive. Twenty-three occurrences of probably authentic microfossils have been described, and twelve of these have been characterized carefully on the basis of thin-section micrography (Hofmann and Schopf 1983). Among these, two types of assemblages are common: (1) shallow water stromatolitic communities preserved by silica permineralization in cratonal carbonate sediments, and (2) relatively deeper water assemblages that are commonly preserved as carbonaceous compressions in offshore black shales.

In early Proterozoic sediments five morphologically distinct groups of microfossils have been recognized: coccoid bodies, septate filaments, tubular microstructures, branched filaments, and "bizarre forms", i.e., fossils of highly atypical morphology. Members of the latter two categories are numerically insignificant, although metallogenium-like remains dominate in nonstromatolitic facies of the 2.0 b.y.-old Gunflint Formation. Figures 8.39 and 8.40 are microphotographs of part of the biota discovered on the north shore of Lake Superior in cherts of this formation (Barghoorn and Tyler 1965).

FIGURE 8.39. Cyanobacterial filaments and spheroidal unicellular prokaryotes from the Gunflint Formation, Ontario, Canada (Barghoorn and Tyler 1965). (Copyright 1965 by the American Association for the Advancement of Science.)

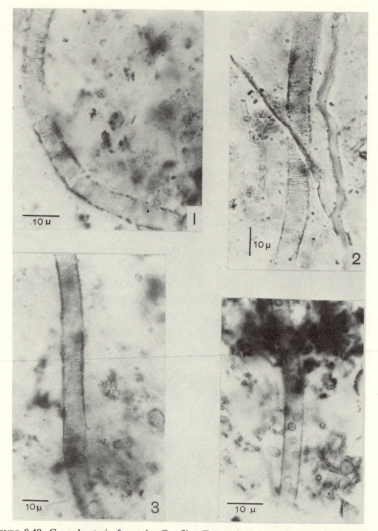

FIGURE 8.40. Cyanobacteria from the Gunflint Formation, Ontario, Canada (Barghoorn and Tyler 1965). (Copyright by the American Association for the Advancement of Science.)

The early Proterozoic biota appear to have been composed largely, and perhaps entirely, of prokaryotic microbes. Many of these organisms may have had noncyanobacterial affinities, but several lines of evidence suggest that oxygen-producing photosynthetic microorganisms of the cyanobacterial type were also present. If cyanobacteria were present during the early part of the Proterozoic Era, systematic aerobic biochemistry could have occurred once the atmospheric oxygen pressure had attained a level of about 10^{-4} atm. The threshold for organismic aerobiosis is about an order of magnitude higher. The evidence for the presence of early eukaryotes 1.4 b.y. ago suggests that this level (ca. 0.005 PAL) was established at least as early as 1.4 b.y.b.p. (Chapman and Schopf 1983). The paleontological evidence available today is therefore consistent with, but does not demand a level of oxygen in the early Proterozoic atmosphere equal to that indicated by the geochemical and sedimentological data.

The ecology of the early Proterozoic biota was apparently complex and adapted to rather variable environments. It seems likely that many cyanobacteria were highly productive both in the presence of free oxygen and under anoxic conditions (see, for instance, Krumbein et al. 1979). Under anoxic conditions these organisms can thrive on the operation of anoxygenic photosystem I based on the reaction

$$CO_2 + 2H_2S \longrightarrow CH_2O + H_2O + 2S \tag{8.13}$$

(Cohen, Padan, and Shilo 1975; Garlick, Oren, and Padan 1977; see also Towe 1978). When oxidizing conditions are established or reestablished, photosynthesis takes place via the operation of photosystem II

$$CO_2 + 2H_2O \longrightarrow CH_2O + H_2O + O_2. \tag{8.14}$$

Such organisms would therefore have been able to live in environments such as those inferred from the deposition of iron formations, where oxygenic and anoxic conditions may well have alternated rapidly in space and time.

Prokaryotic organisms also have excellent defense mechanisms against solar ultraviolet light, and these may have been important early in Earth history. When P_{O_2} is very low, the concentration of ozone is smaller than at present and is insufficient to shield the surface of the Earth from solar ultraviolet radiation at wavelengths between ca. 205 and 300 nm. The relationship between P_{O_2} and the solar UV flux at the Earth's surface has been explored in a number of progressively more refined models. Berkner and Marshall (1965) treated the evolution of O_3 in O_2-deficient paleoatmospheres in a qualitative manner. Ratner and Walker (1972), Hesstvedt, Henriksen, and Hjartarson (1974), and Blake and Carver (1977) developed quantitative photochemical models of increasing complexity; Levine, Boughner, and Smith (1980) have published the most complete treatment to date, but even their results should be treated with a little caution, because they included neither tropospheric photochemical nor chemical processes, nor the effects of the chemistry of carbon and chlorine species on the concentration of ozone in paleoatmospheres. Figure 8.41 shows the computed intensity of solar ultraviolet radiation between 184.6

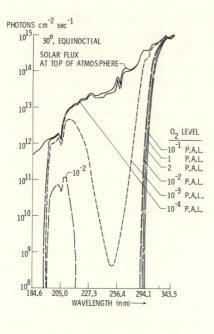

FIGURE 8.41. The flux of solar radiation between 184.6 and 342.5 nm at the surface of the Earth as a function of values of the oxygen pressure between 10^{-4} and 2 PAL (Levine, Boughner, and Smith 1980). (Copyright © 1980 by D. Reidel Publishing Company.)

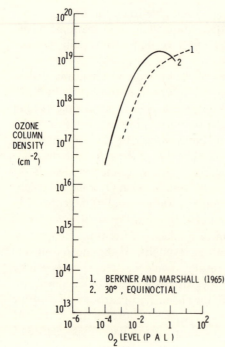

FIGURE 8.42. The total O_3 column above the Earth's surface as a function of the atmospheric O_2 level. Curve 1 is from Berkner and Marshall (1965); curve 2 is from Levine, Boughner, and Smith (1980) for 30° latitude under equinoctial conditions. (Copyright © 1980 by D. Reidel Publishing Company.)

and 342.5 nm at the surface of the Earth at 30° latitude under equinoctial conditions as a function of the oxygen content of the atmosphere. The computed UV intensities at a given value of P_{O_2} are slightly higher at the equator and somewhat lower at higher latitudes. The wavelength window between 200 and 220 nm is closed only at values of $P_{O_2} > 10^{-2}$ PAL. The window between 220 and 294 nm closes between 10^{-3} and 10^{-2} PAL.

The total O_3 column at the Earth's surface seems to go through a maximum at an O_2 pressure of 0.1 PAL. This effect is shown in figure 8.42 together with the earlier curve of Berkner and Marshall (1965), which was based on a much less complete analysis of the problem. Ozone is biologically quite toxic. Chameides and Walker (1975) have pointed out that the present concentration of the gas in the atmosphere is close to toxic levels, and that many varieties of plant life are damaged extensively when exposed to ozone densities two or three times greater than the average ambient density.

The use of the data in figure 8.42 as an indicator of near-surface ozone levels at O_2 pressures different from that of the present day is still somewhat uncertain and will continue to be so until chemical and photo-chemical processes in the troposphere are included in the computational model. The computations indicate very clearly, however, that prokaryotic organisms may have needed defense mechanisms against solar ultraviolet radiation during the Proterozoic Era. Microbial mats probably offered some protection against solar UV light. The presence of nitrate and nitrite in the solutions surrounding prokaryotes in such settings would have offered additional protection, and the operation of enzymatic light-insensitive repair systems as well as repair systems mediated by visible light might have permitted prokaryotic organisms to flourish even in the absence of an effective atmospheric screen against ultraviolet radiation (Margulis, Walker, and Rambler 1976).

Productivity and the Preservation of Organic Matter

Locally, the productivity of the Archean and Proterozoic oceans could have been very high. Bunt (1975) listed values of 0.65–2.15 gm $C/m^2/day$ for the primary productivity of intertidal cyanobacteria at Eniwetok, and Krumbein, Cohen, and Shilo (1977) recorded productivities as high as 12.0 gm $C/m^2/day$ for stromatolites from Solar Lake in the Sinai. These productivities are considerably greater than the mean productivity of the oceans today, ca. 0.27 gm $C/m^2/day$ (see, for instance, Holland 1978, pp. 211–213). However, the areal extent of the quiet, shallow water settings in which these high productivities have been observed is quite limited. More typical Archean shallow marine environments were charac-terized by high rates of clastic and pyroclastic influx (Knoll 1979); in the equivalent modern environments, benthonic productivity is a good deal lower: 0.08–0.53 gm $C/m^2/day$ in northern U.S. estuaries, 0.02–0.22 gm $C/m^2/day$ in tropical sediments, and 0.01–0.03 gm $C/m^2/day$ in Scottish shallow marine sediments (Bunt 1975).

It is still uncertain to what extent the open oceans were populated during the Precambrian. The oldest, assuredly planktonic microfossils have been found in early Proterozoic sediments (Hofmann 1976; Knoll, Barghoorn, and Awramik 1978). In the present-day ocean, prokaryotic photoautotrophs probably account for only a small fraction of the total primary production of open-ocean ecosystems. Until recently it was thought that marine planktonic cyanobacteria were restricted to a few nostocalean genera, of which only *Trichodesmium* was capable of forming extensive water blooms. However, Waterbury et al. (1979) have demonstrated the widespread open-ocean occurrence of a small chroococcalean cyanobacterium belonging to the genus *Synechococcus*. These cyanobacteria can achieve considerable cell densities; however, their importance as primary producers in the world's oceans remains to be determined. It is not known whether the apparent, relatively low productivity of the cyanobacteria in the open oceans today reflects their competitive exclusion from pelagic environments, or whether the oceanic habitat was underexploited prior to the evolution of nucleated organisms (Knoll 1979).

Underexploitation of the open oceans during the early part of the Proterozoic Era would serve to explain the suboxic state that was proposed for deep ocean water in the course of the discussion of the origin of iron formations in the previous section. It would also go far to explain the low oxygen pressure in the atmosphere more than 2.0 b.y. ago. The total rate of photosynthesis would then have been limited largely by the available area and by the environmental conditions in near-shore settings where prokaryotic mats flourished. The fraction of the annual photosynthetic yield of organic carbon that had to be buried with sediments to maintain the redox balance of the crust would almost certainly have been larger than today. It has been observed that the destruction of organic matter is more complete in well-oxygenated than in anoxic settings (see, for instance, Demaison and Moore 1980). The oxygen content of the atmosphere during late Archean and early Proterozoic time may therefore have been limited by the necessity to preserve and bury a relatively large fraction of the organic matter generated in near-shore settings. If this interpretation is correct, the high oxygen pressure today is largely a consequence of the increase in the total rate of photosynthesis, particularly in the open oceans, that has been made possible by a series of biological innovations during the past 2 b.y. of Earth history.

This explanation for the evolution of atmospheric oxygen is attractive, if for no reason other than that it is simple. It may, however, be wrong.

Marine photosynthesis today is almost invariably limited by the availability of nutrients; it is therefore worth exploring the possibility that nutrient limitation rather than evolutionary factors limited photosynthesis in the early oceans, and that nutrient limitation was somehow responsible for the lower oxygen content of the early Proterozoic atmosphere. Among the necessary nutrients, phosphate and nitrate are most commonly limiting today (see, for instance, Holland 1978, 211–224). In an atmosphere containing 0.02 PAL of oxygen, the inorganic fixation of nitrogen was probably a good deal less rapid than today (Yung and McElroy 1979). However, the ability of cyanobacteria to fix nitrogen (see, for instance, Wolk 1973; Schindler 1977) may well have compensated for a smaller supply of fixed nitrogen from the atmosphere. Phosphate rather than nitrate may therefore have been the limiting nutrient in most Archean and Proterozoic marine settings.

It is instructive to see how the marine geochemistry of phosphorus might have exercised a dominant control on biological productivity and on the oxygen content of the Precambrian atmosphere. Today, phosphate cycles rapidly through the marine biosphere (see figure 8.43). Much of the phosphate incorporated into organisms in the photic zone is released

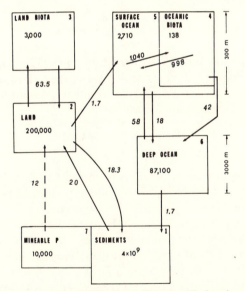

FIGURE 8.43. Aspects of the cycle of phosphorus (Lerman, Mackenzie, and Garrels 1975); reservoir numbers are given in the upper right corner of each box. Roman figures (such as 3000 in reservoir 3) refer to the phosphorus content in units of 10^{12} gm. Italic figures adjacent to arrows refer to interreservoir fluxes in 10^{12} gm/yr. The surface ocean reservoirs are taken to be 300 m deep; the depth of the deep ocean reservoirs is ca. 3000 m. See also Froelich et al. 1982. (Reproduced by permission of the Geological Society of America.)

there during decay and is recycled rapidly in situ (see, for instance, McCarthy and Goldman 1979). A very significant quantity of phosphate is, however, removed with sinking organic matter from the upper 300 m of the oceans. Much of this phosphate is regenerated in the water column, and some is regenerated at and below the water-sediment interface. The remainder is lost as a constituent of organic matter buried with sediments and probably by a variety of other mechanisms as well. Phosphate regenerated in the water column or within sediments is returned to the photic layer by upward mixing of deep water, and then participates in additional biological cycling. The intensity of phosphorus cycling and the nature of the inorganic phosphate sinks are related to the oxygen content of seawater and hence to the oxygen content of the atmosphere. Under some circumstances the ocean-atmosphere system might therefore have operated with an atmospheric oxygen pressure much lower than its present value.

Let us consider an ocean containing $M_{\Sigma P}$ moles of dissolved phosphate. Let a fraction, ϕ, of this phosphate be used annually in photosynthesis, and let the molar C/P ratio of photosynthetic organisms be equal to the Redfield ratio, R. The total marine rate of photosynthesis, $(dM_C/dt)_{\text{Phot.}}$, will then be equal to

$$\left(\frac{dM_C}{dt}\right)_{\text{Phot.}} = M_{\Sigma P}\phi R. \tag{8.15}$$

Now let the fraction of the photochemically generated organic matter that is buried with marine sediments be ξ. If we let the C/P ratio of the buried organic matter be α, then the marine loss rate of phosphorus accompanying the burial of organic matter will be

$$\left(\frac{dM_{\Sigma P}}{dt}\right)_{\text{C}^0,\,\text{Sed.}} = -\frac{M_{\Sigma P}\phi R\xi}{\alpha}. \tag{8.16}$$

Phosphorus sinks other than organic matter are almost certainly important, although there is still considerable disagreement concerning the relative importance of phosphate incorporation in calcium carbonate, the adsorption of phosphate on $Fe(OH)_3$-rich sediments, and the precipitation of phosphate minerals removes ca. 20% of the river flux of phosphate. The effect of these removal mechanisms can be introduced formally into the model by adding a term y to the mass balance equation for phosphorus in the oceans. From equation 8.16 it follows that at steady state

$$\left(\frac{dM_{\Sigma P}}{dt}\right)_{\text{R.W.}} = \frac{M_{\Sigma P}\phi R\xi}{\alpha} + y, \tag{8.17}$$

where y is the rate of removal of phosphorus from the oceans by mechanisms other than the burial of organic matter.

At steady-state the number of moles of phosphorus in the oceans is therefore

$$M_{\Sigma P} = \frac{\alpha}{\phi R \xi} \left[\left(\frac{dM_{\Sigma P}}{dt} \right)_{\text{R.W.}} - y \right].$$ (8.18)

The functional relationship between y and $M_{\Sigma P}$ is not known. The two must be related, because at high phosphate concentrations the precipitation of apatite, and probably of other phosphate minerals as well, becomes inevitable. For the sake of simplicity, we shall assume that

$$y = aM_{\Sigma P}.$$ (8.19)

The use of a power function rather than a simple proportionality relating y and $M_{\Sigma P}$ does not alter the conclusions, which are qualitative in any case.

From equation 8.17 it follows that

$$M_{\Sigma P} = \left(\frac{1}{\dfrac{\phi R \xi}{\alpha} + a} \right) \left(\frac{dM_{\Sigma P}}{dt} \right)_{\text{R.W.}}.$$ (8.20)

The total rate of photosynthesis in such an ocean will be equal to

$$\left(\frac{dM_{C^0}}{dt} \right)_{\text{Phot.}} = \left(\frac{\phi R}{\dfrac{\phi R \xi}{\alpha} + a} \right) \left(\frac{dM_{\Sigma P}}{dt} \right)_{\text{R.W.}},$$ (8.21)

and the number of mole of C^0 buried annually with sediments

$$\left(\frac{dM_{C^0}}{dt} \right)_{\text{Sed.}} = \left(\frac{1}{\dfrac{1}{\alpha} + \dfrac{a}{\phi R \xi}} \right) \left(\frac{dM_{\Sigma P}}{dt} \right)_{\text{R.W.}}.$$ (8.22)

During late Archean and early Proterozoic time the rate of degassing, tectonism, and chemical weathering were all probably somewhat more rapid than they are today. Since the C^0 content of sediments from this period of Earth history is comparable to that of more recent sediments, the value of $(dM_{C^0}/dt)_{\text{Sed.}}$ was also larger then than now. It seems likely, therefore, that the ratio $(dM_{C^0}/dt)_{\text{Sed.}}/(dM_{\Sigma P}/dt)_{\text{R.W.}}$ has remained roughly constant; if so, then the term $[1/\alpha + a/\phi R \xi]$ has also remained roughly constant. For the sake of convenience, this term can be recast in the form,

$$\frac{1}{R} \left(\frac{R}{\alpha} + \frac{a}{\phi \xi} \right) \simeq \text{constant}.$$ (8.23)

The question can now be asked whether there is more than one value of the oxygen content of the atmosphere for which $(R/\alpha + a/\phi\xi)(1/R)$ has its present value, and if so, whether one of these corresponds to the most likely value of P_{O_2} during early Proterozoic time.

The C/P ratio of organic matter in modern marine sediments is generally several times larger than the Redfield ratio (see, for instance, Hartmann et al. 1973; Aller 1980). R/α is therefore less than unity. The very low P_2O_5 content of the highly carbonaceous, black, early Proterozoic schists of the Outokumpu region (see table 8.6) suggests that this was also true early in the Proterozoic Era. There are, however, no reliable data to show how much smaller than unity the ratio R/α was between 2.0 and 2.5 b.y.b.p.

The parameter ξ in equation 8.23 depends in part on biological factors, in part on the oxidation state of the oceans. At present the rate of production of organic carbon in the oceans is ca. 35×10^{15} gm/yr; the rate of burial of organic carbon with marine sediments is ca. 0.1×10^{15} gm/yr (see, for instance, Holland 1978, chap. 5). The value of ξ, the fraction of carbon buried, is therefore

$$\xi \approx \frac{0.1 \times 10^{15} \text{ gm/yr}}{35 \times 10^{15} \text{ gm/yr}} = 0.003. \tag{8.24}$$

As pointed out previously, the decomposition of organic matter in anoxic basins is much less efficient than in well-oxygenated areas; as much as several percent of the organic matter produced in the photic zone of anoxic basins tends to become buried (see, for instance, Demaison and Moore 1980). It is likely that ξ decreases progressively with increasing P_{O_2}; however, at any given value of P_{O_2} the value of ξ almost certainly depends on the distribution of photosynthetic levels in the oceans, on the circulation of seawater in oceanic basins, and hence both on world geography and on world climate (see chapter 9). The nature of the organic matter produced in the oceans and the nature of the distribution of scavenging organisms must also affect the value of ξ. It is clear, therefore, that no single value of ξ can be assigned to a particular value of P_{O_2}.

The fraction, ϕ, of dissolved phosphorus that is used annually in the photosynthetic production of organic matter is largely a function of the mixing time of the oceans. This, in turn, depends on a large number of variables that include biologic and climatic parameters as well as the geometry of the ocean basins. The oxidation state of the system per se probably has little influence on ϕ.

The effect of atmospheric oxygen on the value of the parameter a in equation 8.23 could be intense and interesting. Under anoxic and suboxic conditions the value of a may be large, because PO_4^{-3} can then be removed more efficiently as a constituent of vivianite, $Fe_3(PO_4)_2 \cdot 8H_2O$; under oxygenated conditions, adsorption on amorphous $Fe(OH)_3$ may be an important marine sink of phosphorus. Apatite, $Ca_5(PO_4)_3(OH,F)$, formation should be essentially independent of the oxidation state of the ocean-atmosphere system, but might have been a strong function of the concentration of cations in seawater that act as inhibitors of nucleation of apatite in marine settings. The value of the parameter a could, therefore, have been a complex function of the state of the ocean-atmosphere system.

From all this it follows that constancy of the term $[R/\alpha + a/\phi\xi]$ in equation 8.23 does not necessarily imply a constant level of atmospheric oxygen. It implies only that changes in any one of the parameters required that P_{O_2} or other environmental parameters change in a fashion such that the term as a whole has remained essentially invariant.

The relationship developed above between the level of atmospheric oxygen and the geochemical cycle of phosphorus can be regarded as an extension of an earlier treatment of controls on atmospheric oxygen (Holland 1978, pp. 284–295). It was proposed there that the oxygen content of the atmosphere is controlled by a feed-back system involving the availability of phosphorus (see figure 8.44). The relationship between P_{O_2}

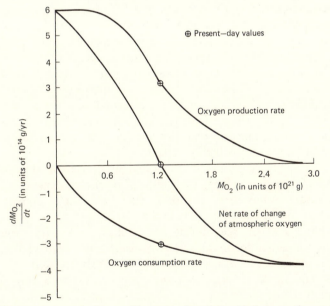

FIGURE 8.44. Semischematic diagram illustrating the relationship between the oxygen content of the atmosphere, the rate of oxygen production, the rate of oxygen consumption, and the net rate of change of atmospheric oxygen (Holland 1978, p. 287). (Reproduced by permission of John Wiley & Sons, Inc.)

and the rate of O_2 generation by the burial of elemental carbon is essentially given by equation 8.22. At a constant rate of phosphate input to the oceans and at constant values of α, a, and ϕ, the rate of C^0 burial with sediments is related to P_{O_2} via the effect of P_{O_2} on the parameter ξ.

As pointed out earlier (Holland 1978, p. 294), variations in biology and geography also affect the rate of C^0 burial. Equation 8.22 is therefore only a first attempt to quantify the effects of some important parameters. What this attempt has shown is that the crossover point of the curve in figure 8.44 depends rather strongly on the evolution of the biology and of the geography of the planet, and that P_{O_2} must have varied during Earth history in response to these evolutionary trends.

6. Summary with Extensions

The evidence that can be extracted from marine sediments regarding the oxidation state of the atmosphere-ocean system during the earlier part of the Precambrian Era supports the inferences in chapter 7, which were based on evidence from terrestrial settings. The mildly oxidizing atmosphere between 2 and 3 b.y.b.p. was in equilibrium with mildly oxygenated ($m_{O_2} \cong 5$ μmol/l) surface seawater, which contained very little dissolved manganese and iron except in areas of active upwelling and of intense injection of hydrothermal solutions. At least some deep waters contained significant quantities of iron, manganese, and phosphate. Biologic productivity was apparently high in areas of upwelling and led to the formation of black shales. Compositionally and isotopically these shales are similar to their younger counterparts. The concentration of sulfide sulfur in these shales is correlated with the concentration of elemental carbon, the normal suite of trace metals is enriched in these sediments, their organic carbon is isotopically light, and the isotopic composition of their sulfur is highly variable.

Iron and SiO_2-rich sediments were apparently precipitated in and close to areas of upwelling where the influx of detrital sediments was very small. Mildly evaporative conditions may have been responsible for much of the precipitation of SiO_2 in these areas.

The biology of the seas during late Archean and early Proterozoic time was dominated by prokaryotic organisms. Some of these almost certainly produced free oxygen. Their preferred habitats were near-shore settings, and it seems likely that the open oceans were only sparsely populated.

A restriction of much of the biota to near-shore areas might account for the low oxygen content of the atmosphere between 2 and 3 b.y.b.p. Other explanations cannot, however, be ruled out. In particular, P_{O_2}

may have been limited by the availability of phosphate, which could have been removed from the oceans by the precipitation of vivianite and/or other phosphate minerals.

The reasons for the rise of P_{O_2} during late Proterozoic time are not known, but it is likely that they are related to biological innovations that led to an increase in the productivity of the oceans. The development of animals and of large plants during the last ca. 100 m.y. of the Precambrian Era (Ediacaran) must have had a dramatic effect on the behavior of organic matter in the water column and in sediments that were subjected to bioturbation. Bioturbation could have been at least partly responsible for the rather respectable oxygen pressure during the early part of the Phanerozoic Eon (see chapter 9).

This view of the Precambrian ocean-atmosphere system has developed rather gradually. In my first paper on the subject (Holland 1962) I suggested that free atmospheric oxygen in appreciable amounts did not exist more than 1.8 b.y.b.p. Since then the work of Grandstaff (1976, 1980) has shown that the survival of detrital uraninite does not require extremely low partial pressures of oxygen in the ambient atmosphere, and numerous circumstances have been described that are best explained by the presence of significant quantities of oxygen during late Archean and early Proterozoic time.

Cloud's views of the oxygen content of the Precambrian atmosphere have undergone a similar evolution. In earlier papers (see, for instance, Cloud 1972), free atmospheric oxygen was considered to be absent or rare more than ca. 1.9 b.y.b.p. In a more recent paper Cloud (1980) suggests that the transition from an anoxic to an oxygenous atmosphere occurred either earlier or was more gradual than previously supposed, and that the transition from an essentially anoxic atmosphere to one in which O_2 was generally present at concentrations greater than 10^{-2} PAL took place between 2.3 and 2.0 b.y.b.p. The evidence for the timing of these changes is not convincing. In particular, it is hard to understand the rarity of highly manganiferous Archean carbonates if oxygen was absent from near-surface seawater more than 2.3 b.y. ago.

The importance of the "Urey effect" (Urey 1959) for regulating oxygen pressures between ca. 10^{-3} and 10^{-2} PAL seems to have been overestimated in the past. This is almost certainly true also of the Pasteur level as a self-regulating mechanism for atmospheric oxygen (Rutten 1970). Both mechanisms seem to fail in the context of the operation of the whole atmosphere-ocean-crust system.

In part, the shift in emphasis toward a more oxygenated Precambrian atmosphere has followed the promptings of those who believe that completely uniformitarian models of chemical sedimentation can be applied to the Precambrian, and that there is no evidence for orders-of-magnitude changes in the average atmospheric or hydrospheric abundances of chemically reactive inorganic species (Dimroth and Kimberly 1976). However, the data presented in this and in the previous chapter indicate that such an extreme position is probably not tenable. Dimroth and Lichtblau's (1978) observation that ferric oxide crusts on shallow-water Archean pillow basalts are very similar to their Cenozoic equivalents demonstrates beyond reasonable doubt that free oxygen was present in the atmosphere and in the shallow parts of the oceans some 2.65 b.y. ago. However, their suggestion that the rate of formation of these crusts is a linear function of the oxygen concentration in seawater may well be wrong. The formation of altered crusts on pillow basalts involves large-scale hydration and carbonation reactions. If these processes rather than oxidation reactions were and are the rate-determining steps, then the thickness of crust enriched in ferric iron has been independent of the oxygen content of the surrounding seawater as long as m_{O_2} was sufficiently high to oxidize Fe^{+2} as it was made available by hydration and carbonation reactions.

The shape of the controversy regarding ferric oxide is symptomatic of the present state of the field. We have arrived at a point where the presence of some oxygen in the atmosphere at least since late Archean time is well established. Additional quantitative studies are now badly needed to set more precise limits on the actual level of atmospheric oxygen during the Precambrian Era.

The definition of ocean water chemistry during the Precambrian is in a similar state. The oceans were surely saturated or supersaturated with respect to calcite and dolomite. P_{CO_2} and the concentration of dissolved CO_2 were almost certainly higher then than now, but by a factor that is still poorly defined. The concentration of dissolved SiO_2 was probably between ca. 20 and 120 ppm, and the concentration of SO_4^{-2} was sufficiently high to produce the observed S^{-2}/C^0 correlation in black shales. The concentration of Fe^{+2} and Mn^{+2} was very low in surface waters, but at least in some near-bottom waters the concentration of these cations was probably on the order of 1 ppm.

The mineralogy of marine evaporites could set significant additional limits on the possible range of seawater composition during the Precambrian Era (see chapter 9). Unfortunately, Precambrian evaporites are rare. Until the early 1960s almost all known marine evaporites were of Phanerozoic age. Since then large masses of gypsum have been discovered in Upper Proterozoic marine sedimentary rocks of the Amadeus Basin in Central Australia (see figure 8.45). Smaller occurrences of Precambrian evaporites or of their replaced residues have been described from northern Australia, North America, and South Africa (for references, see Stewart

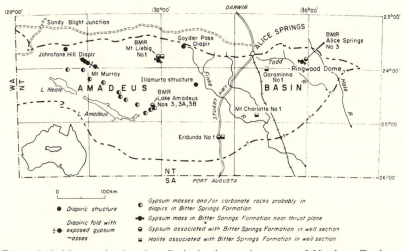

FIGURE 8.45. Map of the Amadeus Basin in the southern part of Northern Territory, Australia, showing the location of gypsum and halite occurrences in the Bitter Springs Formation (Stewart 1979, modified from Wells et al. 1970). (Reproduced by permission of Blackwell Scientific Publications, Ltd.)

1979). The Amadeus Basin occurrences remained among the world's largest deposits of Precambrian evaporites until extensive beds of marble in the mid-Proterozoic (1.8–1.4 b.y.b.p.) McArthur Basin were recognized as replacements after evaporites (Walker et al. 1977).

The evaporites of the Amadeus Basin occur in the Bitter Springs Formation, which is about 900 m.y. old (Marjoribanks and Black 1974). The formation was deposited under relatively stable, shallow marine conditions. As sedimentation proceeded, parts of the sea became partly or totally landlocked, and shales and carbonates interspersed by evaporites were deposited in barred basins and lagoons (Stewart 1979). The known occurrences of gypsum and halite in the Amadeus Basin are too few to define a concentric arrangement; but they suggest that gypsum occurs in the east, south, and west of the basin, whereas halite only occurs in its center and in the northern part. Figure 8.46 is a generalized lithologic log and figure 8.47 a diagrammatic representation showing the sequence of the major rock types of the Ringwood evaporite near Alice Springs at the eastern end of the Amadeus Basin. Evaporation in this part of the basin did not, apparently, extend to the deposition of halite.

The sequence from carbonates to gypsum ± anhydrite to halite in the Bitter Springs Formation is identical to that in Phanerozoic marine evaporites (see, for instance, Braitsch 1962). The sequence could therefore have

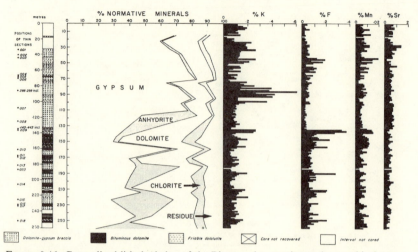

FIGURE 8.46. Generalized lithologic log of the Ringwood evaporite, and a graphical representation of partial rock norms and results of analyses for trace elements (Stewart 1979). (Reproduced by permission of Blackwell Scientific Publications, Ltd.)

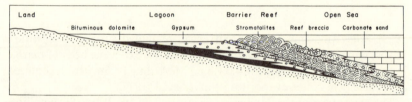

FIGURE 8.47. Diagrammatic representation of a barred basin showing the major rock types of the Ringwood evaporite behind an algal stromatolitic reef of the Bitter Springs Sea (Stewart 1979). (Reproduced by permission of Blackwell Scientific Publications, Ltd.)

been deposited from seawater of present-day composition. It could also have been deposited from seawater of somewhat different composition, because the sequence of minerals during the early phases of seawater evaporation is not very sensitive to changes in the composition of seawater. The limits imposed on excursions in seawater composition by the sequence carbonates $\rightarrow CaSO_4 \pm 2H_2O \rightarrow NaCl$ were explored some time ago (Holland 1972), and are reexamined in chapter 9. It seems likely that the concentration of the major constituents of seawater during the formation of the Bitter Springs evaporites was not much more than twice nor much less than half their concentration in present-day seawater. Attempts such as Ronov's (1968) to define the proportion of the major cations in Precambrian seawater more precisely seem to be rather poorly based.

The evidence for the presence of large marine evaporites much older than those of the Bitter Springs Formation is largely circumstantial. Casts that have been interpreted as replacements of gypsum and halite are reasonably common (see, for instance, Hoffman, 1969; Bell and Jackson, 1974; Badham and Stanworth 1977). The oldest of these replacements comes from the 3.5 b.y.-old Warrawoona Group in Australia (see fig. 8.48).

The interpretation of casts is somewhat uncertain, but the available data do suggest that the evaporation of seawater during the past 3.5 b.y. has always progressed through the sequence carbonate → gypsum ± anhydrite → halite. If this turns out to be correct, then the concentration

FIGURE 8.48. Gypsum rosettes replaced by chert, Warrawoona Group, North Pole, Australia; photo by R. Buick, courtesy of M. Muir.

of the major ions has probably not varied greatly during much of Earth history.

The presence of casts of gypsum with casts of halite crystals has implications not only for the chemistry of seawater but for the temperature during evaporite formation in the Precambrian. Figure 8.49 shows Hardie's (1967) data for the temperature at which gypsum and anhydrite are in equilibrium as a function of the activity of water in surrounding solutions. The activity of water in NaCl solutions decreases with increasing NaCl concentration; in a solution saturated with respect to NaCl, gypsum and anhydrite coexist stably at ca. 18°C. At higher temperatures anhydrite is stable in the presence of a saturated NaCl solution; at lower temperatures the stable phase is gypsum. If the occurrence of gypsum and halite casts within the same hand specimens (see, for instance, Bell and Jackson 1974) can safely be interpreted to mean that gypsum and halite coexisted stably during or shortly after sedimentation, then temperatures during the formation of such sediments could not have been more than a few degrees higher than 18°C. The uncertainty in the temperature limit is due to the tendency of gypsum to crystallize metastably in the anhydrite field; fortunately, the uncertainty is not large. The agreement between the expected and the observed mineralogy of sulfates in the sabkhas of the Trucial Coast is very satisfactory (Kinsman 1966); this implies that the gypsum-anhydrite conversion in such settings is sufficiently rapid to

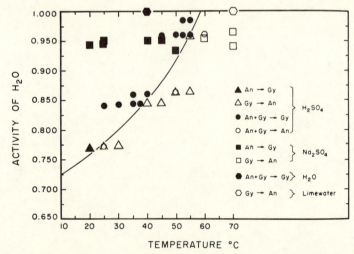

FIGURE 8.49. The stability of gypsum and anhydrite determined experimentally as a function of temperature and activity of H_2O at atmospheric pressure. Only runs are plotted in which a conversion was achieved (Hardie 1967). (Copyright by the Mineralogical Society of America.)

prevent the metastable persistence of gypsum under conditions that are far removed from equilibrium.

The gypsum-halite association is therefore an indicator of temperatures $\leq 25°C$. The association of gypsum, anhydrite, and halite casts in units of the McArthur Group in the Northern Territory of Australia (Muir 1979) suggests that temperatures during the deposition of evaporites in this area 1.6 b.y. ago were similar to those along the Trucial Coast of the Persian Gulf today. These temperature estimates are not in conflict with estimates based on the $\delta^{18}O$ value of massive Precambrian cherts (see, for instance, Knauth and Epstein 1976) during mid-Proterozoic time. At present there are no casts of comparable abundance from Archean rock units to serve as a check on the suggestion that ambient surface temperatures were considerably higher more than 2.5 b.y. ago.

All of the data assembled in chapters 5, 6, 7, and 8 seem to indicate that the Precambrian history of the atmosphere and oceans was dominated by the interplay of the gradual decrease in the volcanic and tectonic activity of the Earth and by the influences of biological evolution. The changes in the rate of continental growth and of continental recycling are still a matter for debate (see, for instance, DePaolo 1980, 1981; Armstrong 1981), but it is likely that the rate of crustal recycling has not decreased by more than a factor of ca. 3 during the past 3 b.y. This decrease per se does not seem to have required any great adjustment in the chemistry of the atmosphere and oceans. The effects of biological evolution have been much more marked. In particular, the O_2 content of the atmosphere has increased dramatically, probably in response to a progressive increase in the photosynthetic production of this gas, and because of the requirement that the overall chemistry of the Earth's crust remain constant. More subtle changes almost certainly remain to be discovered. The concentration of N_2O in the present atmosphere is determined largely by the production of this gas during nitrification in soils and in the oceans, particularly in low-O_2 environments. N_2O destruction seems to be occurring primarily via photolysis in the stratosphere (see, for instance, Elkins et al. 1979; Goreau 1981). In the absence of a dense plant cover on land during the Precambrian Era, the terrestrial N_2O production rate was probably lower than today. On the other hand, the lower O_2 concentration in surface seawater might have led to a considerably higher marine N_2O production. Since N_2O can exert a major influence on the ozone content of the atmosphere and has a strong greenhouse coefficient, its concentration in the Precambrian atmosphere is of considerable interest. Other trace gases have similar functions. Our understanding of the production and consumption of trace gases in the present atmosphere has increased remarkably since 1970; these new insights bid fair to yield important advances in the study of atmospheric evolution. It may ultimately be possible to develop models of the Precambrian atmosphere that contain a sufficiently large number of cross-checks so that they can be accepted with confidence.

References

Aller, R. C. 1980. Diagenetic processes near the sediment-water interface of Long Island Sound: I. Decomposition and nutrient element geochemistry (S, N, P). *Advances in Geophys.* 22:237–350.

Aller, R. C., and Benninger, L. K. 1981. Spatial and temporal patterns of dissolved ammonium, manganese, and silica fluxes from bottom sediments of Long Island Sound, U.S.A. *Jour. Marine Res.* 39:295–314.

Appel, P.W.U. 1983. Rare earth elements in the early Archean Isua iron-formation, West Greenland, *Precamb. Res.* 20:243–258.

Armstrong, R. L. 1981. Comment on "Crustal growth and mantle evolution: Inferences from models of element transport and Nd and Sr isotopes," *Geochim. Cosmochim. Acta* 45:1251.

Awramik, S.M., Schopf, J.W., and Walker, M.R. 1983. Filamentous fossil bacteria from the Archean of Western Australia. *Precamb. Res.* 20:357–374.

Badham, J.P.N., and Stanworth, C. W. 1977. Evaporites from the lower Proterozoic of the East Arm, Great Slave Lake. *Nature* 268:516–517.

Barghoorn, E. S., and Tyler, S. A. 1965. Microorganisms from the Gunflint chert. *Science* 147:563–577.

Baur, M. E., Hayes, J. M., Studley, S. A., and Walter, M. R. 1983. Millimeter scale variations of stable isotopic abundances in carbonates from banded iron formations in the Hamersley Group of Western Australia. *Econ. Geol.*, submitted for publication.

Becker, R. H., and Clayton, R. N. 1972. Carbon isotopic evidence for the origin of a banded iron-formation in Western Australia. *Geochim. Cosmochim. Acta* 36:557–595.

Bell, R. T., and Jackson, G. D. 1974. Aphebian halite and sulfate indications in the Belcher Group, Northwest Territories. *Can. J. Earth Sci.* 11:722–728.

Bencini, A., and Turi, A. 1974. Mn distribution in the Mesozoic carbonate rocks from Lima Valley, Northern Apennines. *Jour. Sed. Petrol.* 44:774–782.

Berkner, L. V., and Marshall, L. C. 1965. On the origin and rise of oxygen concentration in the Earth's atmosphere. *J. Atm. Sciences* 22:225–261.

Berner, R.A. 1980. *Early Diagenesis: A Theoretical Approach.* Princeton, N.J.: Princeton University Press.

Blake, A. J., and Carver, J. H. 1977. The evolutionary role of atmospheric ozone. *J. Atm. Sciences* 34:720–728.

Bodine, M. W., Jr., Holland, H. D., and Borcsik, M. 1965. Coprecipitation of manganese and strontium with calcite. In *Problems of Postmagmatic Ore Deposition*, vol. 2, edited by M. Štemprok, 401–406. Proc. Sym. Prague, Geol. Survey of Czechoslovakia.

Borchert, H. 1952. Die Bildungsbedingungen mariner Eisenerzlagerstätten. *Chemie d. Erde* 16:49–74.

———. 1960, Genesis of marine sedimentary iron ores. *Trans. Inst. Min. Met.* 69:261–279.

Braitsch, O. 1962. *Entstehung und Stoffbestand der Salzlagerstätten.* Vol. 3 of *Mineralogie und Petrographie in Einzeldarstellungen.* Berlin: Springer-Verlag.

Brand, U., and Veizer, J. 1980. Chemical diagenesis of a multicomponent carbonate system: 1. Trace elements. *Jour. Sed. Petrol.* 50:1219–1236.

Braterman, P. S., Cairns-Smith, A. G., and Sloper, R. W. 1983. Photooxidation of hydrated Fe^{+2} enhanced by deprotonation: Significance for banded iron formations. *Nature*, in press.

Brewer, P. G. 1975. Minor elements in seawater. In *Chemical Oceanography*, 2d ed. vol. 1, edited by J. P. Riley and G. Skirrow, chap. 7. New York: Academic Press.

Broecker, W. S. 1970. A boundary condition on the evolution of atmospheric oxygen. *J. Geophys. Res.* 75:3553–3557.

Bunt, J. S. 1975. Primary productivity of marine ecosystems. In *Primary Productivity of the Biosphere*, edited by H. Lieth and R. H. Whittaker, 169–183. New York: Springer-Verlag.

Button, A. 1976. Transvaal and Hamersley Basins: Review of basin development and mineral deposits. *Minerals Sci. Engng.* 8:262–293.

Cairns-Smith, A. G. 1978. Precambrian solution photochemistry, inverse segregation, and banded iron formations. *Nature* 276:807–808.

Cameron, E. M. 1982. Sulfate and sulfate reduction in early Precambrian oceans. *Nature* 296:145–148.

Cameron, E. M., and Garrels, R. M. 1980. Geochemical compositions of some Precambrian shales from the Canadian Shield. *Chem. Geol.* 28:181–197.

Cannon, W. F., and Force, E. R. 1983. Potential for high-grade shallow-marine manganese deposits in North America. In *Cameron Symposium on Unconventional Mineral Sources*, edited by P. Shanks.

Chameides, W., and Walker, J.C.G. 1975. Possible variation of ozone in the troposphere during the course of geologic time. *Amer. Jour. Sci.* 275:737–752.

Chapman, D. J., and Schopf, J. W. 1983. Biological and biochemical effects of the development of an aerobic environment. In *Origin and Evolution of Earth's Earliest Biosphere: An Interdisciplinary Study*, edited by J. W. Schopf, chap. 13. Princeton, N.J.: Princeton University Press.

Cloud, P. 1972. A working model of the primitive Earth. *Amer. Jour. Sci.* 272:537–548.

———. 1980. Early biogeochemical systems. In *Biogeochemistry of Ancient and Modern Environments*, edited by P. A. Trudinger, M. A. Walter, and B. J. Ralph, 7–27. New York: Springer-Verlag.

Cohen, Y., Padan, E., and Shilo, M. 1975. Facultative anoxygenic photosynthesis in the cyanobacterium oscillatoria limnetica. *Jour. Bact.* 123:855–861.

Craig, H. 1953. The geochemistry of the stable carbon isotopes, *Geochim. Cosmochim. Acta* 3:53–92.

Davison, W. 1979. Soluble inorganic ferrous complexes in natural waters. *Geochim. Cosmochim. Acta* 43:1693–1696.

Degens, E. T., and Epstein, S. 1962. Relationship between $^{18}O/^{16}O$ ratios in coexisting carbonates, cherts and diatomites. *Bull. Am. Assoc. Petrol. Geologists* 46:534–542.

Deines, P. 1980. The carbon istotopic composition of diamonds: Relationship to diamond shape, color, occurrence and vapor composition. *Geochim. Cosmochim. Acta* 44:943–961.

Demaison, G. J., and Moore, G. T. 1980. Anoxic environments and oil source bed genesis. *Bull. Am. Assoc. Petrol. Geologists* 64:1179–1209.

DePaolo, D. J. 1980. Crustal growth and mantle evolution: Inferences from models of element transport and Nd and Sr isotopes. *Geochim. Cosmochim. Acta* 44:1185–1196.

———. 1981. Crustal growth and mantle evolution: Inferences from models of element transport and Nd and Sr isotopes (Reply to a comment by R. L. Armstrong). *Geochim. Cosmochim. Acta* 45:1253–1254.

Deuser, W. G. 1975. Reducing environments. In *Chemical Oceanography*, 2d ed. vol. 3, edited by J. P. Riley and G. Skirrow, chap. 16. New York: Academic Press.

Dimroth, E. 1973. Geology of the Central Labrador Trough between Lat. 56°30' and the Height-of-Land. Quebec Dep. Nat. Res., Open File Report GM 28619.

———. 1976. Aspects of the sedimentary petrology of cherty iron-formations. In *Handbook of Strata-bound and Strataform Ore Deposits*, vol. 7, edited by K. H. Wolf, chap. 5. New York: Elsevier.

Dimroth, E., and Chauvel, J. 1973. Petrography of the Sokoman Iron Formation in part of the central Labrador Trough, Quebec, Canada. *Bull. Geol. Soc. Amer.* 84:111–134.

Dimroth, E., Côté, R., Provost, G., Rocheleau, M., Tassé, N., and Trudel, P. 1975. Third progress report on the stratigraphy, volcanology, sedimentology, and tectonics of Rouyn-Noranda area. Quebec Dep. Nat. Res., Open File Report DP 300.

Dimroth, E., and Kimberley, M. M. 1976. Precambrian atmospheric oxygen: Evidence in the sedimentary distributions of carbon, sulfur, uranium, and iron. *Can. J. Earth Sci.* 13:1161–1185.

Dimroth, E., and Lichtblau, A. P. 1978. Oxygen in the Archean Ocean: Comparison of ferric oxide crusts on Archean and Cainozoic pillow basalts. *N. Jb. Mineral Abh.* 133:1–22.

Drever, J. I. 1974. Geochemical model for the origin of Precambrian banded iron formations. *Bull. Geol. Soc. Amer.* 85:1099–1106.

Elderfield, H., Hawkesworth, C. J., Greaves, M. J., and Calvert, S. E. 1981. Rare earth element zonation in Pacific ferromanganese nodules. *Geochim. Cosmochim. Acta* 45:1231–1234.

Elkins, J. W., Wofsy, S. C., McElroy, M. B., Kolb, C. E., and Kaplan, W. A. 1979. Aquatic sources and sinks for nitrous oxide. *Nature* 275:602–606.

Emerson, S., Grundmanis, V., and Graham, D. 1982. Carbonate chemistry in marine pore waters: MANOP Sites C and S. *Earth Planet. Sci. Lett.* 61:220–232.

Eugster, H. P., and Chou, I-Ming, 1973. The depositional environments of Precambrian banded iron-formations. *Econ. Geol.* 68:1144–1168.

Ewers, W. E. 1980. Chemical conditions for the precipitation of banded iron-formations. In *Biogeochemistry of Ancient and Modern Environments*, edited by P. A. Trudinger, M. R. Walter, and B. J. Ralph, 83–92. New York: Springer-Verlag.

Fripp, R.E.P., Donnelly, T. H., and Lambert, I. B. 1979. Sulfur isotope results for Archean banded iron-formations, Rhodesia. *Spec. Publ. Geol. Soc. South Afr.* 5:205–208.

Froelich, P. N., Klinkhammer, G. P., Bender, M. L., Luedtke, N. A., Heath, G. R., Cullen, D., Dauphin, P., Hammond, D., Hartman, B., and Maynard, V. 1979. Early oxidation of organic matter in pelagic sediments of the eastern equatorial Atlantic: Suboxic diagenesis. *Geochim. Cosmochim. Acta* 43:1075–1090.

Froelich, P. N., Bender, M. L., Luedtke, N. A., Heath, G. R., and DeVries, T. 1982. The marine phosphorus cycle. *Amer. Jour. Sci.* 282:474–511.

Fryer, B. J. 1977. Rare earth evidence in iron-formations for changing Precambrian oxidation states. *Geochim. Cosmochim. Acta* 41:361–367.

Galimov, E. M. 1978. Problem of the origin of diamonds in the light of new data on the carbon isotopic composition of diamonds. Abs., 13. VII Symp. Isotope Geochem. Moscow, 1978.

Garlick, S., Oren, A., and Padan, E. 1977. Occurrence of facultative anoxygenic photosynthesis among filamentous and unicellular cyanobacteria. *Jour. Bact.* 129:623–629.

Garrels, R. M., Perry, E. A., Jr., and Mackenzie, F. T. 1973. Genesis of Precambrian iron-formations and the development of atmospheric oxygen. *Econ. Geol.* 68:1173–1179.

Gehman, H. M., Jr. 1962. Organic matter in limestones. *Geochim. Cosmochim. Acta* 26:885–897.

Gole, M. J., and Klein, C. 1981. Banded iron-formations through much of Precambrian time. *Jour. Geol.* 89:169–183.

Goodwin, A. M., Monster, J., and Thode, H. G. 1976. Carbon and sulfur isotope abundances in Archean iron-formations and early Precambrian life. *Econ. Geol.* 71:870–891.

Goreau, T. J. 1981. Biogeochemistry of nitrous oxide. Ph.D. diss., Harvard University.

Graf, J. L., Jr. 1978. Rare earth elements, iron formations and seawater. *Geochim. Cosmochim. Acta* 42:1845–1850.

Grandstaff, D. E. 1976. A kinetic study of the dissolution of uraninite. *Econ. Geol.* 71:1493–1506.

————. 1980. Origin of uraniferrous conglomerates at Elliot Lake, Canada and Witwatersrand, South Africa: Implications for oxygen in the Precambrian atmosphere. *Precamb. Res.* 13:1–26.

Gross, G. A. 1965. Geology of iron deposits in Canada. Can. Geol. Surv. Econ. Geol. Rept. 22.

Gross, G.A. 1983. Tectonic systems and the deposition of iron-formations. *Precamb. Res.* 20:171–187.

Harder, H. 1978. Synthesis of iron layer silicate minerals under natural conditions. *Clays and Clay Minerals* 26:65–72.

Hardie, L. A. 1967. The gypsum-anhydrite equilibrium at one atmosphere pressure. *Amer. Mineral.* 52:171–200.

Hartmann, M., Müller, P., Suess, E., and van der Weijden, C. II. 1973. Oxidation of organic matter in recent marine sediments. "Meteor" Forsch-Ergebnisse, series C, no. 12, 74–86.

Hesstvedt, E., Henriksen, S.-E., and Hjartarson, H. 1974. On the development of an aerobic atmosphere: A model experiment. *Geophys. Norv.* 31, no. 1.

Hill, T. P., Werner, M. A., and Horton, M. J. 1967. Chemical composition of sedimentary rocks in Colorado, Kansas, Montana, Nebraska, North Dakota, South Dakota and Wyoming. U.S. Geol. Survey Prof. Paper 561. Washington, D.C.: U.S. Govt. Printing Office.

Hill, T. P., and Werner, M. A. 1972. Chemical composition of sedimentary rocks in Alaska, Idaho, Oregon, and Washington. U.S. Geol. Survey Prof. Paper 771. Washington D.C.: U.S. Govt. Printing Office.

Hoffman, P. 1969. Proterozoic paleocurrents and depositional history of the East Arm fold belt, Great Slave Lake, Northwest Territories. *Can. J. Earth Sci.* 6:441–462.

Hofmann, H. J. 1976. Precambrian microflora, Belcher Islands, Canada: Significance and systematics. *Jour. Paleont.* 50:1040–1073.

Hofmann, H. J., and Schopf, J. W. 1983. Early Proterozoic microfossils. In *Origin and Evolution of Earth's Earliest Biosphere: An Interdisciplinary Study*, edited by J. W. Schopf, chap. 14. Princeton, N.J.: Princeton University Press.

Holeman, J. N. 1968. The sediment yield of major rivers of the world. *Water Resour. Res.* 4:737–747.

Holland, H. D. 1962. Model for the evolution of the Earth's atmosphere. In *Petrologic Studies: A Volume in Honor of A. F. Buddington*, edited by A.E.J. Engel et al., 447–477. Geol. Soc. Amer., Boulder, Colo.

———. 1972. The geologic history of sea water: An attempt to solve the problem. *Geochim. Cosmochim. Acta* 36:637–651.

———. 1973. The oceans: A possible source of iron in iron formations. *Econ. Geol.* 68:1169–1172.

———. 1978. *The Chemistry of the Atmosphere and Oceans.* New York: Wiley.

———. 1979. Metals in black shales: a reassessment. *Econ. Geol.* 74:1676–1680.

Ichikuni, M. 1973. Partition of strontium between calcite and solution: Effect of substitution by manganese. *Chem. Geol.* 11:315–319.

Jacobs, L., and Emerson, S. 1982. Trace metal solubility in an anoxic fjord. *Earth Planet. Sci. Lett.* 60:237–252.

Jahnke, R., Heggie, D., Emerson, S., and Grundmanis, V. 1982. Pore waters of the central Pacific Ocean: Nutrient results. *Earth Planet. Sci. Lett.* 61:233–256.

James, H. L. 1966. Chemistry of the iron-rich sedimentary rocks. In *Data of Geochemistry*, 6th ed., edited by M. Fleischer, chap. W. U.S. Geol. Survey Prof. Paper 440-W. Washington, D.C.: U.S. Govt. Printing Office.

James, H. L., and Sims, P. K. 1973. Precambrian iron-formations of the world. *Econ. Geol.* 68:913–914.

James, H. L., and Trendall, A. F. 1982. Banded iron-formation: Distribution in time and paleoenvironmental significance. In *Mineral Deposits and the Evolution of the Biosphere*, edited by H. D. Holland and M. Schidlowski, 199–217. New York: Springer-Verlag.

Johnson, K. S. 1982. Solubility of rhodochrosite ($MnCO_3$) in water and seawater. *Geochim. Cosmochim. Acta* 46:1805–1809.

Junge, C. E., Schidlowski, M., Eichmann, R., and Pietrek, H. 1975. Model calculations for the terrestrial carbon cycle: Carbon isotope geochemistry and evolution of photosynthetic oxygen. *J. Geophys, Res.* 80:4542–4552.

Keith, M. L., and Weber, J. N. 1964. Carbon and oxygen isotopic composition of selected limestones and fossils. *Geochim. Cosmochim. Acta* 28:1787–1816.

Kimberley, M. M. 1974. Origin of iron ore by diagenetic replacement of calcareous oolite. *Nature* 250:319–320.

———. 1978. Paleoenvironmental classification of iron formations. *Econ. Geol.* 73:215–229.

Kinsman, D.J.J. 1966. Gypsum and anhydrite of Recent age, Trucial Coast, Persian Gulf. In *Second Symposium on Salt*, vol. 1, edited by J. L. Rau, 302–326. The Northern Ohio Geol. Soc., Cleveland.

Klein, C., Jr. 1973. Changes in mineral assemblages with metamorphism of some banded Precambrian iron-formations. *Econ. Geol.* 68:1075–1088.

———. 1974. Greenalite, stilpnomelane, minnesotaite, crocidolite and carbonates in a very low-grade metamorphic Precambrian iron-formation. *Canadian Mineral.* 12:475–498.

Klein, C., Jr., and Fink, R. P. 1976. Petrology of the Sokoman Iron Formation in the Howells River area at the western edge of the Labrador Trough. *Econ. Geol.* 71:453–487.

Klein, C., and Gole, M. J. 1981. Mineralogy and petrology of parts of the Marra Mamba Iron Formation, Hamersley Basin, Western Australia. *Amer. Mineral.* 66:507–525.

Klinkhammer, G. P. 1980. Early diagenesis in sediments from the eastern equatorial Pacific: II. Pore water metal results. *Earth Planet. Sci. Lett.* 49:81–101.

Klinkhammer, G. P., and Bender, M. L. 1980. The distribution of manganese in the Pacific Ocean. *Earth Planet. Sci. Lett.* 46:361–384.

Klinkhammer, G. P., Heggie, D.T., and Graham, D. W. 1982. Metal diagenesis in oxic marine sediments. *Earth Planet. Sci. Lett.* 61:211–219.

Knauth, L. P., and Epstein, S. 1976. Hydrogen and oxygen isotope ratios in nodular and bedded cherts. *Geochim. Cosmochim. Acta* 40:1095–1108.

Knoll, A. H. 1979. Archean photoautotrophy: Some alternatives and limits. *Origins of Life* 9:313–327.

Knoll, A. H., and Barghoorn, E. S. 1977. Archean microfossils showing cell division from the Swaziland System of South Africa. *Science* 198: 396–398.

Knoll, A. H., Barghoorn, E. S., and Awramik, S. M. 1978. New microorganisms from the Aphebian Gunflint Iron Formation, Ontario. *Jour. Paleont.* 52:976–992.

Koljonen, T. 1975. Behavior of selenium and sulfur in Svecokarelian sulfide-rich rocks. *Bull. Geo. Soc. Finland* 47:25–31.

Krumbein, W. E., Cohen, Y., and Shilo, M. 1977. Solar Lake (Sinai): 4. Stromatolitic cyanobacterial mats. *Limnol. Oceanogr.* 22:635–656.

Krumbein, W. E., Buchholz, H., Franke, P., Giani, D., Giele, C., and Wonneberger, K. 1979. O_2 and H_2S coexistence in stromatolites: A model for the origin of mineralogical laminations in stromatolites and banded iron formations. *Naturwiss.* 66:381–389.

Laajoki, K. 1975. Rare-earth elements in Precambrian iron formations in Väyrylänkylä, South Puolanka area, Finland. *Bull. Geol. Soc. Finland* 47:93–107.

Laajoki, K., and Saikkonen, R. 1977. On the geology and geochemistry of the Precambrian iron formations in Väyrylänkylä, south Puolanka area, Finland. Bulletin 292, Geological Survey of Finland.

LaBerge, G. L. 1973. Possible biological origin of Precambrian iron-formation. *Econ. Geol.* 68:1098–1109.

Landing, W. M., and Bruland, K. W. 1980. Manganese in the North Pacific. *Earth Planet. Sci. Lett.* 49:45–56.

Lepp, H., and Goldich, S. S. 1964. Origin of Precambrian iron formations. *Econ. Geol.* 59:1025–1060.

Lerman, A., Mackenzie, F. T., and Garrels, R. M. 1975. Modeling of geochemical cycles: Phosphorus as an example. In Mem. 142 of the Geol. Soc. Amer., 205–218.

Levine, J. S., Boughner, R. E., and Smith, K. A. 1980. Ozone, ultraviolet flux and temperature of the paleoatmosphere. *Origins of Life* 10: 199–213.

Leytes, A. M., Miller, A. D., and Aleksandrova, L. B. 1977. Rhenium content of Archean rocks of the Aldan Shield as a possible indicator of sedimentation conditions in the early Precambrian. *Doklady Akad. Nauk SSSR* 237:235–237.

McCarthy, J. J., and Goldman, J. C. 1979. Nitrogenous nutrition of marine phytoplankton in the nutrient-depleted water. *Science* 203:670–672.

Margulis, L., Walker, J.C.G., and Rambler, M. 1976. Reassessment of roles of oxygen and ultraviolet light in Precambrian evolution. *Nature* 264:620–624.

Marjoribanks, R. W., and Black, L. P. 1974. Geology and geochronology of the Arunta Complex north of Ormiston Gorge, central Australia. *J. Geol. Soc. Aust.* 21:291–299.

Marmo. V. 1960. On the sulfide and sulfide-graphite schists of Finland. Bull. 190 de la Comm. Géol. de Finlande, Helsinki.

Maxwell, J. A., Dawson, K. R., Tomilson, M. E., Pocock, D.M.E., and Tetreault, D. 1965. Chemical analyses of Canadian rocks, minerals, and ores compiled from the records of the Geological Survey of Canada for the period 1846–1955. *Canada Geol. Surv. Bull.* 115.

Mel'nik, Iu. P. 1973. *Physiochemical Conditions of Formation of Ferruginous Quartzites.* Kiev Akad. Nauk Ukrainskoi SSR Institut Geokhimii i Fiziki Minerolov (Izdate'stro "Naukova Dumka"), in Russian.

Michard, G. 1968. Coprecipitation de l'ion manganeaux avec le carbonate de calcium. *C. R. Acad. Sci. Paris, Ser. D* 267:1685–1688.

Monster, J., Appel, P.W.U., Thode, H. G., Schidlowski, M., Carmichael, C. M., and Bridgwater, D. 1979. Sulfur isotope studies in early Archean sediments from Isua, West Greenland: Implications for the antiquity of bacterial reduction, *Geochim. Cosmochim. Acta* 43:405–413.

Muir, M. D. 1979. A sabkha model for the deposition of part of the Proterozoic McArthur Group of the Northern Territory, and its implications for mineralization. *BMR Jour. Austr. Geol. and Geophys.* 4:149–162.

Münnich, K. O. 1963. Der Kreislauf des Radiokohlenstoffs in der Natur. *Naturwiss.* 50 (6):211–218.

Münnich, K. O., and Roether, W. 1967. Transfer of bomb ^{14}C and tritium from the atmosphere to the ocean: Internal mixing of the ocean on the basis of tritium and ^{14}C profiles. In *Radioactive Dating and Methods of Low-Level Counting,* 93–104. Int. Atom. Energy Agency, Vienna.

Oehler, D. Z., Schopf, J. W., and Kvenvolden, K. A. 1972. Carbon isotopic studies of organic matter in Precambrian rocks. *Science* 175:1246–1248.

Ohmoto, H., and Kerrick, D. 1977. Devolatilization equilibria in graphitic systems. *Amer. Jour. Sci.* 277:1013–1044.

Ohmoto, H., and Rye, R. O. 1979. Isotopes of sulfur and carbon. In *Geochemistry of Hydrothermal Ore Deposits,* edited by H. L. Barnes, 509–567. New York: Wiley.

Peltola, E. 1960. On the black schists of the Outokumpu region in eastern Finland. Bull. 192 de la Comm. Géol. de Finlande, Helsinki.

———. 1968. On some geochemical features in the black schists of the Outokumpu area, Finland. *Bull. Géol. Soc. Finland* 40:39–50.

Peng, T.-H., Broecker, W. S., Mathieu, G. G., Li, Y.-H., and Bainbridge, A. E. 1979. Radon evasion rates in the Atlantic and Pacific Oceans as determined during the Geosecs Program. *J. Geophys. Res.* 84: 2471–2486.

Perry, E. C., Jr., and Tan, F. C. 1973. Significance of carbon isotopic variations in carbonates from the Biwabik Iron Formation, Minnesota. In *Genesis of Precambrian Iron and Manganese Deposits*, 299–305. Proc. Kiev Symp., 1970. Paris: UNESCO.

Piper, D. Z. 1974. Rare earth elements in the sedimentary cycle: A summary. *Chem. Geol.* 14:285–304.

Prasada Rao, C., and Naqvi, I. H.·1977. Petrography, geochemistry and factor analysis of a lower Ordovician subsurface sequence, Tasmania, Australia. *Jour. Sed. Petrol.* 47:1036–1055.

Ratner, M. I., and Walker, J.C.G. 1972. Atmospheric ozone and the history of life. *Jour. Atm. Sci.* 29:803–808.

Ripley, E. M., and Nicol, D. L. 1981. Sulfur isotopic studies of Archean slate and graywacke from northern Minnesota: Evidence for the existence of sulfate reducing bacteria. *Geochim. Cosmochim. Acta* 45: 839–846.

Ronov, A. B. 1968. Probable changes in the composition of sea water during the course of geologic time. *Sedimentology* 10:25–43.

Rutten, M. G. 1970. The history of atmospheric oxygen. *Space Life Sci.* 2:5–17.

Sangster, D. F. 1976. Sulfur and lead isotopes in strata-bound deposits. In *Handbook of Strata-Bound and Stratiform Ore Deposits*, vol. 2, edited by K. H. Wolf, chap. 8. New York: Elsevier.

Schidlowski, M. 1979. Antiquity and evolutionary status of bacterial sulfate reduction: Sulfur isotope evidence. *Origins of Life* 9:299–311.

———. 1982. Content and isotopic composition of reduced carbon in sediments. In *Mineral Deposits and the Evolution of the Biosphere*, edited by H. D. Holland and M. Schidlowski, 103–122. New York: Springer-Verlag.

———. 1983. $^{13}C/^{12}C$ ratios as indicators of biogenicity. In *Biological Markers*, edited by R. B. Johns. Amsterdam: Elsevier.

Schidlowski, M., Eichmann, R., and Junge, C. E. 1975. Precambrian sedimentary carbonates: Carbon and oxygen isotope geochemistry and implications for the terrestrial oxygen budget. *Precamb. Res.* 2:1–69.

———.1976. Carbon isotope geochemistry of the Precambrian Lomagundi carbonate province, Rhodesia. *Geochim. Cosmochim. Acta* 40: 449–455.

Schidlowski, M., Appel, P.W.U., Eichmann, R., and Junge, C. E. 1979. Carbon isotope geochemistry of the 3.7×10^9-yr-old Isua sediments,

West Greenland: Implications for the Archaean carbon and oxygen cycles. *Geochim. Cosmochim. Acta* 43:189–199.

Schindler, D. W. 1977. Evolution of phosphorus limitation in lakes. *Science* 195:260–262.

Schopf, J. W., and Walter, M. R. 1983. Archean microfossils: New evidence of ancient microbes. In *Origin and Evolution of Earth's Earliest Biosphere: An Interdisciplinary Study*, edited by J. W. Schopf, chap. 9. Princeton, N.J.: Princeton University Press.

Skyring, G. W., and Donnelly, T. H. 1982. Precambrian sulfur isotopes and a possible role for sulfite in the evolution of biological sulfate reduction. *Precamb. Res.* 17:41–61.

Spencer, D. W., and Brewer, P. G. 1971. Vertical advection, diffusion and redox potentials as controls on the distribution of manganese and other trace metals dissolved in waters of the Black Sea. *J. Geophys. Res.* 76:5877–5892.

Spencer, D. W., Bacon, M. P., and Brewer, P. G. 1981. Models of the distribution of ^{210}Pb in a section across the north equatorial Atlantic Ocean. *J. Mar. Res.* 39:119–138.

Stewart, A. J. 1979. A barred-basin marine evaporite in the Upper Proterozoic of the Amadeus Basin, central Australia. *Sedimentology* 26:33–62.

Stumm, W., and Morgan, J. J. 1981. *Aquatic Chemistry*, 2d ed., New York: Wiley-Interscience.

Sunda, W. G., Huntsman, S. A., and Harvey, G. R. 1983. Photoreduction of manganese oxides in seawater and its geochemical and biological implications. *Nature* 301:234–236.

Thode, H. G., and Goodwin, A. M. 1983. Further sulfur and carbon isotope studies of late Archean iron-formations of the Canadian Shield and the rise of sulfate-reducing bacteria. *Precamb. Res.* 20:337–356.

Towe, K. M. 1978. Early Precambrian oxygen: A case against photosynthesis. *Nature* 274:657–661.

Trask, P. D., and Patnode, H. W. 1942. *Source Beds of Petroleum*. Tulsa, Oklahoma: American Association of Petroleum Geologists.

Trefry, J. H., and Presley, B. J. 1982. Manganese fluxes from Mississippi Delta sediments. *Geochim. Cosmochim. Acta* 46:1715–1726.

Trendall, A. F. 1972. Revolution in Earth history. *J. Geol. Soc. Aust.* 19:287–311.

———. 1973. Precambrian iron-formations of Australia. *Econ. Geol.* 68:1023–1034.

Trendall, A. F., and Blockley, J. C. 1970. The iron formations of the Precambrian Hamersley Group, Western Australia, with special reference to the associated crocidolite. Bull. 119 of the Geol. Surv. of West. Austr.

Trüper, H. G. 1982. Microbial processes in the sulfur cycle through time. In *Mineral Deposits and the Evolution of the Biosphere*, edited by H. D. Holland and M. Schidlowski, 5–30. New York: Springer-Verlag.

Tyler, S. A., Barghoorn, E. S., and Barrett, E. P. 1957. Anthracitic coal from Precambrian Upper Huronian black shale of the Iron River District, Northern Michigan. *Bull. Geol. Soc. Amer.* 68:1293–1304.

Urey, H. C. 1959. Primitive planetary atmospheres and the origin of life. In *The Origin of Life on the Earth*, vol. 1, 16–22. Proc. of First International Symposium of the International Union of Biochemistry. London: Pergamon Press.

Veizer, J. 1971. Chemical and strontium isotopic evolution of sedimentary carbonate rocks in geologic history. Ph.D. diss., Australian National University, Canberra, Australia.

———. 1978. Secular variations in the composition of sedimentary carbonate rocks: II. Fe, Mn, Ca, Mg, Si and minor constituents. *Precamb. Res.* 6:381–413.

Veizer. J., and Hoefs, J. 1976. The nature of $^{18}O/^{16}O$ and $^{13}C/^{12}C$ secular trends in sedimentary carbonate rocks, *Geochim. Cosmochim. Acta* 40:1387–1395.

Veizer, J., and Garrett, D. E. 1978. Secular variations in the composition of sedimentary carbonate rocks: I. Alkali metals. *Precamb. Res.* 6:367–380.

Veizer, J., Holser, W. T., and Wilgus, C. K. 1980. Correlation of $^{13}C/^{12}C$ and $^{34}S/^{32}S$ secular variations. *Geochim. Cosmochim. Acta* 44:579–587.

Veizer, J., Compston, W., Hoefs, J., and Nielsen, H. 1980. Mantle buffering of the early oceans (abstract). Int'l Geol. Congress, Abstract no. 26, vol. 2, 619.

Vine, J. D., and Tourtelot, E. B. 1970. Geochemistry of black shale deposits: A summary report. *Econ. Geol.* 65:253–272.

Vinogradov, A. P., Kropotova, O. I., Orlov, Yu. L., and Grinenko, V. A. 1966. Isotopic composition of diamond and carbonado crystals. *Geochemistry International* 3:1123–1125.

Visser, J.N.J. 1964. *Analyses of Rocks, Minerals, and Ores.* Handbook 5, Geol. Surv. of South Africa.

Walker, R. N., Muir, M. D., Diver, W. L., Williams, N., and Wilkins, N. 1977. Evidence of major sulfate evaporite deposits in the Proterozoic McArthur Group, Northern Territory, Australia. *Nature* 265:526–529.

Walsh, J. J. 1977. A biological sketchbook for an eastern boundary current. In *The Sea*, vol. 6, edited by E. D. Goldberg, I. N. McCave, J. J. O'Brien, and J. H. Steele, chap. 24. New York: Wiley.

Waterbury, J. B., Watson, S. W., Guilland, R.R.L., and Brand, L. E. 1979. Widespread occurrence of a unicellular, marine, planktonic cyanobacterium. *Nature* 277:293–294.

Weber, J. N. 1964. Trace element composition of dolostones and dolomites and its bearing on the dolomite problem. *Geochim. Cosmochim. Acta* 28:1817–1868.

Wells, A. T., Forman, D. J., Ranford, L. C., and Cook, R. J. 1970. Geology of the Amadeus Basin, Central Australia. Aust. Bur. Miner. Resour. Geol. Geophys. Bull. 100.

Wickman, F. E. 1956. The cycle of carbon and stable carbon isotopes. *Geochim. Cosmochim. Acta* 9:136–153.

Wolk, C. P. 1973. Physiology and cytological chemistry of blue-green algae. *Bacteriol. Rev.* 37:32–101.

Yung, Y. L., and McElroy, M. B. 1979. Fixation of nitrogen in a prebiotic atmosphere. *Science* 203:1002–1004.

Zelenov, K. K. 1960. Transportation and accumulation of iron and aluminum in volcanic provinces of the Pacific. *Akad. Nauk SSSR Ser. Geol. 1960*, 8, 47–59.

9

The Composition of the Atmosphere and Oceans during the Phanerozoic Eon

1. Introduction

The four previous chapters were devoted largely to the compositional evolution of the atmosphere and of seawater during the period between ca. 3.8 b.y. ago and the end of the Proterozoic Eon ca. 0.6 b.y. ago. A number of observations were made in the course of these chapters concerning the history of the atmosphere and oceans during the Phanerozoic Eon, and it seems worthwhile to summarize these here. The near-constancy of the fraction of $(MgO + CaO)_{sil. corr.}$ in sedimentary rocks during much of geologic time was described in chapter 5; this near-constancy implies that the rate of supply of CO_2 by the degassing of the crust and mantle has been matched rather closely by the rate of weathering of igneous rocks during the past 3 b.y. This match does not define the CO_2 pressure in the atmosphere during the past 600 m.y.; it only requires that P_{CO_2} was such that approximately the same fraction of the $(MgO + CaO)$ content of igneous and high grade metamorphic rocks exposed to weathering was converted to carbonate minerals during this, as during earlier periods of Earth history.

The mineralogy and chemical composition of carbonates has changed considerably during the Phanerozoic Eon. The dolomite/limestone ratio is dramatically smaller in carbonates of the last 200 m.y. than in those of the preceding 400 m.y.; this difference is due in part to the dolomitization of limestones long after deposition, but much of the change in the dolomite/limestone ratio is almost certainly due to a decrease in the degree of penecontemporaneous dolomite formation during the last 200 m.y. It was argued in chapter 6 that this decrease is largely the result of biological changes that increased the difficulty of dolomitizing $CaCO_3$ sediments during the Mesozoic and Cenozoic eras. As a result, the Mg/Ca ratio in seawater has probably increased during the past 200 m.y., Mg-Ca exchange during seawater cycling through mid-ocean ridge basalts has increased, and much of the Mg^{+2} that was previously removed from seawater during dolomite formation has been removed at mid-ocean ridges. There seem to have been no major changes in the ratio of the rate of seawater cycling through MORs to the rate of sedimentation during the Phanerozoic Eon. This was shown most clearly in chapter 6 by the rough constancy of the lithium excess in sedimentary rocks. The low Mn and Fe content of most Phanerozoic marine carbonates (see chapter 8) and the absence of large accumulations of detrital uraninite or of other minerals that are readily oxidized (see chapter 7) is surely in agreement with the high oxygen requirement of Phanerozoic animals.

The sum of all these observations does not go very far toward defining the chemical composition of seawater and of the atmosphere during the Phanerozoic Eon. Fortunately, a variety of other data are available for this part of Earth history, and these allow a much more satisfactory definition of the composition of the atmosphere and oceans during the Phanerozoic Eon than during the Precambrian Era. Marine evaporites are particularly useful for defining the composition of seawater in the past. The next two sections of this chapter are devoted to these sediments.

2. Marine Evaporites and the Composition of Seawater

In principle it should be possible to define the relative proportions of all cations and anions in seawater of any period of Earth history by determining the chemical composition of a contemporary marine evaporite. This requires, however, that marine evaporites are residues of the complete evaporation of batches of seawater, and this is seldom, if ever, the case. During the formation of marine evaporites, periods of net evaporation alternate with periods of net dilution. Evaporation rarely goes to completion, the processes of brine addition and removal are usually complex, chemical equilibrium is not always maintained, and postdepositional reactions frequently modify the primary mineralogy and chemical composition of marine evaporites.

Nevertheless, the mineralogy of marine evaporites sets rather useful limits on possible excursions in the composition of seawater since late Precambrian time. During the last ca. 900 m.y., marine evaporites have normally begun with a significant carbonate section; this has been followed by a gypsum-anhydrite section and a halite section. The sequence is illustrated in figure 9.1. Differences between the mineralogy of marine evaporites tend to be pronounced only during the crystallization of sulfates and chlorides from the final bitterns (see, for instance, Borchert and Muir 1964; Braitsch 1971). The evaporation of present-day seawater yields a sequence of minerals very similar to that of most marine evaporites. Understandably, students of marine evaporites have tended to show little interest in possible variations in the composition of seawater with time, since no changes in seawater composition are needed to explain the mineralogy of marine evaporites.

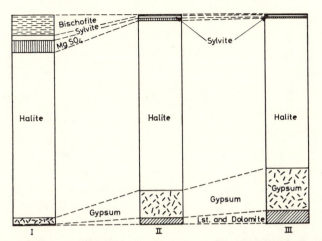

FIGURE 9.1. Comparative precipitation profiles from I, the experimental evaporation of seawater; II, Zechstein evaporites (Permian); and III, the average of numerous other marine salt deposits (Borchert and Muir 1964). (From *Salt Deposits, the Origin, Metamorphism and Deformation of Evaporites,* by Herman Borchert and Richard Muir, copyright © 1964, D. Van Nostrand Co., Ltd. Reprinted by permission of Wadsworth Publishing Co., Belmont, Calif. 94002.)

    Historians of seawater who wish to use the mineralogy of marine evaporites must attempt to solve the inverse problem; they must define the limits on possible excursions in the composition of seawater that are set by the condition that the deposition sequence of the major evaporite minerals has been essentially constant. The problem for the major constituents of seawater can be stated rather easily in geometric terms. Let us imagine a hyperspace in which the concentration of the four major cations (Na^+, Mg^{+2}, Ca^{+2}, K^+) and of the three major anions (Cl^-, SO_4^{-2}, HCO_3^-) are plotted along each of seven axes. The present composition of seawater is a point in this hyperspace. Neighboring points define the composition of solutions that yield the same sequence of minerals on evaporation. However, solutions whose composition plots some distance away from present-day seawater are apt to yield different sequences of evaporite minerals. What is needed, then, is a mapping of the surfaces that bound the volume containing all seawater compositions that yield the present sequence of minerals on evaporation. To simplify matters, we will consider first a two-dimensional section through the seven-dimensional volume. $m_{Ca^{+2}}$ and $m_{SO_4^{-2}}$ have been selected as axes for this section, rather than log $m_{Ca^{+2}}$ and log $m_{HCO_3^-}$, which were used as axes in a previous treatment of this problem (Holland 1972). The proposed section is shown in figure 9.2. The composition of present-day

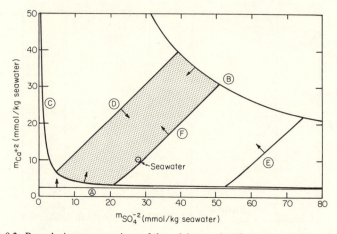

FIGURE 9.2. Boundaries on excursions of the calcium and sulfate concentration of seawater imposed by the sequence of minerals in marine evaporites.

seawater plots within a shaded area, which will be shown to contain the most likely values of the calcium and sulfate concentration of Phanerozoic seawater.

THE CARBONATE MINERALS

Surface seawater today is supersaturated with respect to calcite; this condition has probably persisted throughout Phanerozoic time. The degree of this supersaturation has almost certainly varied, but it is likely that the variations have not been large. The equilibrium constant for the reaction,

$$CaCO_3 + CO_2 + H_2O \rightleftharpoons Ca^{+2} + 2HCO_3^- \tag{9.1}$$

is

$$K_{9.1} = \frac{m_{Ca^{+2}} m_{HCO_3^-}^2 \gamma_{Ca^{+2}} \gamma_{HCO_3^-}^2}{P_{CO_2}}.$$

If we let

$$K'_{9.1} = \frac{K_{9.1}}{(\gamma_{Ca^{+2}})(\gamma_{HCO_3^-}^2)}, \tag{9.2}$$

then today

$$K'_{9.1} = \frac{(10.2 \times 10^{-3})(2.4 \times 10^{-3})^2}{3.3 \times 10^{-4}} \frac{(mol/kg)^3}{atm}, \tag{9.3}$$

where the units of concentration are moles per kg of seawater. It follows that at constant P_{CO_2}, temperature, and degree of supersaturation,

$$m_{Ca^{+2}} \cdot m^2_{HCO_3^-} \cong 58.8 \ (mmol/kg)^3. \tag{9.4}$$

The concentration of Ca^{+2} in present-day seawater (10.2 mmol/kg) is so much greater than that of HCO_3^- (2.4 mmol/kg) that during the early stages of the evaporation of seawater nearly all of the HCO_3^- is removed by $CaCO_3$ precipitation, whereas only ca. 10% of the Ca^{+2} is precipitated. This process insures that all of the significant carbonates in recent marine evaporites are precipitated early in the evaporation sequence, and that sodium carbonate minerals, such as trona, are absent or extremely rare. In this respect marine evaporites differ strongly from many nonmarine evaporites, in which the presence of sodium carbonates is the rule rather than the exception (see, for instance, Eugster and Jones 1979).

Sodium carbonates and bicarbonates are absent not only from recent marine evaporites but from all Phanerozoic marine evaporites. This absence implies that in Phanerozoic seawater $m_{Ca^{+2}}$ has always exceeded $m_{HCO_3^-}/2$. If P_{CO_2}, T, and the degree of supersaturation have remained approximately constant, equation 9.4 can be used to show that $m_{Ca^{+2}}$ has always been in excess of ca. 2.45 mmol/kg s.w., and $m_{HCO_3^-}$ less than 4.90 mmol/kg s.w. I will argue later in this chapter that P_{CO_2} was higher during the earlier part of the Phanerozoic than it is at present. However, the effect of changes in P_{CO_2} on the minimum concentration of Ca^{+2} in seawater is quite modest. An increase of P_{CO_2} by a factor of 10 raises the minimum value of $m_{Ca^{+2}}$ to only 5.2 mmol/kg seawater.

One could argue that these calculated minimum values are too high, because additional HCO_3^- can be removed from seawater as a consequence of dolomitization. If evaporation takes place in an area of prior carbonate accumulation, Ca^{+2} can be released by dolomitization via the reaction,

$$2CaCO_3 + Mg^{+2} \longrightarrow CaMg(CO_3)_2 + Ca^{+2}. \tag{9.5}$$

Ca^{+2} released in this fashion can then be precipitated as $CaCO_3$. Typical marine evaporite sequences are found, however, even in areas where prior limestone accumulation and dolomitization did not take place. Furthermore, the large extent of gypsum and/or anhydrite deposition in typical marine evaporites indicates that a good deal of Ca^{+2} has always remained in evaporating seawater after the end of carbonate precipitation. The minimum Ca^{+2} concentration proposed above and indicated by line A in figure 9.2 is therefore almost certainly a conservatively low figure for the minimum concentration of Ca^{+2} in seawater during the Phanerozoic Eon.

GYPSUM AND ANHYDRITE PRECIPITATION

A strong upper limit for the Ca^{+2} and SO_4^{-2} concentration in Phanerozoic seawater is set by the solubility of gypsum. Although surface seawater is and has probably been supersaturated with respect to calcite and aragonite during the entire Phanerozoic Eon, it is considerably undersaturated with respect to gypsum. Apparently this condition has prevailed throughout this Eon. Gypsum and anhydrite have been restricted to evaporites, because their precipitation in evaporite basins and in mid-ocean ridges, together with the removal of sulfur by sulfate reduction followed by pyrite precipitation, has always been able to dispose of the river flux of sulfate to the oceans.

The solubility of gypsum in seawater has been studied repeatedly (see, for instance, Posnjak 1940) and has been summarized by K. S. Johnson (personal communication, 1980). Johnson's (1980) recommended values for the solubility of gypsum are shown in table 9.1; the recommended values for the product $m_{Ca^{+2}} \cdot m_{SO_4^{-2}}$ at saturation are shown in table 9.2. The solubility of gypsum in seawater passes through a maximum with increasing salinity and temperature. At 298.2°K and a salinity of 35 gm/kg solution,

$$m_{Ca^{+2}} \cdot m_{SO_4^{-2}} \cong 10^{-2.77} \ (mol/kg)^2$$
$$\cong 1700 \ (mmol/kg)^2.$$

The use of the solubility product of gypsum in present-day seawater to calculate the concentration of Ca^{+2} in seawater saturated with respect to gypsum as a function of the SO_4^{-2} concentration is complicated by the presence of significant quantities of the complexes $CaSO_4^0$, $NaSO_4^-$, and

TABLE 9.1.

Solubility of gypsum ($CaSO_4 \cdot 2H_2O$) in seawater at 1 atm; note that concentrations are in units of molality (mmol/kg H_2O).

Salinity	Temperature (°K)				
Gm/kg H_2O	273.2	298.2	323.2	348.2	368.2
15	22.3	26.6	27.3	26.1	24.9
35	28.6	32.8	33.6	33.0	32.6
50	31.5	35.4	36.3	36.3	36.6
75	33.8	37.3	38.5	39.3	40.7
100	33.8	36.9	38.4	40.0	42.2
150	29.0	31.7	33.7	36.4	39.5
200	21.2	23.7	26.0	28.9	32.1

TABLE 9.2.

Solubility product of gypsum ($CaSO_4 \cdot 2H_2O$) in seawater at 1 atm; note that concentrations are in units of molality.

Salinity	$-log(m_{Ca^{+2}} \cdot m_{SO_4^{-2}})(mol/kg\ H_2O)^2$ Temperature (°K)				
Gm/kg H_2O	273.2	298.2	323.2	348.2	368.2
15	3.17	3.04	3.02	3.05	3.09
35	2.87	2.77	2.76	2.77	2.78
50	2.73	2.65	2.64	2.64	2.63
75	2.59	2.53	2.51	2.50	2.48
100	2.51	2.46	2.44	2.41	2.38
150	2.46	2.41	2.38	2.34	2.29
200	2.50	2.44	2.40	2.34	2.29

$MgSO_4^0$ (see, for instance, Garrels and Thompson 1962). If the subscript f denotes the concentration of free Ca^{+2} and SO_4^{-2} ions in solution, then

$$m_{Ca^{+2}} \cong {}_f m_{Ca^{+2}} + m_{CaSO_4^0} \qquad (9.6)$$

and

$$m_{SO_4^{-2}} = {}_f m_{SO_4^{-2}} + m_{CaSO_4^0} + m_{MgSO_4^0} + m_{NaSO_4^-}. \qquad (9.7)$$

At saturation with respect to gypsum at 25°C, the concentration of the complex $CaSO_4^0$ is $10^{-2.14}$ moles/kg H_2O at an ionic strength of 0.477 mol/kg H_2O (Kalyanaraman, Yeatts, and Marshall 1973). The concentration of the neutral complex is not a strong function of the ionic strength of the solutions, and must have nearly the same value in seawater saturated with respect to gypsum at 25°C.

The concentration of the complex $MgSO_4^0$ is determined by the equilibrium constant of the equation,

$$ {}_f Mg^{+2} + {}_f SO_4^{-2} \Longleftrightarrow MgSO_4^0 \qquad (9.8)$$

$$K'_{9.8} = \frac{m_{MgSO_4^0}}{{}_f m_{Mg^{+2}} \cdot {}_f m_{SO_4^{-2}}},$$

and the concentration of the complex $NaSO_4^-$ by the equilibrium constant of the equation,

$$ {}_f Na^+ + {}_f SO_4^{-2} \Longleftrightarrow NaSO_4^- \qquad (9.9)$$

$$K'_{9.9} = \frac{m_{NaSO_4^-}}{{}_f m_{Na^+} \cdot {}_f m_{SO_4^{-2}}}.$$

Thus,

$$m_{SO_4^{-2}} = {}_{f}m_{SO_4^{-2}}[1 + K'_{9.8}\,{}_{f}m_{Mg^{+2}} + K'_{9.9}\,{}_{f}m_{Na^+}] + 7.2 \text{ mmol/kg } H_2O. \tag{9.10}$$

In present-day seawater approximately 56% of the sulfate is present as $MgSO_4^0 + NaSO_4^-$ and rather little as $CaSO_4^0$ (Holland 1978, p. 162). It follows that in seawater of the same Mg^{+2} and Na^+ concentration but saturated with respect to gypsum,

$$m_{SO_4^{-2}} \cong 2.27\,{}_{f}m_{SO_4^{-2}} + 7.2. \tag{9.11}$$

The solubility product of gypsum expresssed in terms of the free ions is therefore

$$\frac{[m_{Ca^{+2}} - 7.2][m_{SO_4^{-2}} - 7.2]}{2.27} = K'_G. \tag{9.12}$$

If gypsum is added to present-day seawater at 25°C, saturation is reached when the Ca^{+2} concentration is 33.0 mmol/kg H_2O and the SO_4^{-2} concentration is 50.5 mmol/kg H_2O. Thus in present-day seawater that is saturated with respect to gypsum at 25°C,

$$[m_{Ca^{+2}} - 7.2][m_{SO_4^{-2}} - 7.2] \cong 1120 \left(\frac{\text{mmol}}{\text{kg } H_2O}\right)^2. \tag{9.13}$$

This equation has been used to calculate the position of the gypsum boundary, curve B, in figure 9.2.

A very strong lower limit for the Ca^{+2} and the SO_4^{-2} concentration in Phanerozoic seawater is set by the observation that gypsum and/or anhydrite have always precipitated from evaporating seawater prior to halite. The data in table 9.1 and 9.2 can be used to calculate the value of the product $m_{Ca^{+2}} \cdot m_{SO_4^{-2}}$ in seawater that reaches saturation at various stages of evaporation. The results are shown as squares in figure 9.3. Seawater in which

$$m_{Ca^{+2}} \cdot m_{SO_4^{-2}} = 1700 \, (\text{mmol/kg } H_2O)^2$$

is, of course, saturated with respect to gypsum without evaporative concentration. Progressively lower values of the product $m_{Ca^{+2}} \cdot m_{SO_4^{-2}}$ are required to reach saturation with respect to gypsum at progressively higher degrees of concentration. The measure of the degree of concentration used in figure 9.3 is that used by Eugster, Harvie, and Weare (1980), and is defined as the ratio of the original quantity of water in a given weight of seawater to the quantity of water remaining after evaporation. Present-day seawater reaches saturation with respect to halite at a concentration factor (C.F.) of 10.82 (Eugster, Harvie, and Weare 1980). If the

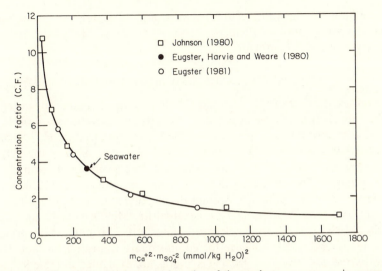

FIGURE 9.3. The relationship between the value of the product $m_{Ca^{+2}} \cdot m_{SO_4^{-2}}$ in seawater and the concentration factor at which seawater becomes saturated with respect to gypsum ($T = 25°C$, $P = 1$ atm).

product $m_{Ca^{+2}} \cdot m_{SO_4^{-2}}$ is ca. 23 (mmol/kg solution)^2, saturation is reached simultaneously with respect to gypsum and halite. This, then, is approximately the minimum value of the product $m_{Ca^{+2}} \cdot m_{SO_4^{-2}}$ in seawater of present-day salinity (35 gm/kg solution) to account for the appearance of gypsum before halite in marine evaporites if no other minerals containing Ca^{+2} or SO_4^{-2} precipitate before halite.

The value 23 (mmol/kg)^2 is only approximate. At 25°C anhydrite rather than gypsum is the stable calcium sulfate mineral in equilibrium with seawater saturated with respect to halite; fortunately, the difference between the solubility product of gypsum and that of anhydrite is not large, and the error is small. Similarly, the error introduced by neglecting the presence of the complex $CaSO_4^0$ is small, because the concentration of this complex in high-ionic-strength solutions is only ca. 1.6. mmol/kg H_2O (Kalyanaraman, Yeatts, and Marshall 1973).

Since $CaCO_3$ precipitates earlier and since the concentration of Ca^{+2} is lowered thereby, a more precise expression for the condition that gypsum precipitates before halite is that

$$(m_{SO_4^{-2}}) \left(m_{Ca^{+2}} - \frac{m_{HCO_3^-}}{2} \right) > 25 \text{ (mmol/kg } H_2O)^2. \qquad (9.14)$$

The large quantity of gypsum and/or anhydrite that normally precedes halite in Phanerozoic marine evaporites indicates that this is a very strong lower limit for the value of the product $(m_{SO_4^{-2}})(m_{Ca^{+2}} - m_{HCO_3^-}/2)$, and that the composition of Phanerozoic seawater has generally been well to the right of curve C in figure 9.2.

PRECIPITATION OF THE LATE SULFATE MINERALS

The concentration of Ca^{+2} and SO_4^{-2} in seawater is limited by several other minerals in marine evaporites. Present-day seawater (see table 9.3) has a composition such that most of the Ca^{+2} remains after HCO_3^- has been exhausted by $CaCO_3$ deposition, and such that most of the SO_4^{-2} still remains after Ca^{+2} has been exhausted by gypsum and/or anhydrite precipitation. The remaining SO_4^{-2} is then removed as a constituent of several late sulfate minerals. Among these, polyhalite, $K_2MgCa_2(SO_4)_4 \cdot 2H_2O$; epsomite, $MgSO_4 \cdot 7H_2O$; hexahydrite, $MgSO_4 \cdot 6H_2O$; and kieserite, $MgSO_4 \cdot H_2O$ are particularly important. This point is illustrated by the mineralogical sequence for equilibrium evaporation in table 9.4, and by the data in figure 9.4. The symbols are explained in table 9.5.

The precipitation of these minerals implies that in ancient as well as in present-day seawater,

$$m_{SO_4^{-2}}^0 > m_{Ca^{+2}}^0 - \frac{m_{HCO_3^-}^0}{2}. \tag{9.15}$$

The abundance of polyhalite, $K_2Mg(Ca)_2(SO_4)_4 \cdot 2H_2O$, in at least some ancient marine evaporites (see for instance Harvie et al., 1980) suggests that in the past

$$m_{SO_4^{-2}}^0 \geq 2\left[m_{Ca^{+2}}^0 - \frac{m_{HCO_3^-}^0}{2} \right]. \tag{9.16}$$

TABLE 9.3

Concentration of major cations and anions in present-day seawater of 35‰ salinity.

	mmol/kg Seawater	mmol/kg H_2O
Na^+	468.0	485.0
Mg^{+2}	53.2	55.1
Ca^{+2}	10.2	10.6
K^+	10.2	10.6
Cl^-	545.0	564.8
SO_4^{-2}	28.2	29.2
HCO_3^-	2.38	2.47

TABLE 9.4.

The mineralogical sequence during the equilibrium evaporation of seawater (Eugster, Harvie, and Weare 1980).

Segment	First appearance of	C.F.	% H₂O left	I	a_{H_2O}	Facies
a	G + sol.	3.62	27.63	2.6	0.929 ⎤	penesaline
b	A + sol.	9.82	10.18	6.6	0.772 ⎦	penesaline
c	A + H + sol.	10.82	9.24	7.2	0.744	saline
d	A + H + Gl + sol.	13.15	7.60	7.5	0.738 ⎤	
e		29.17	3.43	9.1	0.714	
f	A + H + Gl + Po + sol.	38.50	2.60	10.1	0.697	
g	A + H + Po + sol.	44.76	2.23	10.7	0.685	
h	A + H + Po + Ep + sol.	73.56	1.36	13.0	0.590	supersaline
	A + H + Po + Hx + sol.	85.05	1.18	13.8	0.567	
	A + H + Po + Ki + sol.	102.40	0.98	14.9	0.498	
i	A + H + Po + Ki + Car + sol.	117.11	0.85	15.15	0.463	
	A + H + Ki + Car + sol.	159.74	0.63	15.33	0.457	
	A + H + Ki + Car + Bi + sol.	246.00	0.41	17.40	0.338 ⎦	

Notes: C.F.: concentration factor. For seawater, C.F. = 1.

I: ionic strength.

a_{H_2O}: activity of H₂O.

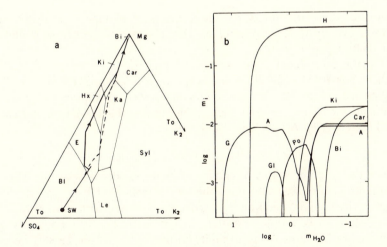

FIGURE 9.4. The equilibrium evaporation path of seawater and the proportion of minerals formed during seawater evaporation (Harvie et al. 1980). (Copyright 1980 by the American Association for the Advancement of Science.)

TABLE 9.5

Abbreviations for minerals in marine evaporites
(Eugster, Harvie, and Weare 1980).

A	anhydrite	$CaSO_4$
Ap	aphthitalite	$NaK_3(SO_4)_2$
Ant	antarcticite	$CaCl_2 \cdot 6H_2O$
Bi	bischofite	$MgCl_2 \cdot 6H_2O$
Bl	bloedite	$Na_2Mg(SO_4)_2 \cdot 4H_2O$
Car	carnallite	$KMgCl_3 \cdot 6H_2O$
Ep	epsomite	$MgSO_4 \cdot 7H_2O$
G	gypsum	$CaSO_4 \cdot 2H_2O$
Gl	glauberite	$Na_2Ca(SO_4)_2$
H	halite	$NaCl$
Hx	hexahydrite	$MgSO_4 \cdot 6H_2O$
Ka	kainite	$KMgClSO_4 \cdot 3H_2O*$
Ki	kieserite	$MgSO_4 \cdot H_2O$
Le	leonite	$K_2Mg(SO_4)_2 \cdot 4H_2O$
Pic	picromerite	$K_2Mg(SO_4)_2 \cdot 6H_2O$
Po	polyhalite	$K_2MgCa_2(SO_4)_4 \cdot 2H_2O$
Syl	sylvite	KCl
Syn	syngenite	$K_2Ca(SO_4)_2 \cdot 6H_2O$
Tc	tachyhydrite	$Mg_2CaCl_6 \cdot 12H_2O$
Th	thenardite	Na_2SO_4

* Braitsch (1971) reports 11/4 H_2O.

This is certainly true today, but the complexities of mineral deposition in marine evaporites are so formidable that no universality can be claimed for relationship 9.16.

On the other hand, boundary E in figure 9.2 seems very well established. Toward the end of marine evaporite formation, SO_4^{-2} is typically exhausted before Mg^{+2}, and the remaining Mg^{+2} is removed as a constituent of chloride minerals such as bischofite, $MgCl_2 \cdot 6H_2O$, and carnallite, $KMgCl_3 \cdot 6H_2O$. This implies that

$$m^0_{SO_4^{-2}} < m^0_{Mg^{+2}} + (m^0_{Ca^{+2}} - \tfrac{1}{2}m^0_{HCO_3^-}). \tag{9.17}$$

Since the position of this boundary in figure 9.2 depends on the Mg^{+2} concentration of seawater, the boundary may have shifted considerably during the Phanerozoic Eon. In particular, if the concentration of Mg^{+2} was lower in Paleozoic seawater than it is today, then boundary E imposed rather more stringent limits on the Ca^{+2} and on the SO_4^{-2} concentration in Paleozoic seawater than on the concentration of these ions in seawater today.

Precipitation of Glauberite

The rather wide expanse of presently permitted Ca^{+2} and SO_4^{-2} concentrations may be limited by boundary F; along this boundary glauberite, $Na_2Ca(SO_4)_2$, begins to crystallize together with halite during the evaporation of seawater. To the left of boundary F, glauberite begins to crystallize after halite; to the right of the boundary, glauberite precedes halite in the evaporation sequence. Glauberite is apparently common in nonmarine evaporites but a rarity in marine evaporites. This is probably due in part to its subsequent replacement by polyhalite (see, for instance, Harvie et al., 1980; Eugster, Harvie, and Weare 1980). Polyhalite is a typical mineral of the halite facies and is not found in the gypsum/anhydrite facies. Thus, if the scarcity of glauberite in marine evaporites is solely due to its replacement by polyhalite, boundary F must be taken very seriously. Other explanations for the apparent rarity of glauberite have been advanced. Glauberite may have been missed in studies of evaporite mineralogy (see, for instance, Braitsch 1962, p. 122), its precipitation may have been kinetically inhibited, or it may have been redissolved during later phases of evaporite deposition. It is also possible that the data for the solubility of glauberite are in error. None of these explanations is satisfactory. Most students of marine evaporites have not looked for glauberite per se, but some have sought to define the entire mineralogy of their marine evaporites; hence the paucity of recorded glauberite occurrences is probably not merely a consequence of faulty observation (W. T. Holser, personal communication, 1981). The precipitation of glauberite is kinetically somewhat difficult, but the mineral has been produced synthetically (see, for instance, Hill and Wills 1938; Block and Waters 1968), and it is unlikely that its extreme rarity can be explained on kinetic grounds. This conclusion is reinforced by the presence of glauberite in evaporites where significant terrestrial contributions have been made to evaporating seawater (Braitsch 1962, p. 80).

The dissolution or replacement of glauberite is predicted by the evaporation sequence of Eugster, Harvie, and Weare (1980). It seems unlikely, however, that the absence of reported glauberite in the gypsum/anhydrite facies can be explained in this manner, since polyhalite always seems to have followed the onset of halite precipitation. The solubility data for glauberite may be in error, but the agreement between the observed and computed solubility trends of glauberite lends support to the validity of the available solubility data (J. H. Weare, personal communication, 1981).

The equilibrium evaporation path of present-day seawater in table 9.4 indicates that glauberite should start to precipitate at a concentration factor (C.F.) of 13.15, shortly after the onset of halite precipitation at a

C.F. of 10.82. Calculations of evaporation paths for seawater compositions containing somewhat different concentrations of Ca^{+2} and SO_4^{-2} (see table 9.6) indicate that glauberite and halite begin to crystallize simultaneously when the initial Ca^{+2} concentration is 15 mmolal and the concentration of SO_4^{-2} is 35 mmolal. This result can be used to define the boundary between the composition of solutions that become saturated with respect to glauberite before halite and those that become saturated with respect to halite before glauberite.

Consider a solution that becomes saturated simultaneously with respect to glauberite and halite. The concentration of Ca^{+2} at that point in the evaporation sequence is

$$m_{Ca^{+2}} \cong (C.F.)(m^1_{Ca^{+2}} - \Delta), \tag{9.18}$$

where

$$m^1_{Ca^{+2}} = (m^0_{Ca^{+2}} - \tfrac{1}{2}m^0_{HCO_3^-})$$

and

$$\Delta = \text{the quantity of } Ca^{+2} \text{ removed by gypsum}$$
$$\text{and anhydrite precipitation in mmol/kg}$$
$$\text{of original } H_2O.$$

Similarly,

$$m_{SO_4^{-2}} = (C.F.)(m^0_{SO_4^{-2}} - \Delta). \tag{9.19}$$

Since the solution is saturated with respect to gypsum or anhydrite,

$$(C.F.)^2(m^1_{Ca^{+2}} - \Delta)(m^0_{SO_4^{-2}} - \Delta) = K'_{Gyp.} \text{ or } K'_{Anh.} \tag{9.20}$$

At 25°C the solubility of gypsum differs only slightly from that of anhydrite in solutions saturated with respect to halite. The value of $K'_{Gyp.}$ at halite saturation is close to 2900 $(mmolal)^2$. Since C.F. is close to 11,

$$(m^1_{Ca^{+2}} - \Delta)(m^0_{SO_4^{-2}} - \Delta) \approx \tfrac{2900}{121} = 24 \ (mmolal)^2. \tag{9.21}$$

Since saturation with respect to glauberite and halite was achieved simultaneously when $m^1_{Ca^{+2}}$ was 15 mmolal and $m^0_{SO_4^{-2}}$ was 35 mmolal,

$$(15 - \Delta)(35 - \Delta) \approx 24 \ (mmolal)^2, \tag{9.22}$$

and hence

$$\Delta \approx 13.9 \text{ mmolal.}$$

At saturation with respect to glauberite,

$$(C.F.)^5 \ (m^0_{Na^+})^2 \ (m^1_{Ca^{+2}} - \Delta)(m^0_{SO_4^{-2}} - \Delta)^2 \approx K'_{Glaub.} \tag{9.23}$$

TABLE 9.6.

Calculated evaporation path of seawater with altered Mg^{+2}, Ca^{+2}, and SO_4^{-2} concentrations (Eugster 1981, personal communication).

First Appearance of	Solution #1		Solution #2		Solution #3	
	C.F.	H/A	C.F.	H/A	C.F.	H/A
G	5.40		5.80		5.80	
A	10.55		10.21		9.70	
A + H	11.68		11.29		10.75	
A + H + Po	57.3	60.2	52.7	60.9	48.05	62.1
A + H + Po + Car	162.7		115		93.7	
A + H + Po + Car + Ki	198		150		106.7	
A + H + Bi + Car + Ki	511		257		183.3	

First Appearance of	Solution #4	
	C.F.	H/A
G	4.29	
A	10.79	
A + H	11.34	
A + H + Gl	24.48	31.1
A + H + Gl + Po	55.65	
A + H + Po	61.8	
H + Ki + Po	183	
A + H + Ki + Po + Car	195	
A + H + Ki + Car + Bi	624	

First Appearance of	Solution #5	
	C.F.	H/A
G	4.29	
A	10.25	
A + H	11.34	
A + H + Po	47.68	45.9
A + H + Po + Car + Ki	135.5	
A + H + Po + Car + Ki	290.7	

First Appearance of	Solution #6	
	C.F.	H/A
G	4.38	
H	9.96	
A + H	11.02	
A + H + Po	44.05	46.5
A + H + Po + Car	101.1	
A + H + Po + Car + Ki	107.7	
A + H + Bi + Car + Ki	200.5	

(*continued*)

TABLE 9.6 (*continued*)

First Appearance of	Solution #7	
	C.F.	H/A
G	1.33	
A + Gl	10.85	0
A + Gl + H	11.9	
A + Gl + H + Po	55.8	
A + Bl + Gl + H + Po	110.5	
Ep + Bl + Po + A + H	159	
Hx + Ep + Po + A + H	325	
Ki + Hx + Po + A + H	397	
Car + Ki + Po + A + H	444	
Car + Ki + Bi + A + H		

First Appearance of	Solution #8	
	C.F.	H/A
G	1.41	
A	10.40	
A + Gl	10.91	0
A + Gl + H	11.49	
A + Gl + H + Po	45.89	
A + Gl + H + Po + Bl	77.4	
A + Bl + Ep + H + Po	104	
A + Ep + Hx + H + Po	168	
A + Ki + H + Po	204	
A + Ki + Car + H + Po	237	
A + Ki + Bi + Car + H	1120	

First Appearance of	Solution #9	
	C.F.	H/A
G	1.48	
A	9.89	
A + Gl	10.92	0
A + Gl + H	11.49	
A + Gl + H + Po	39.35	
A + Gl + H + Po + Bl	59.7	
A + H + Po + Ep	75.2	
A + H + Hx + Po	112	
A + H + Ki + Po	136	
A + H + Ki + Po + Car	155	
A + H + Bi + Ki + Car	398	

TABLE 9.6   (*continued*)

First Appearance of	Solution #10	
	C.F.	H/A
G	2.12	
A	10.96	
A + Gl	11.50	0
A + Gl + H	12.11	
A + Gl + H + Po	53.6	
A + Gl + Bl + H + Po	111.7	
A + Bl + Ep + H + Po	141.6	
A + Hx + H + Po	223.7	
A + Ki + H + Po	272	
A + Ki + H + Po + Car	312	
A + Ki + Bi + H + Car	> 5000	

First Appearance of	Solution #11	
	C.F.	H/A
G	2.12	
A	10.41	
A + Gl + H	11.51	0
A + Gl + H + Po	45.78	
A + H + Bl + Po	77.4	
A + H + Ep + Po	91.1	
A + H + Hx + Po	131.6	
A + H + Ki + Po	163.0	
A + H + Ki + Po + Car	184.5	
A + H + Ki + Bi + Car	594.5	

First Appearance of	Solution #12	
	C.F.	H/A
G	2.19	
A	9.92	
A + H	11.29	
A + H + Gl	11.87	2.52
A + H + Gl + Po	40.67	
A + H + Po	52.8	
A + H + Po + Ep	76.2	
A + H + Po + Hx	96.9	
A + H + Po + Ki	114	
A + H + Po + Ki + Car	135	
A + H + Ki + Bi + Car	279	

Note: Symbols are those used in earlier tables.

The initial concentration of Na^+ was held essentially constant in the computations, and C.F. was not a strong function of the initial Ca^{+2} and SO_4^{-2} concentration of the solutions. Thus,

$$(m_{Ca^{+2}}^1 - \Delta)(m_{SO_4^{-2}}^0 - \Delta)^2 \approx \text{const.} \approx (24)(21.1)$$
$$\approx 506 \text{ (mmolal)}^3, \tag{9.24}$$

when glauberite and gypsum ($\pm$ anhydrite) are in equilibrium with the same solution,

$$\frac{(m_{Ca^{+2}}^1 - \Delta)(m_{SO_4^{-2}}^0 - \Delta)^2}{(m_{Ca^{+2}}^1 - \Delta)(m_{SO_4^{-2}}^0 - \Delta)} = \frac{506}{24}. \tag{9.25}$$

Therefore,

$$(m_{SO_4^{-2}}^0 - \Delta) = 21.1 \tag{9.26}$$

and

$$\Delta = m_{SO_4^{-2}}^0 - 21.1; \tag{9.27}$$

but since

$$m_{Ca^{+2}}^1 - \Delta = 1.1, \tag{9.28}$$

it follows that

$$m_{Ca^{+2}}^1 = m_{SO_4^{-2}}^0 - 20.0 \tag{9.29}$$

for seawater compositions that become saturated simultaneously with glauberite and halite during evaporation.

This expression has been used to draw line F in figure 9.2. The lower limit of the boundary is set by the intercept of line F with boundary C. Its upper limit is set by the intersection with boundary B. As expected, the composition of present-day seawater lies just to the left of boundary F. The most Ca^{+2}- and SO_4^{-2}- rich composition for which an evaporation path has been computed and listed in table 9.6. ($m_{Ca^{+2}}^1 = 20$ mmolal and $m_{SO_4^{-2}}^0 = 45$ mmolal) lies somewhat to the right of boundary F.

The proximity of the present composition of seawater to boundary F is intriguing. In seawater today,

$$m_{Ca^{+2}}^1 = m_{SO_4^{-2}}^0 - 19. \tag{9.30}$$

An increase of only 1 mmol/kg H_2O in the SO_4^{-2} content of present-day seawater would be required at the present Ca^{+2} concentration to reach boundary F. The scarcity of glauberite in marine evaporites therefore suggests that the difference ($m_{SO_4^{-2}}^0 - m_{Ca^{+2}}^1$) was never much more than it is today, and that the Ca^{+2} and SO_4^{-2} concentration of Phanerozoic seawater has been confined to the area between boundaries *B, C, D,* and *F,* if the effects of changes in the other components of seawater have been minor.

THE EFFECT OF CHANGES IN THE CONCENTRATION
OF Mg^{+2}, K^+, Na^+, AND Cl^-

The other seawater components of importance are Mg^{+2}, K^+, Na^+, and Cl^-. The results of the exploratory calculations in table 9.6 of the solutions of table 9.7 show that no significant differences between the observed sequence of evaporites develop as the Mg^{+2} concentration of seawater is reduced from its present level of 56 mmolal to 30 mmolal. This implies that the hypothesis developed in chapter 6 for the evolution of the dolomite/limestone ratio in Phanerozoic carbonates is permitted by the mineralogy of Phanerozoic marine evaporites. An absolute lower limit for the Mg^{+2} concentration in seawater that contains the present amount of SO_4^{-2}, Ca^{+2}, and HCO_3^- is ca. 20 mmolal. At that concentration, boundary *E* passes through the point in figure 9.2 representing the composition of present-day seawater.

TABLE 9.7.

Composition of solutions whose equilibrium evaporation sequence is listed in table 9.6.

Solution Number	Na_2^+	K_2^+	Mg^{+2}	Ca^{+2}	Cl_2^-	SO_4^{-2}
1			30.0	7.5		15.0
2			40.0	7.5		15.0
3			50.0	7.5		15.0
4			30.0	10.0		20.0
5			40.0	10.0		20.0
6			50.0	10.0		20.0
7			30.0	20.0		45.0
8			40.0	20.0		45.0
9			50.0	20.0		45.0
10			30.0	15.0		35.0
11			40.0	15.0		35.0
12			50.0	15.0		35.0
S.W.		5.17	56.0	9.40		29.44

Notes: Concentrations are in units of mmol/kg H_2O. The subscript 2 implies that the listed concentration is half that of the individual ion. The mmolality of Ca^{+2} is the true mmolality minus one-half the mmolality of HCO_3^-.

The effect of increasing the Mg^{+2} concentration of seawater on the sequence of minerals in evaporite minerals has not been explored in detail. A doubling of the Mg^{+2} concentration in seawater today would double the rate of Mg^{+2} removal by reaction of seawater at mid-ocean ridges (see chapter 6) and would almost certainly increase the rate of Mg^{+2} loss from seawater by dolomitization in mildly evaporative settings. If the oceans are close to steady-state with respect to Mg^{+2} today, such a high Mg^{+2} concentration could only be sustained if the rate of input of Mg^{+2} to the oceans were increased by more than a factor of two or if the rate of seawater cycling through mid-ocean ridges were decreased by more than a factor of two.

None of the parameters that determine the Mg^{+2} balance of the oceans are known well enough to rule out such variations, and a more careful exploration of evaporite sequences from seawater in which the concentration of Mg^{+2} is higher than in present-day seawater is surely warranted. It seems likely, however, that the Mg^{+2} content of seawater today is as high as or higher than the Mg^{+2} content of seawater during earlier parts of the Phanerozoic Eon (see chapter 6).

The effects of lower and higher K^+ concentrations in seawater on the mineral sequence in marine evaporites has been explored by Eugster, Harvie, and Weare (1980). Their starting compositions are shown in table 9.8. Solution A has bloedite, $Na_2Mg(SO_4)_2 \cdot 4H_2O$, as a primary phase; solutions B and C intersect the kainite field during equilibrium evaporation, and solutions D and E have sylvite as well as kainite as primary precipitates. Bloedite, when it is present, is almost entirely a secondary mineral in marine evaporites (Braitsch 1962, p. 45). Seawater containing only half the present concentration of K^+ therefore yields a most unusual mineral assemblage on evaporation. Kainite is not uncommon in marine evaporites, where its presence is predictable (see, for instance,

TABLE 9.8.

Starting compositions to test effect of variations in concentration of K^+ on sequence of evaporite minerals (Eugster, Harvie, and Weare 1980).

	A	SW	B	C	D	E
K_2^+	2.65	5.17	7.55	9.82	11.98	14.04
Mg^{+2}	57.48	55.99	54.58	53.24	51.97	50.75
Ca^{+2}	9.65	9.40	9.16	8.94	8.72	8.52
SO_4^{-2}	30.22	29.44	28.70	28.00	27.33	26.69

Note: Concentration conventions are those of table 9.7.

Braitsch 1962, p. 69; and table 1 in Harvie et al. 1980). However, sylvite in marine evaporites is apparently formed largely by secondary processes involving the incongruent solution of carnallite (see, for instance, Braitsch 1962, pp. 92–94). Concentrations of K^+ more than twice that of present-day seawater therefore also lead to the development of unusual evaporite sequences. It would be helpful to have this conclusion confirmed by calculations of mineral assemblages in marine evaporites derived from starting solutions in which not only the K^+ concentration but also those of the other major constituents differ significantly from their concentration in present-day seawater.

The data presented above support the contention that the concentration of many of the major constituents of seawater could have varied only modestly during the Phanerozoic Eon. This conservatism extends to the concentration of Na^+ and Cl^- as well. Zharkov (1981) has compiled the available data for the quantity of sulfate and halite rocks in Paleozoic strata, and has proposed that the volume of "salt rocks" in Paleozoic evaporite basins is 2.944×10^6 km^3 (Zharkov 1981, p. 157). If all of these rocks consist of pure halite, Paleozoic evaporite basins contain 6.4×10^{21} gm NaCl. This corresponds to approximately 15% of the NaCl content of the present-day oceans. Holser (personal communication, 1981) has estimated that the entire inventory of halite in sedimentary rocks of all ages amounts to ca. 30% of the NaCl content of the oceans. There is no disagreement between the two estimates.

If all of the estimated Na^+ and Cl^- in halite deposits were added to the oceans, the salinity of seawater would increase by only 30%. It can be argued that the present inventory of halite in sedimentary rocks is incomplete, and that the maximum increase in the salinity of seawater has been underestimated; however, the distribution of halite in Phanerozoic rocks indicates that there was never a time during the past 600 m.y. when all of the halite that is now present in evaporites was dissolved in the oceans. It is therefore unlikely that the salinity of average seawater during the past 600 m.y. was ever more than 30% higher than at present.

A lower bound on the salinity of seawater is more difficult to extract from the data for the distribution of marine evaporites in Phanerozoic rocks. These evaporites are probably a small fraction of the halite that was deposited during the Phanerozoic Eon. One or more periods of intense evaporite formation could have decreased the salinity of seawater significantly. The salt removed during a period of intense evaporite formation could have been restored subsequently to the oceans, and no useful trace of the NaCl loss and of its restoration might have been left in the sedimentary record. Fortunately, a tracer for such events has been discovered in the bromine content of basal halite in marine evaporite successions.

Br⁻ substitutes for Cl⁻ in halite, but its distribution coefficient is much less than unity (see, for instance, Herrmann et al. 1973). The Br⁻/Cl⁻ ratio therefore rises in the residual brines. Since complete evaporation of seawater is rare, the rate of Br⁻ removal with evaporites is much less than the removal rate of Cl⁻. During a period of intense evaporite formation the concentration of Cl⁻ in seawater therefore tends to be drawn down much more than the concentration of Br⁻. Hence the Br⁻/Cl⁻ ratio in seawater increases during such periods, and the Br⁻ content of basal halite in evaporites formed from such Cl⁻-depleted seawater is higher than the Br⁻ content of basal halites of earlier evaporites. There is at present no evidence for periods during which the Br⁻/Cl⁻ ratio of seawater was significantly greater than today (Holser 1979). The available data are summarized in figure 9.5. Unfortunately, these show a good deal of scatter. This is probably related in part to variations in the value of the distribution coefficient of Br⁻ between brines and halite due to differences in the rate of halite precipitation, and to the effects of other dissolved salts in evaporite brines. Part of the scatter is probably also due to the later recrystallization of primary halite. The interpretation of the Br⁻ content of basal halites is therefore still uncertain, and conclusions based on this criterion alone must be treated with caution. However, the

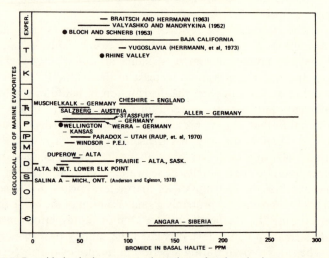

FIGURE 9.5. Bromide in the lowermost salt sections of various basins, as a measure of the Br/Cl ratio of ancient seawater; data modified from tabulation by Kühn (1969). The upper section indicates the range of values expected from modern seawater, calculated from various experimental results (Holser 1979). (Copyright by the Mineralogical Society of America.)

data in figure 9.5 lend little support to the proposition that there were periods during the Phanerozoic Era when the salinity of seawater was much lower than 35‰.

Additional support for the proposition that excursions in the composition of seawater have been minor is supplied by the composition of fluid inclusions in evaporite minerals. Holser (1963) extracted brine from a number of inclusions in Permian halite and from a single inclusion in Silurian halite. His partial chemical analyses of these brines are summarized in table 9.5 and plotted in figure 9.6. The Mg/Cl and Br/Cl ratios in the inclusion brines are quite close to the evaporation trend line of modern seawater. The Mg/Cl ratio of only one brine differs by more than a factor of two from the expected ratio for modern seawater. The errant brine is the only one in which the Mg/Cl ratio is higher than expected. Mg/Cl ratios that fall somewhat below the modern seawater trend line could be due to Mg loss during dolomitization. This interpretation is certainly consistent with the single, rather low SO_4/Cl ratio reported in table 9.9. It is also possible, however, that Mg/Cl ratios below the trend line of present-day seawater reflect a somewhat lower Mg/Cl ratio in the Permian and Silurian seawater whose evaporation residues were trapped in the analyzed fluid inclusions. The Mg^{+2} deficit, if indeed there was a deficit, was not large. The lowest Mg/Br ratio

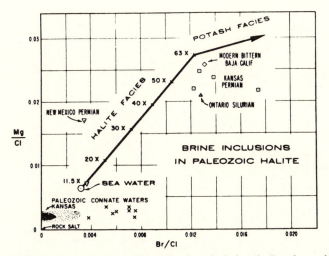

FIGURE 9.6. The Mg/Cl and Br/Cl ratio in brines from inclusions in Permian and Silurian halites and in a modern bittern from Baja California. The trend lines show the increase in the Mg/Cl and Br/Cl ratio during the halite and potash deposition facies of modern seawater (Holser 1963). Reproduced with permission of the Northern Ohio Geological Society.

TABLE 9.9.

Chemical composition of brine inclusions in Permian and Silurian halites (Holser 1963).

| Sample No. | Locality | Formation and Age | Composition by Weight | | | | | Crystal |
| | | | Brine Inclusions | | | | | |
			Mg/Cl	Br/Cl	Mg/Br	Ca/Cl	SO₄/Cl	Br (ppm)
—	Salt flats, Laguna Ojo de Liebre, Baja California	Seawater	0.0671	0.0034	19.7	0.0211	0.139	—
17008-20		Recent salt	0.258	0.0133	19.4	0.001	n.d.	92
17167-3	Carey Mine; Hutchinson, Kans.	Wellington; Permian	0.219	0.0122	18.0	0.000	n.d.	
17167-4	do	do	0.237	0.0138	17.2	0.000	n.d.	92
17167-5	do	do	0.247	0.016	15	0.00	0.034	
16947-20	NAS Core 546.5 ft.; Hutchinson, Kans.	do	0.22	0.017	13	0.00	n.d.	n.d.
16933-3-1	U.S. Borax Mine, Carlsbad, N.M.	Salado, 1st ore zone; Permian	0.068	0.0034	20	0.000	n.d.	n.d.
16933-3-2	do	do	0.171	0.0035	49	0.000	n.d.	
17570-3	Core #2, 1505 ft; Goderich, Ontario	Salina "B" salt; Silurian	0.21	0.013	16	0.13	n.d.	88

Note: n.d. = not determined.

observed in any of the analyzed brines ($\#$ 16947-20) is 0.65 that of present-day seawater. This could mean that the Mg^{+2} concentration in the seawater from which this brine was derived was 35% less than the Mg^{+2} concentration of modern seawater if the concentration of Br in Permian seawater was equal to that of present-day seawater. The Br content of Permian halite indicates that the Br content of Permian seawater was not very different from that of present-day seawater. Much more data will be needed before the departures of brine analyses such as those in table 9.9 from the modern seawater trend can be interpreted with confidence.

Somewhat bothersome questions regarding the history of inclusion fluids in halite will also have to be answered. Halite and the other evaporite minerals are highly soluble; they are therefore particularly subject to recrystallization. Holser (1963) has commented on this problem in connection with the Br/Cl ratio of brines in halite from the Salado salt at Carlsbad, New Mexico. Sabouraud-Rosset (1974) has shown that diagenetic solutions have filled secondary cavities along cleavage planes or have replaced initial liquid inclusions in primary cavities in many of the gypsum crystals in which she has studied the Cl/Br ratio of contained fluids. It should, however, be possible to find brine inclusions whose pedigree is above suspicion. Analytical data for such fluids should prove the basis for an excellent and exacting test of the hypothesis advanced in chapter 6 that the Mg^{+2} content of Paleozoic seawater was significantly lower than that of present-day seawater.

3. The Isotopic Composition of Sulfur and Oxygen in Seawater Sulfate

The chemical and mineralogical data discussed in the previous section all indicate that fluctuations in the composition of seawater during the Phanerozoic Eon have been minor. It therefore comes as somewhat of a surprise that the isotopic composition of sulfur in seawater sulfate has been highly variable during the same period of time, and that the isotopic composition of oxygen in seawater sulfate has definitely not been constant.

VARIATIONS IN THE $\delta^{34}S$ VALUE OF SEAWATER

The evidence for these variations can be read from the isotopic record of sulfur and oxygen in gypsum and anhydrite from marine evaporites. The body of observations for the changes in the value of $\delta^{34}S$ in gypsum and anhydrite is large and convincing. The broad outlines of the $\delta^{34}S$-time curve have been known since the publication of papers by Ault and

Kulp (1959) and by Thode, Monster, and Dunford (1961). Since then a large number of investigators have contributed a mass of data that has confirmed these outlines and that has begun to define the details of the time variation of $\delta^{34}S$ in marine evaporite minerals. Among these more recent contributions, those of Nielsen (1965), Holser and Kaplan (1966), and Claypool et al. (1980) have been particularly important.

The isotopic composition of sulfur in gypsum and anhydrite is nearly the same as the isotopic composition of sulfur in seawater sulfate. The value of $\delta^{34}S$ in gypsum and anhydrite is only ca. $1.65 \pm 0.12‰$ more positive than that of solutions from which they precipitate (Thode and Monster 1965); this difference is smaller than the usual spread of $\delta^{34}S$ values in gypsum and anhydrite from single marine evaporites.

The isotopic composition of sulfur in the brine of an evaporite basin is not necessarily identical to that of contemporary seawater. Sulfate reduction within evaporite basins can lead to more positive $\delta^{34}S$ values in basin brines, and the addition of sulfate by rivers that enter the basin can either increase or decrease the $\delta^{34}S$ value of basin brines. The $\delta^{34}S$ value of nonmarine evaporite brines need bear no relationship at all to the $\delta^{34}S$ value of contemporary seawater. For all these reasons the reconstruction of the time variation of $\delta^{34}S$ in seawater from the isotopic composition of sulfur in evaporites has required a very large body of observational data combined with a great deal of good judgment. Although neither ingredient has been lacking, there are still uncertainties in some portions of the $\delta^{34}S$-time curve of seawater, where a sufficient number of marine evaporites has not been studied or where the origin of sulfate in the evaporites that have been studied is in doubt. Fortunately, the uncertain sections of the $\delta^{34}S$-time curve during the Phanerozoic Eon are short. Considerable reliance can therefore be placed on the $\delta^{34}S$-time plots in figures 9.7 to 9.11 and on the Phanerozoic portion of the summary $\delta^{34}S$-time plot in figure 9.12. The Proterozoic data are limited to the latest few hundred million years of this Eon, because we know of no marine evaporites older than ca. 1200 m.y. Even the data for the last 600 m.y. of the Proterozoic Era are somewhat unsatisfactory, because the number of marine evaporites from this period of Earth history is small, and because the age of the few late Proterozoic evaporites is much more uncertain than that of Phanerozoic evaporites.

During the past billion years $\delta^{34}S$ in seawater sulfate has reached values slightly above $+31‰$ and slightly below $+10‰$ relative to the Cañon Diablo sulfur standard. Both the maximum in the $\delta^{34}S$-time curve in the Cambrian and its minimum in the Permian are well documented. Relatively rapid excursions in the $\delta^{34}S$ value of seawater sulfate occurred in the Devonian, Triassic, and early Cretaceous. Additional, similarly rapid shifts may be found as the $\delta^{34}S$-time curve is defined more precisely.

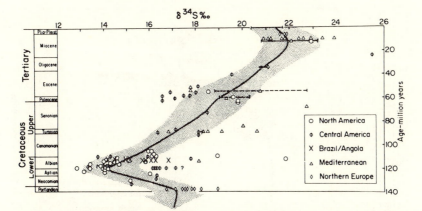

FIGURE 9.7. Sulfur isotopes in evaporites of Recent to Cretaceous age (Claypool et al. 1980). Evaporite suites that may be nonmarine are identified with a question mark; only the maximum value of $\delta^{34}S$ in the lower Oligocene Rhine Valley evaporites is shown; horizontal lines represent literature data for which only a range of $\delta^{34}S$ values is indicated; the shaded area indicates the estimated uncertainty of the $\delta^{34}S$-time curve; see Claypool et al. (1980) for sources of data. (Reproduced by permission of Elsevier Scientific Publishing Company.)

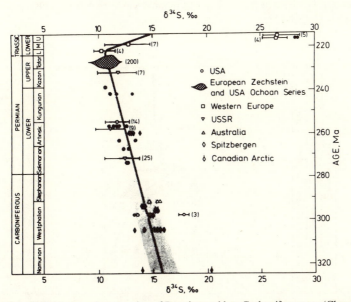

FIGURE 9.8. Sulfur isotopes in evaporites of Permian and late Carboniferous age (Claypool et al. 1980); filled symbols indicate new analyses by Claypool et al. (1980); open symbols represent data from other sources; lines indicate ranges and means where only these are available; the large number of analyses for the Zechstein and for the Ochoan Series are combined into a single shaded area; the number of analyses where these are large is indicated in parentheses; for the significance of other symbols see the legend of figure 9.7. (Reproduced by permission of Elsevier Scientific Publishing Company.)

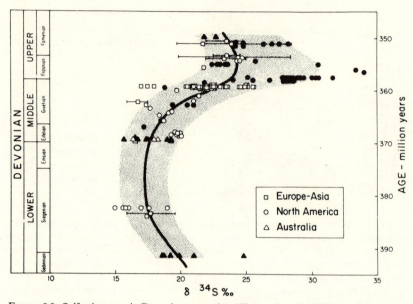

FIGURE 9.9. Sulfur isotopes in Devonian evaporites (Claypool et al. 1980); the symbols have the same significance as in figures 9.7 and 9.8. (Reproduced by permission of Elsevier Scientific Publishing Company.)

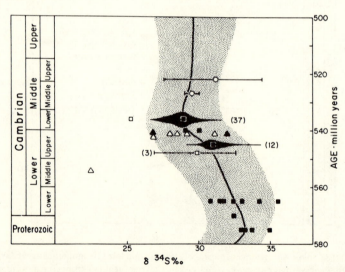

FIGURE 9.10. Sulfur isotopes in Cambrian evaporites (Claypool et al. 1980); the symbols have the same significance as in the previous figures. (Reproduced by permission of Elsevier Scientific Publishing Company.)

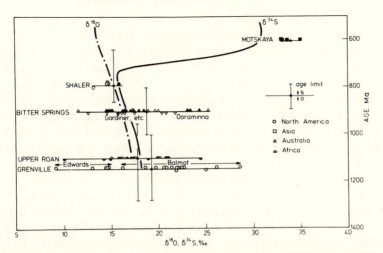

FIGURE 9.11. Sulfur and oxygen isotopes in Precambrian evaporites (Claypool et al. 1980); the limits of the probable age of the several evaporites are indicated by vertical bars; $\delta^{34}S$ values are plotted above the median line for each evaporite, $\delta^{18}O$ values below the median line; the other symbols have the same meaning as in the previous figures. (Reproduced by permission of Elsevier Scientific Publishing Company.)

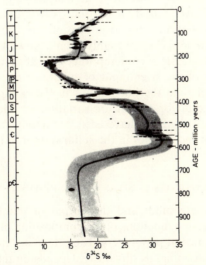

FIGURE 9.12. Summary of sulfur isotope data for marine evaporites (Claypool et al. 1980). Horizontal dashed lines indicate the range of $\delta^{34}S$ in evaporites for which relatively few analyses are available; the heavy curve is the best estimate of Claypool et al. (1980) for the $\delta^{34}S$ of sulfate minerals in equilibrium with the contemporary sulfate in world ocean surface waters; the shaded area indicates the estimated uncertainty of the $\delta^{34}S$-time curve. (Reproduced by permission of Elsevier Scientific Publishing Company.)

The major processes that contribute to the observed variations in the $\delta^{34}S$ value of marine sulfate are well understood. The precipitation of gypsum and anhydrite is accompanied by little isotopic fractionation of sulfur, but the bacterial reduction of sulfate to sulfide in carbonaceous sediments can produce very large isotopic fractionations. Sulfide sulfur is always isotopically lighter than the sulfate sulfur from which it is formed. The precise difference between the isotopic composition of sulfur in these two valence states in any given sediment depends on a host of environmental variables. Fortunately most of the reduction of marine sulfate to sulfide takes place in a single environment: the upper few centimeters of rapidly accumulating marine sediments that are rich in organic matter (Kaplan, Emery, and Rittenberg 1963; Hartmann and Nielsen 1969; Goldhaber and Kaplan 1974, 1980; Sweeney and Kaplan 1980; Berner 1980; and Westrich 1983). In this type of setting the fractionation of sulfur isotopes is approximately 40‰. Sulfate reduction therefore drives the isotopic composition of seawater toward more positive values, and one of the major factors that determines the $\delta^{34}S$ value of seawater is the fraction of sulfur that leaves the oceans as a constituent of sulfide minerals. The other major influence on the isotopic composition of seawater sulfate is exercised by the $\delta^{34}S$ value of world-average river sulfate. This in turn is determined by the relative contribution of the weathering of sulfate and sulfide minerals to world river runoff and by the spectrum of $\delta^{34}S$ values in the sulfur of the weathered sulfates and sulfides. Rivers sample a wide range of evaporites and sulfides. They probably do not present the oceans with a perfectly unbiased sample of contemporary crustal sulfur, but fluctuations in the $\delta^{34}S$ value of the river input of sulfate to the oceans must have been much smaller than the fluctuations in the isotopic composition of marine sulfur. The major reason for variations in the value of $\delta^{34}S$ in marine sulfate must therefore have been fluctuations in the relative proportion of sulfides and sulfates in the sulfur output of the oceans (see, for instance, Holland 1973).

EVAPORITES AS SINKS FOR MARINE SULFATE

The output of sulfate sulfur and the output of sulfide sulfur from the oceans are complicated functions of the composition of seawater and of a variety of environmental parameters. The rate of sulfate output as a constituent of evaporite minerals obviously depends on the concentration of SO_4^{-2} in contemporary seawater. If evaporation in evaporite basins were or generally complete, the sulfate output in these settings would simply be the product of the sulfate concentration in seawater times the annual rate of evaporation in evaporite basins. Evaporation is, however, rarely complete. The abnormally great abundance of gypsum and anhydrite in marine evaporites was described in the previous section of this chapter, and it is clear that the early phases of evaporation determine the loss of sulfate from the oceans by this mechanism. The removal of sulfate during this stage depends on the Ca^{+2} as well as on the SO_4^{-2} concentration in seawater. It depends also on the presence or absence of $CaCO_3$ sediments which can be dolomitized by evaporating seawater. Dolomitization is a potentially important intermediate step, since it can lead to the removal of SO_4^{-2} via the reaction,

$$2CaCO_3 + Mg^{+2} + SO_4^{-2} \longrightarrow CaMg(CO_3)_2 + CaSO_4. \quad (9.31)$$

The volume of seawater that is annually evaporated to the extent that gypsum and/or anhydrite precipitate in significant quantities depends on rather fine details of contemporary world geography. Marine evaporite basins owe their existence to unlikely circumstances. Contact with the open oceans must be sufficiently restricted, so that evaporative concentration can proceed apace, yet sufficiently unimpeded, so that large volumes of seawater are evaporated during the history of the basin. Such settings can occur both on cratons and in areas of rifting. It seems intuitively reasonable that the intensity of evaporite formation has varied during geologic time.

The data in figure 9.13 for the abundance of sulfate rocks in Paleozoic sediments are based on Zharkov's (1981) extensive compilation, and seem

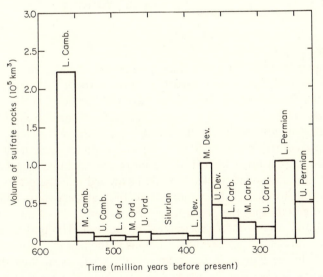

FIGURE 9.13. The inventory of sulfate rocks in Paleozoic sedimentary rocks; data from Zharkov (1981).

to confirm this geological intuition. Lower Cambrian, middle Devonian, and lower Permian sedimentary rocks contain particularly large quantities of gypsum and anhydrite. It is worth asking, however, whether the exceptional abundance of evaporites in these sedimentary rocks is due to an abnormally rapid rate of evaporite formation or to an exceptional degree of evaporite preservation.

The total estimated quantity of lower Cambrian "sulfate rock" is 2.25×10^5 km^3. If we assume that all of these sulfate rocks consist of pure anhydrite of density 3.05 gm/cc, and if all of this anhydrite was deposited between 575 and 550 m.y.b.p., then the formation of these evaporites demands a mean CaSO$_4$ deposition rate of

$$\frac{2.25 \times 10^5 \text{ km}^3 \times 10^{15} \text{ cc/km}^3 \times 3.0 \text{ gm/cc}}{25 \times 10^6 \text{ yrs}}$$

$$= 0.27 \times 10^{14} \text{ gm CaSO}_4/\text{yr}.$$

The present-day river input of SO$_4^{-2}$ to the oceans can be computed by multiplying the SO$_4^{-2}$ content of average river water (11 ppm) by the flux of river water to the oceans (4.6×10^{19} cc/yr). If all of this SO$_4^{-2}$ were precipitated as anhydrite, the annual rate of deposition of anhydrite would be ca.

$$11 \times 10^{-6} \frac{\text{gm SO}_4^{-2}}{\text{cc river water}} \times 4.6 \times 10^{19} \text{ cc river water/yr} \times \frac{136}{96}$$

$$= 7.2 \times 10^{14} \text{ gm CaSO}_4/\text{yr}.$$

This figure is surely an upper limit; perhaps as much as one-half of the present flux of river water sulfate is anthropogenic (see, for instance, Holland 1978, pp. 97–98), and on average only one-half of the river flux of sulfide has left the oceans as a constituent of sulfate minerals. However, even after the appropriate reductions are made, the estimated preindustrial steady-state rate of CaSO$_4$ precipitation under present-day conditions is some six times the calculated mean rate required to produce the lower Cambrian sulfate rocks. It could be argued that the present rate of SO$_4^{-2}$ input to the oceans is more rapid than average, but even so it is virtually certain that the preserved evaporites are a small fraction of the gypsum and/or anhydrite precipitated in the past. The pattern in figure 9.13 therefore tells us less about variations in the rate of evaporite formation during the Paleozoic Era than about variations in their degree of preservation, unless CaSO$_4$ removal by a mechanism other than evaporite formation has been important.

The Oceanic Crust as a Sink for Marine Sulfate

Such a mechanism does exist. Major quantities of anhydrite could have been precipitated in the oceanic crust during the cycling of seawater through hot basalts within and near mid-ocean ridges. Seawater becomes saturated with respect to anhydrite on heating to temperatures in the vicinity of 110°C, and virtually all of the SO_4^{-2} in seawater that is cycled through hot basalt can be removed by precipitation with Ca^{+2} released during the reaction of seawater with basalt (see, for instance, Bischoff and Dickson 1975; Hajash 1975; Mottl and Holland 1978). The presence of anhydrite in basalts of the Reykjanes Ridge in Iceland and in the vicinity of kuroko deposits in the greenschist region of Japan shows that this process does take place.

If the rate of seawater cycling through the axial regions of MORs is $(0.4 \pm 0.2) \times 10^{17}$ cc/yr (see chapter 6), then the potential annual loss of seawater sulfate due to the precipitation of anhydrite in MORs is

$$2800 \times 10^{-6} \frac{gm\ SO_4^{-2}}{gm\ seawater} \times (0.4 \pm 0.2) \times 10^{17} \frac{gm\ S.W.}{yr}$$

$$= (1.1 \pm 0.6) \times 10^{14} \frac{gm\ SO_4^{-2}}{yr}.$$

This is ca. $30 \pm 15\%$ of the present-day nonanthropogenic flux of river sulfate. Anhydrite is rare in the altered MOR basalts that have been recovered to date; but the mineral has been found in the deepest part of the hole drilled during DSDP Leg 83, where temperatures were sufficiently high to prevent resolution of anhydrite (R. N. Anderson et al. 1982). If the distribution of anhydrite in this hole turns out to be representative, the oceanic crust could well turn out to be an important sink for marine sulfate.

The Isotopic Composition of Oxygen in Seawater Sulfate

The isotopic composition of oxygen in seawater sulfate is an intriguing piece of this incomplete puzzle. The present value of $\delta^{18}O_{SO_4^{-2}}$, $+8.6‰$, is far out of equilibrium with the $\delta^{18}O$ value of ocean water. Holser et al. (1979) have shown that this can be explained by the persistence of nonequilibrium between oxygen in seawater and oxygen in the river input of sulfate, which owes its $\delta^{18}O$ value to the combined effect of the solution of evaporite minerals and the oxidation of sulfides during weathering.

An alternative explanation calls on equilibration of the $\delta^{18}O$ value of marine sulfate with seawater oxygen at elevated temperatures. Lloyd's (1968) data show that at temperatures in the vicinity of 200°C, $\delta^{18}O_{SO_4^{-2}}$ is ca. 9‰ more positive than $\delta^{18}O_{H_2O}$. Somewhat lower equilibration temperatures are required by the data of Chiba et al. (1981). If as much seawater cycles through MORs as was suggested above, the reaction between seawater and dissolved sulfate could act as an effective buffer for the $\delta^{18}O$ value of sulfate. However, the operation of this mechanism demands that seawater sulfate that cycles through MORs be returned to the oceans. This in turn implies the later resolution of anhydrite precipitated during the heating of seawater above ca. 110°C. The apparent rarity of anhydrite in altered MOR basalts was noted above and is in agreement with the requirements for the operation of this buffer mechanism. Reasons for the wholesale resolution of anhydrite precipitated in MORs can surely be advanced, but there are still considerable grounds for withholding judgment regarding the importance of this mechanism as a control on $\delta^{18}O_{SO_4^{-2}}$ in seawater today.

The $\delta^{18}O$ values of late Proterozoic and Phanerozoic anhydrites is much less variable than their $\delta^{34}S$ value. Figure 9.14 shows that $\delta^{18}O_{SO_4^{-2}}$ has been virtually independent of $\delta^{34}S_{SO_4^{-2}}$ except during the Permian period when the δ value of both isotopes reached minimum values. The relative constancy of $\delta^{18}O_{SO_4^{-2}}$ during the past 1 b.y. is probably more consistent with a hydrothermal control system than with the explanation proposed by Holser et al. (1979).

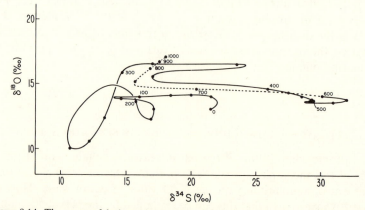

FIGURE 9.14. The course of the isotopic composition of sulfur and oxygen in sulfates of marine evaporites. Numbers along the evolution line are in millions of years; the dashed line for late Proterozoic evaporites is uncertain (Claypool et al. 1980). (Reproduced by permission of Elsevier Scientific Publishing Company.)

THE SULFUR CONTENT OF CARBONACEOUS SEDIMENTS

The reduction of sulfate to sulfide and the subsequent removal of sulfur as a constituent of iron sulfides was discussed briefly in chapter 8, where the correlation between the concentration of pyrite and of organic carbon was used to provide proof of bacterial sulfate reduction in sediments at least as old as the early part of the Proterozoic Era. Figure 9.15 shows a similar correlation between the sulfide sulfur and the organic carbon content of some modern marine sediments. The data compiled by Berner fall close to a trend line along which the S^{-2}/C^0 ratio is 0.36 gm S/gm C^0. The data for anoxic basins falls close to a line of slope 0.71 gm S^{-2}/gm C^0. If these lines are reasonably typical, the worldwide S^{-2}/C^0 ratio today falls between 0.36 and 0.71. Such an intermediate value is consistent with other geochemical data. The best estimate of the rate of present-day C^0 sedimentation is probably $(1.2 \pm 0.3) \times 10^{14}$ gm/yr (Holland 1978, p. 272). The input of nonanthropogenic river sulfate to the oceans today was estimated above. If one-half of this quantity is removed as a constituent of sulfides, the sulfide sulfur removal rate is ca. 0.4×10^{14} gm/yr (see also Berner 1982). The average S^{-2}/C^0 ratio should therefore be 0.33 ± 0.15, i.e., within the observed range. The agreement between the observed and the predicted value of the S^{-2}/C^0 ratio in marine sediments is reassuring. However, the uncertainties in the estimated worldwide S^{-2}/C^0 ratio are so large that this ratio can only be used as a rough measure of

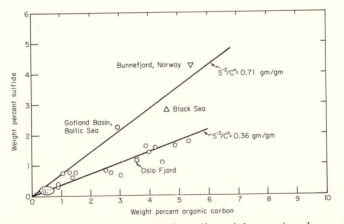

FIGURE 9.15. The correlation between the sulfide sulfur and the organic carbon content of marine sediments; the open circles are data for anoxic marine sediments compiled by Berner (1982); the remaining data are taken from Calvert's (1976) compilation of data for the chemistry of near-shore sediments.

the current partitioning of river sulfate between the ocean output of sulfate and sulfide minerals.

Figure 9.16 summarizes the C^0 and sulfide analyses of 116 samples of upper Devonian shales from the North American craton. Most of the sets of analyses from the several sampling areas fall close to a line for which the S^{-2}/C^0 ratio is 0.86. Two of the data sets plot along a line for which S^{-2}/C^0 is 0.53. The weighted S^{-2}/C^0 ratio of the entire set is 0.73. Leventhal (personal communication 1983) has pointed out that this data set can be fitted quite well by a single line with intercept near 1.3 weight percent sulfur. However, Vyas, Aho, and Robl (1981) have studied the chemical composition of ten cores through the Mississippian Sunbury shale and the Cleveland and Huron members of the Devonian Ohio shale in Lewis and Fleming counties, Kentucky. All three shales are highly carbonaceous. Their average carbon content is 11.42, 9.04, and 5.61%, respectively; their sulfur content 3.54, 2.42, and 3.15%; and their S^{-2}/C^0 ratio is 0.31, 0.27, and 0.56. The range of S^{-2}/C^0 ratios in Mississippian and Devonian black shales of the eastern United States is therefore large, similar to that of their modern equivalents, and cannot all fit on a single line.

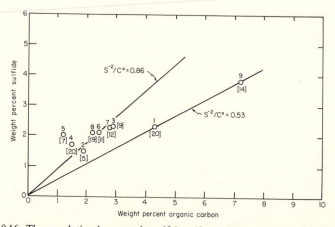

FIGURE 9.16. The correlation between the sulfide sulfur and the organic carbon content of upper Devonian shales from the North American craton (Leventhal 1978a,b, 1979; and Leventhal and Hosterman 1982). The unbracketed numbers indicate the area from which the samples were taken; the bracketed numbers indicate the number of samples in each data set. (1) Perry County, Kentucky; (2) Jackson County, West Virginia; (3) Lincoln County, West Virginia; (4) Cattaraugus County, New York; (5) Washington County, Ohio; (6) Carroll County, Ohio; (7) Wise County, Virginia; (8) Martin County, Kentucky; (9) Overton County, Tennessee.

Hence, the time variation of the isotopic composition of sulfur in seawater sulfate is by far the most reliable indicator of the relative proportion of sulfides and sulfates in the output of sulfur from the oceans. The relative magnitude of these outputs has varied a great deal during the Phanerozoic Eon. The reasons for these variations are complex (see, for instance, Berner and Raiswell 1983), and reliable information regarding variations in the SO_4^{-2} concentration of seawater in the past can probably not be extracted from the $\delta^{34}S$-time curve in figure 9.12. It is clear, however, that the variation of $\delta^{34}S$ in seawater sulfate during the past 1 b.y. is an indicator of considerable gains and losses of oxygen due to the operation of the sulfur cycle. During periods when the rate of sulfide weathering on land greatly exceeded the rate of sulfide deposition from the oceans, the rate of O_2 use by sulfide oxidation to sulfate greatly exceeded the rate of "O_2" return by sulfate reduction to sulfide. O_2 was therefore lost from the ocean-atmosphere system due to the operation of the sulfur cycle during such periods. The effect of these imbalances on the O_2 content of the atmosphere has been potentially large. It is impossible to calculate the magnitude of the effect precisely, because the validity of the calculations depends on the validity of a number of reasonable but unproven assumptions (see, for instance, Holland 1973). Schidlowski, Junge, and Pietrek (1977) have proposed that O_2 consumption due to imbalances in the sulfur cycle since the late Proterozoic Era probably amounted to several times the present O_2 content of the atmosphere. Claypool et al.'s (1980) calculations indicate that the O_2 demand due to excess sulfide oxidation in the Permian amounted to about 95% of the present quantity of atmospheric O_2.

The Sulfur-Carbon-Oxygen System

It was pointed out some years ago (Holland 1973; Garrels and Perry 1974; see also Garrels and Lerman 1981) that such oxygen deficits were probably balanced or nearly balanced by O_2 gains due to the operation of the carbon cycle. When this proposition was made, the supporting evidence was quite weak. Recently, however, Veizer, Holser, and Wilgus (1980) have presented convincing data in favor of the proposed approximate balance; Lindh et al. (1981) and Holser (1984) have further strengthened the case in favor of this proposition. Both sets of authors have shown that the isotopic composition of sulfur in marine evaporite sulfate minerals has been correlated negatively with the isotopic composition of carbon in contemporaneous marine carbonates. The data of Veizer, Holser, and Wilgus are reproduced in figure 9.17. Lindh et al. have shown that the $\delta^{13}C$ value of Cambrian carbonates may have been closer to

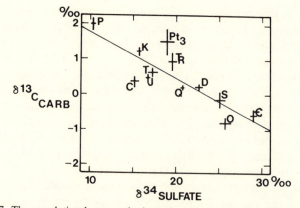

FIGURE 9.17. The correlation between the isotopic composition of carbon in marine carbonates and of sulfur in marine evaporites during the Phanerozoic Eon (Veizer, Holser, and Wilgus 1980). (Reproduced by permission of Pergamon Press, Ltd.)

$-1‰$, and that of Permian carbonates closer to $+3‰$. These two periods of extreme $\delta^{13}C_{carb.}$ and $\delta^{34}S_{sulf.}$ values anchor a $\delta^{34}S$-$\delta^{13}C$ correlation line, which passes conveniently close the value of the isotopic composition of sulfur and carbon in recent marine sulfates and carbonates. Veizer, Holser, and Wilgus (1980) showed that the regression line for their data is

$$\delta^{13}C_{carb.} = 3.074 - 0.131\delta^{34}S_{sulf.}.$$ (9.32)

The equation for Lindh et al.'s (1981, and personal communication, 1982) regression line, which is based on a good deal more data, is approximately

$$\delta^{13}C_{carb.} = (5.0 \pm 2.0) - (0.20 \pm 0.10)\delta^{34}S_{sulf.}$$ (9.33)

The difference between these equations is a measure of the rather large uncertainties in the $\delta^{13}C_{carb.}$-$\delta^{34}S_{sulf.}$ correlation; these are related in part to the large variety of rock types on which the $\delta^{13}C$ data base rests.

The negative correlation between $\delta^{13}C_{carb.}$ and $\delta^{34}S_{sulf.}$ is best explained by a coupling between the carbon cycle and the sulfur cycle, so that oxygen gains due to the operation of one cycle are balanced by oxygen losses due to the operation of the other. Figure 9.18 is a flow diagram for a part of the geochemical cycle of the two elements. Carbon is present in the crust as organic carbon, C^0, and as carbonate carbon, C^{+4}; sulfur is present largely as sulfate sulfur, S^{+6}, and as sulfide sulfur, S^{-2}. Both elements are released during weathering and enter the ocean-atmosphere system, whence they are removed in geologically rather short periods of time and restored to the crustal reservoirs. The total number

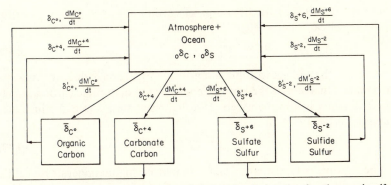

FIGURE 9.18. A simplified representation of the geochemical cycle of carbon and sulfur. Juvenile contributions have been neglected.

of moles of carbon and sulfur added to the ocean-atmosphere system per unit time at steady state is equal to the number of moles of these elements removed from the ocean-atmosphere system. The proportion of carbon and sulfur added from the oxidized and reduced crustal reservoirs of these elements does not, however, need to be equal to the proportion in which they are returned to these reservoirs. All that is required is that any imbalance in the oxygen content of the system due to shifts in the occupancy of the carbon reservoirs is balanced by an equal and opposite shift in the occupancy of the sulfur reservoirs, so that there is no net gain or loss of oxygen in the ocean-atmosphere system.

The mass balance in the carbon cycle implies that

$$\frac{dM_{C^0}}{dt} + \frac{dM_{C+4}}{dt} = \frac{dM'_{C^0}}{dt} + \frac{dM'_{C+4}}{dt}. \tag{9.34}$$

The symbols for the several fluxes in this equation are defined in figure 9.18. Similarly, the requirement of mass balance in the sulfur cycle implies that

$$\frac{dM_{S^{-2}}}{dt} + \frac{dM_{S+6}}{dt} = \frac{dM'_{S^{-2}}}{dt} + \frac{dM'_{S+6}}{dt}. \tag{9.35}$$

By analogy, the isotopic balance of the two elements demands that

$$\delta C^0 \frac{dM_{C^0}}{dt} + \delta C^{+4} \frac{dM_{C+4}}{dt} = \delta' C^0 \frac{dM'_{C^0}}{dt} + \delta' C^{+4} \frac{dM'_{C+4}}{dt}, \tag{9.36}$$

and that

$$\delta S^{-2} \frac{dM_{S^{-2}}}{dt} + \delta S^{+6} \frac{dM_{S+6}}{dt} = \delta' S^{-2} \frac{dM'_{S^{-2}}}{dt} + \delta' S^{+6} \frac{dM'_{S+6}}{dt}. \tag{9.37}$$

There is virtually no fractionation of carbon isotopes during the precipitation of carbonates. Hence,

$$\delta'_{C^{+4}} \simeq {}_0\delta_C \quad , \tag{9.38}$$

where ${}_0\delta_C$ is the average isotopic composition of carbon in dissolved marine carbonate. On average, the isotopic composition of organic carbon in marine sediments is approximately 25‰ lighter than ${}_0\delta_C$ (see, for instance, Claypool et al. 1980); thus,

$$\delta'_{C^o} \simeq {}_0\delta_C - 25. \tag{9.39}$$

The isotopic composition of sulfur in marine sulfate minerals is slightly more positive than the isotopic composition of sulfur in marine sulfate (see above), but neglecting this small effect,

$$\delta'_{S^{+6}} \simeq {}_0\delta_S. \tag{9.40}$$

The isotopic composition of sulfide sulfur is, on average, circa 40‰ lighter than marine sulfate, so that

$$\delta'_{S^{-2}} \simeq {}_0\delta_S - 40. \tag{9.41}$$

If equations 9.38 and 9.39 are substituted into equation 9.36, and if equations 9.40 and 9.41 are substituted into equation 9.37, we obtain

$$\delta_{C^o}\frac{dM_{C^o}}{dt} + \delta_{C^{+4}}\frac{dM_{C^{+4}}}{dt} = {}_0\delta_C\left[\frac{dM'_{C^o}}{dt} + \frac{dM'_{C^{+4}}}{dt}\right] - 25\frac{dM'_{C^o}}{dt}, \tag{9.42}$$

and

$$\delta_{S^{-2}}\frac{dM_{S^{-2}}}{dt} + \delta_{S^{+6}}\frac{dM_{S^{+6}}}{dt} = {}_0\delta_S\left[\frac{dM_{S^{-2}}}{dt} + \frac{dM_{S^{+6}}}{dt}\right] - 40\frac{dM_{S^{-2}}}{dt}. \tag{9.43}$$

It seems likely that during most geological periods the mean isotopic composition of weathering organic carbon has been roughly related to that of weathering carbonate carbon by the expression,

$$\delta_{C^o} \simeq \delta_{C^{+4}} - 25. \tag{9.44}$$

Similarly it is likely that the isotopic composition of weathering sulfide sulfur is related to the isotopic composition of weathering sulfate sulfur by the approximate expression,

$$\delta_{S^{-2}} \simeq \delta_{S^{+6}} - 40. \tag{9.45}$$

If these relationships are substituted into equations 9.42 and 9.43, respectively, the resulting equations can be rearranged to yield

$$\delta_{C+4} - 25 \left[\frac{\dfrac{dM_{C^0}}{dt} - \dfrac{dM'_{C^0}}{dt}}{\dfrac{dM'_{C^0}}{dt} + \dfrac{dM'_{C+4}}{dt}} \right] \simeq {}_0\delta_C \tag{9.46}$$

and

$$\delta_{S+6} - 40 \left[\frac{\dfrac{dM_{S-2}}{dt} - \dfrac{dM'_{S-2}}{dt}}{\dfrac{dM'_{S-2}}{dt} + \dfrac{dM'_{S+6}}{dt}} \right] \simeq {}_0\delta_S. \tag{9.47}$$

Equations 9.46 and 9.47 are coupled by the requirement that there is no net oxygen gain or loss due to the combined operation of the two cycles. Oxygen losses for the ocean-atmosphere system due to a net oxidation of C^0 to C^{+4} via the reaction

$$C^0 + O_2 \longrightarrow CO_2 \tag{9.48}$$

are equal to

$$-\left(\frac{dM_{O_2}}{dt}\right)_C = \frac{dM_{C^0}}{dt} - \frac{dM'_{C^0}}{dt}. \tag{9.49}$$

Oxygen losses due to a net oxidation of sulfur in pyrite to seawater sulfate via the reaction

$$S_2^{-2} + 2OH^- + \tfrac{7}{2}O_2 \longrightarrow 2SO_4^{-2} + H_2O \tag{9.50}$$

are equal to

$$-\left(\frac{dM_{O_2}}{dt}\right)_S = \frac{7}{4}\left[\frac{dM_{S-2}}{dt} - \frac{dM'_{S-2}}{dt}\right]. \tag{9.51}$$

Thus, when there is no net gain or loss of oxygen,

$$-\left[\frac{dM_{C^0}}{dt} - \frac{dM'_{C^0}}{dt}\right] = -\frac{7}{4}\left[\frac{dM_{S-2}}{dt} - \frac{dM'_{S-2}}{dt}\right]. \tag{9.52}$$

If this equation is combined with equations 9.47 and 9.48, one obtains the expression

$$-\left(\frac{25}{40}\right)\left(\frac{7}{4}\right)\left[\frac{\dfrac{dM_S}{dt}}{\dfrac{dM_C}{dt}}\right] = \frac{{}_0\delta_C - \delta_{C+4}}{{}_0\delta_S - \delta_{S+6}}, \tag{9.53}$$

where

$$\frac{dM_S}{dt} = \frac{dM_{S-2}}{dt} + \frac{dM_{S+6}}{dt} \tag{9.54}$$

and

$$\frac{dM_C}{dt} = \frac{dM_{C^0}}{dt} + \frac{dM_{C+4}}{dt}. \tag{9.55}$$

Equation 9.53 is equivalent to Veizer, Holser, and Wilgus's (1980) equation 5. Rearranging equation 9.53, one obtains

$$_0\delta_C = \left[1.09 \left(\frac{\dfrac{dM_S}{dt}}{\dfrac{dM_C}{dt}}\right)\delta_{S+6} - \delta_{C+4}\right] - 1.09 \left[\frac{\dfrac{dM_S}{dt}}{\dfrac{dM_C}{dt}}\right] {}_0\delta_S. \tag{9.56}$$

The form of this equation is, of course, equivalent to that of equations 9.32 and 9.33. The value of the sulfur input to the oceans today has already been discussed above. The value uncorrected for anthropogenic inputs is approximately 0.53×10^{13} mol S/yr. This is reduced to $(0.27 \pm 0.10) \times 10^{13}$ mol S/yr if the anthropogenic contribution to the estimated river flux is 50%. The carbon input to the oceans is not equal to the HCO_3^- content of river water, because some of the HCO_3^- is derived from atmospheric CO_2 which has participated in chemical weathering of carbonates and which will shortly be returned to the atmosphere. The most likely value for the carbon input rate by weathering (see Holland 1978, p. 272) is $(2.1 \pm 0.4) \times 10^{13}$ mol/yr. Hence the ratio of the two fluxes is

$$\frac{dM_S/dt}{dM_C/dt} = \frac{0.27 \pm 0.10}{2.1 \pm 0.4} = 0.13 \pm 0.09. \tag{9.57}$$

If we use the grand mean values of δ_{S+6} and δ_{C+4} for sulfates and for carbonates of the Phanerozoic Era as proposed by Veizer, Holser, and Wilgus, equation 9.56 becomes

$$_0\delta_C = [1.09(0.13 \pm 0.09)(20.0) - 0.46] - (1.09)(0.13 \pm 0.09) {}_0\delta_S$$
$$= (2.8 \pm 1.8) - (0.14 \pm 0.10) {}_0\delta_S. \tag{9.58}$$

The agreement between epuation 9.58 and equations 9.32 and 9.33 is remarkably good, considering the uncertainties in the derivation of both equations. Their agreement indicates that the losses and gains of oxygen by the operation of the sulfur cycle have indeed been roughly balanced by gains and losses of oxygen in the operation of the carbon cycle.

The approximate agreement between the observed relationship of $\delta^{13}C_{carb.}$ to $\delta^{34}S_{sulf.}$ during the Phanerozoic Eon and the relationship predicted on the basis of the simplified model outlined above also suggests that the simplifications in the modeling were largely justified. The rather minor part that redox reactions of iron have played in the oxygen balance of the atmosphere during the past 600 m.y. was pointed out some time ago (Holland 1978, p. 285). However, the fact that variations in the flux of juvenile volatiles can be neglected and, in particular, that changes in the ratio of the flux of sulfur to the flux of carbon $dM_S/dt/dM_C/dt$, have been small during the past 600 m.y. is somewhat unexpected.

The mechanism by which the carbon cycle is coupled to the sulfur cycle is probably the feedback mechanism proposed earlier (Holland 1978, pp. 284–295). This mechanism does not provide a P_{O_2}-stat, but it minimizes excursions in the atmospheric oxygen pressure, and it assures that the total rate of O_2 usage in weathering is essentially balanced by the rate of O_2 production via the burial at sea of C^0, S^{-2}, and Fe^{+2}.

Even if this feedback mechanism has been controlling the redox balance of the system, the level of atmospheric oxygen could well have varied during the Phanerozoic Eon. It is tempting to use deviation from the regression line in figure 9.17 to reconstruct the history of atmospheric P_{O_2}. This temptation should be resisted. The deviations from the regression line could be due to a large number of factors, including the scatter in the $\delta^{34}S_{sulf.}$ and $\delta^{13}C_{carb.}$ data. Other criteria for the evolution of P_{O_2} during the Phanerozoic Eon will have to be developed before a credible P_{O_2}-time curve can be constructed for this portion of Earth history.

4. The Composition of Phanerozoic Black Shales

As pointed out in chapter 8, the concentration of elements scavenged from seawater into carbonaceous sediments can be used to monitor the geochemical behavior of these elements in the exogenic cycle. The enrichment of many trace elements in lower Proterozoic metasediments of the Outokumpu and Talvivaara regions of Finland is quite similar to their enrichment in Phanerozoic sediments; this suggests that the processes that control the enrichment of trace elements in carbonaceous sediments have not changed dramatically during the past 2 billion years. The body of analytical data for carbonaceous sediments from the Phanerozoic Eon is so much larger than for Precambrian sediments that it seems worthwhile to see what changes in the chemistry of seawater and in the operation of the exogenic cycle are indicated by the observed variations in the composition of these sediments during the past 600 m.y.

The Black Sea is the best-studied of the modern anoxic basins. In some ways this is a pity, because the Black Sea is a rather isolated body of water, which communicates with the open ocean only via the Mediterranean Sea, the Dardanelles, the Sea of Marmara, and the Bosporus. On reaching the Bosporus, seawater is abnormally saline due to evaporation en route from the Atlantic Ocean, and flows into the Black Sea as a bottom-current of ca. 38.5‰ salinity (Gunnerson and Özturgut 1974). The surface return flow through the Bosporus has a salinity of ca. 17.5%. Dilution of the saline inflow to the Black Sea is due to the flux of fresh water into the Black Sea, largely via the Danube and the Dnepr. The trace-metal input to the Black Sea is determined by the composition and flux of the somewhat altered seawater from the Mediterranean and by their influx as a constituent of river waters and river sediments.

The upper 200 m of the Black Sea are oxygenated (see chapter 8). Oxygen is absent below this level, and H_2S is present. Organic matter is destroyed much less completely in such settings than in areas overlain by oxygenated water columns (see, for instance, Demaison and Moore 1980), and the sediments that have accumulated in the Black Sea during periods of seawater influx are unusually carbonaceous. The composition of these sediments has been studied in considerable detail, and the results have been summarized in the monograph edited by Degens and Ross (1974). The sediments are enriched in all of the usual elements. Figure 9.19 shows the correlation between the concentration of molybdenum, vanadium, copper, and nickel in these carbonaceous sediments as a function of their organic carbon content. The concentration of all four metals increases significantly with increasing $C_{org.}$ content. The scatter in the data is due at least in part to variations in the concentration of the metals in the noncarbonaceous sediment fraction. Their Ni content, for instance, is well correlated with their Cr content, and it is likely that some of this correlation is a consequence of a similar detrital provenance for both elements.

Part of the scatter in the correlations of figure 9.19 may be due to analytical problems. Figure 9.20 is a plot of the molybdenum concentration in carbonaceous Black Sea sediments reported by Volkov and Fomina (1974) together with the data reported by Hirst (1974). The difference in the slope of the Mo/C^0 correlation lines through the two sets of data is significant and is not easily explained in terms of differences between the sample sets.

The nonanalytical scatter in the data is presumably due to the operation of the reactions that have combined to enrich these sediments in Mo and in a rather large number of other elements: their precipitation or coprecipitation with organic compounds, oxides, hydroxides, and/or sulfides, reactions during early diagenesis and, possibly, recycling into the overlying water column. The considerable complexity of these processes was discussed in the previous chapter in connection with the behavior of iron and manganese in suboxic sediments; at present these processes are quite poorly understood, and the relationship between the

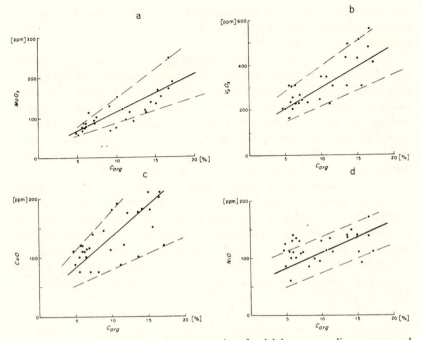

FIGURE 9.19. Correlation between the concentration of molybdenum, vanadium, copper, and nickel and the organic carbon content in sapropelic muds from the Black Sea (Volkov and Fomina 1974). (Reproduced by permisssion of the American Association of Petroleum Geologists.)

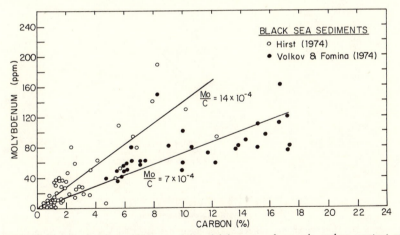

FIGURE 9.20. The correlation between the molybdenum and organic carbon content of sediments from the Black Sea; data from Hirst (1974) and Volkov and Fomina (1974).

concentration of metals in seawater and their concentration in carbonaceous sediments is still largely a matter of empirical observation (see, for instance, Elderfield 1981).

For Black Sea sediments this relationship can be extracted from the available analytical data by plotting the concentration of each trace metal as a function of the Mo content of the sediments, determining the trace metal concentration at a given Mo content, and plotting this concentration against the concentration of the trace metal in seawater. This has been done for the analytical data of Hirst (1974) and Volkov and Fomina (1974). The results for Co, Cu, Ni, V, and Mo are shown in figure 9.21 normalized to a Mo concentration of 200 ppm. There is a good deal of scatter in the diagrams from which the enrichment figures were extracted. Hence, the position of the data points in figure 9.21 is reasonably uncertain. Points that fall on the solid line in this figure define the locus of elements whose concentration in the sediments is 160,000 times their concentration in average seawater. Such an enrichment is not excessive; enrichments up to ca. 500,000 can be explained rather easily by the scavenging of elements from seawater masses into sediments in anoxic basins (Holland 1979).

The scatter of the data points in figure 9.21 around the enrichment line is probably due to a combination of analytical error, the differences between the trace element concentration of average seawater and the Black

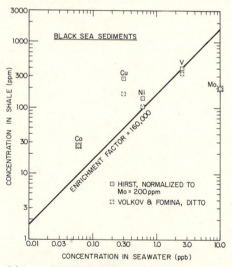

FIGURE 9.21. The enrichment of Black Sea sediments in Co, Cu, Ni, V, and Mo as a function of the concentration of these metals in average seawater; data from Hirst (1974) and Volkov and Fomina (1974); see text for description of method of construction of the diagram. (Reproduced by permission of *Economic Geology*.)

Sea, and the method by which the diagram was constructed. Molybdenum is almost invariably concentrated in anoxic sediments less strongly than vanadium and most of the other enriched elements (see, for instance, Holland 1979). The reasons for the greater enrichment in Cu and Co than in V are not known; they may be related to the behavior of these elements during very early diagenesis.

The enrichment of black shales in these and other metals is also rather variable. The trace element suite of the highly carbonaceous Eocene Suzak Formation has been studied by Poplavko et al. (1977). Their data have been recast in figure 9.22 to show the rough relationship between the enrichment of these sediments in a number of metals and the concentration of these metals in modern seawater. Mo is again enriched less strongly than V; many of the metals scatter mildly around an enrichment line of 500,000; the concentration of a few elements deviates widely from this line. It is not known whether this is because their concentration in the seawater from which the trace metals were extracted differed greatly from their concentration in present-day seawater or whether the deviations are a consequence of depositional and postdepositional processes. The concentration of trace metals is quite variable within the oceans today. This variability is due in large part to biological removal processes in shallow waters, and to the release of biologically sequestered metals during the oxidation of organic matter and the dissolution of tests in mid-water horizons. The data in figure 9.23 show that the release patterns

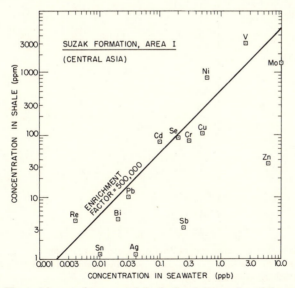

FIGURE 9.22. The enrichment of trace metals in the Suzak Formation of Central Asia; data from Poplavko et al. (1977). (Reproduced by permission of *Economic Geology*.)

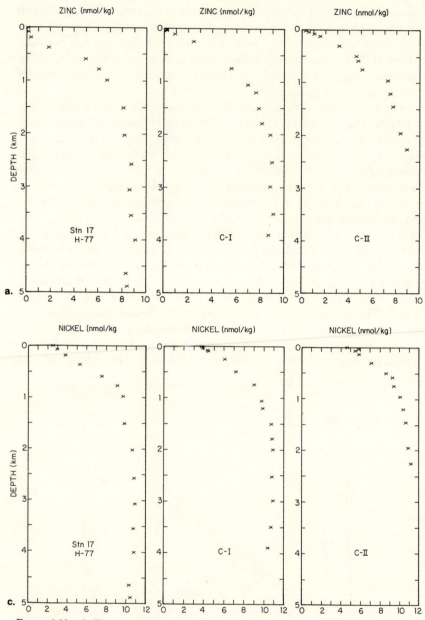

FIGURE 9.23a–d. The concentration of Zn, Cd, Ni, and Cu as a function of depth in waters of the central North Pacific Ocean (Bruland 1980). (Reproduced by permission of Elsevier Scientific Publishing Company.)

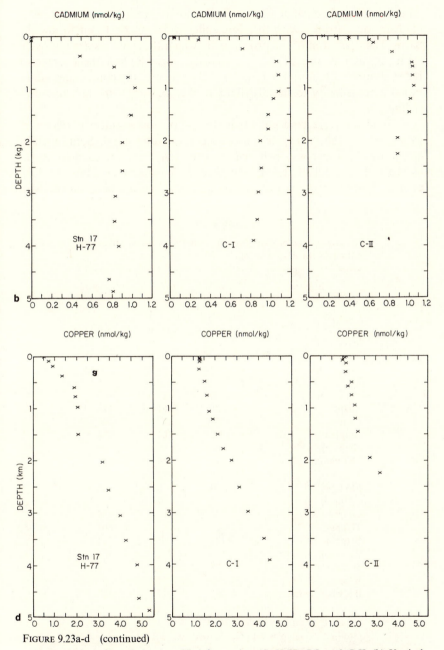

FIGURE 9.23a-d   (continued)

(a) Vertical profiles of zinc (nmol/kg) for station 17, H-77, C-I, and C-II; (b) Vertical profiles of cadmium (nmol/kg) for station 17, H-77, C-I, and C-II; (c) Vertical profiles of nickel (nmol/kg) for station 17, H-77, C-I, and C-II; (d) Vertical profiles of copper (nmol/kg) for station 17, H-77, C-I, and C-II.

differ from metal to metal. The concentration of the metals and the ratio of their concentration can therefore differ from anoxic basin to anoxic basin. The operation of the processes by which these metals are removed from seawater is also probably somewhat variable, and the diagenetic redistribution of material in sediments almost certainly contributes additional heterogeneity to the distribution of trace metals in carbonaceous sediments.

It could therefore be argued that the use of trace-element enrichment as a guide to the history of seawater is a quixotic quest. Some metals are, however, rather well behaved. As expected, these are elements such as Mo and U, which are greatly enriched in carbonaceous sediments. The data in figure 9.20 show that the best correlation line between the Mo

TABLE 9.10.

Analytical data for some Atlantic black shales and neighboring sediments (Brumsack 1980).

Element	Group 1	Group 2	Group 3
C_{org} (%)	6.18	0.69	0.20
S_{total} (%)	2.19	1.44	0.76
$S_{sulfide}$ (%)	1.27	0.65	0.42
$S_{sulfate}$ (%)	0.90	0.80	0.35
Fe_{total} (%)	3.54	4.15	4.43
Fe_{total}–Fe_{py} (%)	2.44	2.89	3.77
Mn (ppm)	910	720	1617
Zn (ppm)	828	161	117
Cu (ppm)	156	117	104
Cr (ppm)	202	164	134
Ni (ppm)	186	90	98
Pb (ppm)	16	14	19
V (ppm)	822	272	207
Mo (ppm)	59	8	3
Ag (ppm)	4.5	0.7	0.8
Cd (ppb)	923	—	933
Tl (ppb)	1757	—	607
Bi (ppb)	141	—	313
Ba (ppm)	530	382	401
Sr (ppm)	278	279	164
B (ppm)	84	73	97
B (carbonate-free) (ppm)	97	117	107

Group 1: 38 samples with C_{org} > 1%
Group 2: 11 samples with C_{org} between 0.5–1%
Group 3: 39 samples with C_{org} < 0.5%
Note that there is some doubt regarding the origin of the $S_{sulfate}$ in these sediments.

and the organic carbon content of Black Sea sediments goes essentially through the origin. Enrichment of Mo in black shales by a factor of 100 over the concentration of this element in average shales (2 to 3 ppm) is quite common, and Mo enrichment by a factor of 1000 has been reported repeatedly (see, for instance, figure 9.22). The Mo/C^0 ratio in these enriched shales is surprisingly constant. Table 9.10 shows the mean composition of black shales deposited in the Atlantic Ocean during the Cretaceous anoxic events, as well as the mean composition of associated, less carbonaceous sediments. The average value of the ratio Mo/C^0 in the most carbonaceous samples (Group 1) is

$$\frac{59 \times 10^{-6} \text{ gm Mo/gm sed.}}{6.18 \times 10^{-2} \text{ gm } C^0/\text{gm sed.}} = 9.5 \times 10^{-4}.$$

This ratio lies between Hirst's (1974) and Volkov and Fomina's (1974) value of the mean Mo/C^0 ratio for Black Sea sediments (see figure 9.20), and is close to the Mo/C^0 ratio of 7.6×10^{-4} in sediments from the Gotland Basin of the Baltic Sea (Manheim 1961; see also Calvert 1976).

Several of the other elements in the Cretaceous sediments of table 9.10 fall close to an enrichment line of 150,000 (see figure 9.24). In fact, the fit to this line is considerably better than the fit of the Black Sea sediment data. Again, Mo is much less enriched than the other trace metals. The data in figure 9.24 are clearly consistent with the proposition that these sediments were enriched in trace metals by processes that extracted metals

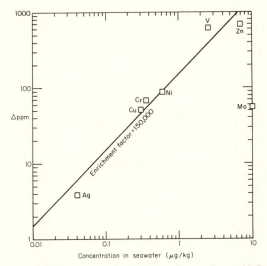

FIGURE 9.24. The metal enrichment of carbonaceous sediments from mid-Cretaceous anoxic events in the Atlantic Ocean; based on the data for Group 1 sediments in table 9.10.

from seawater, and that the relative concentration of the several metals in seawater at the time of enrichment was similar to their relative concentration in seawater today.

The conservative behavior of Mo extends at least as far back in time as the Devonian. The Devonian black shales in the eastern and midwestern United States have been analyzed rather thoroughly, particularly by the U.S. Geological Survey. Table 9.11 summarizes a good deal of the

TABLE 9.11.

Organic carbon, sulfur, uranium, and molybdenum content of Devonian black shales from the eastern United States.

Group	Location	Number of Samples	Organic C (%)	S (%)	U (ppm)	Mo (ppm)
A. Data from Leventhal, Crock, and Malcolm (1981)						
1	Johnson Co., Ky.	10	2.96	2.13	15.3	
2	Wetzel Co., W. Va.	9	0.91	1.14	5.0	
3	McKean Co., Pa.	13	1.44	0.82	4.4	
4	Allegheny Co., Pa.	9	3.01	2.47	15.2	
5	Erie Co., Pa.	10	2.04	1.24	7.1	
6	Indiana Co., Pa.	17	0.61	1.04	3.6	
7	Knox Co., Ohio	13	3.04	1.83	13.9	
8	Ashtabula Co., Ohio	9	2.34	2.10	8.0	
9	Lorain Co., Ohio	12	2.11	1.85	10.6	
10	Grainger Co., Tenn. (Core 7)	17 5	1.83 7.76	1.53 3.37	8.9 30.4	
11	Grainger Co., Tenn. (Core 6)	9 34	3.66 3.42	3.65 2.27	13.2 12.4	
12	Hancock Co., Tenn.	34	3.44	1.52	10.1	
13	Hancock Co., Tenn.	34	1.49	1.08	7.0	
14	Cuyahoga Co., Ohio	35	1.78	0.93	7.7	
15	Tompkins Co., N.Y.	8	0.75	0.77	4.3	
16	Livingston Co., N.Y.	5	4.81	1.76	24.0	
17	Allegany Co., N.Y. (N.Y. 1)	3	3.56	1.32	7.3	
18	Wayne Co., Ill.					
19	Hardin Co., Ill.					
B. Data from Leventhal (1978a,b) and Leventhal and Hosterman (1982)						
20	Perry Co., Ky.	21	4.12	2.28	16.5	52
21	Jackson Co., W. Va.	25	1.94	—	9.1	33
22	Lincoln Co., W. Va.	9	2.9	2.3	11.8	48
23	Cattaraugus Co., N.Y.	20	1.5	1.7	6.7	21
24	Washington Co., Ohio	7	1.2	2.0	6.6	12
25	Carroll Co., Ohio	10	2.4	2.1	10.6	29
26	Wise Co., Va.	12	2.8	2.3	13.4	26
27	Martin Co., Ky.	19	2.2	2.1	10.8	30
28	Overton Co., Tenn.	14	7.2	3.8	43.8	78

data bearing on the organic carbon, sulfur, uranium, and molybdenum content of these shales. The figures are averages of quantitative analyses of drill core samples. They scatter a good deal less strongly than plots of individual analyses along each core; some of this decrease in scatter is probably due to the smoothing effect of the postdepositional redistribution of these elements.

The average Mo concentration in cores for which quantitative Mo analyses are available has been plotted against their average organic carbon content in figure 9.25. The slope of the Mo-C^0 line is close to 12.5×10^{-4}; this falls within the permitted range of the Mo-C^0 slope for Black Sea sediments and is slightly higher than the Mo/C^0 ratio in the Cretaceous carbonaceous sediments from the Atlantic Ocean.

The correlation between the uranium and the carbon content of Devonian black shales of the eastern United States is also excellent (see also Leventhal 1981). Figure 9.26 shows that only one data set (#28, Overton Co., Tennessee) falls decisively above the correlation line. The samples in this set contain significantly more P_2O_5 than those in the other sample sets (Leventhal 1979). Most of the P_2O_5 in these sediments is almost certainly present as a constituent of apatite, which is a fine host for uranium but not for molybdenum. The plot in figure 9.27 of the U vs the Mo concentration of the Devonian black shales emphasizes the presence of a uranium excess and the lack of a Mo excess in sample set 28.

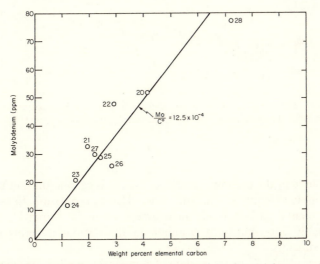

FIGURE 9.25. The correlation between the concentration of molybdenum and organic carbon in Devonian black shales from the eastern United States; data from table 9.11.

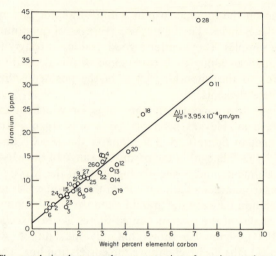

FIGURE 9.26. The correlation between the concentration of uranium and organic carbon in Devonian black shales from the eastern United States; data from table 9.11.

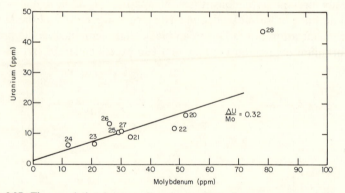

FIGURE 9.27. The correlation between the concentration of uranium and molybdenum in Devonian black shales from the eastern United States; data from table 9.11.

The slope of the uranium enrichment curve is ca. one-third of the slope of the molybdenum enrichment curve. The extrapolation to zero percent elemental carbon yields an intercept of about 1 ppm U, which is presumably contributed by uranium in the detrital minerals of these these sediments.

Unfortunately, there are no data of comparable quality for the uranium content of Black Sea sediments or in sediments of other modern anoxic basins. The data in chapter 8 for the uranium content of the lower Proterozoic black schists of the Talvivaara area (see table 8.8) are not very satisfactory; however their mean U/C^0 ratio falls between 1.6×10^{-4} and 4.0×10^{-4}, and is therefore consistent with the analyses of the Devonian black shales. The reported U/C^0 ratio for the black schists from the Outokumpu Region is approximately 10×10^{-4}.

The sulfur content of the Devonian black shales of the eastern United States is reasonably well correlated with their content of elemental carbon. Figure 9.28 shows, however, that the correlation between the concentration of these two elements is not as good as the correlation between the concentration of Mo, U, and C^0. There is sufficient scatter in the data of figure 9.28 and in the unaveraged S-analyses to cast some doubt on the proper value of the intercept on the sulfur axis and on the interpretation of this intercept in terms of the paleogeography of the basins in which these shales accumulated.

The position of the S-C^0 curve approximately fits the chemistry of modern carbonaceous sediments in the Black Sea ($\bar{C}^0 = 1.32\%$, $\bar{S} = 1.63\%$; Hirst 1974); the Gotland Basin of the Baltic Sea ($\bar{C}^0 = 4.6\%$, $\bar{S} = 2.8\%$); Bunnefjord, Norway ($\bar{C}^0 = 5.44\%$, $\bar{S} = 4.26\%$); and Oslofjord ($\bar{C}^0 = 3.58\%$, $\bar{S} = 1.15\%$) (see Calvert 1976). However, the Cretaceous anoxic sediments of table 9.10 from the Atlantic Ocean seem to contain a good deal less sulfide sulfur than modern and Devonian sediments with the same organic carbon content. There is a good deal of uncertainty regarding the fraction of sulfate sulfur that was derived from the oxidation of sulfide

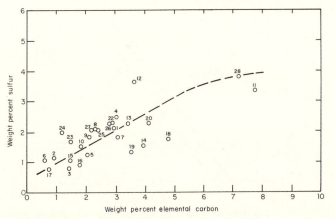

FIGURE 9.28. The correlation between the concentration of sulfur and organic carbon in Devonian black shales of the eastern United States; data from table 9.11.

sulfur prior to the analysis of these samples. This matter requires clarification before much weight can be attached to the apparent, relatively low sulfide sulfur content of these sediments.

A good deal of Leventhal's data for the concentration of trace elements other than Mo and U in Devonian black shales is semiquantitative. Shaffer et al. (1981) have reported quantitative analyses of several trace metals in the highly carbonaceous Henryville bed of the Mississippian-Devonian New Albany shale in the southwestern parts of Indiana. Figure 9.29 summarizes their averages for the concentration of these metals in samples from two drill holes, SDH-273 in Maricon County, Indiana, and SDH-290 in Clark County, Indiana. Both drill cores intersected several black shale horizons; the correlation between the C^0 content and the concentration of the several metals in these horizons is so variable that the rather spectacular correspondence between the data in figure 9.29 and those for more recent sediments comes as a mild surprise. The anomalously high concentration of Pb in the New Albany shale is intriguing. Its meaning will presumably become clearer when there are comparable data for the trace-metal content of contemporary shales from other areas, and when more is known about the relative enrichment of Pb compared to that of the trace metals that fall close to the enrichment lines in figure 9.29.

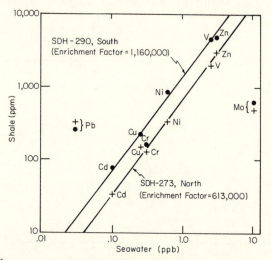

FIGURE 9.29. The average concentration of a number of trace metals in the Henryville bed of the Mississippian-Devonian New Albany shales in southwestern Indiana, U.S.A., plotted against their concentration in average present-day seawater (Shaffer et al. 1981).

It is perhaps significant that the Pb content of the carbonaceous upper Cambrian Alum shale in Sweden is as abnormally high as the lead content of the New Albany shale. The correlation in figure 9.30 of the trace-metal content of the Alum shales with their concentration in present-day seawater is rather poor. It is difficult to say whether this is due to differences in the proportion of these elements in upper Cambrian seawater or whether the efficiency of the extraction processes of these elements into the Alum shale was rather more variable than it is today. The Mo/C^0 ratio in the upper Cambrian Alum shales is ca. 22×10^{-4}; their U/C^0 ratio is ca. 17×10^{-4}. Both of these ratios are well in excess of their value in Black Sea sediments and in the Devonian black shales that were discussed above. Values much lower than average Mo/C^0 ratios have been reported for the upper Jurassic Kimmeridge Shale in England. The C^0 content of these highly carbonaceous sediments ranges from ca. 15 to 70%. Mo is well correlated with C^0, but the mean value of the Mo/C^0 ratio is only $(1.5 \pm 0.2) \times 10^{-4}$ gm/gm. These sediments are enriched in Br and I as well as in Mo, but—rather curiously—their reported concentrations of Ni, Cu, and Zn are essentially equal to the concentration of these metals in average, noncarbonaceous shales (Cosgrove 1970).

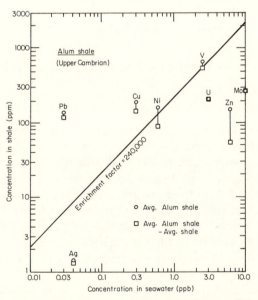

FIGURE 9.30. Correlation between the trace metal content of the Swedish upper Cambrian Alum shale and the concentration of these trace metals in present-day seawater; data from Armands (1972).

The data that have been presented in this section, together with the semiquantitative data of Vine and Tourtelot (1970) confirm the conclusion reached in chapter 8 that the removal of trace metals and sulfur from seawater into carbonaceous sediments has been relatively uniform during the past 2.2 b.y. or so. The ratio of Mo and U to carbon in Phanerozoic carbonaceous shales is not highly variable, and the ratio of these elements to others that are also strongly concentrated in black shales does not vary by orders of magnitude; there are, however, well-documented, sizable variations in all of these ratios in Phanerozoic black shales. The significance of these variations is still unclear. It is impossible to tell which of the variations are due to changes in the composition of average seawater during the Phanerozoic Eon and which are due to changes in the other parameters that influence the trace metal content of black shales. We seem to be stymied in our quest for a precise definition of the trace-metal chemistry of Phanerozoic seawater, and we will probably remain so until we know more about the processes that control the trace metal enrichment of black shales (see, for instance, Elderfield 1981). This is unfortunate; elements such as Mo are enriched both in reduced and in oxidized sediments (Bertine and Turekian 1973); they are therefore potentially important indicators of the oxidation state of the oceans. To date, their concentration in black shales has given indications that are little better than qualitative. Variations in the concentration of such elements in other sediment types are surely worth exploring.

5. Trace Elements in Phosphates and Carbonates

THE DISTRIBUTION OF RARE-EARTH ELEMENTS IN MARINE PHOSPHATE

Phosphates and carbonates are sediment types whose trace-element content is of particular interest in studies of the chemical evolution of seawater. The rare-earth element (REE) distribution in lower Proterozoic apatites has already been described in chapter 8. The depletion of Ce relative to La and Pr in these apatites is quite similar to that in present-day seawater, and indicates that the processes that deplete seawater preferentially in Ce were also operating ca. 2.1–2.3 b.y. ago. The data for the distribution of the REE in Phanerozoic phosphates is somewhat more abundant than for Precambrian phosphates; perhaps for this reason it is also somewhat more confusing.

Wright-Clark et al. (1982) and Wright-Clark, Seymour, and Shaw (1983) have determined the concentration of a number of REE in Phanerozoic conodonts and icthyoliths. The data from the first of these papers are summarized in table 9.12 and in figures 9.31 and 9.32. The Cambrian,

TABLE 9.12.

REE Content of Conodonts, Icthyoliths and the NASC Standard (Wright-Clark et al. 1982)

Age of sample	Location	La	Ce	Sm	Eu	Tb	Yb	Lu
Cambrian (Dresbachian)*	Ash Meadows, Nev.	501 ± 10	1200 ± 20	197 ± 2	17.5 ± 1.6	24 ± 0.6	11.4 ± 0.9	1.38 ± 0.10
Ordovician (Trentonian)*	Dane Co., Wisc.	98 ± 2	206 ± 8	20.6 ± 0.3	3.5 ± 0.4	—	—	—
Ordovician (Trentonian)†	Dane Co., Wisc.	95 ± 2	194 ± 17	14.52 ± 0.19	—	—	—	—
Silurian (Llandoverian-Wenlockian)*	Death Valley, Calif.	179 ± 3	373 ± 11	60 ± 0.7	6.2 ± 0.4	5.8 ± 0.2	2.6 ± 0.4	0.38 ± 0.03
Silurian (Llandoverian-Wenlockian)*	Cap Island, Alaska	25.4 ± 0.5	79 ± 3	20.6 ± 0.2	3.2 ± 0.3	—	—	1.02 ± 0.05
Silurian (Llandoverian-Wenlockian)*	Carnic Alps, Austria	69.2 ± 1.2	210 ± 18	24.7 ± 0.3	6.9 ± 0.8	—	—	0.37 ± 0.05
Devonian (Gedinnian)*	Klamath Mtns, Calif.	44.9 ± 0.8	72 ± 3	5.43 ± 0.07	2.04 ± 0.19	0.50 ± 0.06	—	—
Mississippian (Osagean)*	Lincoln Co., Wyo.	65 ± 2	47 ± 5	22.4 ± 0.3	4.3 ± 0.8	—	—	1.29 ± 0.13
Pennsylvanian (Desmoinesian)*	White Pine Co., Nev.	103 ± 3	52 ± 4	14.38 ± 0.19	6.5 ± 0.8	2.2 ± 0.2	3.9 ± 0.4	0.56 ± 0.06
Pennsylvanian (Desmoinesian)*	White Pine Co., Nev.	90 ± 17	—	13.7 ± 0.2	—	—	3.2 ± 0.2	0.54 ± 0.03
Pennsylvanian (Desmoinesian)†	White Pine Co., Nev.	970 ± 20	444 ± 11	122.3 ± 1.4	27 ± 2	14.5 ± 0.4	39 ± 2	6.6 ± 0.3
Permian (Guadalupian/Capitanian)*	Culbertson Co., Tex.	35 ± 3	—	8.9 ± 0.3	—	—	—	—
Permian (Guadalupian/Capitanian)†	Culbertson Co., Tex.	201 ± 4	159 ± 10	46.8 ± 0.6	8.2 ± 1.1	4.5 ± 0.7	—	—
Triassic (Smithian-Spathian)*	Summit Co., Utah	24.2 ± 0.5	58 ± 5	6.14 ± 0.07	0.90 ± 0.90	0.85 ± 0.06	—	—
North American Shale Composite‡		39	76	7.0	2.0	1.30	3.4	0.60

* Conodont sample.
† Icthyolith sample.
‡ Haskin et al. (1968).

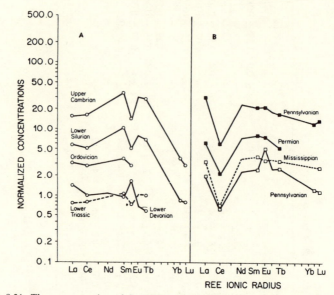

FIGURE 9.31. The concentration of the REE in conodonts (open circles and squares) and in icthyoliths (closed squares) normalized to the North American shale composite (Wright-Clark et al. 1982). (Copyright 1982 by the American Association for the Advancement of Science.)

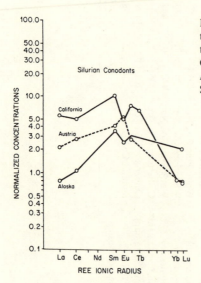

FIGURE 9.32. The concentration of the REE in three Silurian conodont samples normalized to the North American shale composite (Wright-Clark et al. 1982). (Copyright 1982 by the American Association for the Advancement of Science.)

Ordovician, Silurian, and Devonian phosphates have no or only an in-significant negative Ce anomaly. The Carboniferous and Permian phos-phates have a large negative Ce anomaly; however, the single lower Triassic sample does not. Figure 9.32 shows that the REE patterns of Silurian conodonts from three widely separated areas are similar in that none has a significant Ce anomaly; however, the rather large differences between the relative abundance of several of the other rare earths indi-cates that the use of REE patterns in conodonts to reconstruct the REE chemistry of seawater during the Phanerozoic Eon is probably not com-pletely straightforward.

Blokh's (1961) roughly comparable data for the REE content of Paleo-zoic fish bones from the Russian Platform are summarized in Table 9.13. With the exception of one specimen (#257), the time trend of the negative Ce anomaly in Blokh's analyses agrees with that found by Wright-Clark et al. (1982). Lower Paleozoic fish remains do not show a Ce anomaly; two of three Carboniferous samples have a significant negative Ce anom-aly; and the single Permian sample has a slight negative Ce anomaly.

The uptake of much of the REE content of these phosphatic samples must have taken place after the death of the organisms of which they are now a part. It is likely that much, if not all, of the REE was taken up not long after death; but the possibility of much later additions of REE cannot be ruled out. It is therefore likely, but not certain, that the REE pattern in these phosphatic remains is related to the REE pattern of the seawater from which the major portion of their REE budget was obtained.

Not a great deal is known about the REE budget of the present-day oceans and about variations in the REE content and spectrum of sea-water. Martin, Høgdahl, and Philippot (1976) have shown that soluble REE supplied by the Gironde River in France are largely removed in the estuarine zone, and that REE in the solid discharge contain the major part of the REE supplied by rivers to the oceans. Figure 9.33 shows the REE patterns of water of the Gironde River, in surface seawater from the Barents Sea, and in one intermediate and two marine deep-water samples. If these analyses are representative, river and some surface ocean waters have REE patterns rather similar to that of average shale, whereas deeper ocean waters have a pronounced negative Ce anomaly and are relatively depleted in all of the light rare earths compared to surface waters (see Elderfield and Greaves 1982, and Klinkhammer, Elderfield, and Hudson, 1983).

The missing Ce has almost certainly been taken up by ferromanganese nodules. The large positive Ce anomaly in many of these nodules is shown by the data in figure 9.34. No believable mass balance has yet been con-structed to show that ferromanganese nodules are the major scavengers of marine Ce, but there seem to be few, if any, likely alternatives.

TABLE 9.13.

Rare-earth element content of remains of Paleozoic fishes of the Russian Platform according to X-ray spectroscopic analysis (Blokh 1961).

No. in seq.	Sp. no.	Total RE in %	Rare-earth elements, %															Locality	Stratigraphic position	Fossils
			La	Ce	Pr	Nd	Sm	Eu	Gd	Tb	Dy	Ho	Er	Tu	Yb	Lu	Y			
1	18	1.50	9.4	28.2	6.3	21.6	5.2	?	6.3	0.7	3.7	0.5	1.3	?	0.7	?	16.1	Luga River	D_2^3, Lower Luga beds	Unidentifiable scales
2	41	1.93	15.8	29.2	7.5	24.7	4.9	?	4.4	0.5	2.3	0.4	0.8	?	0.3	0.1	9.1	Os'mino	Same	Homostius bones
3	73	1.71	11.0	27.2	7.9	25.8	7.5	?	6.0	0.8	2.7	0.8	0.9	?	0.4	0.3	8.7	Plyussa River, village of Bereshki	"	"
4	85	1.93	12.7	30.7	8.7	16.6	9.6	?	7.9	0.8	1.7	0.4	1.0	?	0.2	0.2	7.5	LugaR., vill. Kleskushi	"	"
5	13	1.25	8.4	29.1	6.7	23.2	4.4	?	5.5	0.6	3.5	0.6	1.4	?	0.6	?	16.0	Lovat'R. vill. Kulakovo	D_3^1, Nadsnezhsk beds	Bothriolepis bones
6	36	0.55	10.8	33.3	6.9	25.6	5.4	?	4.6	0.6	2.2	0.5	0.6	?	0.3	0.2	9.0	Msta R. between vill. of Navoloki and Solodka	"	Holoptychius scales
7	64	1.01	9.7	26.5	6.7	21.3	4.0	?	5.1	0.6	2.8	0.6	1.2	?	0.8	?	20.7	Lovat'R. below Kulakovo	"	Undeterminable bones

No.																		Locality	Horizon	Material
8	67	0.67	8.4	28.1	6.3	23.6	5.6	?	6.3	0.8	3.3	0.6	1.1	?	0.8	—	15.1	Lovat'R. at Kulakovo	" "	Bothriolepis bones
9	69	0.23	Not determined															Lovat'R., Luka	"	Aspidosteus bones
10	43	0.30	Not determined															Msta., village of B. Svetitsy	D_3^3, Middle Famennian	Phyllolepis bones
11	1	1.16	Not determined															Village of Lyubanovo (Oryol district)	D_3^2, Oryol-Saburov sequence	Indeterminable debris
12	192	0.13	Not determined															Livenka R., Livny	D_3^2, Livny beds	Antiarchi bones
13	300	0.33	17.6	19.3	4.8	13.8	5.2	?	4.6	1.1	5.0	1.1	2.6	0.4	1.0	?	23.3	Myachkovo	C_2 Myachkovo horizon	Teeth Polychizodus concavis
14	301	0.19	25.3	19.7	4.4	17.1	4.5	?	5.1	0.9	4.3	0.8	1.9	0.3	0.7	?	15.0	"	To Same	Lagorodus angustus bones
15	257	0.64	3.5	18.5	2.2	17.6	2.7	?	4.8	0.4	6.6	0.9	2.7	?	0.4	?	39.7	Oka R., village of Spas-Teshevo	C_2, Vereyan horizon	Gyracanthusspines
16	254	0.44	19.1	26.4	6.5	22.9	4.4	?	5.1	0.5	3.1	0.5	1.0	?	0.5	0.1	9.9	Village of Petrino near Vyazniki	P_2, Tatarian horizon	Bones of Palaeoniscidae
18			13.6	31.1	4.5	19.4	4.5	?	5.0	1.3	2.9	?	1.3	?	0.9	?	15.5			
257			4.0	9.9	?	13.2	4.6	?	6.1	1.3	6.6	?	4.4	?	2.1	?	47.7			

Results of Turanskaya's X-ray Spectrographic analysis

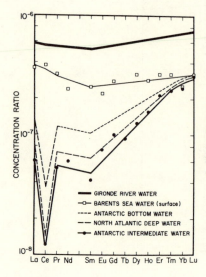

FIGURE 9.33. The shale-normalized concentration of REE in seawater from the Atlantic Ocean (Høgdahl, Melsom, and Bowen 1968) and the Gironde River (Høgdahl 1970). Figure from Piper (1974). (Reproduced by permission of Elsevier Scientific Publishing Company.)

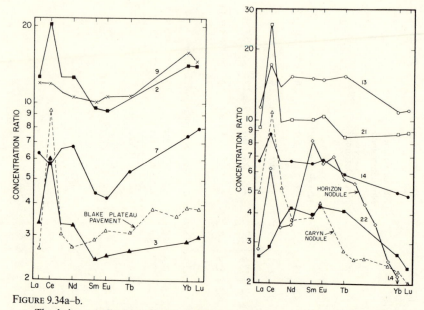

FIGURE 9.34a–b.

a. The shale-normalized concentration of REE in ferromanganese nodules from depths less than 3000 m (Piper 1974).

b. The shale–normalized concentration of REE in ferromanganese nodules from depths greater than 3500 m (Piper 1974).

(Reproduced by permission of Elsevier Scientific Publishing Company.)

The REE pattern of ferromanganese nodules is rather variable; the reasons for this variability are still a matter of some debate. Elderfield and Greaves (1981) have shown that deep-water ferromanganese nodules from the Bauer Basin in the south equatorial Pacific are actually depleted in Ce (see figure 9.35); they propose that the usual REE pattern of these nodules can be explained by the input of hydrothermal iron oxyhydroxides and associated REE to the Bauer Basin and by the transfer of REE to the nodules via diagenetic reactions in the surrounding sediments. These reactions are thought to involve the solution of REEs and their reprecipitation in phosphorus- and iron-rich phases. Ce may, however, be added directly from seawater both to deep sea nodules and to sediments (Elderfield, Hawkesworth, et al. 1981).

The variety of the REE patterns that has been observed in present-day seawater and in modern sediments shows that the marine chemistry of the REE is reasonably complicated. The REE pattern of components of modern sediments, such as foraminifera, seems to reflect the REE pattern

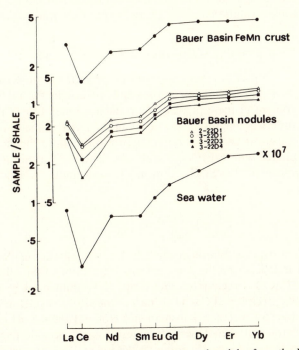

FIGURE 9.35. REE patterns in ferromanganese crusts and nodules from the Bauer Basin (Elderfield and Greaves 1981). (Reproduced by permission of Elsevier Scientific Publishing Company.)

of the seawater in which they were formed. The REE patterns of components such as ferromanganese nodules tend to reflect the processes by which the REE pattern of shallow ocean water is converted into the REE pattern of deep ocean water. It seems likely that conodonts and fish bones are indicators of the REE pattern of ambient seawater, but these objects can be used to define the operation of the oceans only after it has been shown when and from what seawater they have obtained their relatively large complement of REE. Data such as those of Blokh (1961) and Wright-Clark et al. (1982) surely suggest that the removal of Ce from the oceans by ferromanganese nodules has varied significantly during the Phanerozoic Era, but more data are needed to establish the validity of this proposition beyond reasonable doubt. If the lingering doubts are removed, the changes in the REE patterns of seawater during the Phanerozoic Era should be able to tell us a good deal about the history of iron and manganese phases on the ocean floor (see Wright-Clark and Holser 1982). Since there is abundant evidence for the presence of free oxygen in the atmosphere during the Phanerozoic Era (see section 6), the absence of a strong negative Ce anomaly in seawater would not constitute proof of a low oxidation state in the contemporary atmosphere.

The Strontium Content of Marine Carbonates

Some of the complexities that cloud the interpretation of the REE patterns in phosphates and carbonates are virtually absent in the interpretation of the Sr content of these phases. The residence time of Sr in the oceans is several orders of magnitude greater than the mixing time of the oceans, and the concentration of Sr is quite uniform throughout the oceans. Sr^{+2} is coprecipitated with calcite and aragonite. The Sr/Ca ratio of a $CaCO_3$ phase is related to the Sr/Ca ratio in the surrounding seawater by the relationship,

$$\left(\frac{x_{SrCO_3}}{x_{CaCO_3}}\right)_{crystal} = k_{Sr}\left(\frac{m_{Sr^{+2}}}{m_{Ca^{+2}}}\right)_{seawater}. \tag{9.59}$$

The value of the distribution coefficient, k_{Sr}, is smaller for calcite than for aragonite; it depends on the degree of supersaturation of the solution from which the $CaCO_3$ precipitates, and is apt to be influenced by biological factors if the precipitated $CaCO_3$ is part of the shell or skeleton of a living organism (see, for instance, Kinsman and Holland 1969; Katz et al. 1972; see also chapter 6). The observed values of k_{Sr} at oceanic temperatures range from ca. 0.04 to ca. 1.1. Values of k_{Sr} for aragonite in marine oölites and corals tend to be close to 1.0; values for many calcitic organisms are ca. 0.16 ± 0.02. T. G. Thompson and Chow (1955), Turekian and Armstrong (1960), and Pilkey and Goodell (1963) have demonstrated this for mollusks; G. Thompson and Bowen (1969) for coccolithophorids; and Bender, Lorens, and Williams (1975) for foraminifera. The value of k_{Sr} for these groups of organisms is reasonably independent of the Sr/Ca, Mg/Ca, and Na/Ca ratios of the seawater in which they grow (Lewin and Chow 1961; Lorens and Bender 1980).

If, as seems likely, the incorporation of Sr into the shells of calcitic organisms has not varied significantly as a result of evolutionary processes, the Sr content of the shells of organisms can be used to set approximate limits on the Sr/Ca ratio of ancient seawater (see, for instance, Turekian 1955; Lowenstam 1961). The most serious uncertainties in the results of this procedure surround the postdepositional alteration of the Sr content of calcitic organisms. Their recrystallization is frequently accompanied by loss of Sr (see, for instance, figure 6.24). Great care must therefore be exercised in the choice of samples and in the interpretation of the analytical data.

The most thorough and careful study of this type has been reported by Graham et al. (1982), who measured the Sr content of a large number of foraminifera in an attempt to define the Sr/Ca ratio of seawater during the course of the Cenozoic Era. Their sampling was worldwide (see figure 9.36). Their data in figure 9.37 show that there was a small but probably real dip in the Sr/Ca ratio of foraminifera between ca. 4 and 8 m.y.b.p. The mean Sr/Ca atom ratio in *Orbulina* and *G. Sacculifer* today is ca. 1.3×10^{-3}. Six m.y. ago their Sr/Ca ratio was ca. 1.1×10^{-3}. The

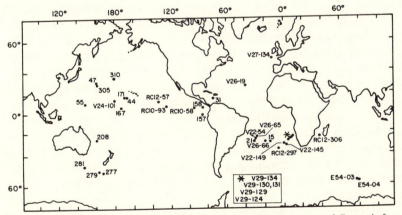

FIGURE 9.36. Distribution of sample sites in the study of the Sr content of Cenozoic foraminifera by Graham et al. (1982). Letters designate Lamont-Doherty collection sites: V = Vema; RC = Robert Conrad; E = Eltanin. (Reproduced by permission of Pergamon Press, Ltd.)

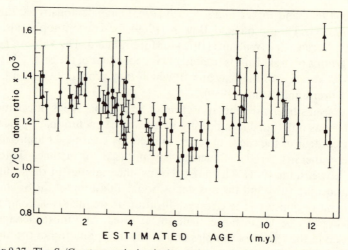

FIGURE 9.37. The Sr/Ca atom ratio in single species of planktonic foraminifera since the middle Miocene (Graham et al. 1982). The length of the bar indicates the statistical counting error. (Reproduced by permission of Pergamon Press, Ltd.)

difference of ca. 15% is probably not due to diagenetic processes, since the Sr/Ca ratio of the foraminifera was observed to return with increasing age to values that are essentially indistinguishable from those of their modern descendants.

Figure 9.38 shows the variation of the Sr/Ca ratio of fossil planktonic foraminifera during the entire Cenozoic Era. The histogram to the left shows the range of Sr/Ca ratios observed in samples of recent foraminifera from the Indian Ocean. This range is comparable to that observed at any given time in the data set plotted in figure 9.37 for samples of foraminifera from the last 10 m.y. of ocean history. The data in figure 9.38 show that there have been no major changes in the Sr/Ca ratio of foraminifera during the past 80 m.y. The small dip during the past 10 m.y. is somewhat obscured; the reality of the suggested dip ca. 50 m.y.b.p. is somewhat questionable. Samples more than 60 m.y. old tend to suffer seriously from the effects of diagenesis, and the overall decrease in the Sr/Ca ratio of foraminifera with increasing age may be due more to progressive alteration than to a progressive decrease in the Sr/Ca ratio of seawater toward the beginning of the Cenozoic Era.

The presence of a slight degree of alteration in samples more than 10 m.y. old is suggested by the trend in the Na/Ca ratios in figure 9.39. The incorporation of Na^+ into the $CaCO_3$ structure is much more complicated than the simple substitution of Sr^{+2} for Ca^{+2}. It is most likely

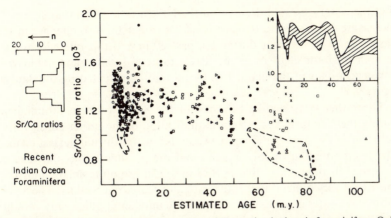

FIGURE 9.38. The Sr/Ca atom ratio in samples of Cenozoic planktonic foraminifera. Only data for samples with counting errors $< \pm 10\%$ in the ratio have been plotted. The dashed line surrounds samples that are altered as shown by SEM observations and $\delta^{18}O$ measurements. The inset is the mean Sr/Ca ratio at the 80% confidence limit of well-preserved samples; these ratios were calculated for 1 m.y. age increments from 0 to 10 m.y.; for 5 m.y. increments from 10 to 60 m.y.; and for 10 m.y. increments from 60 to 80 m.y. (Graham et al. 1982). The histogram to the left shows the variations in the Sr/Ca ratio in Recent Indian Ocean foraminifera. (Reproduced by permission of Pergamon Press, Ltd.)

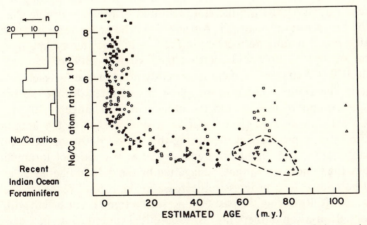

FIGURE 9.39. The Na/Ca ratio in Cenozoic fossil planktonic foraminifera. Only samples for which the counting error in the Na/Ca ratio was $< \pm 10\%$ are plotted. The dashed line surrounds samples for which alteration is indicated by SEM evidence. A histogram of Na/Ca ratios in present-day Indian Ocean foraminifera is shown for comparison (Graham et al. 1982). (Reproduced by permission of Pergamon Press, Ltd.)

that the observed rapid decrease in the Na/Ca ratio of foraminifera with increasing age is due to diagenetic effects rather than to changes in the Na/Ca ratio of seawater during the past 60 m.y. If this inference is correct, then minor changes in the Sr/Ca ratio of foraminifera could well have accompanied the major loss of Na that is indicated by the data of figure 9.39.

The rather conservative behavior of the Sr/Ca ratio in foraminifera during the Cenozoic Era is another indication of the rather conservative behavior of the composition of seawater as a whole during the last 60 m.y. The fluctuations in the Sr/Ca ratio and the possible trends that can be discerned in the data of figures 9.37 and 9.38 are quite small, and their causes are very difficult to identify with any degree of certainty. The Sr/Ca ratio of seawater depends on a rather large number of potentially variable parameters; among these the Sr/Ca ratio of mean river water, the ratio of aragonite to calcite in the $CaCO_3$ output from the oceans, and the rate of seawater cycling through MORs are probably the most important.

The Sr content of seawater as a whole is not particularly sensitive to the rate of seawater cycling through MORs, because the Sr content of seawater is virtually unchanged during cycling through MORs (see chapter 6). However, the Ca concentration of seawater increases dramatically during reaction with MOR basalts at elevated temperatures; hence this process tends to increase the input of Ca to the oceans. At present the flux of Ca from MORs probably accounts for roughly $(10 \pm 5)\%$ of the total flux of Ca to the oceans. Sizable changes in the cycling of seawater through MORs are therefore required to produce sizable changes in the Sr/Ca ratio of seawater. A decrease in the Sr/Ca ratio of seawater of 10% might require an increase in the rate of seawater cycling by ca. a factor of three. This would decrease the $^{87}Sr/^{86}Sr$ ratio of seawater by ca. 0.0020 (see chapter 6); there is presently no evidence for such a large shift in the isotopic composition of Sr in seawater during the period between 6 and 10 m.y.b.p. On the other hand, a change of 10% in the Sr/Ca ratio of average river water or a similar shift in the aragonite/calcite ratio of marine $CaCO_3$ during this period does not seem unreasonably large. The long-term trend in the Sr/Ca ratio of foraminifera during the Cenozoic Era that is suggested by the data in figure 9.38 could well be related to a progressive decrease in the rate of seawater cycling through MORs during the past 60 m.y.; such a trend is consistent with the observed variation of the $^{87}Sr/^{86}Sr$ ratio of seawater during the Cenozoic Era (see figure 6.18) and of the variation of $\delta^{34}S$ in seawater during the same time interval (see figure 9.12). Other interpretations of both curves are, however, also permitted, and the matter is presently unsettled.

6. The O_2 and CO_2 Content of the Phanerozoic Atmosphere

INTRODUCTION

The partial pressure of oxygen and carbon dioxide are among the most interesting parameters of the ocean-atmosphere system. The literature on the subject, particularly that dealing with the evolution of atmospheric oxygen, is voluminous and frequently contradictory (see, for instance, Holland 1962; Berkner and Marshall 1965; Rutten 1970; Walker 1977; and Budyko and Ronov 1979). It is very likely that the level of atmospheric oxygen was lower and the level of atmospheric CO_2 higher in mid-Precambrian time than it is at present (see chapters 7 and 8), but there is little agreement concerning the path by which the partial pressure of the two gases reached their present-day values.

It could be argued that the correlation of the isotopic composition of sulfur in seawater SO_4^{-2} and of carbon in seawater HCO_3^- demonstrates that the oxygen budget of the ocean-atmosphere system has been balanced during the course of the Phanerozoic Eon, and hence that P_{O_2} has not varied significantly during the past 600 m.y. However, the scatter of the data points about the line of best fit in figure 9.17 is quite significant, and could be due to real imbalances between the oxygen demand and the oxygen supply of the combined sulfur and carbon cycles during significant portions of the Phanerozoic Eon. Even rather minor imbalances of this sort can alter the oxygen content of the atmosphere very significantly in geologically rather short periods of time. The residence time of oxygen in the ocean-atmosphere system with respect to use by weathering and production by sedimentation is approximately 4 m.y. (Holland 1978, p. 284). An imbalance of only 5% between O_2 use and O_2 production can therefore increase or decrease P_{O_2} by 50% in ca. 40 m.y. The signature of a 5% imbalance in the sulfur-carbon cycle would be a deviation smaller than the deviation of several of the data points from the line of best fit in figure 9.17. This does not necessarily imply that there were such imbalances, since at least some of the scatter in figure 9.17 is due to inadequacies in the $\delta^{34}S$ and $\delta^{13}C$ data. Some of the scatter may also be due to compensating imbalances in the oxygen budget of the ocean-atmosphere system caused by the operation of the iron cycle. The data in figure 9.17 can therefore do rather little to define the course of the oxygen content of the atmosphere during the Phanerozoic Eon. The same lack of conviction must also greet invocations of oxygen control mechanisms such as those proposed earlier (Holland 1978, pp. 284–287) to maintain P_{O_2} at the present level for geologically long periods of time. The proposed feed-back system almost certainly serves to stabilize P_{O_2}, but it is unlikely to prevent significant variations in the level of atmospheric oxygen during periods of time as long as the Phanerozoic Eon. We must therefore turn to more direct evidence for the level of atmospheric oxygen in the atmosphere during the past 600 m.y.

P_{O_2} During Late Precambrian Time

The geological evidence for defining the level of P_{O_2} during late Precambrian time is still fragmentary. There are no known deposits of detrital uraninite from this period, and there are few major late Precambrian banded iron formations. P_{O_2} could therefore have been well in excess of 0.02 PAL, the value proposed in chapters 7 and 8 for P_{O_2} during early Proterozoic time. The chemistry of late Precambrian paleosols could do much to settle the question whether P_{O_2} was much higher than 0.02 PAL during late Precambrian time, but at present we know virtually nothing about paleosols from this period.

The available paleontological evidence is similarly unsatisfactory. Eukaryotic organisms are basically aerobic. The origin of obligate aerobiosis must therefore predate this cell type. Although the time of origin of the eukaryotes is not well known, they were certainly present 700 m.y. ago, and may have developed as early as 1400 m.y. ago (Schopf and Oehler 1976; Schopf 1978). The oxygen requirements of obligate aerobes have been reviewed by Chapman and Schopf (1983). The available data are still quite limited, but they point to two important conclusions: systemic aerobic biochemistry can operate at oxygen pressures as low as 0.002 PAL O_2 (4×10^{-4} atm); organismic aerobiosis can operate at oxygen pressures as low as 0.01–0.02 PAL O_2 ($2 \times 10^{-3} - 4 \times 10^{-3}$ atm). The latter values of P_{O_2} are similar to those proposed in chapter 7 for the level of atmospheric oxygen during the period between 2.0 and 2.4 b.y.b.p. Eukaryotes could therefore have developed more than 1400 m.y.b.p. Their effect on the level of atmospheric oxygen is not known but could have been significant. Prokaryotic organisms are common and diverse in freshwater environments today, but they are relatively rare compared to eukaryotes in the oceans (see page 414 and Waterbury et al. 1979). McCarthy and Carpenter (1979) have shown that Oscillatoria (Trichodesmium) Thibautii is not apt to be a successful competitor for inorganic nutrients in the company of eukaryotic marine phytoplankton, and that at ambient phosphate levels in oceanic regions the growth of Oscillatoria may be restricted by the availability of phosphate. Evidence is increasing in favor of the notion that cyanobacteria required elevated levels of phosphate to grow at rates equivalent to those of green algae and diatoms (Lehman, Botkin, and Likens 1975).

Eukaryotes may therefore have replaced cyanobacteria in the Precambrian oceans in part because they were more efficient users of the available phosphate. This increased efficiency was probably translated into a higher net rate of photosynthesis in the oceans, and ultimately required a higher value of atmospheric oxygen to assure that the total rate of burial of reduced carbon and sulfur compounds just compensated their rate of weathering on land (see Holland 1978, pp. 284–288). This hypothesis supplies a rationale for the existence of a higher value of P_{O_2} toward the close of the Phanerozoic Eon, but it does not supply proof thereof.

The most direct indication for the level of atmospheric oxygen between 500 and 800 m.y.b.p. is given by the probable physiology of late Precambrian and early Cambrian faunas. Rhoads and Morse (1971) have compiled data for the distribution of benthic invertebrates in low oxygen regions of the Black Sea, the Gulf of California, and the basins off southern California. They have shown that their faunas can be grouped into three major biofacies assemblages, and that these are associated with different concentration ranges of dissolved oxygen. Tunnicliffe (1981) has observed similar relationships in Saanich Inlet on Vancouver Island. When $m_{O_2} \leq 0.1$ ml O_2 NTP/l, marine sediments are essentially devoid of benthic metazoans; when 1.0 ml/$1 > m_{O_2} > 0.3$ ml/l, benthic faunas are composed chiefly of small, soft-bodied infaunal species; when $m_{O_2} > 1.0$ ml/l, the faunas tend to be relatively diverse and to consist of many species that secrete calcareous hard parts.

Figure 9.40 summarizes these findings. Rhoads and Morse (1971) have proposed that the observed progression of faunas in the biofacies of modern oxygen-deficient basins can be regarded as an analogue of the development of metazoan assemblages during the final stages of the Proterozoic Eon and during the early stages of the Phanerozoic Eon. Webby (1970) has described a variety of traces produced by small, wormlike infaunal deposit feeders from the Late Proterozoic of New South Wales, and Goldring and Curnow (1967) have described a diverse assemblage of traces that is associated with the Late Proterozoic (Cowie 1981) Pound Quartzite (Ediacara, South Australia).

The first appearance of a diverse assemblage of unequivocal body fossils has also been reported from the Pound Quartzite (Glaessner 1961; see also Cloud and Glaessner 1982). This fauna consists of approximately 25 species in 15 genera belonging to three phyla: Coelenterata, Annelida, and *incertae sedis*. This large assemblage of soft-bodied species occurs

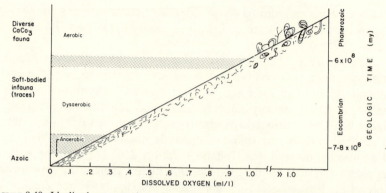

FIGURE 9.40. Idealized cross-section of a Recent marine basin showing the relationship between levels of dissolved oxygen (x-axis) and the benthic fauna (y-axis, left-hand side) (Rhoads and Morse 1971). Data based on the Black Sea (Bacescu 1963), Gulf of California (Parker 1964), Santa Barbara Basin (Emery and Hülsemann 1961), and San Pedro Basin (Hartman 1955, 1966). The right-hand axis relates organism-oxygen relationships in the basin model to the sequential appearance of trace fossils near the Eocambrian glaciation and a calcified fauna at the beginning of the Cambrian. (Reproduced by permission of the editor of *Lethaia*.)

about 200 m stratigraphically below the oldest known shelly fossils of the area, and is separated from them by an unconformity (Stanley 1976). The first diverse benthic assemblage of calcareous fossils occurs in the Lower Cambrian with the appearance of archaeocyathids, brachiopods, and molluscs (see figure 9.41). The faunal changes that accompany an increase in the concentration of dissolved O_2 in the waters of modern marine basins therefore parallel the sequential faunal changes during late Proterozoic and early Phanerozoic time. If the physiology of present-day metazoa is a reliable guide to the physiology of their earliest forebears, the concentration of dissolved O_2 in seawater at the Precambrian-Cambrian boundary must have been ≥ 1 ml O_2/l (45 μmol/kg). The solubility of O_2 in seawater depends both on temperature and salinity, but a concentration of 1 ml O_2/l demands an ambient O_2 pressure ≥ 0.02 atm (≥ 0.1 PAL).

As Rhoads and Morse (1971) have pointed out, the use of faunal sequences in Recent low-oxygen basins as a recapitulation model for the fossil record must be made with reservations, but if these reservations turn out to be unnecessary, the value of P_{O_2} at the opening of the Phanerozoic Era must have been ≥ 0.1 PAL. An upper limit is hard to set. It is possible that P_{O_2} was already considerably higher than 0.1 PAL, and that the evolution of metazoans was not limited by or related to the concentration of oxygen in the contemporary atmosphere.

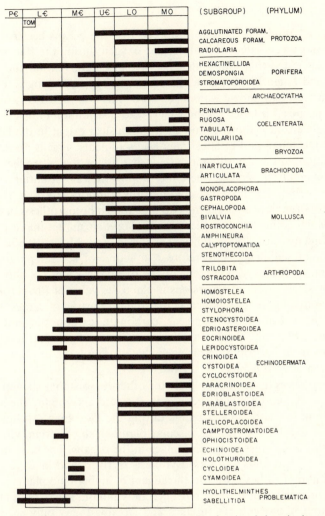

FIGURE 9.41. Currently recognized stratigraphical ranges of major skeletonized taxa of the lower Paleozoic. Minor groups of questionable taxonomic validity are excluded. Only Tommotian groups with significant records in the post-Tommotian strata are included; most others are of uncertain taxonomic affinity. Where sources have specified only the series during which a group appeared, the origin of the group is shown in the middle of the series. Tom = Tommotian stage of Lower Cambrian (Stanley 1976). (Reproduced by permission of the *American Journal of Science*.)

P_{O_2} AND P_{CO_2} DURING THE PALEOZOIC ERA

Berry and Wilde (1978) have proposed that the oceans became progressively more ventilated with molecular oxygen during the Paleozoic Era. They suggested that following successive glaciations in the late Precambrian and Paleozoic, deep, ventilated waters expanded progressively, and that the residual anoxic waters at mid-depths contracted concomitantly. Their proposal is based on the widespread distribution of black shales in lower Paleozoic sediments and their diminished extent in the latter part of the Devonian and in later Paleozoic sediments. This is an interesting suggestion, but it is hardly more than that. Black shales are the products of sedimentation in areas where the ratio of the rate of burial of organic matter to that of noncarbonaceous detritus is abnormally high. The rate of burial of organic matter depends on the rate of photosynthesis in the overlying water column and on the fraction of organic matter that is preserved in sediments (see, for instance, Demaison and Moore 1980). The C^0 content of sediments is therefore a function of several parameters in addition to the oxygen content of the overlying water column. The relatively small extent of excursions in the $\delta^{13}C$ value of Phanerozoic carbonates (see figure 9.17) indicates that the proportion of carbon removed from the oceans as C^0 and as carbonate carbon did not change dramatically during the past 600 m.y. and it is not yet clear just how the changes in the distribution of black shales are related to changes in the dissolved oxygen content of seawater.

Nevertheless it seems very likely for other reasons that a significant increase in P_{O_2} and a major decrease in P_{CO_2} did accompany the conquest of the continents by higher plants during the Devonian (McLean 1978). Prior to the development of vascular plants in the Silurian, virtually all photosynthesis must have taken place in the oceans. The contribution of terrestrial algal mats and photosynthesis in rivers and lakes to the total rate of photosynthesis prior to the advent of higher land plants is hard to define precisely, but it was almost certainly minor. The data in table 9.14 show that photosynthesis in lakes and streams today only accounts for ca. 2% of the net primary production in aquatic environments.

All the best estimates indicate that the photosynthetic productivity on land presently exceeds that in the oceans, perhaps by as much as a factor of two. It seems likely, therefore, that the invasion of the land by vascular plants increased the global rate of photosynthesis significantly, and that the level of atmospheric oxygen increased as a consequence to insure that the total rate of organic carbon burial remained essentially equal to the rate prior to the development of higher land plants.

This development must also have had a considerable effect on the chemistry of weathering. Today the decay of organic matter in soils can increase

TABLE 9.14.

Relationships between ecosystem types, showing area ($\times$ 10^6 km^2) and percentage of the Earth's surface occupied by each, their primary productivity ($\times$ 10^9 metric tons of carbon per year), and their productivity/area ratios (McLean 1978).

Ecosystem type	Area (A) and percentage of earth's surface	Total net primary productivity (P)	P/A
Terrestrial ecosystems			
Tropical rain forest	17.0 (3.4%)	15.3	0.90
Tropical seasonal forest	7.5 (1.5%)	5.1	0.68
Temperate evergreen forest	5.0 (1.0%)	2.9	0.58
Temperate deciduous forest	7.0 (1.4%)	3.8	0.54
Boreal forest	12.0 (2.4%)	4.3	0.36
Savanna	15.0 (3.0%)	4.7	0.31
Woodland and shrub land	8.0 (1.6%)	2.2	0.28
Temperate grassland	9.0 (1.8%)	2.0	0.22
Swamp and marsh	2.0 (0.4%)	2.2	1.10
Tundra and alpine meadow	8.0 (1.6%)	0.5	0.06
Desert scrub	18.0 (3.62%)	0.6	0.03
Rock, ice, and sand	24.0 (4.84%)	0.04	0.002
Marine ecosystems			
Lake and stream	2.5 (0.50%)	0.6	0.24
Continental shelf	26.6 (5.4%)	4.3	0.16
Open ocean	332.0 (66.9%)	18.9	0.06
Upwelling zones	0.4 (0.08%)	0.1	0.25
Algal bed and reef	0.6 (0.12%)	0.5	0.83
Estuaries	1.4 (0.28%)	1.1	0.79

Note: Area and productivity data are from Whittaker and Likens (1975).

the CO_2 content of soil air by as much as factor of 100 (see for instance Holland 1978, pp. 21–22). This in turn increases the rate of decomposition of silicate and carbonate minerals in soil zones, and hence speeds the rate at which CO_2 is removed from the atmosphere and is buried as a constituent of carbonate minerals.

It could be argued that prior to the development of a cover of higher land plants, the CO_2 pressure in the atmosphere had to be comparable to that of soil horizons today in order to effect an equal rate of CO_2 removal from the ocean-atmosphere systems. This, however, turns out to be an overestimate of P_{CO_2} in the pre-Devonian atmosphere. A study of the rate of chemical weathering in a part of Iceland that is virtually free of vegetation today (Cawley, Burruss, and Holland 1969; see also Holland 1978, pp. 103–106) has shown that the required difference between P_{CO_2} prior to and since the advent of higher land plants is probably much less than a factor of 50. There are still too few quantitative data to define the actual difference; the available information suggests that the difference may have been no more than a factor of 5.

A decrease of this magnitude in P_{CO_2} since the Silurian is also much more consistent with the lack of evidence for a major difference in global temperature between the early and the late parts of the Paleozoic Era. Glacial periods seem to have occurred both before and after the advent of higher land plants. If current estimates of the effect of atmospheric CO_2 on world climate are correct, a doubling of P_{CO_2} produces an increase of ca. 2°C in the mean global temperature (see, for instance, Hummel and Reck 1981, and Manabe 1983); it is hard to understand, then, why a dramatic cooling did not accompany the introduction of higher land plants if that introduction decreased P_{CO_2} by as much as a factor of 50.

This analysis is, however, considerably oversimplified. The introduction of a cover of higher land plants almost certainly affected the Earth's climate (Shukla and Mintz 1982) as well as the concentration of atmospheric trace gases that can influence the Earth's greenhouse effect. Nitrous oxide, N_2O, is probably one of these gases. Goreau (1981) has shown that most of the present global inventory of N_2O is probably produced in soils (see also Lipschutz et al. 1981). The greenhouse coefficient of N_2O is large, and its effect on the ozone content of the atmosphere is potentially significant. A thorough analysis of the effect of the development of land plants on the trace constituents of the atmosphere is therefore badly needed. Until this analysis has been done, we can only guess at the climatic consequences of the greening of the land.

A closer look at notions regarding the effects of the development of higher land plants on atmospheric P_{O_2} reveals much the same degree of uncertainty. Before the advent of higher land plants, PO_4^{-3} released during chemical weathering must have made its way directly into rivers and thence to the oceans. Some PO_4^{-3} was surely sequestered by freshwater plants, but these were either reoxidized locally or flushed into the oceans, where they were mixed with and presumably shared the fate of marine organic matter.

Vascular land plants retain a significant portion of the PO_4^{-3} released in soils during chemical weathering. Nearly all of this plant material was probably reoxidized then, as now, in less than 1000 years (see Holland 1978, pp. 262–264). Nevertheless, the residual organic matter brought to the oceans by rivers may well have been a significant fraction of the total organic matter buried with marine sediments. This would not have mattered if the residual terrestrial organic matter were as easily oxidized as average marine organic matter, but that seems unlikely; residual terrestrial organic matter in the oceans is, and presumably has always been particularly resistant to bacterial decay, since it is already a survivor of prolonged bacterial attack prior to entry into the sea (see, for instance, Summerhayes 1981).

A fascinating and still largely unexplained avenue of research into the abundance of atmospheric oxygen was opened by the papers of Heinzinger, Junge, and Schidlowski (1971) and Heinzinger, Schidlowski, and Junge (1974); these authors addressed the evolution of the isotopic composition of atmospheric oxygen. The $\delta^{18}O$ value of atmospheric oxygen is apparently recorded by the oxide phases (mainly magnetite) of cosmic spherules, which are ablation droplets from iron meteorites that are oxidized in the atmosphere at high temperatures (see, for instance, Schidlowski and Ritzkowski 1972). The $\delta^{18}O$ value of magnetite in four recent meteorite falls was found to be $+17.6 \pm 0.4‰$. The $\delta^{18}O$ value of oxide crusts in finds was found to be lower and quite variable, presumably because later oxidation products were included in the analyses together with earlier, high-temperature magnetite.

Atmospheric oxygen today has a $\delta^{18}O$ value of $+23.5‰$ relative to SMOW. Oxygen in the meteorite falls has therefore experienced a fractionation of some $-5.9 \pm 0.4‰$ during the high-temperature formation of magnetite. Table 9.15 reproduces these data and the $\delta^{18}O$ value of Oligocene and Upper Devonian magnetites which are probably meteoritic in origin. The Oligocene spherules have a $\delta^{18}O$ value essentially identical to that of the modern falls. The Devonian magnetite flakes have a considerably lower $\delta^{18}O$ value. If the Devonian sample has retained its original $\delta^{18}O$ value, the $\delta^{18}O$ value of oxygen in the Devonian atmosphere was probably close to $+17.3‰$, some 6‰ lower than that of present-day atmospheric oxygen. If true, this indicates that the difference between the $\delta^{18}O$ value of atmospheric oxygen and the $\delta^{18}O$ value of seawater was considerably smaller then than now (see chapter 6). Since this difference is probably related to fractionation during photosynthesis, the change may have been related to the evolution of plants since the Devonian. Unfortunately, the $\delta^{18}O$ value of the Devonian sample is similar to that of some of the modern finds that have suffered weathering at ground level. It therefore remains to be shown that the lower $\delta^{18}O$ value of the Devonian sample is not simply the result of oxygen exchange between the sample and its environment during the past 350 m.y.

All of the presently available data are still clearly insufficient to permit a meaningful estimate of the effect of the first green revolution on the level of atmospheric O_2. It is likely that P_{O_2} increased significantly during the Devonian, and it is reasonable to believe that the diversification of plants, particularly since the late Cretaceous, has continued to demand

TABLE 9.15.

$\delta^{18}O$ values (‰, SMOW) of recent fusion crusts and fossil ablation products of iron meteorites (Heinzinger, Schidlowski, and Junge 1974).

Sample, age	$\delta^{18}O$ (magnetite)	$\delta^{18}O$ of atmospheric oxygen	Remarks
Recent			
Fusion crust of iron meteorites	17.6 ± 0.4	23.5 ± 0.3	δ-value for crusts represents average of 4 falls (Braunau, N'Goureyma, Treysa, Sikhote Alin)
Tertiary			
Magnetite spherules (aver. 300 μm)	17.4	23.3	Oligocene of Niederrhein Basin (26–38 × 10⁶ yrs ago); value based on 2 aliquots (17.1‰ and 17.6‰)
Devonian			
Magnetite flakes (>50 μm)	11.4	17.3	Nehden Stage of Upper Devonian, Kellerward Mts. (345–360 × 10⁶ yrs ago); value based on 2 aliquots (10.7‰ and 12.1‰)

an increase in atmospheric P_{O_2}; but the evidence and the arguments in favor of these propositions are still decidedly "soft."

OCEANIC ANOXIC EVENTS DURING THE MESOZOIC ERA

The Cenozoic and a good deal of the Mesozoic history of the oceans has gradually been revealed in cores raised by the Deep Sea Drilling Project. Among these revelations, the existence of apparently widespread oceanic anoxic conditions during the Jurassic and Cretaceous periods are surely among the most intriguing. Figure 9.42 shows the location of the approximate position of DSDP sites at the time of deposition of abnormally organic-rich sediments between ca. 90 and 110 m.y.b.p. Jenkyns (1980) has related these oceanic data quite convincingly to the stratigraphy of marine sediments that are now exposed on land. The oceanic anoxic events clearly involved an increase in the intensity and extent of the present midwater oxygen minimum. Late Cretaceous anaerobic environments developed under the influence of an oceanic mid-water oxygen minimum at moderate water depths (500–2500 m; see figure 9.43).

There seems to be a mild correlation between the occurrence of oceanic anoxic events (OAEs) and the high stand of sea level during parts of

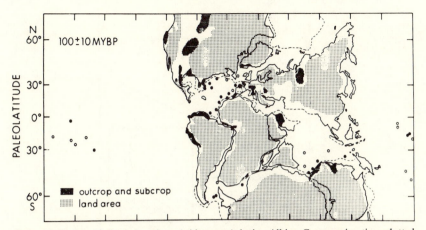

FIGURE 9.42. Land (stippled) and sea (white areas) during Albian-Cenomanian time plotted on a continental reconstruction at 100 ± 10 m.y.b.p. (Arthur and Schlanger 1979; reconstruction from Smith, Briden, and Drewry 1973). Lined areas are regions on continents from which Aptian-Albian or Cenomanian organic carbon-rich marine sediments are known. Circles are Deep Sea Drilling Project sites where sediments of Early to Mid-Cretaceous age were recovered. Filled circles indicate recovery of dark organic-rich marine sediments; empty circles signify no recovery of such sediments. (Modified from Fischer et al. 1977; compiled from numerous sources.) From Arthur and Schlanger 1979. (Reproduced by permission of the American Association of Petroleum Geologists.)

the Cretaceous. However, OAE 1 near the Albian-Aptian boundary does not seem to coincide with a particularly high stand of sea level, and the maximum stand of Cretaceous sea level during the Campanian is not accompanied by an OAE (see figure 9.44). It seems likely that OAEs are produced by a combination of factors, many of which are related to low climatic gradients between the equator and the polar regions. During periods of low climatic gradients, higher temperatures in high latitudes lead to a lower oxygen concentration in surface waters. The rate at which oxygenated surface waters are mixed throughout the oceans also tends to decrease. Deep waters tend to have relatively high temperatures during these periods, and it is likely that this increased the metabolic rates of organisms. The net result of all three factors is to reduce the oxygen content of mid-level and deeper water masses, intensify the oxygen minimum, and produce anoxia in parts of the ocean floor that are normally overlain by mildly oxygenated water masses (Fischer et al. 1977).

The most direct indication of the importance of these events for the carbon imbalance of the atmosphere-ocean system and for the oxygen content of the atmosphere can be read from the isotopic signature of OAEs. A careful study of these signatures has been made by Scholle and Arthur

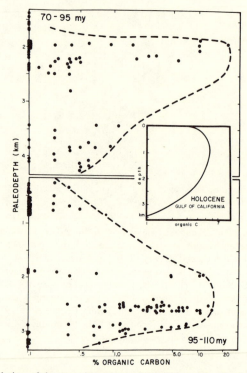

FIGURE 9.43. Variation of the organic content of sediments with paleodepth for the 70–95 and 95–110 m.y.b.p. periods in the South Atlantic (Thiede and van Andel 1977). Dashed line is an envelope to facilitate comparison with the inserted curve representing the relationship between organic carbon content of the surface sediments and water depth of deposition (Holocene) in the Gulf of California (Van Andel 1964); the laminated, diatomaceous continental slope sediments in the Gulf of California have been described in detail (Calvert 1964) and can be related to the mid-water oxygen minimum (Roden 1964) in this region. (Reproduced by permission of Elsevier Scientific Publishing Company.)

(1980), who have analyzed the $\delta^{13}C$ value of more than 1000 closely spaced samples of carbonates from Cretaceous pelagic limestones. Figure 9.45 summarizes their $\delta^{13}C$ data for the Peregrina Canyon section on the eastern coast of Mexico. The $\delta^{13}C$ value of these carbonates rises gradually from 120 m.y.b.p. to ca. 108 m.y.b.p.; it then decreases toward the end of the Cretaceous Period, although there may be an intermediate peak in $\delta^{13}C$ ca. 90 m.y.b.p.

The data in figure 9.46 show that there are similarities in the pattern of $\delta^{13}C$ variations in Cretaceous limestones from several parts of the world; however, the correspondence between the several patterns is by no means

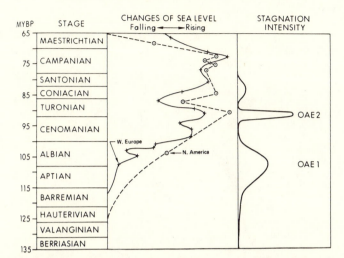

FIGURE 9.44. Relative sea-level stands (Hancock 1975), and timing and intensity of oceanic anoxic events during Cretaceous time (Arthur and Schlanger 1979). Relative intensity of oceanic anoxic events is based on information from amounts of organic carbon preserved, amounts of marine versus terrigenous organic matter, and from carbon isotope studies. (Reproduced by permission of the American Association of Petroleum Geologists.)

perfect. This comes as no surprise, because the magnitude of the observed variations in $\delta^{13}C$ is on the same order as fluctuations due to a combination of geographic variations in the $\delta^{13}C$ value of seawater, differences in the biologic fractionation of carbon isotopes, and the effects of diagenetic processes on the distribution of $\delta^{13}C$ values in limestones.

In the Peregrina Canyon section, the value of $\delta^{13}C$ ranges from $-1.2‰$ to $+4.6‰$ (see figure 9.45); the $\delta^{13}C$ values from most of the other sections (see figure 9.46) range from ca. $+1.0‰$ to more than $+4.0‰$. The positive excursions in $\delta^{13}C$ values are almost certainly the signature of periods in which abnormally large quantities of organic carbon were buried with marine sediments. The stratigraphic coincidence of these periods with OAEs indicates that more organic carbon was indeed buried per unit time during OAE than during non-OAE parts of the Cretaceous Period. Quantitative estimates of the difference in the rate of organic carbon burial during OAEs are difficult to make, because their $\delta^{13}C$ signature differs from section to section and because the intensity of carbon fluxes during the Cretaceous is not well known.

A rough maximum of the differences in the burial rate of organic carbon can probably be obtained by using the data from the Peregrina Canyon section. Prior to the OAE near the Albian-Aptian boundary, $\delta^{13}C_{carb.} \approx +1.0‰$. The difference between the $\delta^{13}C$ value of carbonates and organic

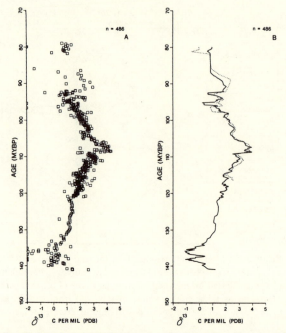

FIGURE 9.45. Detailed carbon isotopic data for the Peregrina Canyon, Mexico section (Scholle and Arthur 1980). (A) Plot of all data points (unfiltered) showing small range of sample scatter. (B) Comparison of smoothed (5 points running average) data from Geochron Labs (dashed line) and Teledyne Isotopes (solid line). These were not analyses of duplicate samples; alternate samples were sent to different laboratories. Thus the plot simulates two independent studies of the section, and indicates a consistent recording of all major isotopic fluctuations in both studies. (Reproduced by permission of the American Association of Petroleum Geologists.)

matter was probably close to the standard value of 25‰ (see Brumsack 1980). From the lever rule it then follows that the fraction, f_1, of carbon leaving the oceans as a constituent of organic matter was

$$f_1 \approx \frac{-\overline{\delta^{13}C_1} + 1.0}{25} \qquad (9.60)$$

where

$$\overline{\delta^{13}C_1} = \text{the isotopic composition of all carbon} \\ \text{deposited from the oceans.}$$

At the peak of the OAE at the Albian-Aptian boundary, the fraction, f, had the value

$$f_2 \approx \frac{-\overline{\delta^{13}C_2} + 4.6}{25}. \qquad (9.61)$$

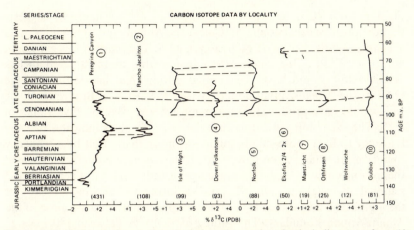

FIGURE 9.46. Carbon isotope ratios of whole-rock samples of pelagic limestone from 10 sections as a function of absolute age (Scholle and Arthur 1980). Carbon isotope data are given as per mil deviation from the PDB standard. Curves represent 5-point running averages with filter to reject points that fall more than 1‰ off average at that level. Numbers in parentheses below each section indicate total number of data points used to construct curve for that locality. (Reproduced by permission of the American Association of Petroleum Geologists.)

If the value of $\overline{\delta^{13}C}$ did not change between ca. 120 and 108 m.y.b.p., then

$$f_2 - f_1 = \Delta f \approx \frac{3.6}{25} = 0.14. \tag{9.62}$$

Thus some 14% more of the total carbon buried per unit time was buried as a constituent of organic matter during the height of the Albian-Aptian OAE than during the preceding non-OAE period. The potential effect of the calculated change in the fraction of carbon buried as a constituent of organic matter could have been sizable. If the total rate of burial of carbon with marine sediments was similar to the present rate, then

$$\Delta C^0 \approx 0.14 \times 36 \times 10^{13} = 5 \times 10^{13} \text{ gm/yr.} \tag{9.63}$$

The corresponding net gain of atmospheric oxygen would have been

$$\Delta O_2 = 5 \times 10^{13} \times \tfrac{32}{12} = 13 \times 10^{13} \text{ gm/yr.} \tag{9.64}$$

If the OAE lasted for 10 m.y., the net gain of O_2 would have been 1.3 $\times$ 10^{21} gm; this is a little more than the entire present inventory of atmospheric oxygen.

The actual effect was probably smaller. The duration of the intense part of the OAE was apparently less than 10 m.y., and comparison of the data for $\delta^{13}C$ in figure 9.45 with those for $\delta^{34}S$ in figure 9.7 suggests that the positive excursion in $\delta^{13}C$ near the Albian-Aptian boundary was accompanied by a negative excursion in $\delta^{34}S$. The sense of this excursion is rather curious. One might expect to find that more rather than less sulfide sulfur would have been buried during than before or after the OAE at the Albian-Aptian boundary (see, for instance, Brumsack 1980). This would have produced an excursion toward more positive values of $\delta^{34}S$. The excursion toward more negative $\delta^{34}S$ values indicates that the tendency to bury more sulfide sulfur was more than offset by a tendency to bury more sulfate sulfur close to the Albian-Aptian boundary. Provided the river flux of SO_4^{-2} did not change significantly during this period, SO_4^{-2} removal in evaporite minerals and/or during the cycling of seawater through MORs must therefore have been particularly intense close to the Albian-Aptian boundary. The ratio $\Delta\delta^{13}C/\Delta\delta^{34}S$ between 140 m.y.b.p. and the Albian-Aptian boundary is ca.

$$\frac{\Delta\delta^{13}C‰}{\Delta\delta^{34}S‰} \approx \frac{4.6 - 0}{14.0 - 17.0} \approx -1.5. \tag{9.65}$$

This is approximately a factor of 10 larger than the slope of the line of best fit in figure 9.17 and the slope for which net oxygen production via the carbon cycle just balances the net loss via the sulfur cycle (see equations 9.32, 9.33, and 9.58). The change in $\delta^{34}S$ therefore indicates that changes in the operation of the sulfur cycle did little to reduce the net gain of atmospheric oxygen via the operation of the carbon cycle during the OAE near the Albian-Aptian boundary. More data are obviously needed from other parts of the oceans to define the true magnitude of the effect of this and other Cretaceous OAEs on atmospheric oxygen, but it is clear that these episodes could have had a significant effect on the level of atmospheric oxygen between 140 and 60 m.y.b.p.

SUDDEN DEATH AT THE END OF THE MESOZOIC

This catchy title introduces the paper by Emiliani, Kraus, and Shoemaker (1981) on the effect of the impact of an asteroid on the Earth's fauna and flora some 65 m.y. ago. The paper claims that the impact of a large extraterrestrial body might have produced a reasonably large transient increase in the global surface temperature, and that this temperature pulse may have been primarily responsible for the observed biologic extinctions at the end of the Cretaceous. The paper adds one more mechanism to the several that have been proposed as important consequences of a major asteroid impact. Darkness, cold, nitrogen oxides, cyanide, and nitric acid have all been discussed as possible consequences and as causes of the major extinction at the Cretaceous-Paleocene boundary (Silver and Schultz 1983).

The history of the impact hypothesis is fascinating. Geologists tend to shy away from explanations involving catastrophes, especially extraterrestrial catastrophes. However, this natural aversion diminished rather quickly when it was discovered that samples taken very close to the time of the Cretaceous-Tertiary extinction, 65 m.y. ago, from deep-sea limestones in Italy, Denmark, and New Zealand contain levels of iridium about 30, 160, and 20 times, respectively, above the normal level (L. W. Alvarez et al. 1980). The analytical data in figure 9.47 can be explained quite well by the impact of an asteroid some 10 ± 4 km in diameter. Other explanations have, of course, also come to mind. The abnormally high Ir concentrations might be local and might be related to normal marine processes that tend to concentrate Ir and other platinum group elements in sediments. Both alternatives are becoming more unlikely. Iridium anomalies have now been reported in 15 marine sites at or very close to the Cretaceous-Tertiary boundary. Sites in Spain, the South Atlantic, the North Pacific, and Texas have been added to the original areas (for references, see Emiliani, Kraus, and Shoemaker 1981). In addition, Orth et al. (1981) have reported the existence of Ir anomalies at the base of a coal layer formed in a swamp in New Mexico at the Cretaceous-Tertiary boundary; this discovery, if it is confirmed, rules out purely marine mechanisms as the cause of the observed Ir anomalies.

Paleontologists have raised a number of objections to the asteroid impact mechanism for extinctions at the Cretaceous-Tertiary boundary. It seemed unlikely to many that the proposed three-year period of darkness following such an impact would produce only the rather limited, though extensive, extinctions observed at the Cretaceous-Tertiary boundary. A reassessment of the length of the "great darkness" following the injection of large quantities of solids into the atmosphere after the proposed asteroid impact has now shown that the darkness might have lasted as little as 3 months, a period that is more consistent with the paleontologic evidence (Pollack et al. 1983).

It is very difficult to develop a detailed description of the consequences of an impact that was sufficiently large to have produced the observed Ir anomalies. The extinctions summarized in table 9.16 can be and have been explained in several ways. Perhaps the most intriguing hypothesis calls on the production of large quantities of nitrogen oxides, their oxidation, and their ultimate rain-out as nitric acid; it has been proposed that the pH of surface seawater was thereby reduced sufficiently to destroy most of the carbonate secreting shallow-water organisms (Lewis et al. 1982).

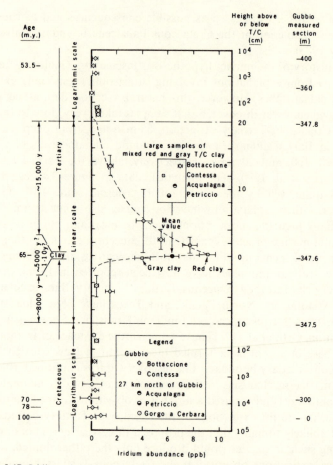

FIGURE 9.47. Iridium abundance per unit weight in 2 N HNO$_3$ acid-insoluble residues from Italian limestone near the Cretaceous-Tertiary boundary (L. W. Alvarez et al. 1980). Error bars on abundances are the standard deviations in counting radioactivity. Error bars on stratigraphic position indicate the stratigraphic thickness of the sample. The dashed line above the boundary is an "eyeball fit" exponential with a half height of 4.6 cm. The dashed line below the boundary is a best-fit exponential (two points) with a half mean and standard deviation of Ir abundance in four large samples of boundary clay from different locations. (Copyright 1980 by the American Association for the Advancement of Science.)

The acceptance of this hypothesis, like that of several others, will have to be preceded by more detailed analyses of their quantitative aspects.

The possible effects of large asteroid impacts on the chemistry of the ocean-atmosphere system are intriguing and potentially important, at least on a short time scale. It will be interesting to see what these effects

TABLE 9.16.

Percentage of Late Cretaceous genera that survived the terminal Cretaceous event (Emiliani, Kraus, and Shoemaker 1981); see this paper for sources of data.

Group	Percentage survival	Group	Percentage survival
Planktonic		*Benthic (neritic)*	
Coccolithophoridae	13	Corals (hermal)	20
Foraminifera	13	Foraminifera	0
Diatoms	31	(Orbitoidiae)	
Dinoflagellates	78	Pelecypods (excl.	
Radiolaria	93	Ostracea and	
		Hippuritacea)	43
Nektonic		Ostracea	32
Ammonoids	0	Hippuritacea	0
Belemnoids	0		
Nautiloids	50	*Terrestrial*	
Elasmobranchii	67	Amphibia	100?*
Osteichthyes	4	Reptilia	
Ichthyosauria	0	Chelonia	23
Pleisosauria	0	Sauropterygia	0
		Squamata	
Benthic		Lacertilia	27
(*bathyal-abyssal*)		Serpentes	0?*
Foraminifera	75–85	Crocodilia	12
		Saurischia	0
		Ornithischia	0
		Pterosauria	0
		Aves	0?*
		Mammalia	52
		Higher plants	69

* Insufficient data.

turn out to be, not only for the event that ended the Mesozoic Eon but for the somewhat smaller impacts that must have occurred both before and after the impact that has given rise to the "lights out" theory of extinction (see, for instance, W. Alvarez et al. 1982 and Turco et al. 1981). It is also surely worth remembering that several nonimpact hypotheses have been advanced during the past several years to explain the biotic crisis at the Cretaceous-Tertiary boundary, and that there is still room for doubt regarding the connection between the impact of extraterrestrial bodies and the history of life on Earth (see, for instance, Rampino and Reynolds 1983) and the interpretation of the observed iridium anomalies (Officer and Drake 1983).

THE LAST 65 MILLION YEARS

The last 65 million years have seen a rather unsteady decrease in the mean annual temperature, culminating in a succession of glacial and interglacial periods. Convincing evidence for the general cooling trend during the Tertiary Era was supplied some time ago by paleobotanical data (see, for instance, Dorf 1964, 1970). The results in figure 9.48 are somewhat impressionistic, but they have been supported rather well by inferences based on the $\delta^{18}O$ value of planktonic and benthic foraminifera. Figure 9.49 shows the inferred decline in the temperature of surface and bottom waters in the North Pacific Ocean (see also Woodruff, Savin, and Douglas 1981); figure 9.50 shows similar data for the subantarctic Pacific Ocean. The $\delta^{18}O$ data for benthic foraminifera show that bottom-water temperatures decreased during the Tertiary, but that this decrease was neither uniform nor monotonic. Tertiary bottom-water temperatures were highest during the Eocene. The Eocene-Oligocene boundary and the mid-Miocene were periods of particularly rapid temperature decline. Typical bottom-water temperatures today are in the vicinity of 1 to 2°C.

The climate of the past 1–2 m.y. has been dominated by alternating glacial and interglacial periods; each cycle has lasted approximately 100,000 years. The pattern of temperature variations during these cycles is well illustrated by the data in figure 9.51 (see also T. F. Anderson and Steinmetz 1981). At the end of each interglacial period the mean annual

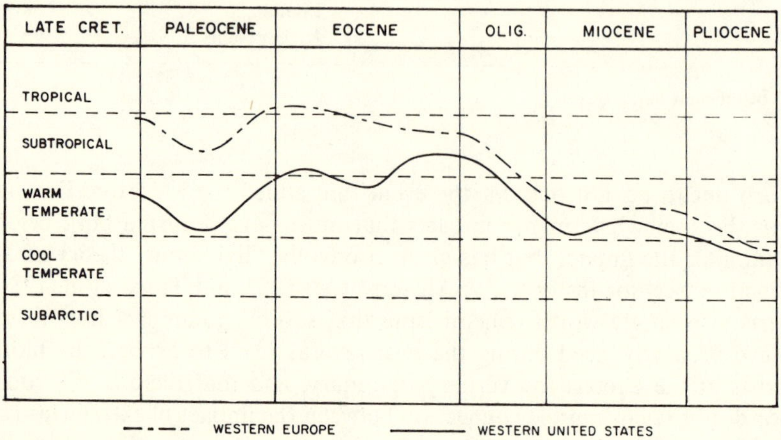

FIGURE 9.48. The climatic regimes of western Europe and the western United States during the Tertiary Era as inferred from paleobotanical data (Dorf 1964). (Reproduced by permission of John Wiley & Sons, Inc.)

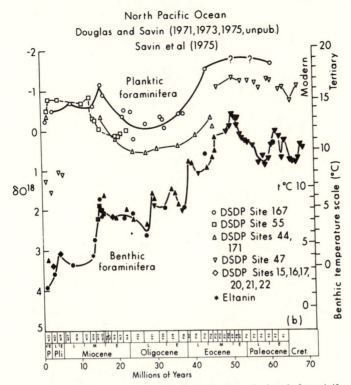

FIGURE 9.49. Isotopic paleotemperature data from Tertiary planktonic foraminifera (open symbols) and benthic foraminifera (closed symbols), primarily from the North Pacific (Savin and Yeh 1981). (Reproduced by permission of John Wiley & Sons, Inc.)

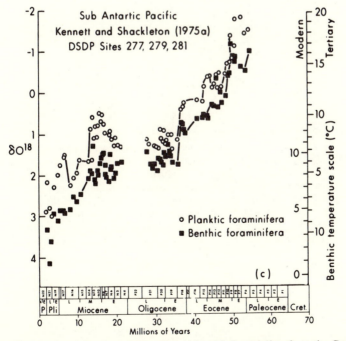

FIGURE 9.50. Isotopic analyses of planktonic and benthic foraminifera from the Campbell Plateau and the Macquarie Ridge (Subantarctic Pacific) by Shackleton and Kennett (Savin and Yeh 1981). (Reproduced by permission of John Wiley & Sons, Inc.)

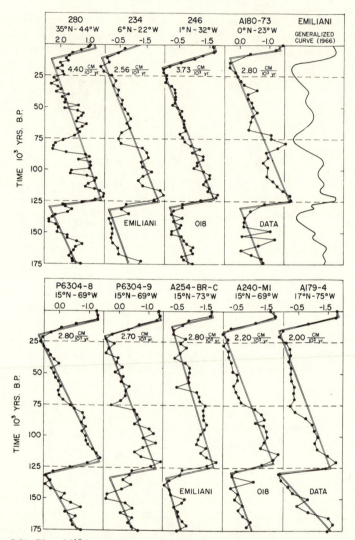

FIGURE 9.51. Plot of $\delta^{18}O$ versus time in the upper part of nine of the eleven North Atlantic and Caribbean cores analyzed by Emiliani (1955, 1964, 1966). The time scale was obtained by assigning an age of 127,000 years to the midpoint of termination II and assuming the sedimentation rate to be constant. Since sedimentation rates during cold periods (i.e., high $\delta^{18}O$) are probably somewhat different from those during warm periods, the age of the point intersected by the dotted line for 75,000 years could be in error by $\pm 10,000$ years. The extent of this distortion would vary from core to core. The idealized curve of Emiliani (1966) is given for comparison (Broecker and van Donk 1970). (Copyright by the American Geophysical Union.)

temperature decreased gradually but somewhat unsteadily; after the minimum temperature was reached, the climate rebounded rapidly to that of an interglacial period.

The causes of these fluctuations are still somewhat obscure. There is rather strong support for the astronomical theory of climate change, which proposes that variations in the seasonal and latitudinal distribution of incoming solar radiation due to long-term variations in the Earth's orbital parameters have exercised a significant effect on global climate during the last 2 m.y. However, the degree to which the climatic record of glacial and interglacial fluctuations is actually a result of orbital controls has not yet been determined. The orbital parameters that are expected to have a significant effect on climate are not always those that are actually reflected in the climatic record. This is particularly true of the important periodicity close to 100,000 years, which should play no more than a minor role in climatic changes if these changes are controlled directly by the Earth's orbital parameters (Kominz and Pisias 1979).

It has always been considered possible that changes in atmospheric chemistry have influenced the alternation of glacial and interglacial periods, but believable data for the composition of the atmosphere during the last glacial cycle have been lacking. Delmas, Ascencio, and Legrand (1980) and Neftel et al. (1982) have shown that the concentration of CO_2 in air bubbles trapped in ice during the height of the last glacial period was almost certainly significantly lower than that of air trapped since the end of this period. Pertinent data are shown in figures 9.52. Approximate age scales are attached to each set of data. The few erratic high CO_2 values are thought to be due to contamination by drilling fluid. The change in $\delta^{18}O$ with depth in these cores is an indicator of the air temperature at the time of accumulation of snow from which the ice in the cores formed. The increase in $\delta^{18}O$ at a depth of about 1100 meters in both cores marks the end of the last glacial period. The nonlinearity of the age-depth relationship is related to foreshortening as the result of ice flow. Although these data are still rather scant, they suggest very strongly that at the height of the last glacial age P_{CO_2} was only ca. 2/3 of its value some 100 years ago, i.e., before fossil-fuel burning had increased P_{CO_2} significantly.

The lower CO_2 pressure during the last glacial period would certainly have served to amplify the effect of orbital and/or other parameters that may have been responsible for the climatic fluctuations of the past million years. It remains to be seen how much of an amplification was supplied by the observed change in P_{CO_2}, why P_{CO_2} was so much lower some 10,000 years ago, and why it rose so rapidly at the end of the last ice age (see, for instance, Broecker 1982a,b).

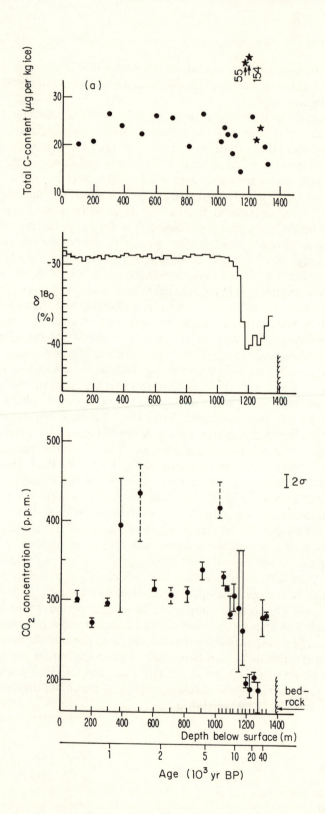

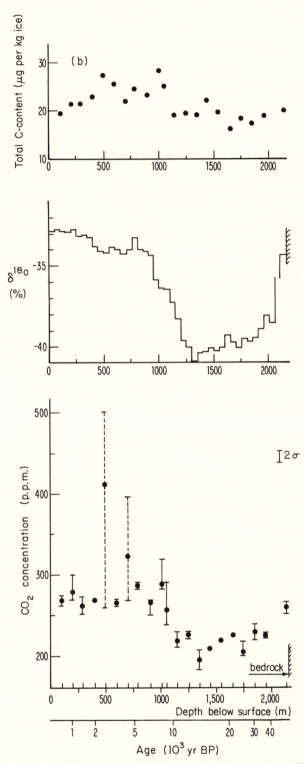

FIGURE 9.52. CO_2 concentration in the bubbles and total carbonate content of (a) the Camp Century core; (b) the Byrd core. The CO_2 concentrations are presented with the lowest, highest, and median values for each depth. The dashed lines indicate depths where drill fluid was observed in the large samples. All samples marked by a star were alkaline, and acid was added during the wet extraction (Neftel et al. 1982). (Reproduced by permission of *Nature*.)

The CO_2 pressure in the atmosphere is potentially quite variable on a rather short time scale, because the quantity of carbon present in atmospheric CO_2 is roughly equal to that of the quantity of carbon in the biosphere and only a few percent of the carbon present in marine HCO_3^-. Carbon in these three reservoirs is in excellent communication and mixes on a time scale of ca. 500 years. Relatively small transfers of carbon within these reservoirs can therefore have a sizable impact on the partial pressure of CO_2 in the atmosphere. This is not true for atmospheric oxygen. Major changes in the size of the biosphere and the dissolved oxygen content of seawater can exercise only a minor influence on P_{O_2}. Changes in P_{O_2} must apparently await the consequences of imbalances in the rate of use and production of this gas during weathering and sedimentation. It is likely, therefore, that P_{O_2} was affected very little by the climatic fluctuations of the past million years, and that P_{O_2} has been close to its present value at least during the past several million years. A backward extrapolation of P_{O_2} beyond the Pliocene-Miocene boundary is probably not warranted. No major excursions in $\delta^{13}C$ and $\delta^{34}S$ have been observed during the Cenozoic (see, for instance, Savin and Yeh 1981; Margolis, Kroopnick, and Showers 1981); however, it was shown above that the Cretaceous OAEs could have had a major effect on atmospheric P_{O_2}, and $\delta^{13}C$ shifts of approximately this magnitude have occurred in the Paleocene, Eocene, and Oligocene (Arthur 1982). The physiology of Tertiary mammals may ultimately set the most stringent limits on both upward and downward excursions of P_{O_2} during their age.

7. Summation

The data assembled in this chapter allow us to say a good deal about the Phanerozoic history of seawater and the atmosphere, but not nearly as much as we might wish. The apparent invariance of mineral sequences in marine evaporites, the relatively small variations in the Br^-/Cl^- ratio of basal halites in marine evaporites, and the composition of fluid inclusions in halites all suggest that the concentration of the major ions in seawater has not varied a great deal during the Phanerozoic Eon. The imposed limits on excursions in the composition of Phanerozoic seawater are, however, fairly broad. They are not in conflict, for instance, with the proposal that the Mg^{+2} content of Paleozoic seawater was only ca. half that of present-day seawater, and they allow considerable latitude for fluctuations in the Ca^{+2} and SO_4^{-2} concentration of Phanerozoic seawater (see figure 9.2).

It seems very likely that such fluctuations have actually occurred. The isotopic composition of sulfur in seawater SO_4^{-2} has varied dramatically during the past billion years (see figure 9.12). These variations must be due in large part to fluctuations in the ratio of the output of sulfide-sulfur to that of sulfate-sulfur from the oceans. The fluctuations of this ratio must have been caused by a combination of geographic, chemical, and biologic factors. Unfortunately it is still impossible to relate the observed variations in $\delta^{34}S$ to variations in the concentration of Ca^{+2} and SO_4^{-2} in Phanerozoic seawater.

The major features of the $\delta^{34}S_{SO_4^{-2}}$-time curve during the last 600 m.y. is similar to that of the variation of the $^{87}Sr/^{86}Sr$ ratio of seawater (figure 6.24) and, possibly, to the variation of the lithium excess in Phanerozoic sediments (figure 6.18) and of the Sr/Ca ratio of seawater (figure 9.37). All four parameters are potentially coupled through the rate of seawater cycling through MORs. The data for all four are consistent with a relatively high rate of seawater cycling during the mid-Phanerozoic compared to that in the early and late parts of this Eon. Although this is an attractive explanation, the similarity of the course of the four curves could be fortuitous, since each parameter depends on so many other unrelated or poorly coupled processes.

The large swings in $\delta^{34}S_{SO_4^{-2}}$ were almost certainly accompanied by large net changes in atmospheric oxygen due to the operation of the sulfur cycle. However, the accumulation of a very large number of $\delta^{13}C$ values for marine limestones has revealed a covariance of $\delta^{13}C_{CO_3^{-2}}$ with $\delta^{34}S_{SO_4^{-2}}$ in seawater during the Phanerozoic Eon (see figure 9.17). This covariance indicates that during periods of net O_2 loss due to the operation of the sulfur cycle, roughly equal gains of O_2 were posted due to the operation of the carbon cycle. This is consistent with the theorem that the overall redox state of the Earth's crust has remained essentially constant. Apparently, imbalances incurred by the sulfur cycle have been balanced, at least roughly, by the cycle of carbon, which is the other element of major importance for the redox state of the atmosphere-ocean-crust system.

Several sediment types yield useful data for the concentration of minor elements in seawater. Black shales concentrate many metals, and the pattern of their enrichment has been surprisingly constant. The data for metals such as Mo and U, for which the largest number of quantitative analyses are available, give the strongest indication of this constancy. The concentration of many of the other metals in black shales is more variable. It remains to be seen whether this variability is due to changes in the concentration of these elements in seawater or to differences in the processes by which they were removed from the oceans in the course of the last 600 m.y.

The distribution of the rare-earth elements in Phanerozoic marine phosphates suggests that the concentration of these elements in seawater has varied significantly with time. The strong negative Ce anomaly in present-day seawater is almost certainly related to the preferential removal of Ce^{+4} with manganese-iron nodules. The apparent absence of a Ce anomaly during the early part of the Phanerozoic Eon and during at least part of the Triassic (see figure 9.31) suggests that this sink for Ce has not always been as important as it is today. More analytical data are, however, needed both for the marine geochemistry of the REE today and for the distribution in ancient marine phosphates.

The chemistry of the atmosphere has almost certainly changed significantly during the Phanerozoic Eon. $CaCO_3$-secreting organisms probably could not have developed during the early part of the Cambrian unless P_{O_2} was greater than ca. 0.1 PAL (0.02 atm). P_{O_2} must therefore have risen significantly between 2.0 and 0.6 b.y.b.p. It is not yet clear whether this increase occurred between 2.0 and 1.0 b.y.b.p. or whether P_{O_2} rose shortly before the opening of the Cambrian.

It seems very likely that the development of the higher land plants and their conquest of the continents led to a significant increase in P_{O_2} and to a significant decrease in P_{CO_2} during the Devonian. A further increase in P_{O_2} may have occurred during the several oceanic anoxic events of the Cretaceous, but it seems likely that P_{O_2} has been roughly equal to its present value of 0.2 atm during the Cenozoic Era. The analyses of air bubbles trapped in Arctic and Antarctic ice indicate that the CO_2 content of the atmosphere was close to 2.0×10^{-4} atm during the height of the last ice age and that P_{CO_2} has remained close to 2.8×10^{-4} atm since the onset of warmer climates some 10,000 years ago.

The partial pressure of N_2 has probably changed rather little during the Phanerozoic Eon. The partial pressure of Ne, Ar, and Kr has probably increased somewhat in response to mantle and crustal degassing, while the He pressure has probably remained roughly constant and adjusted so that the input of He from degassing of the crust and mantle has been balanced by He escape into interplanetary space.

It may be well to close with a retrospective look at the processes and mechanisms that have shaped the evolution of the atmosphere and oceans. The two driving forces have clearly been heat generation within the Earth and energy transfer to the Earth from the Sun. The effects of the Sun have been modified strongly by the biosphere. This has led Lovelock and Margulis (1974a, 1974b) to suggest that the atmosphere of the Earth flows in a closed system controlled by and for the biosphere, and that the Earth is a planet whose surface physical and chemical state is in homeostasis at an optimum set by the contemporary biota, which expends energy to maintain the optimum.

Lovelock and Margulis present this as the "Gaia" hypothesis. I find the hypothesis intriguing and charming, but ultimately unsatisfactory.

The geologic record seems much more in accord with the view that the organisms that are better able to compete have come to dominate, and that the Earth's near-surface environments and processes have accomodated themselves to the changes wrought by biological evolution. Many of these changes must have been fatal or near-fatal to parts of the contemporary biota. We live on an Earth that is the best of all possible worlds only for those who are well adapted to its current state.

One curious aspect of Earth history is the continuity of life during the past 3.8 b.y. I believe that this continuity is a consequence of the relative dullness of Earth history, of the rarity and relatively small magnitude of disruptive events such as asteroid impacts, of the variety of physical and chemical control mechanisms that have tended to maintain the status quo and to damp out rather than to amplify fluctuations, and of evolutionary processes that did not destroy the ecology of the Earth as a whole. It remains to be seen whether this long record will be sustained in the presence of modern man.

References

Alvarez, L. W., Alvarez, W., Asaro, F., and Michel, H. V. 1980. Extraterrestial cause for the Cretaceous-Tertiary extinction. *Science* 208: 1095–1108.

Alvarez, W., Asaro, F., Michel, H. V., and Alvarez, L. W. 1982. Evidence for a major meteorite impact on the Earth 34 million years ago: Implications for Eocene extinctions. *Science* 216:885–888.

Anderson, R. N., Honnorez, J., Becker, K., Adamson, A. C., Alt, J. C., Emmermann, R., Kempton, P. D., Kinoshita, H., Laverne, C., Mottl, M. J., and Newmark, R. L. 1982. DSDP Hole 504B, the first reference section over 1 km through layer 2 of the oceanic crust. *Nature* 300: 589–594.

Anderson, T. F., and Steinmetz, J. C. 1981. Isotopic and biostratigraphical records of calcareous nannofossils in a Pleistocene core. *Nature* 294: 741–744.

Armands, G. 1972. Geochemical studies of uranium, molybdenum, and vanadium in a Swedish alum shale. *Stockholm Contr. Geology* 27: 1–148.

Arthur, M. A. 1982. The carbon cycle: Controls on atmospheric CO_2 and climate in the geologic past. In *Climate in Earth History*, chap. 4. Geophysics Study Committee, National Research Council. Washington, D.C.: National Academy Press.

Arthur, M. A., and Schlanger, S. O. 1979. Cretaceous "Oceanic Anoxic Events" as causal factors in development of reef-reservoired giant oil fields. *Bull. Amer. Assoc. Petr. Geol.* 63:870–885.

Ault, W. U., and Kulp, J. L. 1959. Isotopic geochemistry of sulfur. *Geochim. Cosmochim. Acta* 16:201–235.

Bacescu, M. 1963. Contribution à la biocoenologie de la Mer Noire: L'étage périozoique et le faciès dreissenifère, leurs charactéristiques. *Proc. Verb. Réun. C.I.E.S.M.* 17:107–122.

Bender, M. L., Lorens, R. B., and Williams, D. F. 1975. Sodium, magnesium and strontium in the tests of planktonic foraminifera. *Micropaleont.* 21:448–459.

Berkner, L. V., and Marshall, L. C. 1965. On the origin and rise of oxygen concentration in the Earth's atmosphere. *Jour. Atm. Sci.* 22:225–261.

Berner, R. A. 1980. *Early Diagenesis: A Theoretical Approach.* Princeton, N.J.: Princeton University Press.

———. 1982. Burial of organic carbon and pyrite sulfur in the modern ocean: Its geochemical and environmental significance. *Amer. Jour. Sci.* 282:451–473.

Berner, R. A., and Raiswell, R. 1983. Burial of organic carbon and pyrite sulfur in sediments over Phanerozoic time: a new theory. *Geochim. Cosmochim. Acta* 47:855–862.

Berry, W.B.N., and Wilde, P. 1978. Progressive ventilation of the oceans: An explanation for the distribution of the lower Paleozoic black shales. *Amer. Jour. Sci.* 278:257–275.

Bertine, K. K., and Turekian, K. K. 1973. Molybdenum in marine deposits. *Geochim. Cosmochim. Acta* 37:1415–1434.

Bischoff, J. L., and Dickson, F. W. 1975. Seawater-basalt interaction at 200°C and 500 bars: Implications for origin of sea-floor heavy-metal deposits and regulation of seawater chemistry. *Earth Planet. Sci. Lett.* 25:385–397.

Block, J., and Waters, O. B., Jr. 1968. The $CaSO_4$-Na_2SO_4-NaCl-H_2O system at 25° to 100°C. *Jour. Chem. Eng. Data* 13:336–344.

Blokh, A. M. 1961. Rare earths in the remains of Paleozoic fishes of the Russian Platform. *Geochemistry no.* 5:404–415.

Borchert, H., and Muir, R. O. 1964. *Salt Deposits: The Origin, Metamorphism, and Deformation of Evaporites.* New York: van Nostrand.

Braitsch, O. 1962. *Entstehung and Stoffbestand der Salzlagerstätten.* Berlin: Springer-Verlag.

———. 1971. *Salt Deposits: Their Origin and Composition.* New York: Springer-Verlag.

Braitsch, O., and Herrmann, A. G., 1963, Zur Geochemie des Broms in salinaren Sedimenten. Teil I: Experimentelle Bestimmung der Br-Verteilung in verschiedenen natürlichen Salzsystemen. *Geochim. Cosmochim. Acta* 27:361–391.

Broecker, W. S. 1982a. Glacial to interglacial changes in ocean chemistry. *Prog. Oceanog.* 11:151–197.

———. 1982b. Ocean chemistry during glacial time. *Geochim. Cosmochim. Acta* 46:1689–1705.

Broecker, W. S., and van Donk, J. 1970. Insolation changes, ice volumes, and the O^{18} record in deep-sea cores. *Rev. Geophys. and Space Phys.* 8 169–198.

Bruland, K. W. 1980. Oceanographic distributions of cadmium, zinc, nickel, and copper in the North Pacific. *Earth Planet. Sci. Lett.* 47:176–198.

Brumsack, H.-J. 1980. Geochemistry of Cretaceous black shales from the Atlantic Ocean (DSDP Legs 11, 14, 36, and 41). *Chem. Geol.* 31:1–25.

Budyko, M. I., and Ronov, A. B. 1979. Chemical evolution of the atmosphere in the Phanerozoic. *Geochemistry International* 16(3):1–9.

Calvert, S. E. 1964. Factors affecting distribution of laminated diatomaceous sediments in Gulf of California. In *Marine Geology of the Gulf of California*, 311–330. Memoir 3, Amer. Assoc. Petrol. Geol., Tulsa, Okla.

———. 1976. The mineralogy and geochemistry of near-shore sediments. In *Chemical Oceanography*, 2d. ed. vol. 6, edited by J. P. Riley and R. Chester, chap. 33. New York: Academic Press.

Cawley, J. L., Burruss, R. C., and Holland, H. D. 1969. Chemical weathering in Central Iceland: An analog of pre-Silurian weathering. *Science* 165:391–392.

Chapman, D. J., and Schopf, J. W. 1983. Biological and biochemical effects of the development of an anaerobic environment. In *Origin and Evolution of Earth's Earliest Biosphere: An Interdisciplinary Study*, edited by J. W. Schopf, chap. 13. Princeton, N.J.: Princeton University Press.

Chiba, H., Kusakabe, M., Hirano, S.-I., Matsuo, S., and Somiya, S. 1981. Oxygen isotope fractionation factors between anhydrite and water from 100 to 550°C. *Earth Planet. Sci. Lett.* 53:55–62.

Claypool, G. E., Holser, W. T., Kaplan, I. R., Sakai, H., and Zak, I. 1980. The age curves of sulfur and oxygen isotopes in marine sulfate and their mutual interpretations. *Chem. Geol.* 28:199–260.

Cloud, P., and Glaessner, M. F. 1982. The Ediacarian Period and system: Metazoa inherit the Earth. *Science* 218:783–792.

Cosgrove, M. E. 1970. Iodine in bituminous Kimmeridge shales of the Dorset Coast, England. *Geochim. Cosmochim. Acta* 34:830–836.

Cowie, J. W. 1981. The Proterozoic-Phanerozoic transition and the Precambrian-Cambrian boundary. *Precamb. Res.* 15:199–206.

Degens, E. T., and Ross, D. A., eds. 1974. *The Black Sea: Geology, Chemistry, and Biology*. Memoir 20, Amer. Assoc. Petr. Geol., Tulsa, Okla.

Delmas, R. J., Ascencio, J.-M., and Legrand, M. 1980. Polar ice evidence that atmospheric CO_2 20,000 yr BP was 50% of present. *Nature* 284:155–157.

Demaison, G. J., and Moore, G. T. 1980. Anoxic environments and oil source bed genesis: *Bull. Amer. Assoc. Petr. Geol.* 64:1179–1209.

Dorf, E. 1964. The use of fossil plants in paleoclimatic interpretations. In *Problems of Paleoclimatology*, edited by A.F.M. Nairn, 13–31, 46–48. New York: Wiley-Interscience.

———. 1970. Paleobotanical evidence of Mesozoic and Cenozoic climatic changes. In *Proceedings of the North American Paleontological Convention*, vol. 1, 323–346. Lawrence, Kansas: Allen Press.

Douglas, R. G., and Savin, S. M. 1971. Isotopic analyses of planktonic foraminifera from the Cenozoic of the northwest Pacific, Leg 6. In *Initial Reports of the Deep Sea Drilling Project*, vol. 6. Washington, D.C.: U.S. Govt. Printing Office.

———. 1973. Oxygen and carbon isotope analyses of Cretaceous and Tertiary foraminifera from the central North Pacific. In *Initial Reports of the Deep Sea Drilling Project*, vol. 7. Washington D.C.: U.S. Govt. Printing Office.

Douglas, R. G., and Savin, S. M. 1975. Oxygen and carbon isotope analyses of Tertiary and Cretaceous microfossils from Shatsky Rise and other sites in the North Pacific Ocean. In *Initial Reports of the Deep Sea Drilling Project*, vol. 32. Washington, D.C.: U.S. Govt. Printing Office.

Elderfield, H. 1981. Metal-organic associations in interstitial waters of Narragansett Bay sediments. *Amer. Jour. Sci.* 281:1184–1196.

Elderfield, H., McCaffrey, R. J., Luedtke, N., Bender, M., and Truesdale, V. W. 1981. Chemical diagenesis in Narragansett Bay sediments. *Amer. Jour. Sci.* 281:1021–1055.

Elderfield, H., Hawkesworth, C. J., Greaves, M. J., and Calvert, S. E. 1981. Rare earth element geochemistry of oceanic ferromanganese nodules and associated sediments. *Geochim. Cosmochim. Acta* 45:513–528.

Elderfield, H., and Greaves, M. J. 1981. Negative cerium anomalies in the rare earth element patterns of oceanic ferromanganese nodules. *Earth Planet. Sci. Lett.* 55:163–170.

———. 1982. The rare earth elements in seawater, *Nature* 296:214–219.

Emery, K. O., and Hülsemann, J. 1961. The relationships of sediments, life and water in a marine basin. *Deep Sea Res.* 8:165–180.

Emiliani, C. 1955. Pleistocene temperatures. *Jour. Geol.* 63:538–578.

———. 1964. Paleotemperature analysis of the Caribbean cores A254-BR-C and CP-28. *Bull. Geol. Soc. Amer.* 75:129–144.

————. 1966. Paleotemperature analysis of Caribbean cores P6304-8 and P6304-9 and a generalized temperature curve for the past 425,000 years. *Jour. Geol.* 74:109–126.

Emiliani, C., Kraus, E. B., and Shoemaker, E. M. 1981. Sudden death at the end of the Mesozoic. *Earth Planet. Sci. Lett.* 55:317–334.

Eugster, H. P., and Jones, B. F. 1979. Behavior of major solutes during closed-basin brine evolution. *Amer. Jour. Sci.* 279:609–631.

Eugster, H. P., Harvie, C. E., and Weare, J. H. 1980. Mineral equilibria in a six-component seawater system. Na-K-Mg-Ca-SO$_4$-Cl-H$_2$O, at 25°C. *Geochim. Cosmochim. Acta* 44:1335–1347.

Fischer, A. G., Arthur, M. A., Herb, R., and Premoli Silva, I. 1977. Middle Cretaceous events. *Geotimes* 22(4):18–19.

Garrels, R. M., and Thompson, M. E. 1962. A chemical model for seawater at 25°C and one atmosphere total pressure. *Amer. Jour. Sci.* 260:57–66.

Garrels, R. M., and Perry, E. A., Jr. 1974. Cycling of carbon, sulfur, and oxygen through geologic time. In *The Sea*, vol. 5, edited by E. D. Goldberg, chap. 9. New York: Wiley-Interscience.

Garrels, R. M., and Lerman, A. 1981. Phanerozoic cycles of sedimentary carbon and sulfur. *Proc. Natl. Acad. Sci. USA* 78:4652–4656.

Glaessner, M. F. 1961. Pre-Cambrian animals. *Sci. Amer.* 204(3):72–78.

Goldhaber, M. B., and Kaplan, I. R. 1974. The sulfur cycle. In *The Sea*, vol. 5, edited by E. D. Goldberg, chap. 17. New York: Wiley-Interscience.

————. 1980. Mechanisms of sulfur incorporation and isotope fractionation during early diagenesis in sediments of the Gulf of California. *Mar. Chem.* 9:81–94.

Goldring, R., and Curnow, C. N. 1967. The stratigraphy and facies of the late Precambrian at Ediacara, South Australia. *J. Geol. Soc. Aust.* 14:195–214.

Goreau, T. 1981. Biogeochemistry of nitrous oxide. Ph.D. diss., Harvard University.

Graham, D. W., Bender, M. L., Williams, D. F., and Keigwin, L. D., Jr. 1982. Strontium-calcium ratios in Cenozoic planktonic foraminifera. *Geochim. Cosmochim. Acta* 46:1281–1292.

Gunnerson, C. G., and Özturgut, E. 1974. The Bosporus. In *The Black Sea: Geology, Chemistry, and Biology*, edited by E. T. Degens and D. A. Ross, 99–114. Memoir 20, Amer. Assoc. Petr. Geol., Tulsa, Okla.

Hajash, A. 1975. Hydrothermal processes along mid-ocean ridges: An experimental investigation. *Contrib. Mineral. Petrol.* 53:205–226.

Hancock, J. M. 1975. The sequence of facies in the Upper Cretaceous of northern Europe compared with that in the Western Interior. Geol. Assoc. Canada Spec. Paper 13, 84–118.

Hartman, O. 1955. Quantitative survey of the benthos of San Pedro Basin, southern California: Part 1. Preliminary results. *Allan Hancock Pacific Expedition 19* (1). Los Angeles: University of California Press.

———. 1966. Quantitative survey of the benthos of San Pedro Basin, southern California: Part 2. Final results and conclusions. *Allan Hancock Pacific Expedition 19* (2). Los Angeles: University of California Press.

Hartmann, M., and Nielsen, H. 1969. δ^{34}S-Werte in rezenten Meeressedimenten und ihre Deutung am Beispiel einiger Sedimentprofile aus der westlichen Ostsee. *Geol. Rundschau* 58:621–655.

Harvie, C. E., Weare, J. H., Hardie, L. A., and Eugster, H. P. 1980. Evaporation of seawater: Calculated mineral sequences. *Science* 208:498–500.

Haskin, L. A., Haskin, M. A., Frey, F. A., and Wildeman, T. R. 1968. Relative and absolute terrestrial abundances of the rare earths. In *Origin and Distribution of the Elements*, edited by L. H. Ahrens, 889–912. New York: Pergamon.

Heinzinger, K., Junge, C., and Schidlowski, M. 1971. Oxygen isotope ratios in the crust of iron meteorites. *Z. Naturforsch.* 26a:1485–1490.

Heinzinger, K., Schidlowski, M., and Junge, C. 1974. Isotopic composition of atmospheric oxygen during the geological past. *Z. Naturforsch.* 29a:964–965.

Herrmann, A. G., Knake, D., Schneider, J., and Peters, H. 1973. Geochemistry of modern seawater and brines from salt pans: Main components and bromine distribution. *Contrib. Mineral. Petrol.* 40:1–24.

Hill, A. E., and Wills, J. H. 1938. Ternary systems: XXIV, Calcium sulfate, sodium sulfate and water. *Jour. Amer. Chem. Soc.* 60:1647–1655.

Hirst, D. M. 1974. Geochemistry of sediments from eleven Black Sea cores. In *The Black Sea: Geology, Chemistry, and Biology*, edited by E. T. Degens and D. A. Ross, 430–455. Memoir 20, Amer. Assoc. Petr. Geol., Tulsa, Okla.

Høgdahl, O. T. 1970. Distribution of lanthanides in the waters and sediments of the river Gironde in France. Paper presented at the IAEA Research Agreement Coordinated Program on Marine Radioactivity Studies, Int. At. Energy Agency, Monaco.

Høgdahl, O. T., Melsom, S., and Bowen, B. T. 1968. Neutron activation analysis of lanthanide elements in seawater. In *Trace Inorganics in Water*, edited by R. F. Gould, 308–325. Advan. Chem. Ser. 73. Washington, D.C.: American Chemical Society Publications.

Holland, H. D. 1962. Model for the evolution of the Earth's atmosphere. In *Petrologic Studies: A Volume in Honor of A. F. Buddington*, edited by A.E.J. Engel, H. L. James, and B. F. Leonard, 447–477. Geol. Soc. Amer.

———. 1972. The geologic history of sea water: An attempt to solve the problem. *Geochim. Cosmochim. Acta* 36:637–651.

———. 1973. Systematics of the isotopic composition of sulfur in the oceans during the Phanerozoic and its implications for atmospheric oxygen. *Geochim. Cosmochim. Acta* 37:2605–2616.

———. 1978. *The Chemistry of the Atmosphere and Oceans.* New York: Wiley.

———. 1979. Metals in black shales: A reassessment. *Econ. Geol.* 74: 1676–1680.

Holser, W. T. 1963. Chemistry of brine inclusions in Permian salt from Hutchinson, Kansas. In *Symposium on Salt (First)*, edited by A. C. Bersticker, 86–95. Cleveland, Ohio: Northern Ohio Geol. Soc.

———. 1979. Mineralogy of evaporites. In *Marine Minerals*, vol. 6 of *Short Course Notes*, edited by R. G. Burns, chap. 8. Washington, D.C.: Mineral. Soc. Amer.

Holser, W. T. 1984. Gradual and abrupt shifts in ocean chemistry during Phanerozoic time. In *Patterns of Change in Earth Evolution*, edited by H. D. Holland and A. F. Trendall. Dahlem Konferenzen, Springer-Verlag.

Holser, W. T., and Kaplan, I. R. 1966. Isotope geochemistry of sedimentary sulfates. *Chem. Geol.* 1:93–135.

Holser, W. T., Kaplan, I. R., Sakai, H., and Zak, I. 1979. Isotope geochemistry of oxygen in the sedimentary sulfate cycle. *Chem. Geol.* 25:1–17.

Hummel, J. R., and Reck, R. A. 1981. Carbon dioxide and climate: The effects of water transport in radiative-convective models. *Jour. Geophys. Res.* 86:12, 035–12, 038.

Jenkyns, H. C. 1980. Cretaceous anoxic events: From continents to oceans. *Jour. Geol. Soc. London* 137:171–188.

Kalyanaraman, R., Yeatts, L. B., and Marshall, W. L. 1973. Solubility of calcium sulfate and association equilibria in $CaSO_4$ + Na_2SO_4 + $NaClO_4$ + H_2O at 273 to 623K. *J. Chem. Thermodyn.* 5:899–909.

Kaplan, I. R., Emery, K. O., and Rittenberg, S. C. 1963. The distribution and isotopic abundance of sulfur in recent marine sediments off southern California. *Geochim. Cosmochim. Acta* 27:297–331.

Katz, A., Sass, E., Starinsky, A., and Holland, H. D. 1972. Strontium behavior in the aragonite-calcite transformation: An experimental study at 40–90°C. *Geochim. Cosmochim. Acta* 36:481–496.

Kennett, J. P., and Shackleton, N. J. 1975. Laurentide Ice Sheet meltwater recorded in Gulf of Mexico deep-sea cores. *Science* 188:147–150.

Kinsman, D.J.J., and Holland, H. D. 1969. The coprecipitation of cations with $CaCO_3$: IV. The coprecipitation of Sr^{2+} with aragonite between 16° and 96°C. *Geochim. Cosmochim. Acta* 33:1–17.

Klinkhammer, G., Elderfield, H., and Hudson, A. 1983. Rare earth elements in seawater near hydrothermal vents. *Nature* 305: 185–188.

Kominz, M. A., and Pisias, N. G. 1979. Pleistocene climate: Deterministic or stochastic? *Science* 204:171–173.

Kühn, R. 1969. Vorkommen und Verteilung des Broms in Salzlagerstätten Mittel-und Westeuropas und Nordamerika nebst einigen Auswertungen. *Kaliforschungs-Inst. Wiss. Mitt.* 111. Unpublished report.

Lehman, J. T., Botkin, D. B., and Likens, G. E. 1975. The assumptions and rationales of a computer model of phytoplankton population dynamics. *Limnol. Oceanogr.* 20:343–364.

Leventhal, J. S. 1978a. Trace elements, carbon and sulfur in Devonian black shale cores from Perry County, Kentucky; Jackson and Lincoln Counties, West Virginia; and Cattaraugus County, New York. U.S. Geol. Survey Open-File Report 78-504.

———. 1978b. Summary of chemical analyses and some geochemical controls related to Devonian black shales from Tennessee, West Virginia, Kentucky, Ohio, and New York. Second Eastern Gas Shales Symposium, METC/SP-78/6.

———. 1979. Chemical analysis and geochemical associations in Devonian black shale core samples from Martin County, Kentucky; Carroll and Washington Counties, Ohio; Wise County, Virginia; and Overton County, Tennessee. U.S. Geol. Survey Open-File Report 79-1503.

———. 1981. Pyrolysis gas chromatography-mass spectrometry to characterize organic matter and its relationship to uranium content of Appalachian Devonian black shales. *Geochim. Cosmochim. Acta* 45: 883–889.

Leventhal. J. S., Crock, J. G., and Malcolm. M. J. 1981. Geochemistry of trace elements and uranium in Devonian shales of the Appalachian Basin: Final report to U.S. Department of Energy for the Eastern Gas Shales Program, U.S. Geol. Survey Open File Report 81-778.

Leventhal, J. S., and Hosterman, J. W. 1982. Chemical and mineralogical analyses of Devonian black-shales from Martin County, Kentucky; Carroll and Washington Counties, Ohio; Wise County, Virginia; and Overton County, Tennessee, U.S.A. *Chem. Geol.* 37:239–264.

Lewin, R. A., and Chow, T. J. 1961. La enpreno de strontio en Koko-litoforoj. *Plant Cell Physiol (Japan)* 2:203–208.

Lewis, J. S., Watkins, G. H., Hartman, H., and Prinn, R. G. 1982. Chemical consequences of major impact events on Earth. In *Geological Implications of Impacts of Large Asteroids and Comets on the Earth,* edited by L. T. Silver and P. H. Schultz, 215–223. Geol. Soc. Amer. Special Paper 190.

Lindh, T. B., Saltzman, E. S., Sloan, J. L. II, Mattes, B. W., and Holser, W. T. 1981. A revised $\delta^{13}C$-age curve (abstract). *Abstracts with Programs* 13(7):498. Geol. Soc. Amer.

Lipschutz, F., Zafiriou, O. C., Wofsy, S. C., McElroy, M. B., Valois, F. W., and Watson, S. W. 1981. Production of NO and N_2O by soil nitrifying bacteria. *Nature* 294:641–643.

Lloyd, R. M. 1968. Oxygen isotope behavior in the sulfate-water system. *Jour. Geophys. Res.* 73:6099–6110.

Lorens, R. B., and Bender, M. L. 1980. The impact of solution chemistry on *mytilus edulis*, calcite and aragonite. *Geochim. Cosmochim. Acta* 44:1265–1278.

Lovelock, J. E., and Margulis, L. 1974a. Atmospheric homeostasis by and for the biosphere: The Gaia Hypothesis. *Tellus* 26:2–10.

———. 1974b. Homeostatic tendencies of the Earth's atmosphere. *Origins of Life* 5:93–103.

Lowenstam, H. A. 1961. Mineralogy, $^{18}O/^{16}O$ ratios, and strontium and magnesium contents of recent and fossil brachiopods and their bearing on the history of the oceans. *Jour. Geol.* 69:241–260.

McCarthy, J. J., and Carpenter, E. J. 1979. Oscillatoria (Trichodesmium) Thiebautii (Cyanophyta) in the central North Atlantic Ocean. *J. Phycol.* 15:75–82.

McLean, D. M. 1978. Land floras: The major late Proterozoic atmospheric carbon dioxide/oxygen control. *Science* 200:1060–1062.

Manabe, S. 1983. Carbon dioxide and climatic change. *Advances in Geophys.* 25:39–82.

Manheim, F. T. 1961. A geochemical profile in the Baltic Sea. *Geochim. Cosmochim. Acta* 25:52–70.

Margolis, S. V., Kroopnick, P. M., and Showers, W. J. 1981. *Paleoceanography: The History of the Ocean's Changing Environments.* Rubey Memorial vol. 2. Newark: Prentice-Hall.

Martin, J. -M., Høgdahl, O., and Philippot, J. C. 1976. Rare earth element supply to the ocean. *Jour. Geophys. Res.* 81:3119–3124.

Mottl, M. J., and Holland, H. D. 1978. Chemical exchange during hydrothermal alteration of basalt by seawater: I. Experimental results for major and minor components of seawater. *Geochim. Cosmochim. Acta* 42:1103–1115.

Neftel, A., Oeschger, H., Schwander, J., Stauffer, B., and Zunbrunn, R. 1982. Ice core sample measurements give atmospheric CO_2 content during the past 40,000 yrs. *Nature* 295:220–223.

Nielsen, H. 1965. Schwefelisotope im marinen Kreislauf und das $\delta^{34}S$ der früheren Meere. *Geol. Rundschau* 55:160–172.

Officer, C. B., and Drake, C. L. 1983. The Cretaceous-Tertiary transition. *Science* 219:1383–1390.

Orth, C. J., Gilmore, J. S., Knight, J. D., Pillmore, C. L., Tschudy, R. H., and Fassett, J. E. 1981. An iridium abundance anomaly at the palynological Cretaceous-Tertiary boundary in northern New Mexico. *Science* 214:1341–1343.

Parker, R. H. 1964. Zoogeography and ecology of some macroinvertebrates, particularly molluscs, in the Gulf of California and the continental slope off Mexico. *Vidensk. Meddr. Dansk Naturh. Foren. Copenhagen* 126:1–178.

Pilkey, O. H., and Goodell, H. G. 1963. Trace elements in recent mollusk shells. *Limnol. Oceanogr.* 8:137–148.

Piper, D. Z. 1974. Rare earth elements in the sedimentary cycle: A summary. *Chem. Geol.* 14:285–304.

Pollack, J. B., Toon, O. B., Ackerman, T. P., McKay C. P., and Turco, R. P. 1983. Environmental effects of an impact-generated dust cloud: Implications for the Cretaceous-Tertiary extinctions. *Science* 219:287–289.

Poplavko, Ye., M., Ivanov, V. V., Loginova, L. G., Orekhov, V. S., Miller, A. D., Nazarenko, I. I., Nishankhozhayev, R. N., Razenkova, N. I., and Tarkhov, Yu. A. 1977. Behavior of rhenium and other metals in Central Asian combustible shales. *Geochem. Internat.* 14(1):172–181.

Posnjak, E. 1940. Deposition of calcium sulfate from sea water. *Amer. Jour. Sci.* 238:559–568.

Rampino, M. R., and Reynolds, R. C. 1983. Clay mineralogy of the Cretaceous-Tertiary boundary clay. *Science* 219:495–498.

Rhoads, D. C., and Morse, J. W. 1971. Evolutionary and ecologic significance of oxygen-deficient marine basins. *Lethaia* 4:413–428.

Roden, G. I. 1964. Oceanographic aspects of Gulf of California. In *Marine Geology of the Gulf of California*, 30–58. Memoir 3, Amer. Assoc. Petrol. Geol., Tulsa, Okla.

Rutten, M. G. 1970. The history of atmospheric oxygen. *Space Life Sci.* 2:5–17.

Sabouraud-Rosset, C. 1974. Determination par activation neutronique des rapports Cl/Br des inclusions fluides de divers gypses: Correlation avec les données de la microcryoscopie et interpretations genetiques. *Sedimentology* 21:415–431.

Savin, S. M., Douglas, R. G., and Stehli, F. G. 1975. Tertiary marine paleotemperatures. *Geol. Soc. Amer. Bull.* 86:1499–1510.

Savin, S. M., and Yeh, H.-W. 1981. Stable isotopes in ocean sediments. In *The Oceanic Lithosphere*, vol. 7 of *The Sea*, edited by C. Emiliani, chap. 3. New York: Wiley-Interscience.

Schidlowski, M., and Ritzkowski, S. 1972. Magnetitkügelchen aus dem hessischen Tertiär: Ein Beitrag zur Frage der "kosmischen Kügelchen." *N. Jb. Geol. Paläont. Mh.*, no. 3:170–182.

Schidlowski, M., Junge, C. E., and Pietrek, H. 1977. Sulfur isotope variations in marine sulfate evaporites and the Phanerozoic oxygen budget. *Jour. Geophys. Res.* 82:2557–2565.

Scholle, P. A., and Arthur, M. A. 1980. Carbon isotope fluctuations in Cretaceous pelagic limestones: Potential stratigraphic and petroleum exploration tool. *Bull. Am. Assoc. Petr. Geol.* 64:67–87.

Schopf, J. W. 1978. The evolution of the earliest cells. *Sci. Amer.* 239(3): 110–138.

Schopf, J. W., and Oehler, D. Z. 1976. How old are the eukaryotes? *Science* 193:47–49.

Shaffer, N. R., Leininger, R. K., Ripley, E. M., and Gilstrap, M. S. 1981. Heavy metals in organic-rich New Albany shale of Indiana (abstract). *Abstracts with Programs* 13(7):551. Geol. Soc. Amer.

Shukla, J., and Mintz, Y. 1982. Influence of land-surface evapotranspiration on the Earth's climate. *Science* 215:1498–1500.

Silver, L. T., and Schultz, P. H., eds. 1982. Geological implications of impacts of large asteroids and comets on the earth. *Geol. Soc. Amer. Spec. Paper 190.* Boulder: Geol. Society of America.

Smith, A. G., Briden, J. C., and Drewry, G. E. 1973. Phanerozoic world maps. In *Organisms and Continents through Time.* Paleont. Assoc. London Spec. Papers in Paleontology 12:1–42.

Stanley, S. M. 1976. Fossil data and the Precambrian-Cambrian evolutionary transition. *Amer. Jour. Sci.* 276:56–76.

Summerhayes, C. P. 1981. Organic facies of Middle Cretaceous black shales in deep North Atlantic. *Bull. Amer. Assoc. Petr. Geol.* 65: 2364–2380.

Sweeney, R. E., and Kaplan, I. R. 1980. Diagenetic sulfate reduction in marine sediments. *Mar. Chem.* 9:165–174.

Thiede, J., and van Andel, T. H. 1977. The paleoenvironment of anaerobic sediments in the late Mesozoic South Atlantic Ocean. *Earth. Planet. Sci. Lett.* 33:301–309.

Thode, H. G., Monster, J., and Dunford, H. B. 1961. Sulfur isotope geochemistry. *Geochim. Cosmochim. Acta* 25:159–174.

Thode, H. G., and Monster, J. 1965. Sulfur-isotope geochemistry of petroleum, evaporites and ancient seas. In *Fluids in Subsurface Environments*. 367–377. Memoir 4, Amer. Assoc. Petrol. Geol., Tulsa, Okla.

Thompson, G., and Bowen, V. T. 1969. Analysis of coccolith ooze from the deep tropical Atlantic. *Jour. Mar. Res.* 27:32–38.

Thompson, T. G., and Chow, T. J. 1955. The strontium-calcium atom ratio in carbonate secreting marine organisms. In Papers in Mar. Bio. Oceanogr. Suppl. to *Deep Sea Res.* 3:20–39.

Tunnicliffe, V. 1981. High species diversity and abundance of the epibenthic community in an oxygen-deficient basin. *Nature* 294:354–356.

Turco, R. P., Toon, O. B., Park, C., Whitten, R. C., Pollack, J. B., and Noerdlinger, P. 1981. Tunguska meteor fall of 1908: Effects on stratospheric ozone. *Science* 214:19–23.

Turekian, K. K. 1955. Paleoecological significance of the strontium-calcium ratio in fossils and sediments. *Bull. Geol. Soc. Amer.* 66:155–158.

Turekian, K. K., and Armstrong, R. L. 1960. Magnesium, strontium, and barium concentrations and calcite-aragonite ratios of some recent molluscan shells. *Jour. Mar. Res.* 18:133–151.

van Andel, T. H. 1964. Recent marine sediments of Gulf of California. In *Marine Geology of the Gulf of California*. 216–310. Memoir 3, Amer. Assoc. Petrol. Geol., Tulsa, Okla.

Veizer, J., Holser, W. T., and Wilgus, C. K. 1980. Correlation of $^{13}C/^{12}C$ and $^{34}S/^{32}S$ secular variations. *Geochim. Cosmochim. Acta* 44:579–587.

Vine, J. D., and Tourtelot, E. B. 1970. Geochemistry of black shale deposits: A summary report. *Econ. Geol.* 65:253–272.

Volkov, I. I., and Fomina, L. S. 1974. Influence of organic material and processes of sulfide formation on distribution of some trace elements in deep-water sediments of the Black Sea. In *The Black Sea: Geology, Chemistry, and Biology*, edited by E. T. Degens and D. A. Ross, 456–476. Memoir 20, Amer. Assoc. Petrol. Geol., Tulsa, Okla.

Vyas, K. C., Aho, G. D., and Robl, T. L. 1981. Synthetic fuels from eastern oil shale. Vol. 1, Final Report, U.S. Dept. of Energy, Report DOE/R4/10185-T1.

Walker, J.C.G. 1977. *Evolution of the Atmosphere*. New York: Macmillan.

Waterbury, J. B., Watson, S. W., Guillard, R.R.L., and Brand, L. E. 1979. Widespread occurrence of a unicellular marine planktonic cyanobacterium. *Nature* 277:293–294.

Webby, B. D. 1970. Late Precambrian trace fossils from New South Wales, *Lethaia* 3:79–109.

Westrich, J. T. 1983. The consequences and controls of bacterial sulfate reduction in marine sediments. Ph.D. diss., Yale University.

Whittaker, R. H., and Likens, G. E. 1975. The biosphere and man. In *Primary Productivity of the Biosphere*, edited by H. Leith and R. H. Whittaker, 305–328. New York: Springer-Verlag.

Woodruff, F., Savin, S. M., and Douglas, R. G. 1981. Miocene stable isotope record: A detailed deep Pacific Ocean study and its paleoclimate implications. *Science* 212:665–668.

Wright-Clark, J., and Holser, W. T. 1982. The significance of iron hydroxide to the cerium anomaly of seawater: Implications for secular variation in redox. Unpublished manuscript.

Wright-Clark, J., Seymour, R. S., Holser, W. T., and Goles, G. G. 1982. Are changes in ancient ocean chemistry recorded in fossil apatite? Unpublished ms.

Wright-Clark, J., Seymour, R. S., and Shaw, H. F. 1983. REE and Nd isotopes in conodont apatite: variations with geological age and depositional environment. Paper presented at the Pander Society Symposium, Madison, Wisc.

Zharkov, M. A. 1981. *History of Paleozoic Salt Accumulation.* Translated by R. E. Sorkina, R. V. Fursenko, and T. I. Vasilieva. New York: Springer-Verlag.

Author Index

Subject Index

Abitibi greenstone belt, 311
Acadian orogeny, 240
Achondrites, 7, 20, 37, 38
Actinolite, 279
Adenine, 116
Adenosine phosphate, 116
Albanene, Canadian Shield, 260
Albian, 221, 252, 257, 398, 521, 523–526
Aldan Shield, 368
Aldehyde, 99
Algae, 195, 513
Algal mats, 195, 257, 285, 516
Allende meteorite, 6, 7, 36, 37
Alum shale, Sweden, 497
Aluminum, 13, 17, 18, 147, 156, 157, 212, 216, 220–223, 262, 284, 311
Aluminum-26, 7
Aluminum content: Archean and Proterozoic shales of the Canadian Shield, 224; metamorphosed Proterozoic shales from the USSR, 224; sedimentary rocks of the Russian Platform, 216; sediments near the Japan Trench, 217
Aluminum oxide, 144, 147, 156, 157, 159, 169, 171, 172, 177, 178, 201, 212, 258, 284, 297, 306, 310, 338, 343, 375
Amadeus Basin, Central Australia, 422, 423; map of, 423
Amazon River, 229, 238, 386
Ameralik dikes, 129
American and Russian Platforms, Late Precambrian and Phanerozoic sediments, 158–175
Amino acids, 99, 112, 114
Amino nitriles, 112
Aminoalkane phosphates, 116
Aminoalkyl phosphates, 116
Amitsoq tonalite gneiss, 129
Ammonia, 49, 97–99, 111–114, 116, 121, 184; fugacity of, 49
Ammonium, 112–114, 390
Ammonium bisulfide, 111
Ammonium oxalate, 120
Amphibolite(s), 129, 135, 360
Amsterdam paleosol, 297, 299; chemical and mineralogical composition of, 299

Amygdaloidal lava, 142
Andesite(s), 53, 54, 211
Andesitic magma, 52, 53
Anhydrite, 82, 423–427, 442, 445, 448–450, 453, 454, 458, 465, 466, 470–474; and gypsum, stability diagram of, 426; precipitation of, 446–450
Ankerite, 135, 146, 379, 384, 387, 398, 399, 401
Annelida, 513
Anorthite, 105
Anorthositic rocks, lunar, 13
Antarctic ice, 538
Anthophyllite, 131
Anthracitic coal, 368
Antigorite, 379
Antimony, 15, 209
Apatite, 115–118, 282, 401, 405, 419, 493, 498
Aphebian, 151, 152, 155–157, 173, 223, 260, 353
Aptian, 521–526
Aragonite, 183, 257, 345, 351, 382, 446, 506, 510; diagenetic trends during stabilization by meteoric water, 345
Archean, 51, 136, 142, 151, 152, 155–157, 173, 181, 185, 192, 193, 195, 199, 202, 223, 226, 227, 251, 255, 260, 265, 278, 279, 296, 299, 308, 312, 346, 352, 353, 355, 366–368, 379, 406, 408, 413–415, 417, 420–422, 427; craton of Greenland, 129; craton of North America, map of, 130; cratons, 136; ocean, "Volcanogenic," 227;
Archaeocyathids, 514
Arenaceous rocks, 139
Arenites, 262
Argillite, 278, 343
Argon, 6, 29, 34, 64, 69, 82, 538;
Argon-36, 30, 31, 34, 66–69, 83, 84, 90
Argon-38, 30
Argon-39, 6
Argon-40, 6, 29, 64
Aristarchus, 15
Arsenic, 15, 209
Arctic ice, 538

Library of Congress Cataloging in Publication Data

Holland, Heinrich D.
    The chemical evolution of the atmosphere and oceans.

    (Princeton series in geochemistry)
    Includes bibliographies and index.
        1. Atmospheric chemistry.   2. Chemical oceanography.
3. Chemical evolution.   I. Title.   II. Series.
QC879.6.H63   1984   551.5   83-43077
ISBN 0-691-08348-7   ISBN 0-691-02381-6 (pbk.)